Molecular astrophysics

A volume honouring Alexander Dalgarno

Molecular astrophysics

A VOLUME HONOURING
ALEXANDER DALGARNO

EDITED BY

T. W. Hartquist

Max Planck Institute for Physics and Astrophysics
Institute for Extraterrestrial Physics, Garching, FRG

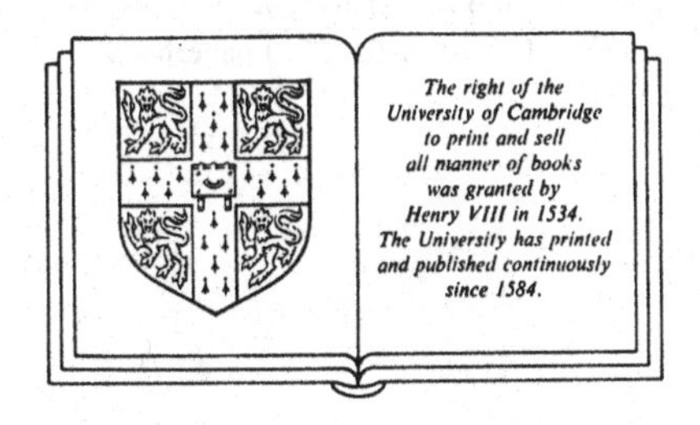

CAMBRIDGE UNIVERSITY PRESS

Cambridge

New York Port Chester

Melbourne Sydney

CAMBRIDGE UNIVERSITY PRESS
Cambridge, New York, Melbourne, Madrid, Cape Town, Singapore, São Paulo

Cambridge University Press
The Edinburgh Building, Cambridge CB2 2RU, UK

Published in the United States of America by Cambridge University Press, New York

www.cambridge.org
Information on this title: www.cambridge.org/9780521363310

First published 1990
This digitally printed first paperback version 2005

A catalogue record for this publication is available from the British Library

Library of Congress Cataloguing in Publication data
Molecular astrophysics: a volume honouring Alexander Dalgarno
edited by T. W. Hartquist.
p. cm.
Includes index.
ISBN 0-521-36331-4
1. Molecular astrophysics. 2. Dalgarno, A. I. Dalgarno, A. II. Hartquist, T. W.
QB462.6.M65 1990
523.01'96--dc20 89-17260 CIP

ISBN-13 978-0-521-36331-0 hardback
ISBN-10 0-521-36331-4 hardback

ISBN-13 978-0-521-01727-5 paperback
ISBN-10 0-521-01727-0 paperback

CONTENTS

DEDICATION

Until about 20 years ago, almost all of our knowledge of astronomical sources derived from observations of the emission and absorption of radiation by neutral and ionized atomic species and by electrons moving in the fields of such species and in large-scale magnetic fields. A large fraction of the astrophysics done up to that time could be described as atomic astrophysics. Of course, nuclear astrophysics was another important area of inquiry, but much of that field's database came from observations of stellar spectra formed by atomic processes.

In 1967 when Alex Dalgarno moved to Cambridge, Massachusetts and began to devote a substantial fraction of his efforts to theoretical molecular astrophysics, CH, CH^+, and CN, which had been observed close to 30 years earlier in optical absorption against stars, and OH, which had been seen recently in emission at 18 cm, were the only molecules to have been detected in the interstellar gas. It was anticipated that H_2 would be observed in ultraviolet absorption against nearby massive O and B stars, and Alex began calculations to investigate quantitatively molecular processes important in the destruction and pumping of molecular hydrogen by the interstellar radiation field. This work made possible the construction of models of diffuse interstellar molecular clouds. Thus began a long term effort to develop molecular astrophysics into a rigorous field of investigation.

Molecules exist in a wide variety of astronomical sources including interstellar clouds, star-forming regions, ionized nebulae, stellar atmospheres, the pregalactic medium, quasar absorption line regions, very young supernova remnants, and the interstellar media and the centers of other galaxies, including many with active nuclei. Though the subjects of this volume are restricted to more distant objects, molecules also exist in extraterrestrial sources, such as planetary atmospheres and comets, in the Solar System. Alex has led efforts to understand the chemistries which form the molecules in this wide variety of sources, to use their abundances

and level populations to diagnose the macroscopic properties of the sources, and to elucidate the roles which they play in regulating those macroscopic properties.

In 1986, he was awarded the Royal Astronomical Society's Gold Medal for his success in these efforts.

Over roughly 20 years, Alex has created and maintained an important school within theoretical astrophysics. Most major active contributors in theoretical molecular astrophysics have been Alex's students or collaborators or have been influenced significantly by working together with one or more of his former students and close associates. Most of Alex's work in theoretical astrophysics has been concerned directly with understanding or using data, and he has interacted with a number of leading observational molecular astronomers as well.

Alex had already led a distinguished career as a theoretical atomic and molecular physicist and as an atmospheric physicist before moving to the States. He has continued to publish prodigiously in those areas and received the Hodgkins Medal of the Smithsonian Institution in 1977 and the Davisson–Germer Award of the American Physical Society in 1980 for his work in them. Indeed, Alex is highly regarded in the four separate communities of physicists, astronomers, chemists, and atmospheric scientists. Alex has profitably used his background in these other areas when conducting his astrophysical investigations. His knowledge of atomic and molecular processes and ability to study them quantitatively, his experience and facility in identifying the dominant physical and chemical mechanisms in complicated plasmas, and his skill in suggesting to laboratory physicists and chemists astrophysically relevant measurements have contributed to his impact as a molecular astrophysicist.

Students constitute a group which has benefitted from Alex's attention. In the mid-1970s, when I studied with him and he was the chairman of the Harvard Department of Astronomy, Alex possessed the reputation of being able to match a project with the constraints of a student's abilities and the available time and of being a faculty member to whom a student facing difficulties could turn for help. He seems to make availability to students a special priority. Despite his own supply of problems, Alex always has been extremely encouraging to his students when they have pursued projects of their own devising. He is an excellent teacher capable of getting students started and supporting their growth as independent research scientists. His help even extends to the teaching of English composition, a useful but sometimes harrowing experience for the instructee.

Alex's personal qualities suit him for more than his role as mentor. They make him a valued companion as well. His humour is dry and playfully interactive and sometimes seems to serve as an affectionate means of communication with friends. Alex combines quiet modesty with a confidence which can reassure others. His words to those of his students and friends who confide in him during moments of real or imagined crisis are sympathetic and wise.

He is an excellent sportsman. The possibility of becoming a professional soccer player existed before he suffered an injury. He is known to be one of the best squash players in astrophysics, and he plays tennis at least as well as any of his former students. When its existence was threatened by a building project, the Observatory tennis court was mentioned to the Harvard-Radcliffe executives as having been one of the benefits which had attracted him to Harvard; happily, the Observatory personnel still have access to a tennis court, and the construction led to the existence of squash courts a short walk from Alex's office.

A book on the subject of molecular astrophysics would be improved by a contribution by Alex. However, one of the two purposes (the other being to provide a self-contained introductory overview of the field which he has played such a dominant role in shaping) of this collection is to honor Alex who on 5 January 1988 celebrated the sixtieth anniversary of his birth in London. In addition, the 1988 volume of *Advances in Atomic and Molecular Physics* consists of papers dedicated to Alex and the January 1988 issue of *Planetary and Space Science* contains contributions written in his honor. These two works should also be consulted to obtain a more complete overview of his scientific influence and other biographical information.

The importance of Alex's research was also recognized when he was elected a Fellow of the American Academy of Arts and Sciences in 1969 and a Fellow of the Royal Society in 1972. He obtained his higher education in University College, London, where he took his Ph.D. in 1951 and from 1951 through 1966 held faculty posts in The Queen's University of Belfast where he was promoted to a chair in 1961. Both of these institutions have honored him; the former by electing him a Honorary Fellow in 1976 and the latter by awarding him an honorary D.Sc. in 1980.

Currently, Alex is a physicist in the Smithsonian Astrophysical Observatory and the Phillips Professor of Astronomy in Harvard University. He is the editor of *The Astrophysical Journal Letters*.

As alluded to above, Alex's theoretical work reflects the approach of an empiricist. In recent years, he has advocated efforts to ensure the future of theoretical atomic and molecular physics in the foremost physics research departments in American universities. He believes 'that physics is embodied not in its equations, but in the solutions to the equations.'

PREFACE

The present volume has been designed to be a self-contained introduction to the field of molecular astrophysics. It can serve as the text for a one semester postgraduate course concerning that subject exclusively or as a supplementary text in a postgraduate course on the interstellar medium. It can also be used by research astronomers, atomic and molecular physicists, chemists, and atmospheric scientists who have interests in weakly ionized plasmas and the physical and chemical processes which occur in them and who wish to become familiar with recent work in molecular astrophysics.

Many of the articles concern theoretical studies and modelling. Part I, consisting of two chapters written by observers, provides a general description of the astronomical context within which much of the remainder of the volume should be considered. The contribution by Per Friberg and Åke Hjalmarson reviews briefly our understanding of the global physical properties of the Galactic interstellar medium, describes the wide range of conditions within the dark molecular clouds (those having visual extinctions greater than about 1) and the distribution in the Galaxy of the dark clouds, and compares the chemical contents of different regions in various dark clouds. Leo Blitz has written about observations of molecules in other galaxies including some with active nuclei.

The study of chemistry in diffuse molecular clouds (those having a visual extinction of about 1 or less) constitutes the subject of Part II and is perhaps the most fundamental area in molecular astrophysics. Before beginning Part II, the reader who is unfamiliar with the structure of and spectra of diatomic molecules may wish to turn first to the article by Kate Kirby in which an introduction to those topics is given. Many of the data for the diffuse clouds have been obtained in absorption studies against background stars.

Direct measurements of the H_2 rotational level population distributions and of the H column densities are used to infer the detailed physical properties in diffuse molecular clouds. The H_2 is formed on the surfaces of grains and is destroyed by a photodissociation process involving excitation of higher electronic states which radiatively decay to the rovibrational continuum of the ground electronic state. Radiative decays of the higher electronic states to the discrete rovibrational levels of the ground electronic state also occur and the subsequent radiative cascades through the rovibrational ladder are generally assumed to be responsible for the production of the H_2 detected in the $v = 0, J = 3$–6 levels. The populations of the lower levels are determined by collisional processes. Diffuse cloud modelling is the most rigorously based theoretical enterprise in molecular astrophysics and is reviewed in the paper by Ewine van Dishoeck who also has described efforts to construct low temperature equilibrium models of the diffuse cloud chemistry to account for the measured column densities of a number of simple diatomic species. The chemistry in diffuse clouds is often initiated by the photoionization of atomic species. Because oxygen has an ionization potential exceeding that of atomic hydrogen, there are no ultraviolet photons which can ionize O (implying on diffuse clouds); rather the oxygen chemistry is driven by cosmic ray induced ionization, but once O^+ is formed it reacts quickly with H_2 and eventually OH, the abundance of which is used to infer cosmic ray ionization rates, is formed.

CH^+ and CO are two species which have measured column densities which are much higher than those explained easily with low temperature equilibrium models. In roughly the past fifteen years, the interstellar medium has come to be viewed as a site of violent activity, as mentioned by Per Friberg and Åke Hjalmarson. Consequently, the effects of shocks on diffuse cloud chemistry have been considered. Bill Langer has reviewed critically the observations which give insight into the dynamics and structures of diffuse clouds and, hence, constrain shock models. David Flower, Guillaume Pineau des Forêts, and I have summarized work on the chemistry in diffuse cloud shocks. Because high temperatures ($T \gtrsim 10^3$ K) and ion–neutral streaming can drive reactions which are slow in cold diffuse clouds, shocks have been considered as the sites of formation of molecules including CH^+. Multifluid hydromagnetic models are appropriate for shocks in weakly ionized clouds. The chemistry affects the ionization and shock structures. Because H_2 is collisionally excited in shocks, the observed H_2 level populations constrain models. Neither current low temperature chemical equilibrium models, existing shock models, nor a combination of them seem compatible with the H_2, CH^+, and CO diffuse cloud data.

Part III concerns the chemistry of dense clouds. Tom Millar has written an introduction to the area in which he has described the basic gas phase chemical networks including the processes which determine the ionization structure and chemical fractionation. Cosmic rays are the sole source in dense clouds of the ionization which drives the chemistry. His paper and the one by Ewine van Dishoeck should provide the reader with sufficient background to appreciate the importance of many of the studies described in Part IV. He also has summarized results of steady state and simple time dependent models of dense cloud chemistry. Dust is also a component of molecular clouds and contains a substantial fraction of the heavy elements in them. Chemistry occurs on dust grains as Victoria Buch has discussed. One of several important ways in which dust grains can affect the gas phase chemistry is by removing electrons. Stephen Lepp's article concerns, in part, the effect of small grains or large molecules on the gas phase ionization structures in dense and diffuse clouds. His contribution could have been included in either Part III or Part II. It also contains a discussion of diffuse cloud heating due to photoabsorption by grains. In Part IV, the paper by Evelyne Roueff addresses some of the collisional processes which give rise to radiative losses which balance the heat input into clouds.

Part IV contains contributions dealing with the molecular processes which must be included in models and which produce the spectral features which originate in the molecular clouds. Kate Kirby has provided a short introduction to the structure of diatomic molecules and their spectra and has discussed the mechanisms by which photoionization and photodissociation of molecules occur. As noted above, absorption data provide much of the information known about diffuse molecular clouds and photoionization and photodissociation are processes which are extremely important in establishing the ionization structure and molecular abundances in those clouds. David Smith, Nigel Adams, and Eldon Ferguson have reviewed the experimental techniques used to obtain data relevant to diffuse cloud and dense cloud modelling and have discussed results for several classes of reactions and their relevance to chemical modelling. Sir David Bates has summarized our theoretical understanding of various types of reactions. The modelling of diffuse and dense cloud chemistries relies very heavily on experimental and theoretical studies of the chemical processes. Evelyne Roueff has described theoretical studies of collisional excitation processes. Rotational excitation of molecules results in the radio and millimeter emissions which have been observed to study molecular clouds. A detailed knowledge of collisional excitation mechanisms is often necessary for the diagnosis of cloud conditions. The internal level population distributions in reactants in the laboratory differ from those in reactants in molecular clouds, and Margaret Graff has considered the ways in which the level population distributions affect some astronomically important reactions.

Parts I–IV acquaint the reader with the basic fundamentals of molecular astrophysics and some of the classic, often unsolved, problems. The subsequent parts deal primarily with special topics and particularly challenging outstanding questions.

The theme of Part V is the presence of atomic species in dense clouds. In his review of the observations of neutral and singly ionized atomic species in dark clouds, Gary Melnick has pointed out that while some of the emission from such species is associated with regions in which ultraviolet radiation from stars dissociates molecules and ionizes atoms, some of the emission is extended. He has noted further that a number of viable explanations have been suggested for the means of producing neutral atomic carbon, the emission of which has been studied with sufficient resolution to gain some idea of its spatial extent in M17 and ρ Oph. One of those suggested explanations is that dissociating stellar radiation is able to penetrate deeply into clouds either because they are clumpy or because the far ultraviolet dust opacity is lower than usually assumed. Wayne Roberge has described the detailed radiative transfer theory necessary for the construction of models in which the photodissociation of CO is responsible for the production of the observed neutral carbon. His paper describes a general formalism which is of importance for diffuse cloud modelling and for the modelling of dense H_2 photodissociation zones as discussed by Amiel Sternberg in Part VI. An additional source of photodissociation and photoionization in dense cloud has been considered by Roland Gredel and arises because ultraviolet photons are emitted following the collisionally induced excitation of H_2 by energetic electrons produced by cosmic ray induced ionization. He has summarized the effects of this process on dark cloud chemistry and has calculated the expected neutral atomic carbon to carbon monoxide abundance ratio. Note that Tom Millar has included a briefer discussion of the same process in his article in Part III. Another class of explanations for the neutral atomic carbon in 'apparently quiescent' regions is based on the assumption that the regions are, in fact, dynamic. Steve Charnley and David Williams have written about the chemistry in a region, which if it were further away would appear quiescent but in which the gas is thought to pass repeatedly through a dynamical cycle. The chemistry in that region, Barnard 5, probably does not attain equilibrium and the dynamical processing may ensure that atomic carbon is abundant. Barnard 5 may be typical of low mass star-forming regions and chemical studies may reveal a great deal about dynamic processes occurring in them, but it is probably not representative of all dark clouds. However, a long term goal in molecular astrophysics is the construction of time dependent chemical models for realistic dynamical scenarios. Phil Myers has provided a review of the current state of knowledge of molecular cloud structure, dynamics, and evolution.

Part VI is primarily about infrared line emission from H_2 in regions of massive star formation. Tom Geballe has reviewed the observations. David Chernoff and Chris McKee have discussed shocks in dense molecular material. Most of the H_2 infrared emission in Orion comes from shock heated gas. David Neufeld has written about higher speed shocks which dissociate the molecules which pass through them and the chemistry in the postshock gas. Amiel Sternberg has described work on static ultraviolet induced dissociation regions which are also sources of H_2 infrared emission. Dense dissociation regions near bright stars can have H_2 infrared spectra which are similar to those originating in shocked gas.

Part VII concerns masers which are very near stars and the chemistry of gas in evolved stellar envelopes including supernova ejecta. Jim Moran's contribution contains discussions of the locations of OH, H_2O, and SiO masers relative to the central stars and the theory of radiative transfer in masers. Some astronomical masers are interstellar and are found very near young bright stars, while others are in the outflowing envelopes of highly evolved stars. Mike Jura has considered the chemistry in more expanded circumstellar envelopes around red giants. Dick McCray has reviewed what is known about the physical conditions in the ejecta of the supernova SN 1987A. A knowledge of those conditions is necessary for understanding the origin of and the relative concentrations of the molecules observed to exist in the ejecta. Detailed study of the chemistry in that environment is just beginning.

In Part VIII, Greg Shields has also written about processes in plasmas with temperatures around 10^4 K. In such plasmas thermal charge transfer, a process which one studies theoretically by first calculating the relevant molecular potential curves for the diatomic system and then solving the scattering problem, plays an important role in determining the ionization balance. Emission line regions in galaxies with active nuclei are amongst those sources in which charge transfer is important. John Black has described pregalactic chemistry and searches for molecular hydrogen in QSO spectra.

As editor, I am grateful to the contributors, all of whom have been cooperative and helpful.

Garching
1988

T. W. Hartquist

I

Molecular clouds and the distribution of molecules in the Milky Way and other galaxies

1

Molecular clouds in the Milky Way

PER FRIBERG

Onsala Space Observatory, Onsala, Sweden

ÅKE HJALMARSON

Onsala Space Observatory, Onsala and The Institute of Theoretical Physics, University of Göteborg, Sweden

1.1. The cloudy structure of our Galaxy

The interstellar medium (ISM) in our Galaxy is in a complex state. The temperature and the density vary by about five and ten orders of magnitude respectively. The medium is exposed to cosmic rays and starlight and contains magnetic fields. Its average density is about 10^{-24} g cm^{-3}, or about one hydrogen atom per cubic centimeter, corresponding to 0.025 $M_{\odot}$ pc^{-3}. Thus on average one solar mass of the medium is contained in 40 pc^3. About 75% of the medium is hydrogen, about 25% is helium, and the remaining part (about 2%) is in heavier atoms. A large fraction of the heavier elements has condensed out as dust grains having an average density of about 0.001 $M_{\odot}$ pc^{-3}.

The local kinetic gas temperatures range from 10 K to 10^6 K. These temperatures correspond to energies which are much lower than those typical of cosmic rays. The lower temperatures correspond to energies which are less than those typical for starlight. Cosmic rays and starlight drive processes including ionization; hence, the ionization structure sometimes is very far from that in a gas in thermal equilibrium at the gas kinetic temperature. The energy density of cosmic rays is $\approx$0.5 eV cm^{-3}, of Galactic magnetic fields $\approx$0.2 eV cm^{-3} ($\approx 10^{-6}$ G), and of diffuse starlight $\approx$0.5 eV cm^{-3}. Hence there is a rough equipartition of energy among these components. The description of the interaction of matter with radiation is complicated by the huge variation of opacities for different wavelengths. For example, consider the formation of HII regions. Here photons in the Lyman continuum are heavily absorbed close to the star causing a HII region to form, while photons of longer wavelengths, which are absorbed by dust, can escape out to great distances.

Dynamical processes also affect the appearance of the interstellar medium. The Galactic rotation provides angular momentum that can hinder or delay local contractions. The Galactic magnetic field can also retard collapse. Rotation produces spiral arms and density waves that compress the medium and may cause cloud–cloud collisions. Density waves may also create shock waves which can heat the gas and even disperse clouds. The arms contain stars which become supernovae, generating strong often very disruptive shock waves. But we do not need such violent events to create shock waves. Stellar winds from massive evolved stars are powerful enough to sweep up matter and create shock waves. Bipolar outflows, mass loss during early stellar evolution, also give rise to shock waves or flows that stir up and possibly can disperse the cloud or cloud core that formed the star.

To reach any understanding of the density structure of atomic gas (HI) we have to consider thermal balance. The interstellar matter is heated by cosmic rays and in less obscured regions by Galactic diffuse starlight. The formation of molecular hydrogen ($H + H \rightarrow H_2 + 4.48$ eV) and the gravitational compression of the gas also contribute to the heating. Close to massive stars or star forming regions the intense radiation will heat the gas directly and, indirectly, through dust grains. Heating by cosmic rays and photons occurs through ionization of atoms and molecules in the gas or by photoejection of electrons from grains. The electrons transfer their excess energy to the gas through collisions before they recombine. Since the heating rate per unit volume is proportional to the ionization cross section it will be proportional to the density and can be written $n\Gamma(T)$, where n is the hydrogen density and $\Gamma(T)$ the heating rate per hydrogen atom. $\Gamma(T)$ is a function of the temperature and composition of the medium as well as of the radiation field and cosmic ray flux. Cooling of the gas occurs through radiation losses. Collisional excitation of atoms and/or molecules causes spontaneous emission of photons that can escape from the cloud. Since the collision rate per unit volume is dependent on the square of the number of colliding particles the cooling rate per unit volume can be written $n^2\Lambda(T)$; $\Lambda(T)$ is a function of temperature and composition as well as of the probability that the generated photon escapes the region. For equilibrium the heating has to be equal to the cooling, i.e., $n^2\Lambda(T) - n\Gamma(T) = 0$. In the case of cosmic ray heating Γ will be proportional to the ionization rate ζ (s^{-1}), per particle. Then the equilibrium requirement becomes $\Lambda(T) = \Gamma(T)/n = H/n$, where $H = \text{const} \times \zeta$. The solution to this equation is determined by the shape of the cooling function $\Lambda(T)$. To obtain it, one may draw a horizontal line corresponding to a constant value of H/n in a diagram of the cooling function and read off the temperature where the line crosses the cooling curve (see Figure 1.1).

In order to investigate the stability of the solutions, consider what happens if we add an amount dQ of heat to a volume of the gas. Adding the energy dQ corresponds to increasing the temperature by dT. Hence, in order for the gas to regain its previous temperature the net cooling rate $L(n,T) = n^2\Lambda(T) - n\Gamma(T)$ must now be greater than zero. More formally the condition for instability can be written $(\partial L/\partial S)_P < 0$, where S is the entropy. Since $dS = dQ/T$ for a reversible process this is essentially what we discussed above. The subscript P indicates that the thermodynamical quantity is evaluated for constant pressure. In the ISM, many perturbations are isobaric. Using the ideal gas law this condition can be transformed to (for constant pressure)

$$(\partial L/\partial T)_P = (\partial L/\partial T)_\rho - (\rho_0/T_0)\cdot(\partial L/\partial \rho)_T < 0$$

where ρ_0 and T_0 are the equilibrium density and temperature, respectively. Constant pressure corresponds to constant value of $nT = C$. Hence, for an ideal gas and cosmic ray heating ($\Gamma = H = \text{const} \times \zeta$, independent of the temperature)

$$(\partial L/\partial T)_P = [\partial[(C^2/T^2)\Lambda - (C/T)\Gamma]/\partial T]_P = -(n^2/T)\Lambda + n^2(\partial\Lambda/\partial T)_{nT} < 0$$

$$(\partial\Lambda/\partial T)_{nT} < \Lambda/T \quad \text{or} \quad (\partial \ln\Lambda/\partial \ln T)_{nT} < 1$$

Hence, the equilibrium temperature is unstable if the cooling function Λ does not rise fast enough with temperature. The unstable temperature regions are marked in Figure 1.1.

Figure 1.1. Radiative cooling curve for optically thin, low density interstellar HI gas. Regions prone to thermal instability are cross-hatched. Dominant coolants are shown. Solutions for a given density are obtained by drawing a horizontal line and reading off the temperature where it intersects the cooling curve. Adapted from Shull (1987).

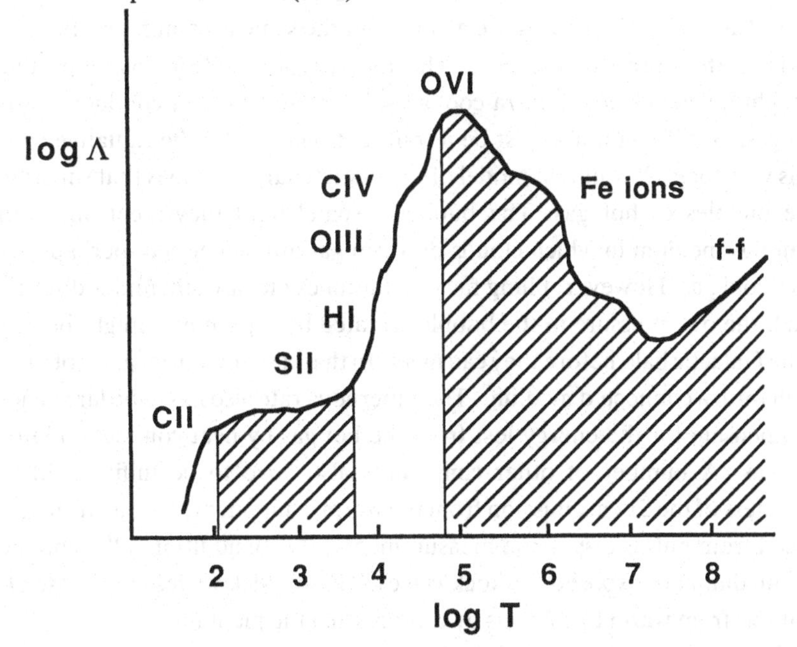

The slope of the cooling function $\Lambda(T)$ is determined by the temperature dependence of the excitation of different atomic lines. Below 100 K, $\Lambda(T)$ increases rapidly due to CII cooling. At 100 K CII cooling saturates and no major coolant turns on until 10^4 K, when Lyman α and some other transitions start to get excited. For temperatures above $\sim10^5$ K the cooling rate decreases since most atoms except iron are completely ionized. Above 10^7 K free–free emission takes over and the cooling rate coefficient increases again. Radiative cooling becomes important in supernova remnants after they have expanded during an adiabatic phase. The immediate postshock temperature is between 5×10^5 and 10×10^5 K when a supernova remnant has radiated away about half of its energy.

In order to have two stable regions coexisting not only thermal balance and stability are required for each region. Pressure balance between the regions must also be maintained. Hence, this requires three conditions to be fulfilled (i) $L(n,T) = n^2\Lambda(T) - n\Gamma(T) = 0$, (ii) $(\partial \ln\Lambda/\partial \ln T)_{nT} < 1$, and (iii) $P \propto nT =$ const. For the case of cosmic ray heating, $\Gamma(T) = H$ is temperature independent. Then by dividing condition (i), the thermal equilibrium equation, with n^2T we get $\Lambda(T)/T = H/(nT)$, where the right hand side is a constant for constant pressure. In a $\log \Lambda - \log T$ diagram the solutions correspond to the intersection of the cooling function with the line $\log [H/(nT)] + \log T$. Remember that $H/(nT)$ is a constant.

Three phase models of the ISM have been proposed. The temperatures which characterize the different components arise naturally from the properties of radiative cooling mentioned above. Cold ($\approx$80 K), dense, neutral gas coexists with warm ($\approx10^4$ K), dilute, neutral or ionized gas and with hot (10^6 K), very dilute gas, see Table 1.1. The phases are all at about the same pressure, and the denser material is, thus, pressure balanced. The hot gas is not stable in this simple theory but might be stabilized by extra cooling sources such as heat conduction to the warm gas. But it might also just be a transient phase since the radiative cooling time is very long. A common model of the interstellar medium is that supernovae create bubbles of hot gas. The bubbles expand until they meet and form a continuous medium in which clouds of warm gas are embedded, perhaps with a core of cold gas. However, filling factors, lifetimes etc. are still under discussion. Indeed, the overpressure in the bubbles created by supernovae might be vented into the Galactic halo before they can meet. In that case the warm, and not the hot, gas will form a connected medium. The supernova rate also has to be large enough to compensate for the energy lost from the hot gas by heat conduction into the warm gas. A number of other constraints also have to be fulfilled, like the smoothness of the 21 cm emission from the warm component, the electron density deduced from pulsar dispersion measurements, etc. In addition, all components of the medium are exposed to shock waves. These shock waves may cause phase transitions from warm to cold gas by compressing the medium.

Table 1.1. *Phases of the interstellar medium*

Name	$n\,(\mathrm{cm}^{-3})$	$T\,(\mathrm{K})$	Filling factor	Diagnostic
Hot intercloud gas	3×10^{-3}	10^6	0.4–0.7	Soft X-ray emission, ultraviolet absorption lines
Warm gas	0.1	10^3–10^4		If ionized – optical emission lines, pulsar dispersion measures
			0.4	If neutral – 21 cm observations
Diffuse clouds	>10	50–100		Optical and ultraviolet absorption lines, 21 cm observations
Cirrus clouds	10–10^3	10–100		Infrared emission, 21 cm observations, molecules
Dark clouds	>10^3	10	0.01?	Dark patches, infrared emission, molecules
GMCs	>10^3	15–40		Infrared emission, molecules
HII regions	>10	10^4		Optical, ultraviolet, infrared, and radio emission lines
SNR	>1	10^4–10^7		Radio to X-ray

Adapted from Jura (1987).

Our general picture is that we have clouds of warm gas with or without cooler cores. When this cooler gas achieves sufficiently high column density the hydrogen molecules (H_2) formed on grains will be protected from destruction by ultraviolet photons by dust grains. An important role in the shielding is also performed by the molecule itself since only ultraviolet photons in lines can dissociate H_2. The ultraviolet photons in these lines are soon exhausted – the molecule is self-shielded. When H_2 has formed, other molecules will start to form and ultimately we have a molecular cloud. Since rotational transitions of molecules can be excited at lower temperatures the cooling will be more efficient and the temperature can decrease towards about 10 K. In these molecular clouds massive stars will form and subsequently replenish the hot phase.

Molecular clouds are, in general, gravitationally bound and not bound by external pressure as are the HI clouds. Therefore, the thermal and pressure balance picture developed for HI clouds is not needed for molecular clouds. Small molecular clouds such as Cirrus clouds may form after compression of HI clouds by shocks.

The remaining part of this chapter will deal exclusively with the molecular clouds. It should be noted that the amount of molecular gas will depend on the metallicity. Lower metallicity means less dust grains shielding the molecules from ultraviolet radiation. Hence a larger column density of gas will be necessary before molecules are protected. This is particularly true for molecules less effectively self-shielded than H_2.

The detection of interstellar NH_3 lines at $\lambda = 1.3$ cm in 1968 was a breakthrough not only since it demonstrated that polyatomic molecules could form and survive in the hostile ISM but, furthermore, since it revealed the existence of a population of dense clouds in the dilute HI medium known from observations of the 21 cm line of atomic hydrogen. The dominant constituents of these clouds, H_2 and He, cannot radiate at the low cloud temperatures. Our knowledge of the physics, distribution and even the very existence of these massive molecular clouds relies on spectral lines from trace molecules, excited by collisions with H_2 and He. This also means that cooling via molecular lines is crucial for cloud energetics and cloud evolution towards protostellar formation.

1.2. Large scale distribution of molecular gas

For more than a decade a number of Galactic CO surveys have been performed. The aim has been to map the spatial, kinematical, mass, and size distributions of the molecular clouds. The total mass of these clouds is estimated to be $\approx 2 \times 10^9$ $M_\odot$, i.e., at least as high as the HI mass.

The average molecular gas density in the Galactic plane is only about one H_2 molecule per cubic centimetre, while the density in the molecular clouds is larger than 100 cm^{-3}. Therefore the volume filling factor of these giant molecular clouds (GMCs) is only 1%, but their 'typical' mass is some 10^5 $M_\odot$. Since the high density and mass facilitates gravitational collapse, this is the active star forming component of the interstellar medium.

The most extensive recent molecular cloud distribution surveys are listed below.

(i) The Massachusetts–Stony Brook survey, contains 40 000 CO spectra in the Galactic longitude/latitude region 8° to 90°/b = −1° to +1°, observed with a 0.8′ antenna beam at a spacing of 3′ for l = 18° to 54° and 6′ spacing outside this interval. At 3′ spacing this survey is expected to detect essentially all clouds of size >20 pc ($M > 10^5$ $M_\odot$) inside the solar circle.

(ii) The Columbia–Cerro Tololo survey, consists of 31 000 CO spectra taken in a region 10–30° wide along the entire Galactic plane, observed with a 0.5° beam and at 0.5° spacing (Dame *et al.* 1987). The same telescopes have also surveyed large regions close the plane with a 8′ beam and 0.125° resolution. According to Dame *et al.* (1986) an isolated cloud of a few times 10^5 $M_\odot$ should be seen at 20 kpc at 8′ resolution, but in the inner Galaxy, where clouds are crowded spatially and in velocity, such a cloud would be difficult to identify as a discrete object. This survey hence is biased towards detection of higher mass clouds.

(iii) The Bell Laboratories survey consists of more than 70 000 ^{13}CO spectra and

4000 CO spectra, covering $b = -1°$ to $1°$ over $l = -5°$ to $120°$. All of a $6'$ grid, most of a $3'$ grid, and small sections at $0.75'$ spacing have been observed with a $1.7'$ antenna beam. This survey, which is still ongoing, is designed ultimately to resolve and detect clouds of mass $>10^3$ $M_\odot$ anywhere in the first quadrant (Stark *et al.* 1988).

1.2.1. Radial and Z distributions

Already the early surveys have revealed that the dense molecular gas is concentrated in a 'ring' between 4 and 8 kpc from the Galactic center (for a solar distance of 10 kpc). While 80–90% of the total H_2 mass is found inside the solar circle only 30–40% of the HI gas is found there (cf. Figure 1.2). Recent

Figure 1.2. Mass of molecular hydrogen versus galactocentric radius, as obtained from azimuthally averaged data. From Clemens *et al.* (1988).

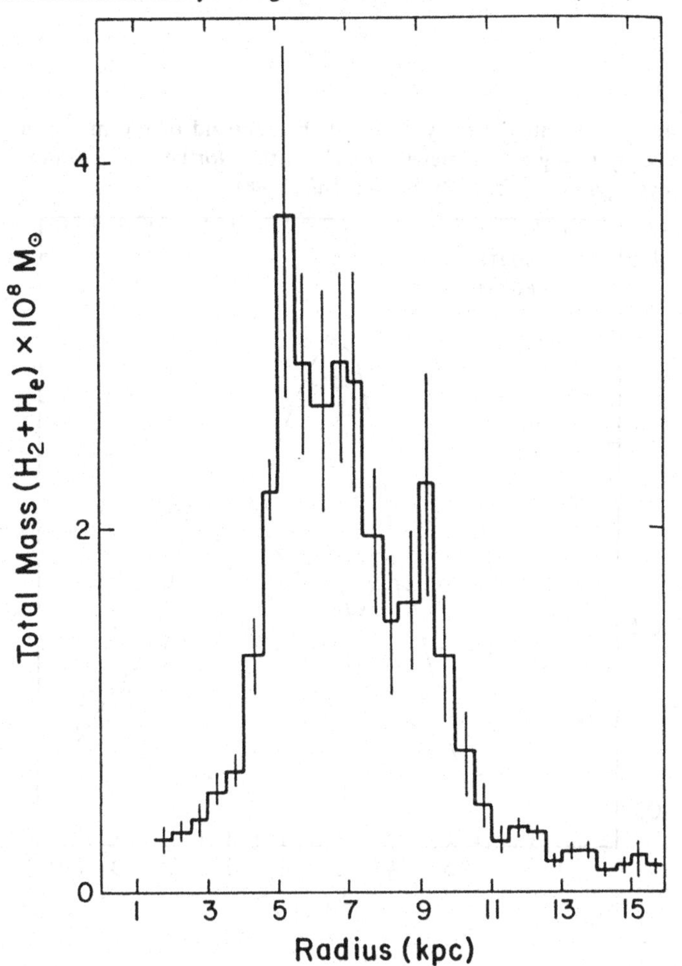

mass estimates for the 2–10 kpc region fall in the surprisingly narrow range (1.5–3) × 10^9 $M_\odot$ (H_2 and He). The proponents of the lower value (Columbia surveys, cf. Bronfman *et al.* (1988)) resolve the discrepancy into differences in instrumental calibration, CO to H_2 conversion, and statistical analyses, while scientists favoring the higher mass (Massachusetts–Stony Brook survey, cf. Clemens, Sanders and Scoville (1988)) suggest that the far-side CO emission from the inner Galaxy may be inadequately accounted for in the Columbia surveys.

Large scale deviations from azimuthal symmetry in the radial distribution between the first and fourth quadrant have been found in the Columbia–Cerro Tololo data (Figure 1.3, Bronfman *et al.* (1988)). This may be a result of spiral arm structures (see Section 1.2.2). The thickness (FWHM) of the molecular disk varies from 80 pc at 3 kpc to 150–300 pc at 10–12 kpc and midplane displacements from the Galactic equator (warps) are apparent (Grabelsky *et al.* 1987, Bronfman *et al.* 1988, Clemens *et al.* 1988). The HI scale height seems to be about twice that of CO.

Figure 1.3. Number density of H_2 at the centroid of the molecular layer derived by fitting axisymmetric models to the Northern and Southern CO Surveys separately. From Bronfman *et al.* (1988).

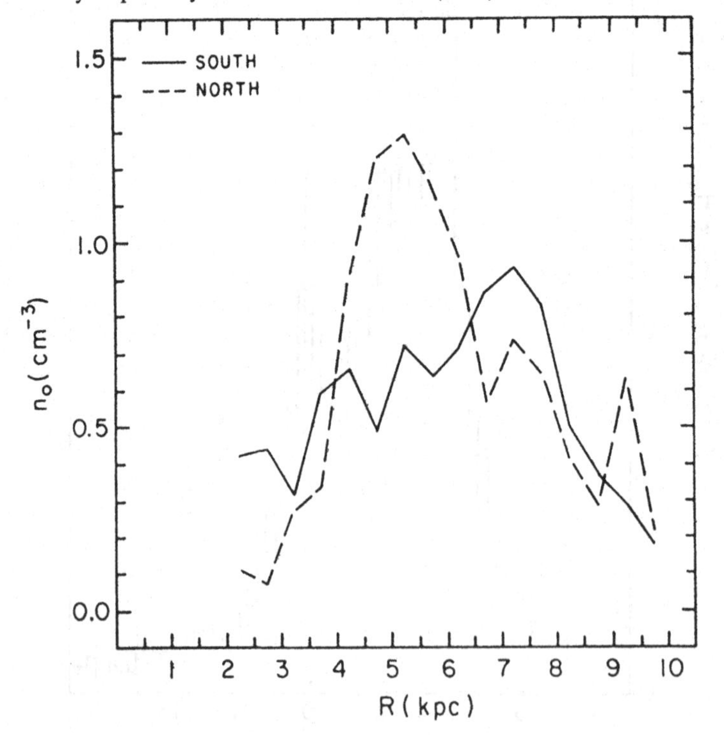

1.2.2. Azimuthal distribution and velocity field

Since spiral arms of stars and HII regions often are conspicuous in external galaxies it is important for our understanding of large scale star and cloud formation processes to find out the distribution of the dense star forming gas. Do spiral arms of giant molecular clouds exist? Can streaming motion be observed across arms?

The Massachusetts–Stony Brook survey has revealed the existence of a spiral arm population of warmer (>10 K) clouds and clouds associated with HII regions, contrasting with a rather uniform disk population of cold (<10 K) clouds (cf. Figures 1.4 and 1.5). The HII regions appear to be associated with systematically warmer, larger and more massive CO clouds (Solomon, Sanders, and Rivolo 1985, Rivolo, Solomon, and Sanders 1986, Scoville *et al.* 1987, Waller *et al.* 1987). The *quadratic* dependence of the number density of HII regions on the local H_2 density may suggest that the most massive stars (OB) form in spiral arms as a result of cloud–cloud collisions (Scoville, Sanders, and Clemens 1986). Solomon *et al.* (1985) note that the warm and cold cloud populations have about the same total emissions, indicating similar total masses. Clemens *et al.* (1988) have tried to delineate the detailed distribution of H_2 in the first quadrant. The resulting face-on map is dominated by the molecular cloud 'ring', and the Sagittarius and Perseus 'arms' – all three being more ring-like than spiral-like. CO emission is present between the arm features and the mean arm–interarm contrast is 3.6:1.

The Columbia–Cerro Tololo surveys, taken with a 8′ antenna beam, tend to emphasize large massive clouds or cloud complexes – especially so since these investigators remove an extended background (of smaller clouds?) to distinguish large complexes clearly. For example, Dame *et al.* (1986) note that in this way they remove 63% of the non-local emission (i.e., emission at velocities corresponding to a non-local kinematic distance). For l = 2–60°, they identified 26 clouds of mass $>5 \times 10^5$ $M_\odot$ and only 7 lower mass clouds. The Sagittarius arm is clearly delineated by 17 large complexes. Using the same data Myers *et al.* (1986) identified 54 cloud complexes with mean mass $\approx 10^6$ $M_\odot$. Clouds associated with radio HII regions tend to have higher mass, as Waller *et al.* (1987) also concluded. The Carina arm in the outer Galaxy is a dominant feature in the Cerro Tololo data for l = 270–300° (Grabelsky *et al.* 1987), and an arm–interarm contrast as large as 13:1 is found. The average contrast ratio between 'rings' at 11.5 kpc and 10.5 kpc is 4.5:1 in the Carina arm.

From the Bell Laboratories ^{13}CO and CO survey Stark *et al.* (1987) conclude that there exist large voids containing few or no GMCs. These voids appear since GMCs are concentrated in spiral arms and are also present in the Columbia survey. In the Bell Laboratories survey numerous lower mass molecular clouds

Figure 1.4. The (l, v) distributions of 4 K clouds, hot cloud cores, and HII regions are shown. The latter two samples exhibit tighter confinement in the (l, v) plane, indicative of a spiral arm population. From Scoville *et al.* (1987).

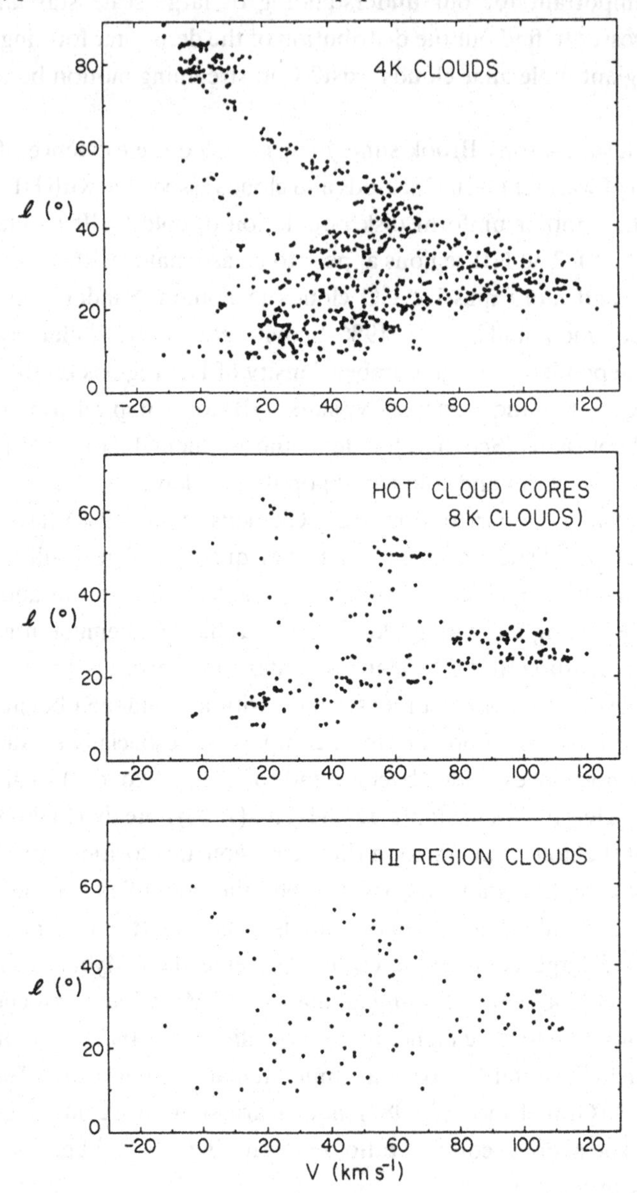

Figure 1.5. Galactic radial distribution of cold molecular cloud cores between longitudes 20 and 50° (COLD ^{12}CO), warm molecular cores between longitudes 20 and 50° (WARM ^{12}CO), and radio HII. Adapted from Solomon *et al.* (1985).

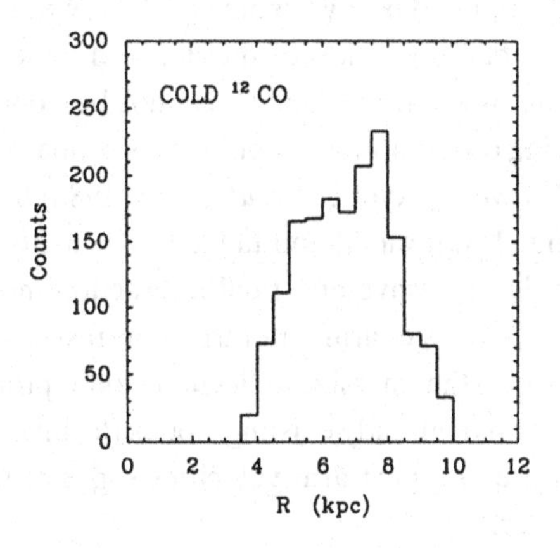

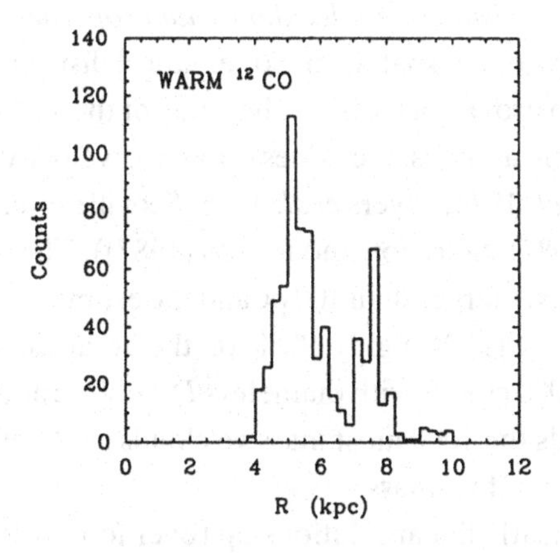

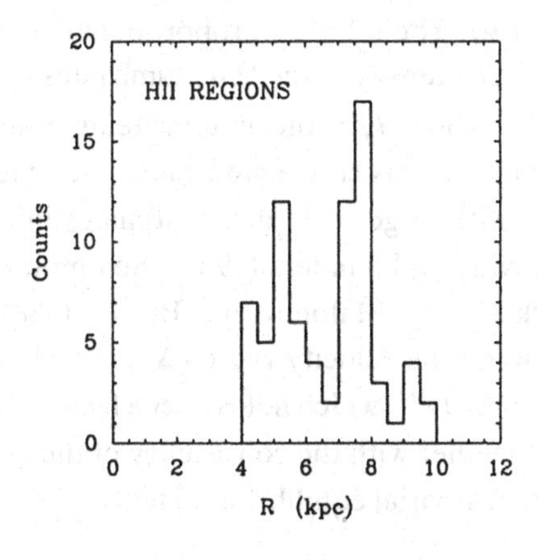

are present throughout the voids/interarm regions, but these clouds are usually not detected in the Columbia survey due to its lower sensitivity for smaller clouds.

Although some streaming motions of 2 kpc length and velocity up to 15 km s^{-1} have been separated out in our Galaxy by Clemens (1985), we are forced to accept the fact that simultaneous determination of the detailed molecular distribution and the streaming motion in the inner Galaxy cannot be done (Clemens *et al.* 1988). The most convincing observations of spiral arms probably are found in the 10 Mpc distant galaxy M51 where Rydbeck *et al.* (1985, 1988) have demonstrated the existence of large and abrupt radial and tangential velocity shifts across CO arms in agreement with density wave and tidal interaction modeling. Rydbeck *et al.* (1988) estimate an average arm–interarm contrast ≥ 4. The observed velocity shifts support the idea that massive molecular cloud spiral arms are caused by density wave organization of already existing clouds ('orbit crowding') and the spatial overlap between pre and post arm velocities suggests that fairly violent cloud–cloud collisions occur.

1.2.3. Statistics on molecular cloud properties

The extensive CO surveys of the spatial and kinematical distribution of molecular clouds in the Galaxy also provide a basis for the study of the statistical distribution of cloud properties such as sizes, densities, masses, velocity dispersions and temperatures (Dame *et al.* 1986, Myers *et al.* 1986, Scoville *et al.* 1987, reviews by Scoville and Sanders (1987), Solomon and Rivolo (1987)). The sizes range from a few parsecs up to complexes larger than 100 pc and the estimated masses vary from some 10^4 $M_\odot$ up to 10^7 $M_\odot$. Roughly 90% of the total Galactic H_2 mass is concentrated in about 5000 clouds with diameters $D > 20$ pc and masses $M > 10^5$ $M_\odot$. Of these 5000 clouds the 1000 most massive clouds ($D > 50$ pc and $M > 10^6$ $M_\odot$) contain 50% of the total H_2 mass.

There seems to be a nearly linear relationship (over four orders of magnitude) between the virial masses of the identified clouds and their observed CO luminosities (see Figure 1.6). The inferred proportionality constant for the conversion of the integrated CO intensity to the H_2 column density is $N(H_2)/\int T_R^* dv = 3 \times 10^{20}$ cm^{-2} (K km $s^{-1})^{-1}$ where T_R^* is the antenna temperate. This is very close to other empirical estimates of this conversion ratio – so extensively used (and discussed) in connection with large scale mass estimates in our own as well as external galaxies (cf. Scoville and Sanders (1987), Solomon and Rivolo (1987), van Dishoeck and Black (1987), Maloney and Black (1988)). There is also a striking relationship between the velocity width (Δv) of a cloud and its diameter (D), *viz.* Δv is proportional to $D^{1/2}$, which holds over a factor of 30 or so in size (see Figure 1.6). The result, together with the consistency of the virial mass estimates shows that the clouds are near virial equilibrium. Hence, the clouds are bound by self-gravity and not confined by pressure equilibrium with the cold or warm HI gas phase of the interstellar medium.

In their progress report on the Bell Laboratories ^{13}CO survey Stark *et al.* (1988) claim that the linewidths of molecular clouds are inversely related to their distance to the Galactic center. They suggest that this may be a consequence of the Galactic tidal field. In this picture, clouds near the Galactic center would be considerably

Figure 1.6. (*a*) The virial mass–CO luminosity relation for molecular clouds. The clouds range in distance from 2 to 15 kpc and in flux over more than two orders of magnitude. The fit is $M_{VT} = 39(L_{CO})^{0.81} M_\odot$. For a given CO luminosity the dispersion in virial mass is 0.13 in the log. (*b*) Molecular cloud velocity dispersion $\sigma(v)$ as a function of size S (harmonic mean of diameter in l and b) for 273 clouds in the Galaxy. The fitted line is $\sigma(v) = S^{0.5}$ km s^{-1}. For virial equilibrium the 0.5 power law requires clouds of constant average surface density. From Solomon *et al.* (1987).

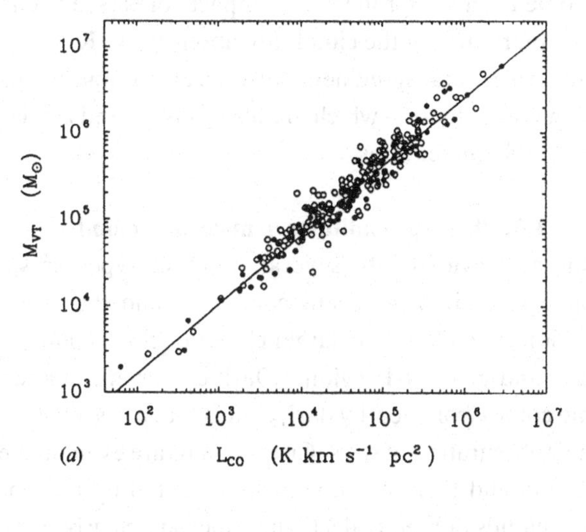

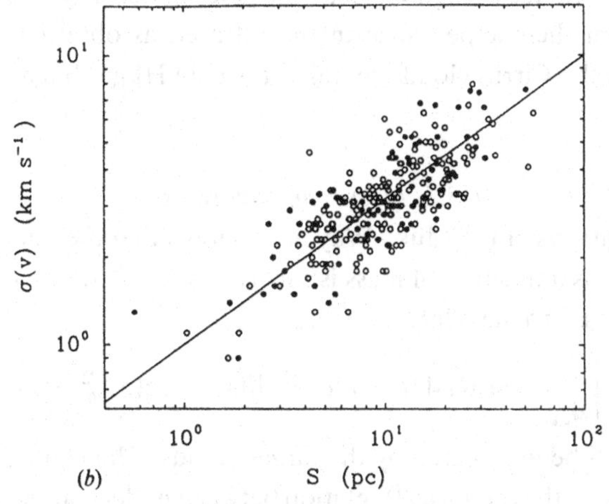

denser than clouds in the solar neighborhood. The mass and mean density of a 'standard' cloud in virial equilibrium are related to the diameter as (cf. Scoville and Sanders (1987)) $M_{vir}/M_{\odot} \approx 3 \times 10^5 (D/40\ \mathrm{pc})^2$ and $\langle n(\mathrm{H_2}) \rangle \approx 180 (D/40\ \mathrm{pc})^{-1}$ cm^{-3}. (The exponent is close to -1; we will use -1 in this paper.) Hence clouds of diameters of 10–100 pc have mean densities of 700 to 70 cm^{-3} and masses of 2×10^4–2×10^6 $M_{\odot}$. As will be discussed in Section 1.3.1 the core densities of these clouds are estimated to be considerably higher, 10^4–10^6 cm^{-3}.

Some comments on the temperature information contained in the Massachusetts–Stony Brook CO survey also seem appropriate here. The catalog by Scoville *et al.* (1987) contains 1427 clouds with peak kinetic temperature $\geq$10 K, 255 warmer cloud cores with peak temperatures $\geq$16 K and 95 clouds associated with 171 radio HII regions (cf. Scoville and Sanders (1987)). Although the peak temperatures of the warmer clouds (as measured by CO) fall in the range 15–60 K, and are known to be even higher in some compact cores (see Section 1.3.2), the average kinetic temperatures of the clouds are about 10–15 K.

There seems to be a general agreement between all groups making surveys that the larger, more massive clouds – which are also warmer and associated with HII regions – outline spiral arm features.

1.3. Physical conditions in molecular clouds

Molecular clouds are divided into several different types as summarized in Table 2. The table also gives the Jeans conditions and the free fall timescale formulae. Giant molecular clouds are larger clouds or cloud complexes associated with massive stars and giant HII regions. Dark clouds are quiescent, low mass clouds. The name comes from nearby dark clouds that are visible as dark areas in the sky due to their obscuration of stars. Cirrus clouds are even smaller clouds with low visual extinction and therefore not easily detected by their obscuration of stars. Only Cirrus clouds at high Galactic latitudes are clearly discernible. Their name comes from their appearance in the infrared as observed by the IRAS satellite. Most of the Cirrus clouds contain only cold HI gas but about 10% have molecular cores.

1.3.1. Masses and densities

The molecular surveys of the Milky Way have shown that the number density of molecular clouds as a function of mass is: $n(M) \approx \alpha M^{-1.5}$, where α is a constant. Since the total mass in a mass interval (M_{up}, M_{low}) is

$$M_{tot} = \int_{M_{low}}^{M_{up}} M n(M)\, \mathrm{d}M = [2\alpha M^{1/2}]_{M_{low}}^{M_{up}} = 2\alpha(M_{up}^{1/2} - M_{low}^{1/2})$$

the total mass will be dominated by the larger clouds. The cloud mass is mainly determined by using the (empirical) relation between molecular hydrogen column

Table 1.2. *Molecular regions in the interstellar medium*

	Giant molecular cloud	Dark cloud
Complex		
Size (pc)	20–80	6–20
Density (cm^{-3})	100–300	100–1000
Mass ($M_\odot$)	8×10^4–2×10^6	10^3–10^4
Linewidth ($km\,s^{-1}$)	6–15	1–3
Temperature (K)	7–15	≈ 10
Cloud		
Size (pc)	3–20	0.2–4
Density (cm^{-3})	10^3–10^4	10^2–10^4
Mass ($M_\odot$)	10^3–10^5	5–500
Linewidth ($km\,s^{-1}$)	3–12	0.5–1.5
Temperature (K)	15–40	8–15
Core		
Size (pc)	0.5–3	0.1–0.4 $\approx \lambda_{\rm Jeans}$
Density (cm^{-3})	10^4–10^6	10^4–10^5
Mass ($M_\odot$)	10–$10^3 \geq M_{\rm Jeans}$	0.3–10 $\approx M_{\rm Jeans}$
Linewidth ($km\,s^{-1}$)	1–3	0.2–0.4 $\approx$ thermal width
Temperature (K)	30–100	≈ 10

$\lambda_{\rm Jeans} = 0.27 \left(\frac{T/10}{n/10^4}\right)^{1/2}$ pc; $M_{\rm Jeans} = 4. \left(\frac{(T/10)^3}{n/10^4}\right)^{1/2} M_\odot$

$t_{ff} = 4.10^5/(n/10^4)^{1/2}$ yr

Adapted from Goldsmith (1987).

density and the CO emissivity: $N(H_2) = (3 \pm 2) \times 10^{20}\, W_{\rm CO}$, where $W_{\rm CO} = \int T_{\rm R}^* v$ (K km s^{-1}). The cloud mass then is obtained by integrating $N(H_2)$ over the cloud surface. In fact, it is not clear that the relation $N(H_2) = 3 \times 10^{20}\, W_{\rm CO}$ is useful for pointwise estimates of the molecular hydrogen column density $N(H_2)$. Dense cores in dark clouds, which locally cause peaks in $N(H_2)$, often are not seen or are just barely seen in the integrated CO intensity, due to the large optical depth in the CO line. The reason why the conversion formula still works can be understood from virial arguments. A more massive cloud in virial equilibrium has a larger velocity dispersion than a lighter cloud ($M_{\rm vir} \approx R\Delta v^2/G$, where R is the cloud radius, Δv is FWHM line width and G is the gravitational constant; this formula can also be directly used to estimate cloud masses). The larger linewidth also causes the integrated CO intensity to be larger for the more massive cloud. Hence the integrated CO intensity should be more affected by the total mass than by the local column density of molecular hydrogen. The large scale relation between mass and integrated CO intensity is also apparent from comparison of CO, HI and γ-rays. The γ-rays are believed to be proportional to the total column density of matter in the observed direction. Due to the crude resolution of the γ-ray observations the correlation between integrated CO intensity and $N(H_2)$ can only be confirmed on large scales (cf. Bloemen *et al.* 1986).

In order to determine the mass in the denser parts of the clouds we must use optically thin emission from rarer isotopic species such as $C^{18}O$. Several authors have established relations between column densities of ^{13}CO and $C^{18}O$ and the visual extinction derived from star counts (Dickman 1978, Frerking, Langer, and Wilson 1982). By using the relation between molecular hydrogen column density and visual extinction we can then compute the mass. However, ^{13}CO and $C^{18}O$ may not be so useful to estimate the total cloud mass since their emissions are too weak for observing large areas, and close to the cloud edges these species may be less excited than CO (Guélin and Cernicharo 1988).

The average cloud or cloud complex density may be computed by dividing the mass by the observed dimensions (the depth has to be estimated from observations of width and height). As already mentioned in Section 1.2 the result is that the average cloud or cloud complex density decreases with increasing size: $\langle n(H_2)\rangle = 180(D/40\ \mathrm{pc})^{-1}\ \mathrm{cm}^{-3}$. Since the mass is proportional to $D^3\langle n\rangle$ we find that it is proportional to D^2. Over the observed sizes molecular clouds have a fractal dimension of 2.

Different molecules and molecular transitions require different densities to be excited. The excitation is controlled by the balance between collisional excitation and radiative decay. Transitions between excited states of molecules with large dipole moments are detected only when the density is high. The high radiative decay rate caused by the large dipole moment must be balanced by a high collision rate in order to get detectable emission. For example, the CS $J = 2–1$ and $J = 7–6$ transitions are observed only if the densities are greater than about $2 \times 10^5\ \mathrm{cm}^{-3}$ and $1 \times 10^7\ \mathrm{cm}^{-3}$ respectively. The detection of such lines permits one to estimate the density directly. It is better to use observations of several transitions and fit them to results from a statistical equilibrium calculation in order to estimate the temperature, density, and abundance.

1.3.2. Cloud temperatures

Since the CO low energy levels are easily collisionally excited and the lines connecting them are very optically thick, their populations are thermalized over large parts of the clouds, i.e., the excitation temperature is equal to the kinetic temperature. From the fact that the $CO/^{13}CO$ $J = 1–0$ intensity ratio is only $\approx 2–10$ while the isotopic ratio is 40–90 it can be seen that the CO line is very optically thick. Thus the brightness temperature derived from the CO emission should be equal to the cloud temperature. However, if the cloud is microturbulent the large optical depth in the CO lines will stop us from looking into the cloud and we will measure the temperature on the front surface at an optical depth of roughly one.

Only if the cloud is composed of many small clumps with random velocities or if we have a large scale velocity gradient will the CO measurement apply to the interior of the cloud. Of course, for more or less isothermal clouds the front temperature will be equal to the cloud temperature (cf. Kwan and Sanders 1986).

Another useful temperature probe is NH_3 (Ho and Townes 1983). The symmetric top molecule NH_3 has many inversion transitions at $\lambda = 1.3$ cm. By comparing the column densities in different rotational states the population distribution (rotation) temperature can be computed. The rotation temperature is equal to the kinetic temperature for low temperatures while corrections are needed for higher temperatures. Since NH_3 is much less abundant than CO and harder to excite we can be sure that its population distribution measures the temperature inside the clouds, irrespective of their density structure and velocity field. An example is the OMC1 molecular cloud, where the NH_3 measurements give 20–50 K, considerably cooler than the temperature measured by CO (T_k(CO) = 60–100 K), consistent with heating of the near side by the HII region. There are, however, also regions close to the star forming KL region where the temperature inside the cloud is high ($\geq$150–250 K). Another example is TMC1, which is a core region in a nearby dark cloud complex. Here, the NH_3 and CO measurements give the same temperature of about 10 K. Other temperature probes are discussed by Hjalmarson and Friberg (1988). As already discussed in Section 1.2, the bulk of the molecular clouds have a mean temperature close to 10 K. This seems to be the characteristic equilibrium temperature determined by cosmic ray heating and molecular line cooling in our Galaxy. However, molecular clouds associated with local heating sources, such as HII regions or massive star formation, have warmer cloud cores. These are the clouds that outline the spiral arms in our Galaxy.

1.3.3. Velocity fields

Many details about the velocity fields still remain uncertain. In the most quiescent cold dark cloud cores the linewidths are close to the width caused by thermal motions of the molecules at 10 K, but, in general, the observed linewidths are highly supersonic. This supersonic velocity dispersion seems to be due to large scale chaotic velocity fields. In some cases we can directly see that the linewidth becomes narrower when the spatial resolution increases. Such an example is found in NGC2024 IRS2, where Black and Willner (1984) observed a linewidth of 1.4 km s^{-1} for vibrational transitions of CO in absorption against the infrared source. The CO mm-wave lines observed by radio telescopes with much coarser spatial resolution have much larger linewidths for all observed isotopes. This tells us that the cloud is built up of several larger regions each having a small velocity dispersion. It is not surprising that the linewidths are caused by macroturbulence since it would be hard energetically to maintain a supersonic microturbulence. Shock waves would in that case dissipate the energy so fast that a powerful energy source would be needed to maintain the microturbulence.

The linewidths for molecular clouds are correlated with their sizes. For a virialized cloud the mass should be proportional to the diameter (D) and the observed line width squared $(\Delta v)^2$: $M \propto D\Delta v^2$. The observed mass–size correlation $M \propto D^2$ then implies that virialized clouds should have a velocity dispersion proportional the square root of the diameter D. Such a relation is also empirically observed (cf. Figure 1.6). This supports the idea that molecular clouds are virialized and hence contained by gravitation and not external pressure. An exception to this might be the high latitude Cirrus clouds that have masses, as derived from CO and extinction observations, which are far too small for their velocity dispersion. Since pressure confinement is also doubtful, the Cirrus clouds may be short-lived condensations behind shockwaves that are not confined (cf. Myers (1987)).

If larger molecular clouds and complexes are created by cloud–cloud agglomeration we would expect a cloud–cloud velocity dispersion which is inversely proportional to the square root of the mass: $\sigma_{cc} \propto M^{-1/2}$, in order to have the same kinetic energy per mass unit in all clouds. The observed cloud–cloud dispersion ranges from 3 to 7 km s^{-1}. This is smaller than expected since the clouds cover three orders of magnitude in mass. However, since the cloud–cloud dispersion can be confused with non-circular orbits and streaming motions along spiral arms, cloud–cloud agglomeration is not ruled out.

1.3.4. Morphology and clumping

It seems to be a general rule that structure is seen in clouds down to the resolution limits of the observations. The general decline of the mean density with cloud size – $\langle n \rangle = 180(\mathrm{D}/40\ \mathrm{pc})^{-1}$ – also makes it clear that the clouds must contain density structure since the average density is too low to excite many of the observed spectral lines. The lower scale for the clumps seems to be 0.1–0.01 pc from observations of dark clouds (cf. Figure 1.8). Dark cloud cores have a NH_3 brightness temperature that is equal to the kinetic temperature. This means that the source is uniform within the antenna beam. It is feasible that larger clouds are built up from such smaller dark cloud cores. However, we cannot make larger clouds just by putting together smaller clouds with a constant mean distance between them. This would lead to a $M \propto D^3$ dependence and not to the observed relationship $M \propto D^2$. A simple way to achieve the observed relationship is to form clouds hierarchically. First medium sized clouds form from the small clouds. Then bigger clouds form from the medium sized clouds, with spaces between them that

Figure 1.7. The black area in the figure is an example of a fractal structure of dimension ln3/ln2 ≈ 1.585. If we double the linear extent of the figure (in a self-similar way) the black area will triple instead of quadruple as normal. This gives the fractal dimension ln3/ln2 ≈ 1.585 instead of the normal ln4/ln2 = 2. To explain the apparent fractal dimension of 2 for the mass of a molecular cloud, instead of 3, it may be built up from smaller clouds in a similar way.

are correspondingly larger – i.e., we should create a fractal with dimension 2, see Figure 1.7 (cf. Falgarone and Perault (1987)).

Self-gravitating systems should be spherical if they have time to relax and have zero total angular momentum and magnetic field. Most larger clouds and cloud complexes are, if not spherical, at least somewhat rounded. Dark cloud complexes like the Taurus complex have a partly filamentary structure. The clouds can still be gravitationally bound but not relaxed. More likely the filamentary structure is supported by magnetic fields. The Cirrus clouds generally have quite irregular shapes and appear to be sheet-like, which fits well with the suspicion that they actually are not gravitationally bound.

1.3.5. Energetics and cloud support against collapse

Most molecular clouds have temperatures between 10 and 15 K, resulting from a balance between heating by cosmic rays and cooling by radiation from CO, H_2O, C, etc. Energetics problems associated with the creation and maintenance of the velocity dispersion and the support against gravitational collapse are not well understood. Gravitationally bound clouds do not seem to evolve on free-fall timescales. While the observed star formation rate in the Galaxy is 3–10 $M_{\odot}$/yr, the total H_2 mass divided by the free-fall time (for $n \approx 100$ cm^{-3}) is about 2×10^9 $M_{\odot}/4 \times 10^6$ yr = 500 $M_{\odot}$/yr. A number of processes have been proposed to explain the discrepancy between these two rates (cf. Downes (1987)):

(i) *Cloud rotation*: not observed in most clouds.

(ii) *Winds*: mass ejection (bipolar flows) from young stars may be energetic enough to maintain the velocity dispersion even if shocks are created. This would mean self-regulating star formation (Norman and Silk 1980).

(iii) *Supersonic turbulence*: the problem is to avoid too much dissipation due to shocks that radiate away the energy. Two types of turbulence have been suggested. Incompressible Kolmogorov type turbulence and compressible 'star-cloud turbulence'. While the origin for the former type of motion is uncertain, the latter type supposedly could be driven by Galactic differential rotation (cf. Scalo (1987)).

(iv) *Magnetic fields*: magnetic fields can by themselves support clouds for some time. Another promising idea is that Alfvén waves can help to spread turbulent energy from intrinsic sources or from bipolar flows without creating shocks. This is possible since the velocities are supersonic but not super-Alfvénic. Support for this idea is provided by the observed relationship between velocity and cloud size (cf. Myers (1987)).

1.4. Chemical content

During the last two decades a growing number of molecules have been identified in the interstellar medium; see Table 1.3. In addition to the more than 70 molecules we may add some 50 isotopic variants. Most of the molecules have been identified through their millimeter wave spectra. It is a challenge to explain the abundances of these often rather complex molecules. They form and survive, at least for some time, in a very dilute cold medium exposed to destructive ultraviolet radiation. Even in dense warm regions near newly formed stars the density is a dozen orders of magnitude less than that in the Earth's atmosphere at sea level and the temperature is still below 200 K. These molecules also provide the only way of studying the physical conditions in these clouds. The line intensity is determined by, apart from the molecular abundance, the rate of collisional excitation, which depends on the temperature and density. When combined with chemical knowledge some abundances also give information about the ionization level. Individual cloud ages may also be possible to determine since the chemical timescale (10^7 yr) is much longer than the dynamic timescale for dense clouds.

We note that of all the molecules in Table 1.3 about 25% have been identified through dedicated spectral surveys of large frequency ranges – spectral scans – and many others by pure accident. Some of the more extensive spectral scans are, in TMC1 22–24 GHz and 36–50 GHz by Kaifu *et al.* (1987); in Sgr B2 72–145 GHz Cummins, Linke, and Thaddeus (1986); in OMC-1 72–91 GHz by Johansson *et al.* (1984), 215–263 GHz by Sutton *et al.* (1985) and Blake *et al.* (1986). Even if a large fraction of the accessible radio frequency range has been covered new molecules

Table 1.3. *Interstellar molecules*

Simple molecules:			
H_2	CO	NH_3	CS
HCl	SiO	SiH_4[a]	SiS
C_2	HNO?	CH_4[a]	OCS
PN	SO_2	H_2O	H_2S
$NaCl$[a]	$AlCl$[a]	KCl[a]	AlF[a]?
Nitriles, acetylene derivatives, and related molecules:			
HCN	$HC{\equiv}C{-}CN$	$H_3C{-}C{\equiv}C{-}CN$	$H_3C{-}CH_2{-}CN$
H_3CCN	$H(C{\equiv}C)_2{-}CN$	$H_3C{-}C{\equiv}C{-}H$	$H_2C{=}CH{-}CN$
$CCCO$	$H(C{\equiv}C)_3{-}CN$	$H_3C{-}(C{\equiv}C)_2{-}H$	HNC
$CCCS$	$H(C{\equiv}C)_4{-}CN$	$H_3C{-}(C{\equiv}C)_2{-}CN$	$HNCO$
$HC{\equiv}CH$[a]	$H(C{\equiv}C)_5{-}CN$	H_3CNC?	$HNCS$
$H_2C{=}CH_2$[a]			
Aldehydes, alcohols, ethers, ketones, amides, and related molecules:			
$H_2C{=}O$	H_3COH	$HO{-}CH{=}O$	H_2CNH
$H_2C{=}S$	H_3CCH_2OH	$H_3C{-}O{-}CH{-}O$	H_3CNH_2
$H_3C{-}CH{=}O$	H_3CSH	$H_3C{-}O{-}CH_3$	H_2NCN
$NH_2{-}CH{=}O$	$H_2C{=}C{=}O$		
HC_2CHO	$(CH_3)_2C{=}O$?		
Ions:			
$CH+$	$HCS+$	H_2D+?	$HCNH+$
HN_2+	$SO+$?	$HOCO+$	$HOC+$
H_3O+?	$HCO+$		
Cyclic molecules:			
C_3H_2	SiC_2[a]	$C{-}C_3H$	
Radicals:			
CH	C_2H	CN	HCO
OH	$l{-}C_3H$	C_3N	NO
	C_4H	NS	SO
	C_5C_6HH	C_2S	H_2CCN

[a] Detected only in envelopes around evolved stars.

are still identified. However, it might be suspected that all major species which have strong radio spectra have been found by now – knock on wood. The last years have witnessed the detection of the very abundant carbon ring molecule C_3H_2 as well as the similarly abundant carbon chain molecule C_2S. We have also started to detect molecules containing less abundant elements such as phosphorus, chlorine, and aluminum. PAHs (polycyclic aromatic hydrocarbons) are other large 'molecules' or small grains whose existence in the ISM is being established.

1.4.1. Abundance determinations

In order to study the chemistry the spectral line intensity information has to be transformed to abundances – often a non-trivial task. To achieve a good determination several different lines of the same species have to be observed. But since these lines often differ in frequency, beam size and beam shape will differ and hence the regions sampled will differ for the lines observed. We can always try to correct for such effects if we know the spatial distribution of the molecules but it introduces an additional uncertainty. In addition, a 20–30% calibration uncertainty due to variations in atmospheric attenuation, errors in antenna efficiencies, and pointing variations etc. are not uncommon. Some of these problems are avoided for molecules for which several transitions, from vastly different energy levels, can be observed simultaneously, e.g., symmetric top molecules such as CH_3CN and CH_3C_2H. Other problems are due to radiative transfer and excitation. The lines may be saturated or even self-absorbed. Self-absorption, where the photons are absorbed by low excitation foreground gas, is only the most extreme case of variations along the line of sight.

In many situations the fractional abundance is computed by dividing the column density of the molecule by that of H_2, i.e., $f[X] = N(X)/N(H_2)$. The H_2 column density is often estimated from the integrated intensity of a CO isotope or from the millimeter-wave dust continuum emission. If the species X *only occurs, or only is excited in parts of a cloud* this method may introduce large errors. A better method would be to use a statistical equilibrium calculation to compute the density necessary to excite the molecule and the fractional abundance needed to achieve the observed intensity. However, this method also has to be based upon assumptions about the velocity structure of the cloud since it affects the radiation field within the cloud. Most often an LVG (large velocity gradient) approximation is used and considerable errors may result since the relevant velocity gradient may be difficult to determine.

It should be noted that many of these problems affect the determination for abundant molecules most. For optically thin emission the observed antenna temperature is proportional to the column density in the upper state. For temperatures much higher than the cosmic background the proportionality constant is independent of the excitation temperature and only the partition function is needed to convert the upper state column density to a total column density. With this in mind it is not surprising that an abundance determination in dense clouds is uncertain to at least a factor of 2–3 and, in some cases, an order of magnitude. Note also that the concentrations of more abundant species often are estimated by observing a rarer isotopic variant and then assuming an isotope ratio – usually the terrestrial one.

1.4.2. Chemical abundances

A relatively thorough knowledge of the abundances of the detected molecules is limited to a few sources. These sources represent physically different regions, which hopefully are representative of rather general interstellar cloud conditions. The regions are Orion KL, which contains at least four subregions of different chemistry; the Galactic center cloud Sgr B2, the most massive giant molecular cloud known in the Galaxy, which also contains several subregions; and to a lesser extent the nearby cold dark clouds TMC1 and L134N. A brief discussion of these regions follows.

(i) *Orion KL*. The giant molecular cloud in Orion is about 500 pc from the sun, making it the nearest region of massive star formation. The cloud itself covers several tens of square degrees, but most of the millimeter-wave observations have focused on the region behind the Orion Nebula (M42). There is a extended ridge of gas and dust with embedded newly formed massive stars, which are detectable in the near infrared but are completely obscured at optical wavelengths. The central core is called OMC-1 and includes the Kleinmann–Low (KL) infrared nebula. Closer observations of Orion KL reveal several subregions which have distinct spatial and spectral distributions. First, the previously mentioned *extended ridge* of gas is elongated in the north–south direction but has a kink at Orion KL. The molecular emissions from this region have linewidths of 1.5–4 km s^{-1}.

The hydrogen density reaches $n \approx 10^5$–10^6 cm^{-3} with a kinetic temperature in the range 25–50 K. Together with the kink at Orion KL there is a sharp velocity gradient such that the ridge has a velocity of 10 km s^{-1} south of Orion KL. Near the interface between these ridge components broad spectral lines are observed, indicating violent activity which seems to be powered by IRc2, the intrinsically most luminous source in Orion KL. The large width of the spectral lines emanating from this region is the reason for the name '*the plateau*'. High resolution single dish and interferometer data resolve the plateau into a somewhat less wide, in velocity space, emission feature from a disk or 'doughnut' and a high velocity bipolar flow. The outflow direction is roughly perpendicular to the plane of the disk or doughnut structure. The *hot core* is a distinct region which is characterized by emission from very highly excited molecules. It may be a large clump of dense material left over from the star forming period. Another region is the *compact ridge* cloud which has the same central velocity and velocity width as the southern ridge cloud but is much warmer ($\approx$150 K). It may represent an area of interaction between the outflow from IRc2 and the southern ridge and has velocity signatures which could be caused by gravitational contraction.

(ii) The *Sgr B2* molecular cloud is also the site of massive star formation in a giant molecular cloud, but on a more extreme scale. The estimated luminosity from the stars formed in Sgr B2 is $\approx 3 \times 10^7$ $L_\odot$. Molecular observations reveal an extended core region 40 pc in diameter and with a molecular hydrogen density of 5×10^5 cm^{-3}. A core region of 5–10 pc diameter contains an extremely rich variety of molecular species. Many of these are detectable because of the very large molecular column density along the line of sight ($N(H_2) \approx (2\text{–}10) \times 10^{23}$ cm^{-2}, with higher values within smaller regions). The mixture of large amounts of cool dense and warm dense gas associated with star forming activity makes this region an excellent fishing ground for new molecules. A variety of physical conditions exists along the line of sight; consequently if a molecule exists in interstellar space this is a likely place to find it. The problem is to establish under which conditions particular molecules are abundant and which molecules are formed together, a program that has just been started.

(iii) *TMC1/L134N*. These are well-studied low mass dark molecular clouds which lack embedded luminous sources. The clouds are very quiescent, exhibiting linewidths close to those expected from thermal broadening alone. The temperature is about 10 K and the molecular hydrogen density 10^4–10^5 cm^{-3}. TMC1 in particular is divided into several small cloudlets along a ridge with position angle $-35°$ (cf. Figure 1.8). L134N has a more rounded appearance and has a less

Figure 1.8. Integrated $C^{18}O$ intensity towards TMC1; $\alpha(1950) = 4^h38^m38^s$, $\delta(1950) = 20° 35' 45''$.

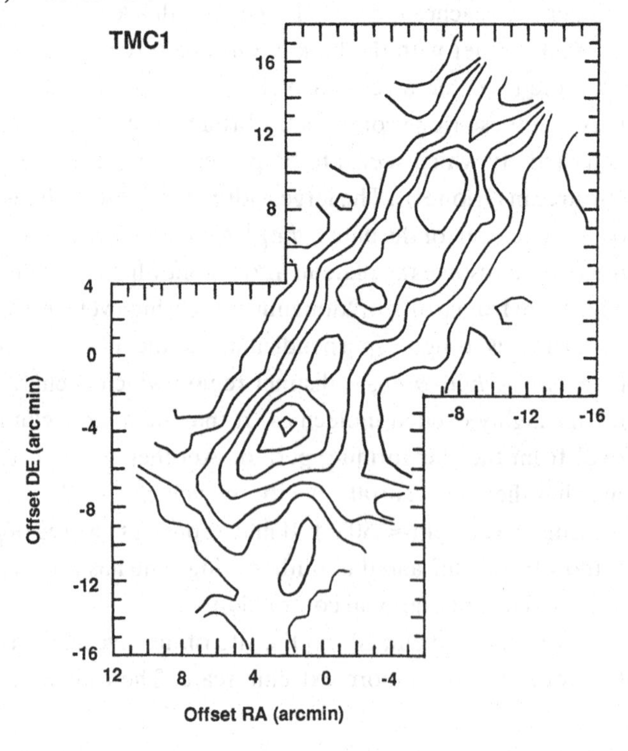

pronounced division into cloudlets but is by no means devoid of such structures. However, the clouds show large chemical differences in spite of all the physical similarities. TMC1 is rich in large carbon chain molecules while L134N has 'normal' abundances of them.

For a comparison of the abundances in the quiescent component of these clouds see Table 1.4. It appears that the abundance variations are not large compared to the rather wide variation in density and temperature. Since molecules in low temperature gas would be expected to stick onto grains upon collisions, a strong inverse abundance dependence versus cloud density might have been expected. Indeed, infrared observations show absorption lines due to solid state CO, H_2O, and NH_3. Evidently, there must be efficient mechanisms to return such material back to the gas phase.

Table 1.4. *Molecular abundances for selected regions*

		Abundance versus H_2 (10^8)		
Species	Name	Orion ridge[a]	TMC1	Sgr B2
Assumed H_2 density ($\times 10^{-22}$)		10	1	20
CO	Carbon monoxide	8000[b]	8000[b]	8000[b]
CH_4	Methane	<80		
C	Atomic carbon	>1000		
C_2	Carbon dimer		5	
OH	Hydroxyl		30	
CH	Methylidyne		2	
C_2H	Ethynyl	1	5–10	>0.5
C_3H	Propynylidyne		0.05	
C_4H	Butadiynyl	<0.03	2	
C_3H_2	Cyclopropenylidene		2	0.1
CH_3C_2H	Methyl acetylene	0.5	0.6	0.4
CN	Cyanogen	0.5	3	2
HCl	Hydrogen chloride	1		
HCN	Hydrogen cyanide	2	2	2
HNC	Hydrogen isocyanide	0.04	2	0.3
$HCNH^+$				0.1
CH_3CN	Methyl cyanide	0.04/0.08	0.1	0.07
HC_3N	Cyanoacetylene	0.04	0.6	0.2
HC_5N	Cyanodiacetylene	0.006	0.3	0.04
CH_2CHCN	Vinylcyanide		0.02	0.02
C_3N	Cyanoethynyl	<0.006	0.1	<0.02
C_3O	Tricarbon monoxide	<0.003	0.01	<0.002
CH_3C_3N	Methylcyanoacetylene		0.05	
CH_3C_4H	Methyldiacetylene		0.2	
HNCO	Isocyanic acid	<0.03	0.02	0.9
N_2H^+		0.2	0.05	
NH_3	Ammonia	20	2	1–10
HCO^+	Formyl ion	0.3	0.8	1

(*continued*)

Table 1.4 *continued*

Species	Name	Abundance versus H_2 (10^8) Orion ridge[a]	TMC1	Sgr B2
Assumed H_2 density ($\times 10^{-22}$)		10	1	20
HCO_2^+		<0.02		0.3
HDO	Deuterated water	<0.04/0.4		
CH_3OH	Methanol	/40	0.4	20
H_2CO	Formaldehyde	3/30	2	2–10
H_2C_2O	Ketene	/0.2	0.01	0.08
$(CH_3)_2O$	Dimethyl ether	/2		0.25
$HCOOCH_3$	Methyl formate	/2	<0.1	0.2
CS	Carbon monosulfide	0.4	1	1
HCS^+	Thioformyl ion	0.02	0.06	0.02
H_2CS	Thioformaldehyde	/0.2		0.3
H_2S	Hydrogen sulfide	<0.1		
OCS	Carbonyl sulfide	/0.9	0.2	1
SO	Sulfur monoxide	0.2/0.5	0.5	0.2
SO_2	Sulfur dioxide	/0.4	<0.1	0.3–2
SiO	Silicon monoxide	<0.1	<0.0005	0.04
SiS	Silicon sulfide	<0.001		0.03
HC_7N	Cyanohexatriyne		0.1	
HC_9N	Cyanooctatetrayne		0.03	
$HC_{11}N$	Cyanodecapentayne		0.01	
HOC^+				
HCO	Formyl	<0.02		
CH_3CHO	Acetaldehyde	<0.02	0.06	0.1
HC_2CHO	Propiolaldehyde		<0.06	
CH_2CHCHO	Acrolein	<0.02		
CH_3CH_2OH	Ethanol	<0.05		0.3
HCOOH	Formic acid	/0.03		~0.2
CH_3COOH	Acetic acid	<0.5		
CH_3NC	Methyl isocyanide	<0.005	<0.01	
CH_3CH_2CN	Ethyl cyanide	<0.03	<0.1	0.03
NH_2CN	Cyanamide	<0.02	<0.01	0.01
NH_2CHO	Formamide	<0.03	<0.2	0.04
CH_3SH	Methyl mercaptan	<0.06		0.08
HCNS	Isothiocyanic acid			0.01
$(NH_2)_2CO$	Urea	<0.07		
NH_2CH_2COOH	Glycine II	<0.05	<0.04	
C_4H_4O	Furan	<0.07		
C_4H_5N	Pyrrole	<0.03	<0.04	
$C_3N_2H_4$	Imidazole	<0.1	<0.03	
CH_3NH_2	Methylamine	<0.1		
PN	Phosphorus nitride	0.01		0.001
PO	Phosphorus monoxide	<0.1		
NO	Nitric oxide	<5		~10

[a] For Orion ridge, values given as x/y refer to the extended and the compact ridge, respectively.
[b] Adopted value.
Adapted from Irvine *et al.* (1987).

Another obvious result is that gas phase chemistry is important. Most abundances for quiescent gas clouds are in reasonable agreement with pure gas phase, mainly ion–neutral chemical models. The rather high abundances of molecular ions such as HCO^+ and the large fractionation of deuterium in several molecules forcefully show the influence of gas phase chemistry in general and ion–neutral reactions in particular. However, one molecule obviously formed on grain surfaces is H_2 which it would be impossible to form efficiently enough in the gas phase. The case for formation of other molecules on grain surfaces is less compelling. Evidence of grain surface 'chemistry' is found in the high abundances of H_2O, NH_3, and CH_3OH in dense warm regions. These increases in abundances are attributed to evaporation of molecules stored in grain mantels (Pauls *et al.* 1983). To what extent the molecules have been processed during the storage period is less certain, however (cf. Brown, Charnley and Millar (1988)).

1.4.3. Chemical differences

Between dark clouds

The cyanopolyyne (HC_3N, HC_5N, HC_7N, . . .) peak in TMC1 stands out relative to the dark cloud L134N as well to GMCs, due to its considerably enhanced abundance of mostly unsaturated carbon chain molecules (cyanopolyynes, C_2H, C_4H, C_6H, C_3N, CH_3C_2H, etc.). Also other carbon species are enhanced here, i.e., C_3H_2, CS, etc. Other abundances are fairly similar to those in L134N, except that NH_3, SO, and SO_2 are more abundant in L134N. This overabundance of carbon chains in TMC1, and in the Taurus region in general, is curious since all known physical conditions are the same for TMC1 and L134N. One proposed explanation is the long time needed to reach chemical equilibrium. Time dependent chemical models predict an abundance peak for carbon containing molecules at about 10^5–10^6 yr, before most of the carbon atoms have been transformed into the stable CO molecule. This might be an explanation for the high abundances of carbon containing molecules in TMC1, but several questions remain. Methanol (CH_3OH) also is predicted to have an abundance peak at an early cloud age since it is formed from hydrocarbons. But the abundances of methanol are equal and large in TMC1 and L134N. If this high carbon molecule abundance is an age effect, why have other carbon molecule rich clouds not been found outside the Taurus region, in spite of several observational attempts? Furthermore, even inside TMC1 it is only the southeastern part that is really rich in carbon chains.

Between GMCs

There is little comprehensive data on chemical abundances in GMCs apart from the spectral surveys of Orion KL and Sgr B2. The nearest and best studied GMC is that in Orion, within which there are striking chemical differences between regions. The dense warm regions close to the newly formed stars differ in many respects from the quiescent cloud; see Figure 1.9. Since such a region would be unresolved at a larger distance observed abundance variations between GMCs could reflect heterogeneities in the regions studied and not any real chemical differences. This problem is particularly troublesome for the Galactic center cloud Sgr B2. Among the many observed differences between Orion KL and Sgr B2 perhaps the differences in chemical selectivity are most apparent. If we compare relative abundances in the two sources it is found that in Orion KL dimethyl ether ($(CH_3)_2O$) is more than forty times as abundant as its isomer ethanol (CH_3CH_2OH) while the two species are equally abundant in Sgr B2. Other such differences are $f[CH_3OH]/f[HCOOCH_3] = 20$ and 100, $f(HCOOCH_3]/f[HCOOH] = 70$ and 1 for Orion KL and Sgr B2, respectively.

Between cloud types

While the most quiescent parts of GMCs and dark clouds are reasonably similar there are clear differences between quiescent matter and regions close to young

Figure 1.9. Abundances relative to H_2 for the Orion ridge and hot core. From Blake (1985).

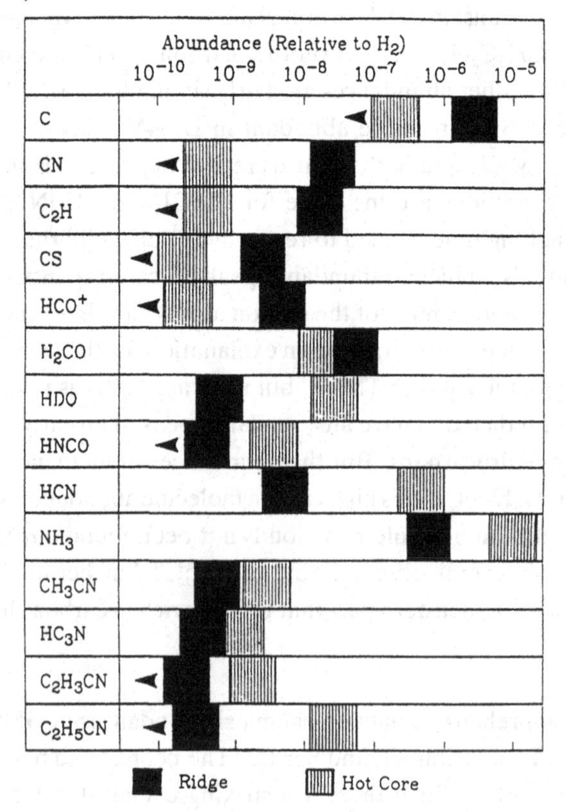

luminous massive stars. These latter regions are much warmer and denser and exhibit much higher abundances of species like methanol (CH_3OH), ammonia (NH_3), and water (H_2O). It is believed that these enhancements depend on evaporation of molecules from grain mantels due to heating by newly formed stars. Support for this idea comes from the observed high fractionation of deuterium in NH_2D and HDO. The high deuterium fractionation implies that the molecules were formed earlier at a much lower temperature (cf. Brown *et al.* (1988)).

Within dark clouds

Several apparent abundance gradients are observed in dark clouds. As already discussed the emission from carbon chain molecules is enhanced to the south-eastern part on the TMC1 ridge. For molecules with high dipole moments this can be explained at least partly by different excitation conditions along the ridge, i.e., a density gradient. But even the carbon chains C_2H and C_4H with low dipole moments decrease in the north-western part, while CH_3OH with a similar dipole moment still shows strong emission. Also in L134N chemical gradients have been observed. As in TMC1 it is not yet obvious what can be explained by excitation and radiative transfer effects and what is really due to chemical gradients. However, the cloud changes its shape and structure depending upon the molecular probe used. NH_3 and $H^{13}CO^+$ are extended in a north–south direction while $C^{18}O$ and CS are extended from east to west.

Within GMCs

Our comparatively detailed knowledge of the physical and chemical heterogeneity of the nearby Orion KL region has been discussed at length by e.g. Irvine, Goldsmith and Hjalmarson (1987) and Blake *et al.* (1987). We note that the abundances of several species are considerably enhanced in the energetic outflow of the *plateau* source (cf. Figure 1.10). Likewise the *hot core* exhibits increased concentrations of many species compared to the ridge (cf. Figure 1.9). Elevated abundances are also apparent in the warm compact ridge cloud. Also in the distant Sgr B2 molecular cloud finer scale physical and chemical structures, including 'hot cores' are just beginning to be resolved (Goldsmith *et al.* 1987, Irvine *et al.* 1987, Vogel, Genzel, and Palmer 1987). But here also it is hard to disentangle the effects of excitation and abundance differences. The peak at Sgr B2 north (N) of HNCO, and vibrationally excited HC_3N emission are most simply explained by infrared excitation due to embedded luminous sources. This, however, does not explain the lack of vibrationally excited HC_3N emission from Sgr B2 middle (M). There both heating from embedded sources and ground state emission from HC_3N are present.

Figure 1.10. Abundances relative to H_2 for the Orion ridge and plateau. From Blake (1985).

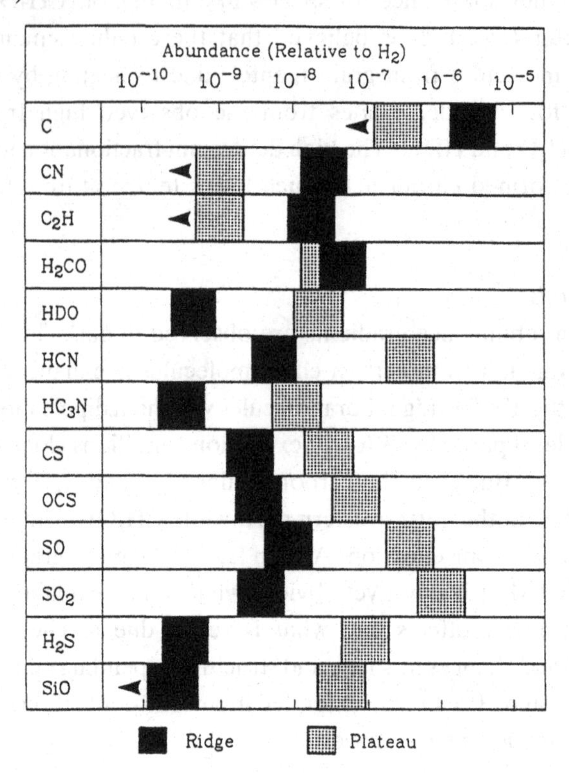

Within Cirrus clouds

About 10% of the Cirrus clouds contain detectable molecules. These molecular Cirrus clouds can be divided into two groups. 'CO-rich' Cirrus which have abundances similar to dark clouds (f[CO] $\approx 4 \times 10^{-5}$) and 'CO-poor' Cirrus which have abundances similar to diffuse clouds (f[CO] $< 10^{-6}$). Both cloud types have low visual extinction ($A_v \approx 0.5$), an order of magnitude lower than for dark clouds but similar to the extinction in diffuse clouds. The enhanced abundances of CO, OH, and H_2CO over the values for diffuse clouds may be caused by decreased photodissociation due to a reduced ambient ultraviolet radiation field, or/and increase in the formation rate of molecules due to non-dissociative shocks. The abundance of OH is, in particular, enhanced over dark cloud values by an order of magnitude, which has been attributed to the shock formation of molecules (Lada and Blitz 1988, Magnani, Blitz, and Wouterloot 1988).

Acknowledgements

We would like to forward our gratitude to colleagues and friends in many countries for stimulating collaborations and discussions, and to the Swedish

Natural Science Research Council (NFR) and Chalmers University of Technology for financial support.

REFERENCES

Black, J. M., and Willner, S. P. 1984, *Ap. J.*, **279**, 673.
Blake, G. A. 1985, Ph.D. Dissertation, California, Institute of Technology.
Blake, G. A., Sutton, E. C., Masson, C. R., and Phillips, T. G. 1986, *Ap. J., Suppl*, **60**, 357.
Blake, G. A., Sutton, E. C., Masson, C. R., and Phillips, T. G. 1987, *Ap. J.*, **315**, 621.
Bloemen, J. B. G. M., Strong, A. W., Blitz, L., Cohen, R. S., Dame, T. M., Grabelsky, D. A., Hermsen, W., Lebrun, F., Mayer-Hasselwander, H. A., and Thaddeus, P. 1986, *Astr. Ap.*, **154**, 25.
Bronfman, L., Cohen, R. S., Alvarez, H., May, J., and Thaddeus, P. 1988, *Ap. J.*, **324**, 248.
Brown, P. D., Charnley, S. B., and Millar, T. J. 1988, *Mon. Not. R. astr. Soc.*, **231**, 409.
Clemens, D. P. 1985, *Ap. J.*, **295**, 422.
Clemens, D. P., Sanders, D. B., and Scoville, N. Z. 1988, *Ap. J.*, **327**, 139.
Cummins, S. E., Linke, R. A., and Thaddeus, P. 1986, *Ap. J. Suppl.*, **60**, 819.
Dame, T. M., Elmegren, B. G., Cohen, R. S., and Thaddeus, P. 1986, *Ap. J.*, **305**, 892.
Dame, T. M., Ungerechts, H., Cohen, R. S., deGeus, E. J., Grenier, I. A., May, J., Murphy, D. C., Nyman, L.-Å., and Thaddeus, P. 1987, *Ap. J.*, **322**, 706.
Dickman, R. L. 1978, *Ap. J. Suppl.*, **37**, 407.
Downes, D. 1987, in *Star Forming Regions*, IAU Symp. No. 115, eds. Peimbert, M., and Jugaku, J. (Dordrecht: D. Reidel).
Falgarone, E., and Perault, M. 1987, in *Physical Processes in Interstellar Clouds*, eds. Morfill, G. B., and Scholer, M. (Dordrecht: D. Reidel).
Field, G. B., Goldsmith, D. W., and Habing, H. J. 1969, *Ap. J.* (*Letters*), **155**, L149.
Frerking, M. A., Langer, W. D., and Wilson, R. W. 1982, *Ap. J.*, **262**, 590.
Goldsmith, P. E. 1987, in *Interstellar Processes*, eds. Hollenbach, D. J., and Thronson, H. A. Jr. (Dordrecht: D. Reidel).
Goldsmith, P. F., Snell, R. L., Hasegawa, T., and Ukita, N. 1987, *Ap. J.*, **314**, 525.
Grabelsky, D. A., Cohen, R. S., Bronfman, L., Thaddeus, P., and May, J. 1987, *Ap. J.*, **315**, 122.
Guélin, M., and Cernicharo, J. 1988, in *Molecular Clouds in the Milky Way and External Galaxies*, eds. Dickman, R. L., Snell, R. L., and Young, J. (New York: Springer).
Hjalmarson, Å., and Friberg, P. 1988, in *Formation and Evolution of Low Mass Stars*, eds. Dupree, A. K., and Lago, M. T. V. T., NATO ASI, (Dordrecht: D. Reidel).
Ho, T. P., and Townes, C. H. 1983, *Ann. Rev. Astr. Ap.*, **21**, 239.
Irvine, W. M., Goldsmith, P. F., and Hjalmarson, Å. 1987, in *Interstellar Processes*, eds. Hollenbach, D. J., and Thronson, H. A. Jr. (Dordrecht: D. Reidel).
Johansson, L. E. B., Andersson, C., Ellder, J., Friberg, P., Hjalmarson, Å., Höglund, B., Irvine, W. M., Olofsson, H., and Rydbeck, G. 1984, *Astr. Ap.*, **130**, 227.
Jura, M. 1987, in *Interstellar Processes*, eds. Hollenbach, D. J., and Thronson, H. A. Jr. (Dordrecht: D. Reidel).
Kaifu, N., Suzuki, H., Ohishi, M., Miyaji, T., Ishikawa, S., Kasuga, T., Morimoto, M., and Saito, S. 1987, *Ap. J.* (*Letters*), **317**, L111.
Kutner, M. L. 1984, *Fund. Cosm. Phys.*, **9**, 233.
Kwan, J., and Sanders, D. B. 1986, *Ap. J.*, **309**, 783.
Lada, E. A., and Blitz, L. 1988, *Ap. J.* (*Letters*), **326**, L69.
Magnani, L., Blitz, L., and Wouterloot, J. G. A. 1988, *Ap. J.*, **326**, 909.
Maloney, P., and Black, J. H. 1988, *Ap. J.*, **325**, 389.

McKee, C. F., and Ostriker, J. P. 1977, *Ap. J.*, **218**, 148.
Myers, P. C. 1987, in *Interstellar Processes,* eds. Hollenbach, D. J., and Thronson, H. A. Jr. (Dordrecht: D. Reidel).
Myers, P. C., Dame, T. M., Thaddeus, P., Cohen, R. S., Silverberg, R. F., Dwek, E., and Hauser, M. G. 1986, *Ap. J.*, **301**, 398.
Myers, P. C., Fuller, G. A., Mathieu, R. D., Beichman, C. A., Benson, P. J., Schild, R. E., and Emerson, J. P. 1987, *Ap. J.*, **319**, 340.
Norman, C. A., and Silk, J. 1980, *Ap. J.*, **238**, 158.
Pauls, T. A., Wilson, T. L., Bieging, J. M., and Martin, R. N. 1983, *Astr. Ap.*, **124**, 123.
Rivolo, A. R., Solomon, P. M., and Sanders, D. B. 1986, *Ap. J.* (*Letters*), **301**, L19.
Rydbeck, G., Hjalmarson, Å., and Rydbeck, O. E. H. 1985, *Astr. Ap.*, **144**, 282.
Rydbeck, G., Hjalmarson, Å., Wiklind, T., and Rydbeck, O. E. H. 1988, in *Molecular Clouds in the Milky Way and External Galaxies,* eds. Dickman, R. L., Snell, R. L., and Young, J. (New York: Springer), and *Astr. Ap.*, submitted.
Scalo, J. 1987, in *Interstellar Processes,* eds. Hollenbach, D. J., and Thronson, H. A. Jr. (Dordrecht: D. Reidel).
Scoville, N. Z., and Sanders, D. B. 1987, in *Interstellar Processes,* eds. Hollenbach, D. J., and Thronson, H. A. Jr. (Dordrecht: D. Reidel).
Scoville, N. Z., Sanders, D. B., and Clemens, D. P. 1986, *Ap. J.* (*Letters*), **310**, L77.
Scoville, N. Z., Yun, M. S., Clemens, D. P., Sanders, D. B., and Waller, W. H. 1987, *Ap. J.* Suppl., **63**, 821.
Shull, J. M. 1987, in *Interstellar Processes,* eds. Hollenbach, D. J. and Thronson, H. A. Jr. (Dordrecht: D. Reidel).
Solomon, P. M., and Rivolo, A. R. 1987, in *The Galaxy,* eds. Gilmore, G., and Carswell, B. (Dordrecht: D. Reidel).
Solomon, P. M., Sanders, D. B., and Rivolo, A. R. 1985, *Ap. J.* (*Letters*), **292**, L19.
Solomon, P. M., Rivolo, A. R., and Yahil, A. 1987. *Ap. J.*, **319**, 730.
Stark, A. A., Bally, J., Knapp, G. R., Krahnert, A., Penzias, A. A., and Wilson, R. W. 1987, in *Star Forming Regions,* IAU Symp. No. 115, eds. Peimbert, M., and Jugaku, J. (Dordrecht: D. Reidel).
Stark, A. A., Bally, J., Knapp, G. R., and Wilson, R. W. 1988, in *Molecular Clouds in the Milky Way and External Galaxies,* eds. Dickman, R. L., Snell, R. L., and Young, J. (New York: Springer).
Sutton, E. C., Blake, G. A., Masson, C. R., and Phillips, T. G. 1985, *Ap. J. Suppl.*, **58**, 341.
van Dishoeck, E. F., and Black, J. H. 1987, in *Physical Processes in Interstellar Clouds,* eds. Morfill, G. E., and Scholer, M. (Dordrecht: D. Reidel).
Vogel, S. N., Genzel, R., and Palmer, P. 1987, *Ap. J.*, **316**, 243.
Waller, W. H., Clemens, D. P., Sanders, D. B., and Scoville, N. Z. 1987, *Ap. J.*, **314**, 397.

Below we list for each section several references which were of particular use during its preparation.

(1) Shull 1987, Field *et al.* 1969, McKee and Ostriker 1977.
(2) Scoville and Sanders 1987, Solomon and Rivolo 1987, Dame *et al.* 1987, Bronfman *et al.* 1988.
(3) Goldsmith 1987, Myers 1987, Kutner 1984, Hjalmarson and Friberg 1988.
(4) Irvine *et al.* 1987.

2

Molecules in galaxies

LEO BLITZ

University of Maryland, Maryland and Institute for Advanced Study, Princeton, USA

2.1. Introduction

The first unambiguous detection of molecules in galaxies other than the Milky Way was that of OH in absorption toward the nuclear continuum sources in NGC 253 and M82 (Weliachew 1971). Subsequently, all of the species detected in the Milky Way that show strong emission or absorption lines have now been detected in other galaxies. Although extragalactic molecules have been observed in regions of the spectrum other than the radio, notably the vibrational–rotational lines of H_2 in the infrared (Thompson, Lebofsky and Rieke 1978), observations remain largely the domain of radioastronomers. Furthermore, the largest fraction of all extragalactic molecular observations have been carried out using the J = 1–0 transition of CO, because it is the easiest transition to detect in most galaxies.

Initial progress in the field was slow, hampered by the simultaneous requirements of large bandwidth, high sensitivity and high angular resolution. The technical requirements were largely overcome in the early 1980s and the field has now burgeoned into one where developments are occurring so fast, on so many fronts that it is impossible in a short review to do justice to all of them. Accordingly, I have chosen a small subset of what I think are among the most interesting recent developments in the study of molecules in other galaxies with the aim of focussing on a few problems of wide astrophysical interest. A good introduction to the field is the comprehensive review by Morris and Rickard (1982). It is still useful although it is now quite dated, only six years after its publication. Alex Dalgarno's seminal work in providing the understanding of basic molecular processes underlies the interpretation of nearly all of the molecular observations and we are all in his debt.

2.2. Finding molecules in galaxies

Molecular hydrogen has been detected directly only through its vibrational–rotational spectrum in the near infrared (Thompson *et al.* 1978), an important probe of the regions of warm molecular gas which in many cases is thought to be shock excited. The largest fraction of molecular gas in a galaxy is, however, quite cold, and the radiation from the H_2 molecules is not detectable. Therefore, as is true for the Milky Way, the detection of the cold molecular gas must be done through trace species, the most abundant of which is CO. In general, the $J = 1–0$ and $J = 2–1$ lines of CO are the strongest transitions of any molecular species; these transitions are excited by the modest densities representative of most of the gas in molecular clouds. The lower transition has been detected in galaxies with redshifts approaching $z = 0.1$ (Sanders *et al.* 1986). More galaxies have been both detected and mapped in CO than in any other molecule.

But how do we find galaxies with CO? The early searches concentrated on nearby galaxies with mixed results. The HI content of a galaxy was found to be an unreliable predictor of CO line strength, but the association of galactic molecular clouds with dust suggested that other tracers of the dust such as far infrared emission might be good CO predictors. Rickard and Harvey (1984) correlated the velocity integrated line strength of CO, I(CO), with 100 μm flux density, S(100 μm), for 19 galaxies measured from the Kuiper Airborne Observatory and found a good correlation between the two quantities. This correlation has subsequently been confirmed by many observers. With the advent of IRAS, 100 μm fluxes are now available for thousands of galaxies, and the IRAS galaxy catalog is perhaps the best predictor of the CO flux from a galaxy. There is, however, considerable scatter in the CO–FIR correlation, the origin of which is not yet understood. Recently, a significant deviation from the I(CO)/S(100 μm) ratio has been found for the 'ultraluminous' galaxies detected by IRAS (Houck *et al.* 1985). These galaxies are found to have substantially more 100 μm emission relative to I(CO) than do their lower luminosity counterparts (Sanders *et al.* 1986, 1988). These authors have interpreted this deviation as being due to a higher efficiency of star formation, based in part on the assumption that I(CO) is directly proportional to H_2 column density, $N(H_2)$ (see the following section for a discussion of this point). Ultraluminous galaxies are discussed further in the section on active galaxies below. A full discussion of how CO correlates with various observational quantities can be found in Verter (1988).

2.3. The abundance of extragalactic CO

There are many problems in extragalactic research that require knowing only how CO is distributed in a galaxy. However, a full understanding of the role that the molecular gas plays in the structure, evolution and star formation in galaxies requires an understanding of how the strength of the detected CO lines is related to the column density of H_2. A great deal of effort has been expended in the Milky Way in trying to determine how to convert CO column densities to H_2 column densities, and an excellent review of the subject has been written by van Dishoeck and Black (1987). Maloney and Black (1988) have also published an extensive treatment of the conversion of CO line strengths to H_2 column densities. This section is a brief treatment of many of the issues considered by these authors with the addition of several new points.

In a seminal paper by Dickman (1978) it was shown that the ^{13}CO column density is linearly proportional to A_v in dark clouds near the Sun over a range of extinction from 1 to 4 magnitudes. In conjunction with the dust-to-gas ratio measured by Bohlin, Savage and Drake (1978), Dickman was able to determine the ratio $N(^{13}CO)/N(H_2)$ in the solar vicinity. Because the CO line strength was found to be roughly proportional to the ^{13}CO line strength in surveys of the Milky Way, (Solomon, Scoville and Sanders 1979), it was suggested that the velocity integrated CO line strength is proportional to $N(H_2)$, even though the CO line is almost always optically thick. That is, because the ^{13}CO line is nearly always optically thin, the rough proportionality (with considerable scatter) between the line strengths of the two isotopic species implies that the CO line is actually a column density tracer despite its generally large optical depth. The usual explanation of this phenomenon is that the optically thick regions in which the CO line is formed are decoupled from one another through their Doppler motions and the integrated CO line intensity is really a measure of the number of these optically thick emitting regions (Zuckerman and Evans 1974). Nevertheless, the CO line emission from cold molecular clouds has never been satisfactorily modeled.

If it could be shown that $I(CO)$ is a reliable tracer of $N(H_2)$, one could then use the CO observations to infer many important properties about the molecular content of galaxies. It is here that the story gets a bit complex. Many observers simply assume that a ratio of $I(CO)/N(H_2)$ determined from galactic observations can be applied to other galaxies to obtain quantities such as the molecular-to-atomic mass fraction, the star formation efficiency, the mass of the nuclear component, etc. Indeed, there is little else that can be done until it becomes possible to calibrate the CO line directly in other galaxies. One must look with great caution, however, at interpretations of extragalactic CO data where $N(H_2)$ is derived from $I(CO)$ because the uncertainties can be quite large. Let us examine the arguments for and against the use of CO as an extragalactic tracer of H_2, starting with the large scale observations of the Milky Way.

Two pieces of evidence within the Milky Way that the $J = 1\text{–}0$ line of CO can be used as a large scale tracer of $N(H_2)$ come from gamma-ray observations of the disk of the Galaxy, and from the assumption that giant molecular clouds are in virial equilibrium. The gamma-rays in the 100–5000 MeV range observed by COS-B are thought to come primarily from interactions of cosmic rays with interstellar gas (e.g. Fazio (1967), Stecker (1971)). The gamma-rays therefore are a tracer of atomic and molecular gas independent of the excitation or density of the gas, and can be used to calibrate the CO line when the contribution of the atomic gas is accounted for. Such an analysis was made by Bloemen *et al.* (1986) who found that a constant $I(CO)/N(H_2)$ ratio can be used inside the solar circle in to a distance of about 2 kpc from the galactic center. The results are sufficiently coarse to allow a gradient of about a factor of 2 in this ratio to exist between the peak of the molecular ring and the solar circle. Such a variation was inferred by Liszt, Burton and Xiang (1984) from variations in the large scale gradient of $I(^{13}CO)/I(CO)$ inside the solar circle. Similar gradients are inferred from the work of the Durham gamma-ray group (Li, Riley and Wolfendale 1983).

The virial theorem also can be used to calibrate the ratio of $I(CO)/N(H_2)$ for molecular clouds that can be isolated from their neighbors in galactic plane surveys. The distance to a cloud can be determined kinematically from the LSR velocity of the cloud if there is some way to resolve the distance ambiguity. From the measured size and velocity dispersion of the cloud, a mass can be derived under the assumption that the cloud is in virial equilibrium. After a correction for helium, the value of $I(CO)/N(H_2)$ can be derived from the measured value of $I(CO)$ across the face of the cloud. The ratio obtained by Sanders, Scoville and Solomon (1985) is within 50% of the ratio derived from the gamma-ray determination, giving one confidence that one can indeed use $I(CO)$ as an H_2 mass tracer. Clouds may be only marginally gravitationally bound, however, in which case the derived value of $I(CO)/N(H_2)$ would have to be raised by a factor of 2 (less H_2 mass for a given value of $I(CO)$). Also, no results have been published on whether the virial theorem value of $I(CO)/N(H_2)$ is a function of galactic radius. A critical review of this method which discusses other uncertainties is given in van Dishoeck and Black (1987).

In spite of the reasonable agreement between the gamma-ray and virial theorem estimates of $I(CO)/N(H_2)$, the ratio of molecular to atomic gas inside the solar circle claimed by various observers far exceeds the variance in the derived CO/H_2 mass ratios. Bloemen *et al.* (1986) derive a ratio of molecular to atomic gas between 2 and 8 kpc from the galactic center of 1.35 (for $R_0 = 10$ kpc). Solomon and Sanders (1980) obtain a ratio of 10 between 4 and 8 kpc, although the $I(CO)/N(H_2)$ ratio they use is only a factor of 2.6 higher than the Bloemen *et al.* value. Most of the discrepancies have now been resolved by Bronfman *et al.* (1988), but if these sorts of difficulties exist in the Milky Way, where good calibrations are possible, what are we to conclude about extragalactic CO observations where they are not?

As an alternative to using the virial theorem, it is possible to compare the ratios of $I(CO)/N(H_2)$ in various galaxies by using the linewidth–size relation for molecular clouds. Various observers doing galactic plane CO surveys find a power-law relation between these two quantities in the inner portion of the Milky Way; all of them so far agree that the linewidth varies nearly as the diameter to the $\frac{1}{2}$ power (e.g. Dame *et al.* 1986, Solomon *et al.* 1987). Currently, there are only three galaxies beyond the Milky Way where molecular clouds have been resolved: the Large Magellanic Cloud (Cohen *et al.* 1988), M31 (Boulanger *et al.* 1984, Ichikawa *et al.* 1985, Casoli, Combes and Stark 1987, Vogel, Boulanger and Ball 1987, Lada *et al.* 1988), and M33 (Israel, personal communication). Of these, only the LMC has been mapped completely in CO. Cohen *et al.* derive a linewidth–size relation for the clouds in the LMC and find a power-law relation for molecular clouds with the same slope as that found for galactic molecular clouds. The constant of proportionality is, however, different from the galactic clouds in the sense that $I(CO)/N(H_2)$ is a factor of 6 lower in the LMC than it is in the Milky Way. Because the metallicity of the LMC is a factor of 4 lower than that of the Milky Way, it is reasonable to suppose that the metallicity difference is reflected in the CO/H_2 ratio. This is the most direct evidence to date that $I(CO)/N(H_2)$ is proportional to the metallicity of the galaxy or the region of a galaxy where the CO lines are being produced. That this is the case has been suggested by numerous authors (e.g. Blitz and Shu (1980), Elmegreen, Elmegreen and Morris (1980), Bhat, Mayer and Wolfendale (1986), Israel (1988)). Furthermore, in M31, where the metallicities are close to those in the solar vicinity, the interferometric data imply a virial theorem value for $I(CO)/N(H_2)$ twice that derived from the galactic molecular clouds (Vogel *et al.* 1987).

Are we then to conclude that if we adjust $I(CO)/N(H_2)$ for metallicity we can obtain a good H_2 calibration from CO? Maybe. The LMC data are not, unfortunately without their difficulties. The telescope beam used for the observations has a diameter of ~50 pc, close to the mean value of the diameter of the giant molecular clouds (GMCs) in the Milky Way (Blitz 1980, Sanders *et al.* 1985). Could it be that the LMC observations do not resolve the GMCs? If so, then the linewidth–size relation for the LMC is not representative of the physical conditions of the GMCs in that galaxy. Yet, if that is the case, why is the slope of that relation the same as it is for galactic GMCs? Furthermore, is there any reason why the molecular clouds in the LMC should not be much larger than those in the Milky Way especially if the galactic tidal forces acting on them are significantly smaller in general (Stark and Blitz 1978, Blitz and Glassgold 1982)? If the Cohen *et al.* linewidth–size relation is an observational artifact in the LMC, then that relation must be suspect in the Milky Way. Fortunately, there are two experiments that can decide the issue. Monto Carlo simulations of molecular clouds in the Milky Way, properly done, can decide whether the observed linewidth–size relation can be obtained from a random collection of molecular clouds. Also, now that there is a 20 m millimeter-wave telescope in the southern hemisphere with a resolution some 17 times better than that of the Cohen *et al.* observations, it will be possible to determine definitively whether their results are affected by their large beam, simply by mapping some of the clouds in their survey.

Extending the question of the H_2 content of galaxies beyond the local group requires evaluating more indirect evidence. Rickard and Blitz (1985) used the argument that established the use of CO as a galactic H_2 mass tracer to evaluate how good CO would be in other galaxies. That is, because CO is almost always optically thick, and ^{13}CO is almost always optically thin, a necessary condition for CO to be a good mass tracer for H_2 is that the *apparent* CO/^{13}CO ratio must be constant from galaxy to galaxy and within a galaxy. It was the near constancy in this ratio when averaged over suitably large areas that suggested that CO could be used as an H_2 mass tracer in the Milky Way in place of ^{13}CO. Small changes in the ratio $I(CO)/I(^{13}CO)$ can mask large changes in the CO/H_2 ratio because of the way large scale metallicity and temperature gradients in a galactic disk are expected to drive the ratio in opposite directions with increasing galactic radius (Kutner and Leung 1985). Rickard and Blitz observed six gas rich spirals and made sensitive CO and ^{13}CO observations of the nuclei and disks of the galaxies. The beam in these galaxies typically subtended 1–2 kpc and therefore only large scale variations were measured. They found significant variations of a factor of 4 in the ratio $I(CO)/I(^{13}CO)$ from galaxy to galaxy, and within a galaxy. The measured ratio was frequently different from that derived in the Milky Way. Similar results had been obtained previously by Encrenaz *et al.* (1979), who made a more limited set of observations. The conclusion therefore seems to be that $I(CO)$ has a rather limited usefulness as an H_2 mass tracer. Subsequent to the Rickard and Blitz study, Young and Sanders (1986) made a similar study of a number of spiral galaxies and confirmed the large variations in the ratio of $I(CO)/I(^{13}CO)$ seen by Rickard and Blitz (1985) and Encrenaz *et al.* (1979), although there are some differences in the values obtained by all three sets of observers when the same galaxies were observed (the beam sizes used in all of the observations was different, however). Nevertheless, Young and Sanders conclude that CO is a good H_2 mass tracer, a result which seems to be contradicted by the data in their tables.

A final point about extragalactic CO observations comes from observations of the Milky Way. Unlike the situation for atomic hydrogen, $I(CO)$ is frequently the strongest in galactic centers (a conspicuous exception is M31, Stark (1979)). In galactic nuclei, the physical conditions in the molecular gas are likely to be very different from those in the disk. How are we to evaluate the use of CO as an H_2

mass tracer in galactic nuclei? Blitz *et al.* (1985) used the gamma-ray emission as a probe of the molecular hydrogen content of the inner 400 pc of the Milky Way using a technique similar to that of Bloemen *et al.* (1986). The strong CO peak in the nucleus should have produced a strong gamma-ray peak, yet no such peak is seen in the COS-B data. They concluded that either the CO/H_2 ratio is vastly different in the center of the Milky Way, or that the cosmic rays that produce the gamma-rays are effectively channeled away from the molecular clouds with a very high efficiency. Either conclusion is rather uncomfortable. If the first is correct, then even in the Milky Way, the ratio of $I(CO)/N(H_2)$ varies by a factor of about 10 and using CO to obtain H_2 column densities can produce large errors. This situation would be exacerbated in the nuclei of other galaxies where the physical conditions are likely to be even more extreme, notably in the centers of active galactic nuclei. On the other hand, since we know from the work of Jansky (1932) that cosmic ray electrons are produced in the galactic center in great profusion, one would need a way to have the magnetic field channel the cosmic ray nuclei away from the molecular clouds over a large range of galactic radius, while still permitting enough of them to penetrate the Sgr B2 molecular cloud to produce the vast profusion of molecules seen there, if the CO/H_2 ratio were close to normal in the galactic center. Solving this conundrum has great importance for understanding the molecular observations in the centers of galaxies.

What are we to conclude then? Understanding the abundance of H_2 relative to HI in a galaxy is important for understanding how the two gas phases are related. In a galaxy in which the interstellar medium is 90% molecular, for example, the atomic gas will probably have little to do with the global star formation in the galaxy. On the other hand, if the relative abundances of the two phases of the ISM are more nearly equal, it is more likely that the production of molecular gas is more intimately related to the atomic phase. One of the major problems in the physics of the interstellar medium is how GMCs form. Although several mechanisms have been proposed, there is, in fact, little hard data on this problem. It appears likely that the star formation in the disks of all spiral galaxies takes place in GMCs, as it does in the Milky Way. All observers seem to agree that in the Milky Way at least, the giant star forming molecular clouds are situated in the spiral arms, perhaps the GMCs *define* the spiral arms. But there are other galaxies where there appears to be abundant star formation, abundant molecular gas, but a total absence of anything resembling spiral arms (Stark, Elmegreen and Chance 1987). How do molecular clouds form in those galaxies without the mediating effects of the spiral arms or of a density wave? Are there even GMCs in these galaxies? The formation of stars appears to be much more vigorous in some places than in others. In M33, for example, the giant HII region NGC 604 is the richest star forming complex in the local group. The object is associated with one or more

GMCs, but why has star formation taken place with such a bang in that object, and why are there no objects which have shown similarly vigorous star formation in the Milky Way or in M31? The nuclei of some galaxies show evidence of star bursts. Is it because they have more molecular gas, or is it that the molecular gas is more efficiently converted into stars? To obtain the answers to these questions we need to know how much molecular gas we are observing when we observe the trace species in these galaxies.

Unfortunately, it appears that the data we have at present cannot adequately assess these questions because we don't yet know how to convert the CO data from these galaxies into H_2 column densities with sufficient accuracy to obtain reliable answers. We note, for example, that for NGC 4321, NGC 6946 and IC 342, Young and Scoville (1982) derive H_2/HI mass ratios of 6.4, 3.5, and 4.3 respectively, whereas Bhat *et al.* (1986) obtained values of 0.9, 0.8 and 0.9 respectively, from the same data, but using different assumptions to convert CO to H_2. Thus, in the disks of spiral galaxies, we need to assign an uncertainty of at least a factor of five in the derived H_2 content, and an even larger uncertainty in galactic nuclei.

2.4. Spiral structure

The question of how concentrated molecular clouds are in the arms of spiral galaxies has been one of the most hotly contested issues in millimeter-wave astronomy. The assumption that GMCs are located in spiral arms (and perhaps *define* the arms) comes from the following chain of reasoning. In the Milky Way, star formation takes place in molecular clouds (Zuckerman and Palmer 1974). Most star formation takes place in clusters and stellar associations (Miller and Scalo 1978), and all local stellar associations form from giant molecular clouds (Blitz 1978, 1980). In spiral galaxies, the spiral arms are most clearly delineated by the population of bright hot stars (OB associations) and HII regions, thus one would expect the star forming molecular clouds either to be located in or to define the arms. On the other hand, Stark, *et al.* (1987) have looked at the CO emission from galaxies that have no clearly delineated spiral structure and have found that both their CO content and the rate of star formation is frequently quite normal. In these galaxies at least, the star forming molecular clouds must be able to form without the intervention of spiral arms. Could it be that the GMCs are, in fact, everywhere and are only 'turned on' by the passage of a spiral density wave?

In the Milky Way, all observers now agree that there is at least some population of molecular clouds that is concentrated in the spiral arms. Cohen *et al.* (1980), showed clearly defined arms, as has Dame *et al.* (1986). After initially claiming that the molecular cloud distribution shows no obvious spiral pattern (Scoville,

Solomon and Sanders 1979), Solomon, Sanders and Rivolo (1985) now agree that there is at least a population of 'hot-centered' clouds, GMCs, that are confined to the spiral arms of the inner galaxy and are the source of at least half of the CO emission. The remainder of the gas, they argued, is distributed more or less uniformly throughout the disk. Furthermore, Blitz, Fich and Kulkarni (1983) showed that the outer galaxy GMCs are closely correlated with the well defined arms seen in HI. When it was then seen that the CO emission in M31 is clearly concentrated in the spiral arms (Combes *et al.* 1977, Stark, Linke and Frerking 1981), it seemed reasonable to conclude that while there might be a substantial population of molecular clouds in the interarm regions, the spectacular spiral arms in galaxies most probably exhibit a strong concentration of molecular clouds, and the arm–interarm contrast is high.

The classic test, of course, is the galaxy M51, the prototype grand design spiral. Single dish observations of that galaxy failed, however, to show a significant concentration of the CO in the spiral arms (Rickard and Palmer 1981, Rydbeck, Hjalmarson and Rydbeck 1985). The latter study showed that the CO flux enhancement in the spiral arms is about 20%, when measured with the 33″ beam of the 20 m Onsala telescope. Why is the flux contrast so low in M51, when it seems to be higher in both the Milky Way and in M31, both of which have more ragged spiral structure than M51?

The answer comes from new observations obtained with the Hat Creek (now BIMA) and Owens Valley interferometers which have synthesized M51 with an angular resolution far surpassing what is possible with single dishes. The maps show that the molecular gas manifests an outstanding spiral structure. In a set of observations made in the central arcminute of the galaxy, observations by Lo *et al.* (1987) show that beautiful spiral structure in CO extends down to within 500 pc of the nucleus and is largely confined to the dusty spiral arms. In a more extensive set of observations (shown in Figure 2.1) which are a mosaic of eight 1′ fields of view with the Owens Valley interferometer at 7″ resolution, Vogel, Kulkarni and Scoville (1988) show that the CO is concentrated in the spiral arms out to at least several kiloparsecs from the nucleus. Both sets of observers infer a minimum arm to interarm brightness ratio of about 3. Furthermore, the Vogel *et al.* observations show that the detected CO is coincident with the dust lanes in the mapped region and that the HII regions are downstream of the detected molecular clouds. The much larger arm–interarm contrast inferred relative to the single dish maps of the galaxy appears to be due to the narrowness of the CO arms compared to the Onsala beam. That is, both sets of observers detect only 30% of the single dish flux, presumably because their observations are not sensitive to the interarm CO component which is relatively uniformly distributed in the disk. Although the

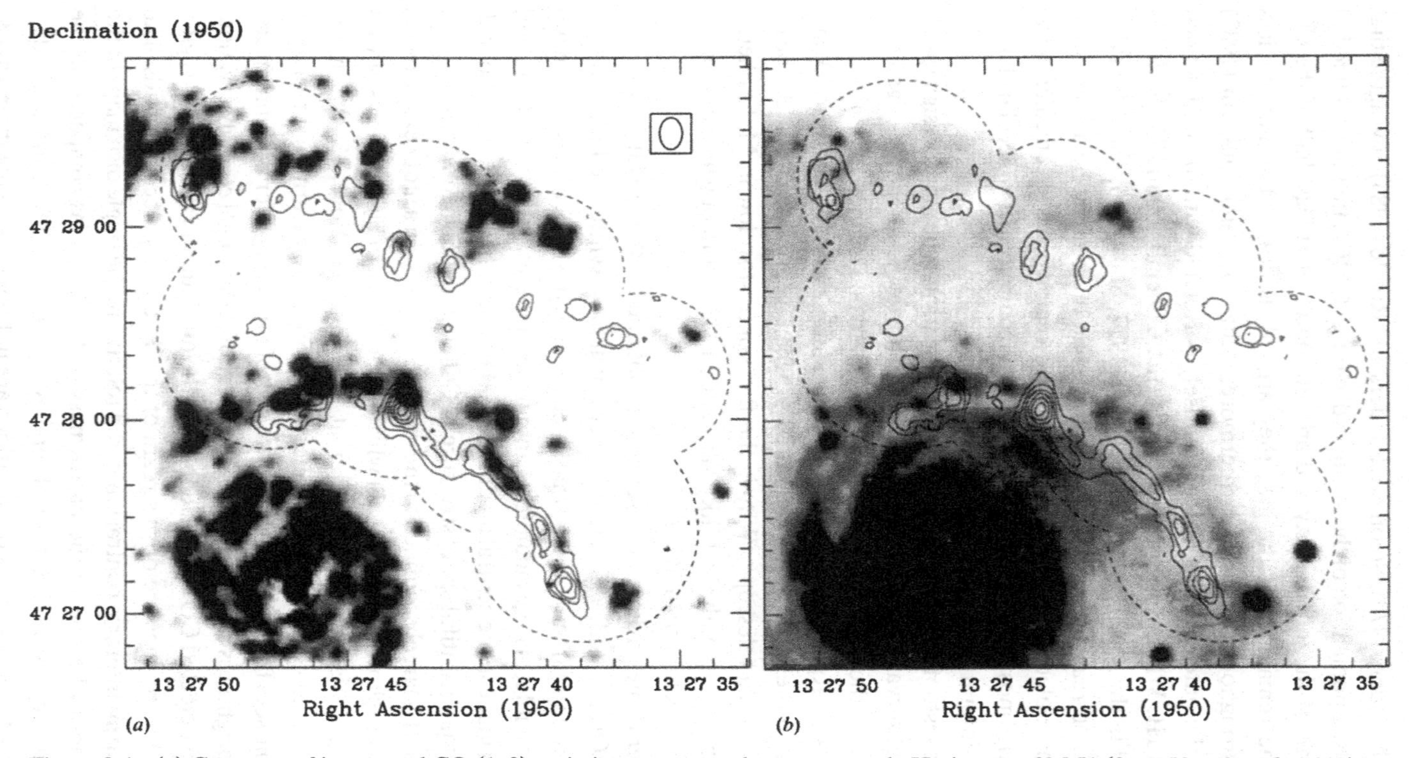

Figure 2.1. (*a*) Contours of integrated CO (1–0) emission superposed on a grey scale Hα image of M 51 (from Vogel *et al.*, 1988). The CO emission was mapped with the OVRO interferometer and is contoured at intervals of 15% of the peak of 45 Jy km^{-1}. The area mapped in CO is outlined by the dashed line. The synthesized beam of the CO observations is indicated in the upper right. The CO arms are seen to lie parallel to the Hα arms but offset toward the nucleus. (*b*) Contours of integrated CO emission superimposed on an R band CCD image. The inner CO arm coincides with the well defined dust lane; the width of the dust lane is about 300 pc.

spiral arm component that they do detect is only 20–30% of the total emission, it fills only a small fraction of the single dish Onsala beam at any particular location, and thus the surface brightness ratio of the arm and interarm gas is very high.

The general picture that emerges is the following. In all galaxies, regardless of spiral arm morphology, there are two components of the molecular emission; one component that is concentrated in the spiral arms and one that is distributed throughout the disk. In the spiral arms, we find a concentration of the giant star forming clouds (GMCs). Evidence for a distributed disk component of molecular gas also comes from M83, where Allen, Atherton and Tilanus (1986) argue that the relative location of the HI and the dark dust lanes indicate that the gas coming *into* the spiral arm segment they observed must be molecular. Since both HI and H_2 are enhanced in the spiral arms, the observational evidence does not yet present a clear picture of whether the GMCs condense from atomic gas or are agglomerations of the molecular clouds entering into the spiral arms. It may be that molecular clouds form in both ways. For example, in the outer disk of the Milky Way and other spirals such as M101, there is surely insufficient interarm molecular gas to be collected into the GMCs observed there. Furthermore, we do not yet know whether there are GMCs in any significant quantity in the interarm regions of spiral galaxies. The limited information available in M31 (Stark *et al.* 1981, Boulanger *et al.* 1984), indicates that there are not. We also do not know whether the star formation in the galaxies without clear spiral structure takes place in GMCs, and if so, how those GMCs form. Preliminary evidence from the LMC (Cohen *et al.* 1988) indicates that at least there, the sites of obvious star formation are indeed the GMCs.

In any event, it now appears to be quite clear that in galaxies which show evidence of spiral structure, the CO, and hence the molecular clouds are substantially enhanced in the spiral arms. We know of no galaxy observed with sufficient angular resolution where this is not the case. It is not yet clear, however, what role the spiral arms or the density waves play in either assembling or triggering the GMCs. An important clue is likely to come from interferometric CO maps of galaxies without discernible spiral structure, and from maps of the spiral arms in galaxies without beautiful grand designs.

2.5. The molecular luminosity function of galaxies

In order to make systematic comparisons between galaxies, and to understand the evolution of the stars in the disks of spiral galaxies, one of the fundamental data we need to have is the H_2 luminosity function of galaxies. However, many obstacles lie in the way and only very recently has the first reliable attempt been made to obtain a molecular luminosity function (Verter 1987). In addition to the normal problems in obtaining a luminosity function for galaxies with which optical

Figure 2.2. (*a*) The mean CO luminosity of the maximum likelihood probability distribution of galaxies for early (E-Sab), intermediate (Sb-Sbc), and late (Sc-Irr) morphological subgroups. For each group the mean is displayed for detections only (diamonds), detections plus upper limits treated as detections (upper plus signs), detections plus upper limits (circles), and detections plus upper limits treated as zero detections (lower plus signs). The true luminosity function of each sample must be bracketed by the plus signs. (*b*) Same as for (*a*) but for CO/HI (from Verter (1987)).

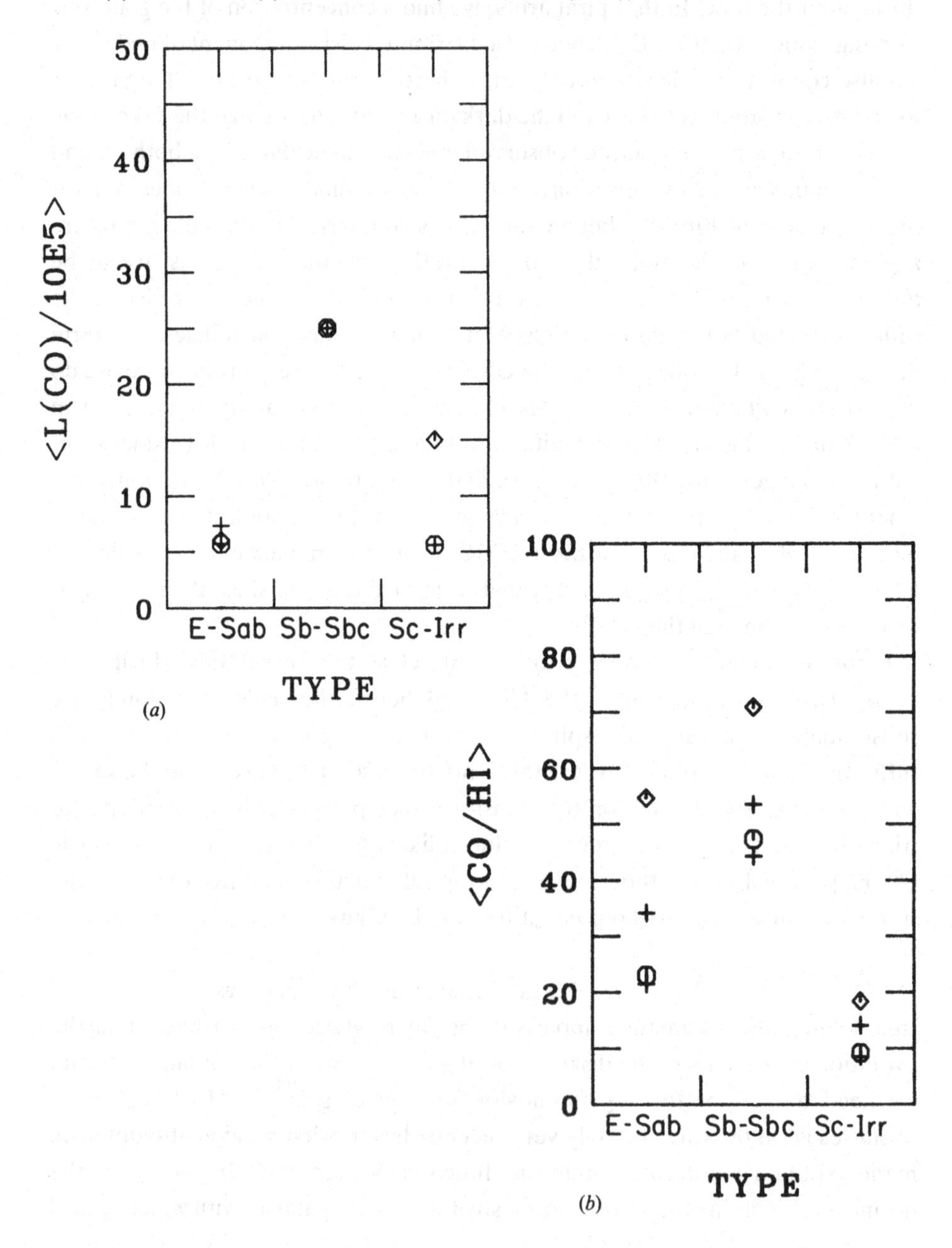

astronomers are familiar, in the millimeter-wave portion of the spectrum the small beams require at least some mapping in order to estimate what the total molecular content of a galaxy is. This is a very time consuming process for galaxies with inherently weak emission to begin with. Furthermore, it is necessary to make careful calibration comparisons between various telescopes, which is exacerbated by continuous upgrades and consequent changes in aperture and beam efficiencies. Such a comparison was made by Verter (1985) and used by her to obtain a molecular luminosity function for galaxies (Verter 1987). Strictly speaking, what she obtained was a CO luminosity function, but she argued that the luminosity function is likely to represent the true H_2 luminosity function because the variation in CO luminosities from galaxy to galaxy is likely to greatly exceed the uncertainty introduced by the problems involving the calibration of CO to H_2. Verter finds that both the total CO content as well as the fractional content (measured as the ratio of the CO/HI flux ratio) peak at around Hubble type Sbc (see Figure 2.2). The fall off in the luminosity function for the early type spirals is not surprising since they have long been known to be gas poor, but the fall off for very late type spirals and irregular galaxies was not obvious *a priori*, especially when the CO is normalized to the HI content. One of the problems for astrophysicists and astrochemists is to understand why it is that the molecular content of galaxies reaches a peak at intermediate morphological types.

Although none of the elliptical galaxies in Verter's sample was detected in CO, it has long been suspected that there might be at least some H_2 trapped in the potential well of the ellipticals from the ejecta of the planetary nebulae and supernovae in the galaxies. A number of elliptical galaxies have been detected in HI, but the question remained about whether any molecular gas exists in the benign environments of low interstellar ultraviolet radiation fields known to exist in ellipticals. Huchtmeier *et al.* (1988) have now detected weak CO from the elliptical galaxy NGC 4472 (M49) in the Virgo cluster using the IRAM 30 m telescope. CO had been found previously toward the peculiar elliptical galaxy NGC 185, a dwarf elliptical companion of M31, but the Huchtmeier *et al.* results are the first CO detections of an apparently normal elliptical. The weakness of the signals insure that work on the molecular phase of the interstellar medium in elliptical galaxies will be difficult for some time to come, but the exciting results open up a new area of research in extragalactic molecular astronomy.

2.6. Active galactic nuclei

Another exciting new area of research for extragalactic molecular observers is the study of active galactic nuclei. This area has benefitted enormously from the increase in sensitivity and stability of millimeter wavelength receivers and from the IRAS survey. The major thrusts have been in Seyfert galaxies and ultraluminous galaxies. There is some overlap in the galaxies being investigated by the two research groups most actively investigating these galaxies, that is, some Seyferts are ultraluminous galaxies, but the research efforts are complementary.

Seyfert galaxies as a class have been the target of molecular line astronomers since the detection of CO in NGC 1068, the brightest member of the class. The first papers searching for CO in Seyferts reported only one or two new detections per observing run. However, new instrumentation, and the use of IRAS fluxes as a search criterion has increased the number of detected Seyferts to 26, enough to obtain some reliable statistical measures (Heckman *et al.* 1988). These authors find that the CO luminosity of Seyfert 2 galaxies is a factor of 4 greater than the luminosity of a carefully selected control sample of galaxies chosen from the Revised Shapley Ames Catalog (Sandage and Tammann 1981). Surprisingly, Seyfert 1 galaxies are not unusual in their CO content. The CO appears to be fueling a burst of star formation in the Seyfert 2 type galaxies suggesting that the molecular gas concentrated in the nucleus is a condition necessary for the formation of a Seyfert. In this picture, the active galactic nucleus is an accompanying byproduct of the formation of a Seyfert, and Seyfert 2s evolve into Seyfert 1s as the molecular gas becomes disrupted by the burst of stars that have formed from it. A similar picture for ultraluminous galaxies has been proposed by Sanders *et al.* (1988), who suggest that these infrared bright galaxies might be the precursors of quasars.

To establish what role the molecular gas plays in the Seyfert nuclei, Blitz *et al.* (1989) have mapped the CO emission in the central 2′ of three galaxies with the BIMA interferometer at Hat Creek. Two of the three show strong central concentrations of CO coincident with the projected emission from the narrow-line emitting region observed optically, thus confirming that at least in some Seyferts the molecular gas is intimately related to the nuclear activity. In NGC 3227, about 30% of the emission seen in the single dish flux detected with a 1′ beam is seen to be located within 1.2″ of the nucleus!

The 'ultraluminous' galaxies are so called because they have luminosities exceeding 10^{12} solar luminosities, making them among the brightest galaxies in the local universe. These galaxies are in most cases galaxies which were known previously from optical observations, but were found by IRAS to be putting out most of their energy in the far infrared (Houck *et al.* 1984, 1985, Soifer *et al.* 1984). Many are associated with active galactic nuclei, and many are known to be morphologically disturbed, suggesting that they are the result of interactions and mergers between galaxies (Sanders *et al.* 1988). These authors have suggested an evolutionary link between quasars and starburst activity similar to that proposed by Heckman *et al.* (1988) for Seyfert galaxies. It is not surprising that these two groups should have come to similar conclusions if indeed quasars are simply more energetic versions of active nuclei.

What is especially intriguing about the Sanders *et al.* hypothesis is that the ultraluminous galaxies would represent the initial dust enshrouded stages of quasars. In their view, once the galactic nuclei shed their obscuring dust, they become optically selected quasars. The formation of quasars then would be a direct consequence of the production or gathering of molecular gas and dust in the centers of merged systems. A complete understanding of the most energetic known phenomenon in the Universe then would require an understanding of the low energy physics of molecular cloud formation.

2.7. Afterword

This chapter has provided only a superficial review of the subject of molecules in galaxies. Because of space limitations, many interesting and important topics are not covered. These include megamasers in the nuclei of galaxies, the chemistry of molecules in other galaxies, observations of species other than CO, the properties of the molecular interstellar medium in other galaxies, and others. Some of these are discussed in the volume *Molecules in Galaxies,* edited by R. L. Dickman, which are the proceedings of a meeting held in Amherst, Massachusetts in November 1987. Millimeter observations of molecules in galaxies have now moved into the mainstream of extragalactic astronomy and will continue to provide an important tool for understanding the structure, evolution and physics of galaxies.

This work is partially supported by funding from NSF grant AST 86-18763.

REFERENCES

Allen, R. J., Atherton, P. D., and Tilanus, R. P. J., 1986, *Nature* **319**, 296.
Bhat, C. L., Mayer, C. J., and Wolfendale, A. W., 1986, *Phil. Trans. Roy. Soc. Lond. A.* **319**, 249.
Blitz, L., 1978, Ph.D. Dissertation, Columbia University.
Blitz, L., 1980, in *Giant Molecular Clouds,* P. M. Solomon and M. G. Edmunds, eds., Pergamon: Oxford, p. 1.
Blitz, L., and Shu, F., 1980, *Ap. J.* **238**, 148.
Blitz, L., and Glassgold, A. E., 1982, *Ap. J.* **252**, 481.
Blitz, L., Fich, M., and Kulkarni, S., 1983, *Science* **220**, 1233.
Blitz, L., Bloemen, J. B. G. M., Hermsen, W., and Bania, T. M., 1985, *Astron. Ap.* **143**, 267.
Blitz, L., *et al.,* 1989, in preparation.
Bloemen, J. B. G. M., *et al.,* 1986, *Astron. Ap.* **154**, 25.
Bohlin, R. C., Savage, B. D., and Drake, J. F., 1978, *Ap. J.* **224**, 132.
Boulanger, F., Bystedt, J., Casoli, F., and Combes, F., 1984, *Astron. Ap.* **140**, 15.

Bronfman, L., *et al.,* 1988, *Ap. J.* **324**, 248.
Casoli, F., Combes, F., and Stark, A. A., 1987, *Astron. Ap.* **173**, 43.
Cohen, R. S., *et al.,* 1988, preprint.
Cohen, R. S., Cong, H-I., Dame, T. M., and Thaddeus, P., 1980, *Ap. J.* (*Letters*) **239**, L53.
Combes, F., Encrenaz, P. J., Lucas, R., and Weliachew, L., 1977, *Astron. Ap.* **55**, 311.
Dame, T. M., *et al.,* 1986, *Ap. J.* **305**, 892.
Dame, T. M., *et al.,* 1987, *Ap. J.* **322**, 706.
Dickman, R. L., 1978, *Ap. J.* (*Suppl.*) **37**, 407.
Elmegreen, B. G., Elmegreen, D. M., and Morris, M., 1980, *Ap. J.* **240**, 455.
Encrenaz, P. J., Stark, A. A., Combes, F., and Wilson, R. W., 1979, *Astron. Ap.* **78**, L1.
Fazio, G. G., 1967, *Ann. Rev. Astron. Ap.* **5**, 481.
Gouguenheim, L., 1979, in *Photometry, Kinematics, and Dynamics of Galaxies,* D. S. Evans, ed., University of Texas: Austin, p. 102.
Heckman, T. M., Blitz, L., Wilson, A. S., Armus, L., and Miley, G. K., 1988, *Ap. J.*, submitted.
Houck, J. R., *et al.,* 1984, *Ap. J.* (*Letters*) **278**, L63.
Houck, J. R., *et al.,* 1985, *Ap. J.* (*Letters*) **290**, L5.
Huchtmeier, W. K., Bregman, J. N., Hogg, D. E., and Roberts, M. S., 1988 preprint.
Ichikawa, T., Nakato, M., Tanaka, Y. O., and Sofue, M., 1985, *PASJ,* **37**, 439.
Israel, F. P., 1988, in *Millimeter and Submillimeter Astronomy,* R. Wolstencroft and W. B. Burton, eds., Reidel: Dordrecht, in press.
Jansky, K., 1932, *Proc. IRE* **20**, 1920.
Kutner, M. L., and Leung, C. M., 1985, *Ap. J.* **291**, 188.
Lada, C. J., Margulis, M., Sofue, Y., Nakai, M., and Handa, T., 1988, *Ap. J.* **328**, 143.
Li, Ti Pei, Riley, P. A., and Wolfendale, A. W., 1983, *MNRAS* **203**, 87.
Liszt, H. S., Burton, W. B., and Xiang, D. L., 1984, *Astron. Ap.* **140**, 303.
Lo, K. Y., Ball, R., Masson, C. R., Phillips, T. G., Scott, S., and Woody, D. P., 1987, *Ap. J.* (*Letters*) **317**, L63.
Maloney, P., and Black, J. H., 1988, *Ap. J.,* in press.
Miller, J., and Scalo, J., *Ap. J.,* 1978, **90**, 506.
Morris, M., and Rickard, L. J., 1982, *Ann. Rev. Astron. Ap.* **20**, 517.
Rickard, L. J., and Harvey, P. M., 1984, *Ap. J.* **89**, 1520.
Rickard, L. J., and Blitz, L., 1985, *Ap. J.* (*Letters*) **292**, L57.
Rickard, L. J., and Palmer, P., 1981, *Astron. Ap.,* **102**, L13.
Rydbeck, G., Hjalmarson, A., and Rydbeck, O. E. H., 1985, *Astron. Ap.* **144**, 282.
Sandage, A., and Tammann, G. A., 1981, *The Revised Shapley Ames Catalog of Bright Galaxies,* Carnegie: Washington.
Sanders, D. B., Scoville, N. Z., and Solomon, P. M., 1985, *Ap. J.* **289**, 373.
Sanders, D. B., *et al.,* 1986, *Ap. J.* **305**, L45.
Sanders, D. B., *et al.,* 1988, *Ap. J.* **325**, 74.
Sanders, D. B., and Mirabel, I. F., 1985, *Ap. J.* **298**, L31.
Scoville, N. Z., Solomon, P. M., and Sanders, D. B., 1979, in *Large Scale Characteristics of the Galaxy,* W. B. Burton, ed., Reidel: Dordrecht, p. 277.
Soifer, B. P., *et al.,* 1984, *Ap. J.* (*Letters*) **283**, L1.
Solomon, P. M., Scoville, N. Z., and Sanders, D. B., 1979, *Ap. J.* (*Letters*) **232**, L89.
Solomon, P. M., and Sanders, D. B., 1980, in *Giant Molecular Clouds,* P. M. Solomon and M. G. Edmunds, eds., Pergammon: Oxford, p. 41.
Solomon, P. M., Sanders, D. B., and Rivolo, A. R., 1985, *Ap. J.* (*Letters*) **292**, L24.

Solomon, P. M., Rivolo, A. R., Barrett, S., and Yahil, A., 1987, *Ap. J.* **319**, 370.
Stark, A. A., 1979, Ph.D. Dissertation, Princeton University.
Stark, A. A., and Blitz, L., 1978, *Ap. J.* (*Letters*), **225**, L15.
Stark, A. A., Linke, R. A., and Frerking, M. A., 1981, *BAAS* **13**, 535.
Stark, A. A., Elmegreen, B. G., and Chance, D., 1987, *Ap. J.* **322**, 64.
Stecker, F. W., 1971, *Cosmic Gamma Rays,* Mono Book Corp., Baltimore.
Thompson, R. I., Lebofsky, M. J., and Rieke, G. H., 1978, *Ap. J.* (*Letters*) **222**, L49.
van Dishoeck, E, and Black, J. H., 1987, in *Physical Processes in Interstellar Clouds,* G. E. Morfill and M. Scholer, eds., Reidel: Dordrecht, p. 241.
Verter, F., 1985, *Ap. J.* (*Suppl.*) **57**, 261.
Verter, F., 1987, *Ap. J.* (*Suppl.*) **65**, 555.
Verter, F., 1988, *Ap. J.,* in press.
Vogel, S. N., Boulanger, F., and Ball, R., 1987, *Ap. J.* (*Letters*) **321**, L145.
Vogel, S. N., Kulkarni, S. R., and Scoville, N. Z., 1988, *Nature* **334**, 402.
Weliachew, L., 1971, *Ap. J.* (*Letters*) **167**, 247.
Young, J. S., and Scoville, N. Z., 1982, *Ap. J.* **258**, 467.
Young, J. S., and Sanders, D. B., 1986, *Ap. J.* **302**, 680.
Zuckerman, B., and Evans, N. J., 1974, *Ap. J.* (*Letters*) **192**, L149.
Zuckerman, B., and Palmer, P., 1974, *Ann. Rev. Astron. Ap.* **12**, 279.

II

Diffuse molecular clouds

3

Diffuse cloud chemistry

EWINE F. VAN DISHOECK

Division of Geological and Planetary Sciences, California Institute of Technology, Pasadena, USA

3.1. Historical perspective

The study of interstellar chemistry started, appropriately, about 60 years ago. In 1926, Eddington discussed in his remarkable Bakerian Lecture the possibility of molecule formation and absorption in dark nebulae. At that time, only atomic species had been identified in interstellar space through their narrow absorption lines superposed on the spectra of background stars. In the next decade, several new interstellar features were detected which could indeed be ascribed to molecules: CH, CH^+ and CN.

In spite of this early success, no other molecule was found in the interstellar gas for the next 25 years, until OH was detected in 1963 by its radio emission lines. In the next two decades, more than 70 different interstellar molecules were identified by centimeter and millimeter wavelength techniques. However, these radio emission line studies were mostly concerned with dense and dark clouds, whereas the early absorption line observations probed much more diffuse gas.

Although a wide variety of interstellar molecules has now been detected in dark clouds, still only a handful of molecules has been found in diffuse clouds. The molecules H_2, HD, OH and CO were discovered in the 1970s by their absorption lines in the ultraviolet through rocket experiments and by the Copernicus satellite. Since the detection of C_2 in 1977 by ground-based techniques, however, no new molecule has firmly been identified in diffuse clouds. The list of molecules sought but not detected is considerably longer and includes such interesting species as NH, HCl, NaH, MgH, H_2O and C_3 (see van Dishoeck and Black (1988*a*) for a recent summary).

Attempts to understand the observed abundances of interstellar molecules were first made by Swings and Rosenfeld (1937) on the basis of a model borrowed from the study of stellar atmospheres, in which thermodynamic equilibrium between the atoms, molecules and a diluted stellar radiation field is assumed. Because of the low densities and temperatures in interstellar clouds compared with the high effective temperatures of the ultraviolet starlight, the assumption of thermodynamic equilibrium is not valid, however. In 1946, Kramers and ter Haar adopted a more realistic model in which the specific rates for the gas-phase formation and destruction of molecules – at that time mainly CH and CH^+ – were equated. This model was greatly improved upon by Bates and Spitzer in 1951, who obtained much more reliable rate coefficients for the various processes. The model failed, however, to reproduce the observed CH and CH^+ abundances with reasonable assumptions for the physical conditions in the cloud.

Because of the difficulties encountered with the early gas-phase reaction schemes, grain surface processes were thought to dominate the formation of interstellar molecules for the next 20 years. Plausible models were made for the formation of the H_2 molecule (Hollenbach and Salpeter 1971), but the production of other molecules appeared difficult to quantify due to uncertainties in the composition of the grain surfaces, the surface reactions that can occur, and the desorption processes.

Stimulated by the detection of new molecules at radio wavelengths, Solomon and Klemperer (1972) and Herbst and Klemperer (1973) recognized that reactions between ions and neutral species can be very rapid at the low temperatures prevailing in interstellar clouds, and reintroduced gas-phase reaction schemes that could account for the abundances of some small polyatomic molecules. At the same time, Black and Dalgarno (1973*a*) proposed that the formation of the carbon-bearing molecules – in particular CH – is initiated by the radiative association reaction between C^+ and H_2, rather than by the previously assumed association of C^+ and H. In addition, Dalgarno, Black and Weisheit (1973), Black and Dalgarno (1973*b*), Watson (1973) and O'Donnell and Watson (1974) identified the reactions important in the deuterium and oxygen chemistries in interstellar clouds, and showed that the observed HD and OH abundances can be used as diagnostics of the cosmic deuterium abundance and the cosmic ray ionization rate. These papers have become the basis of the present interstellar gas-phase chemistry networks. Because of their relative simplicity, diffuse clouds may provide the best environment in which to test these basic concepts.

Interstellar chemistry consists of an intricate and detailed interplay between the study of various molecular processes under interstellar conditions and the actual modeling of interstellar clouds. No one has a better overview of the problems occurring in both areas than Alexander Dalgarno, and his many contributions will be emphasized throughout this chapter.

3.2. Chemistry networks

The gas-phase processes responsible for the formation of the simplest carbon-, oxygen- and nitrogen-bearing molecules were identified about 15 years ago and have been discussed extensively e.g. by Dalgarno and Black (1976), Dalgarno (1976*a*) and Watson (1978). Recent reviews of diffuse cloud chemistry with a large number of references have been given by Crutcher and Watson (1985), Dalgarno (1988) and van Dishoeck and Black (1988*a*). Only limited references will be given throughout this chapter.

Although many different reactions can occur in interstellar clouds, usually only few routes dominate. This results from the fact that interstellar clouds consist mostly of hydrogen with all other species present in trace amounts. Table 3.1 lists the current best estimates of the solar abundances of the various elements relative to hydrogen. Some of the heavier elements are depleted from the gas phase in interstellar clouds, although for diffuse clouds, this depletion is minor and tends to exceed a factor of 4 only for heavier metals like Ca, Ti, Mn, and Fe. The fraction of the solar abundance of element X in the gas phase is denoted by the depletion factor δ_X, with $\delta_X \leq 1$. Except for hydrogen, most elements exist primarily in atomic or ionic form in diffuse clouds, with the most abundant molecule, CO, containing at most 1% of the total carbon. Only H_2 may be comparable in abundance to H. Thus reactions with H and H_2 dominate the networks if they are exothermic, and small hydride molecules are readily formed. The clouds also contain small solid particles, called grains, which account for approximately 1% of the mass.

The amount of molecules in diffuse clouds is small because they are rapidly photodissociated by the interstellar radiation field. This process also severely limits the building-up of more complex molecules. The mostly atomic content and

Table 3.1. *Solar elemental abundances*

Element	Abundance	Element	Abundance
H	1.00	Si	4.3(−5)
He	0.075	S	1.7(−5)
O	8.3(−4)	P	2.8(−7)
C	4.7(−4)	Cl	1.1(−7)
N	1.0(−4)	K	1.3(−7)
Na	2.1(−6)	Ca	2.2(−6)
Mg	4.2(−5)	Fe	4.3(−5)
Al	3.1(−6)		

Note: In this and subsequent tables the notation $a(b)$ indicates $a \times 10^b$.

the importance of photodissociation make diffuse cloud chemistry significantly different from the dark cloud chemistry discussed e.g. in Chapters 6 and 8. The distinction between diffuse and dark clouds is most conveniently described in terms of the extinction A_V at visible wavelengths. Diffuse clouds have A_V less than about 2 magnitudes, whereas dark clouds often have $A_V > 10$ mag, so that ultraviolet photons do not penetrate to the centers of the clouds.

In the following, the individual networks for species containing hydrogen, carbon, oxygen, nitrogen or sulfur will be considered in more detail. An overview of the chemistries of silicon- and metal-bearing molecules can be found e.g. in Dalgarno and Black (1976).

3.2.1. Hydrogen chemistry

There are two basic processes by which atoms can be converted into molecules at the low densities in diffuse clouds. The first one is radiative association

$$X + Y \rightarrow XY + h\nu \tag{3.1}$$

and the second one is through surface reactions on grains

$$X + Y{:}g \rightarrow XY + g \tag{3.2}$$

For the homonuclear molecule H_2, the radiative association reaction of H(1s) + H(1s) is in first order forbidden, and is thus much too slow to account for its large abundance. However, if two hydrogen atoms stick to a grain surface, the grain can absorb (part of) the energy that is liberated upon formation of the molecule. The surface area of dust grains is such that the H_2 formation rate is about

$$r_f \approx 3 \times 10^{-18} T^{1/2} y_f n_H n(\mathrm{H}) \ \mathrm{cm}^{-3}\ \mathrm{s}^{-1} \tag{3.3}$$

where y_f is a formation efficiency factor which is of order unity if every atom that sticks forms a molecule, and if every molecule is released back into the gas phase (see Chapter 7). Here $n_H = n(\mathrm{H}) + 2n(\mathrm{H_2})$ is the density in the cloud and T is the kinetic temperature.

H_2 is destroyed mostly by photodissociation through the two-step process of spontaneous radiative dissociation, first recognized by Solomon (1965) and illustrated in Figure 3.1 (see also Chapter 9). In the first step, an H_2 molecule in the $v = 0$ vibrational level of the ground $X^1\Sigma^+$ electronic state absorbs an ultraviolet photon into an excited electronic state. For photon energies less than the cutoff energy of the interstellar radiation field of 13.6 eV, only the $B^1\Sigma_u^+$ and $C^1\Pi_u$ states can be accessed through electric dipole allowed transitions. The corresponding series of discrete absorptions into the various vibrational levels are called the Lyman and Werner systems, respectively, and oscillator strengths have been computed by Allison and Dalgarno (1970). The electronically excited levels subsequently decay rapidly by spontaneous emission into the vibrational continuum of the ground state, leading to dissociation of the molecule. Dalgarno and Stephens (1970) and Stephens and Dalgarno (1972) performed accurate calculations of the radiative transition probabilities for this process, and found that the average fraction of absorptions leading to dissociation is about 0.1 in the interstellar radiation field.

The remaining 90% of the fluorescent transitions populate various bound excited vibration–rotation levels of the ground $X^1\Sigma_g^+$ state. Because the homonuclear H_2 molecule does not possess a permanent dipole, electric dipole transitions between the various rovibrational levels within the ground electronic state are forbidden. The excited levels can therefore decay only through slow quadrupole transitions on time scales of a few months. Radiative rates have been computed by Turner, Kirby-Docken and Dalgarno (1977). The transitions occur in the infrared, and observations of them will be discussed in Chapter 22.

The main reason for the large abundance of H_2 in interstellar clouds is the fact that its photodissociation is initiated by discrete line absorptions. These lines

Figure 3.1. Potential energy curves of the H_2 molecule illustrating the processes of spontaneous radiative dissociation and ultraviolet pumping.

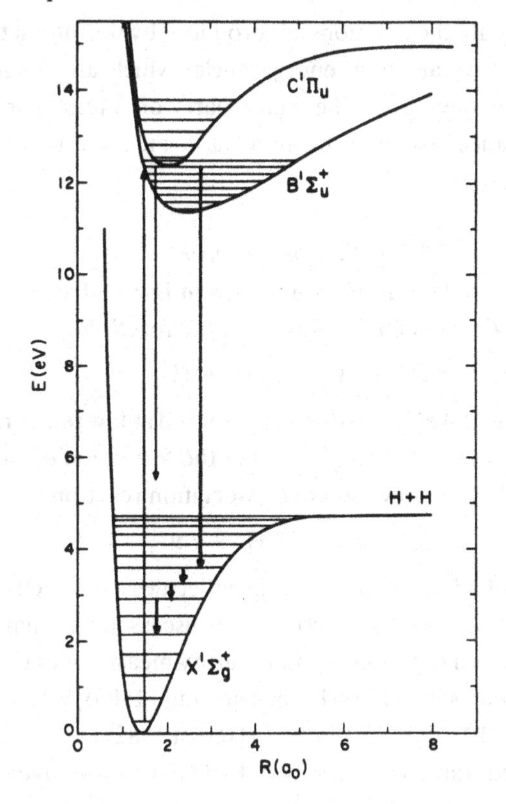

become optically thick, so that H_2 molecules deeper in the cloud experience a greatly reduced flux. The steady-state abundance of H_2 can be computed at each depth by equating the formation and destruction rates, and H_2/H ratios of unity are readily obtained in the centers of diffuse clouds with densities of the order of a few hundred per cubic centimeter.

3.2.2. Deuterium chemistry

If the molecule HD, like H_2, were formed only on grain surfaces, its abundance with respect to H_2 would be expected to be much smaller than the overall deuterium abundance, $[D]/[H] \approx 10^{-5}$. This is because the self-shielding of H_2 reduces its photodissociation rate by more than a factor of 10^4 in the centers of diffuse clouds, whereas the column densities of HD are too small for appreciable self-shielding. Thus abundance ratios $HD/H_2 \approx 10^{-9}$ would be expected. In contrast, the observed ratio is $HD/H_2 \approx 10^{-6}$. This large abundance of HD can be explained only if gas-phase reactions are taken into account, in particular the sequence

$$H^+ + D \rightleftarrows H + D^+ \tag{3.4}$$

$$D^+ + H_2 \rightleftarrows H^+ + HD. \tag{3.5}$$

The H^+ ions which initiate the reactions are produced by the interaction of cosmic rays with H. Cosmic rays are high energy nuclei which are ubiquitous in the Galaxy and which can penetrate the inner parts of interstellar clouds. The observed HD abundance can thus be used as a measure of the cosmic ray ionization rate of hydrogen.

3.2.3. Carbon chemistry

Because the ionization potential of atomic carbon is less than 13.6 eV, carbon exists mostly as C^+ in diffuse clouds. However, the reaction

$$C^+ + H_2 \rightarrow CH^+ + H \tag{3.6}$$

is endothermic by about 0.4 eV, and does not proceed at low temperatures. Black and Dalgarno (1973*a*) therefore suggested that the formation of carbon-bearing molecules is initiated by the slow radiative association reaction

$$C^+ + H_2 \rightarrow CH_2^+ + h\nu \tag{3.7}$$

Thus the abundances of all carbon-containing molecules are directly proportional to the rate coefficient k_7 of this reaction. As discussed in Chapter 11, k_7 is unfortunately still not well determined. Laboratory measurements give an upper limit $k_7 < 1.5 \times 10^{-15}$ cm^3 s^{-1} at 13 K (Luine and Dunn 1985), whereas theoretical estimates suggest $k_7 \approx 10^{-16}$–10^{-15} cm^3 s^{-1} (Herbst 1982).

Once CH_2^+ is formed, rapid reactions with H_2 lead to larger hydrocarbon ions.

Dissociative recombination subsequently produces small neutral hydrides like CH. The network is illustrated in Figure 3.2. Note that the rates for reactions such as $CH_2^+ + H_2 \rightarrow CH_3^+ + H$ and $CH_3^+ + e \rightarrow$ products need not be known with very high precision as long as they are rapid. Much more important in the prediction of some abundances are the branching ratios for dissociative recombination, which are highly uncertain (Bates (1986); see also Chapters 10 and 11). For example, the branching ratio of CH_3^+ to form CH_2 directly affects the CH_2 abundance. On the other hand, the CH abundance is only slightly modified, since CH_2 photodissociates rapidly to form CH.

The formation of molecules containing more than one carbon atom can take place by reactions of the carbon hydrides with the abundant C^+ ion. Many of these ion–molecule reactions are rapid at low temperatures and produce ions such as C_2^+, C_2H^+ and $C_2H_2^+$, which can subsequently recombine to form e.g. C_2 and C_2H. The predicted amount of C_2 in the models is also sensitive to the branching ratio for the CH_3^+ dissociative recombination. If the CH_2 abundance is large, substantial amounts of C_2 are formed through the reaction of CH_2 with C^+, in addition to that produced through the CH with C^+ reaction.

The dominant destruction process of the neutral molecules in diffuse clouds is

Figure 3.2. The most important reactions involving carbon-bearing molecules. M stands for metal.

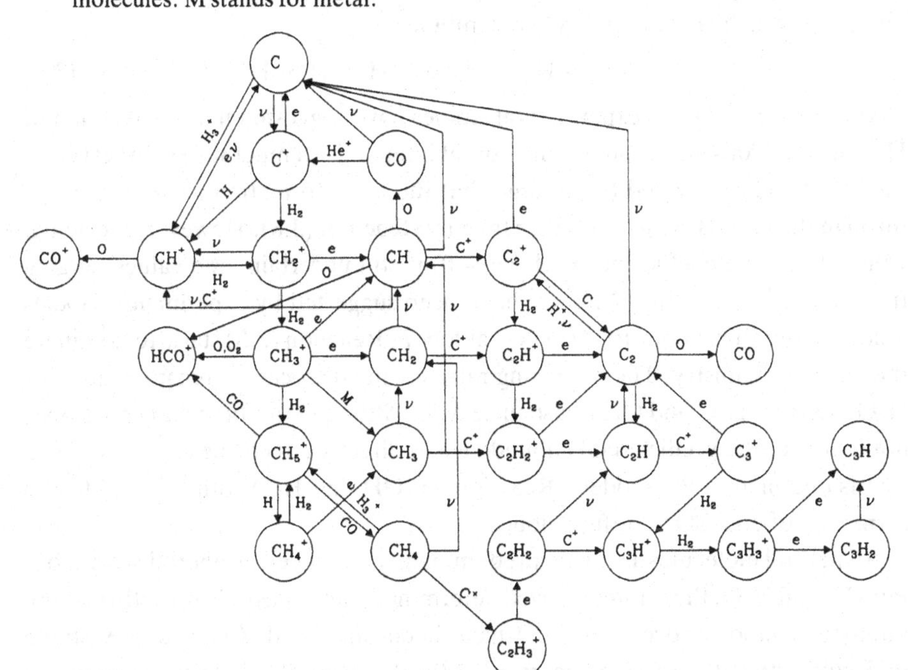

photodissociation. Thus, accurate rates for all photoprocesses are of paramount importance for the study of diffuse cloud chemistry. The photo rates for the various species have recently been evaluated critically by van Dishoeck (1988) (see also Chapter 9).

3.2.4. Oxygen chemistry

Atomic oxygen cannot be photoionized by the interstellar radiation field, so that oxygen is mostly neutral in diffuse interstellar clouds. The source of ionization which drives the ion–molecule chemistry is provided in this case by the cosmic rays which ionize H and H_2 to form H^+ and H_2^+. H_2^+ subsequently reacts very rapidly with H_2 to form H_3^+. The rate at which atomic hydrogen is ionized is denoted by ζ_0. Since the ionization potential of O is accidentally close to that of H, the charge transfer reaction

$$O + H^+ \rightarrow O^+ + H \tag{3.8}$$

requires only a small energy input, and takes place efficiently at low temperatures. Once O^+ is formed, it reacts rapidly with H_2 to form OH^+. Alternatively, OH^+ can be formed by the fast reaction between H_3^+ and O. The formation of neutral oxygen-bearing molecules such as OH and H_2O results again from a series of rapid hydrogen abstraction reactions followed by dissociative recombination, as is illustrated in Figure 3.3. Thus, the abundances of all oxygen-containing molecules are directly proportional to ζ_0. Another crucial reaction in the oxygen network is the rate at which H_3^+ dissociatively recombines

$$H_3^+ + e \rightarrow H_2 + H \quad \text{or} \quad H + H + H \tag{3.9}$$

Various theoretical and experimental studies have suggested that the reaction of H_3^+ with electrons may be unusually slow at interstellar temperatures. In that case, the H_3^+ abundance would be so large that the main formation route of OH^+ is through the O + H_3^+ reaction rather than by Reaction (3.8). However, the actual value of the rate coefficient for $H_3^+ (v = 0)$ is still uncertain, and values ranging from $<10^{-11}$ to 2×10^{-7} cm ^{3} s^{-1} have been suggested by experiments. If k_9 is indeed larger than about 10^{-9} cm^3 s^{-1} at low T, Reaction (3.8) starts to dominate the oxygen chemistry. The branching ratio for the dissociative recombination of H_3O^+ to form H_2O and OH is also uncertain. Although this branching ratio does not influence the predicted OH abundance, it directly affects the amount of H_2O that is obtained in the models. Reactions of OH and H_2O with C^+ lead to the formation of e.g. CO in diffuse clouds.

All neutral molecules are again predominantly destroyed by photodissociation. For OH and H_2O, these rates are well determined, but it was only recently that the photodissociation processes of CO have been unraveled. CO is a very stable molecule with a dissociation energy of 11.09 eV so that its photodissociation can

Figure 3.3. The most important reactions involving oxygen-bearing molecules.

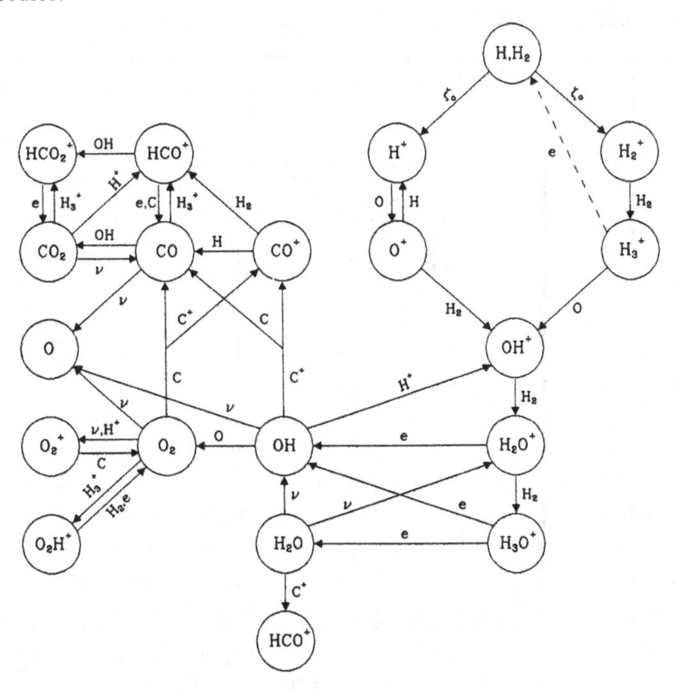

take place only at $\lambda < 1118$ Å. Recent experiments by Letzelter *et al.* (1987) and Yoshino *et al.* (1988) have confirmed earlier speculations that the CO photodissociation occurs mostly by line absorptions into a large number of predissociating bands. If the cross sections derived from these new measurements are adopted, the unshielded rate is about a factor of 40 larger than that used e.g. by Black and Dalgarno (1977).

The depth dependence of the CO photodissociation rate is complicated by several factors. Just as for H_2, the dissociating lines can become optically thick inside the cloud. In addition to this self shielding, dust attenuation and shielding by coincident lines of H and H_2 play a role: the ultraviolet lines of H_2 through which its photodissociation and excitation occur happen to lie in exactly the same wavelength region. In order to account for these effects, van Dishoeck and Black (1988*b*) have simulated the full absorption spectrum of CO at each depth into the cloud. Figure 3.4 contains a small portion of the absorption that would be produced by CO and by H and H_2 at the center of a diffuse cloud such as that toward ζ Oph. It is clear that some bands, such as the one at 1063 Å, are coincident with strong H and H_2 features and are effectively blocked by them, whereas other bands are relatively free. Because of the shielding, the CO photodissociation rate decreases rapidly with depth into the cloud, although the decrease is not as drastic

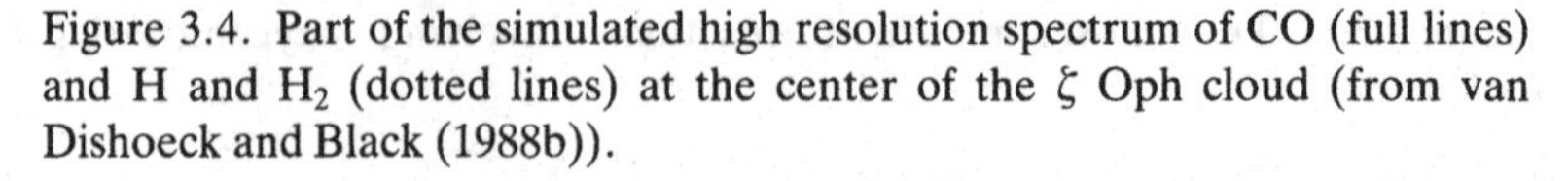

Figure 3.4. Part of the simulated high resolution spectrum of CO (full lines) and H and H_2 (dotted lines) at the center of the ζ Oph cloud (from van Dishoeck and Black (1988b)).

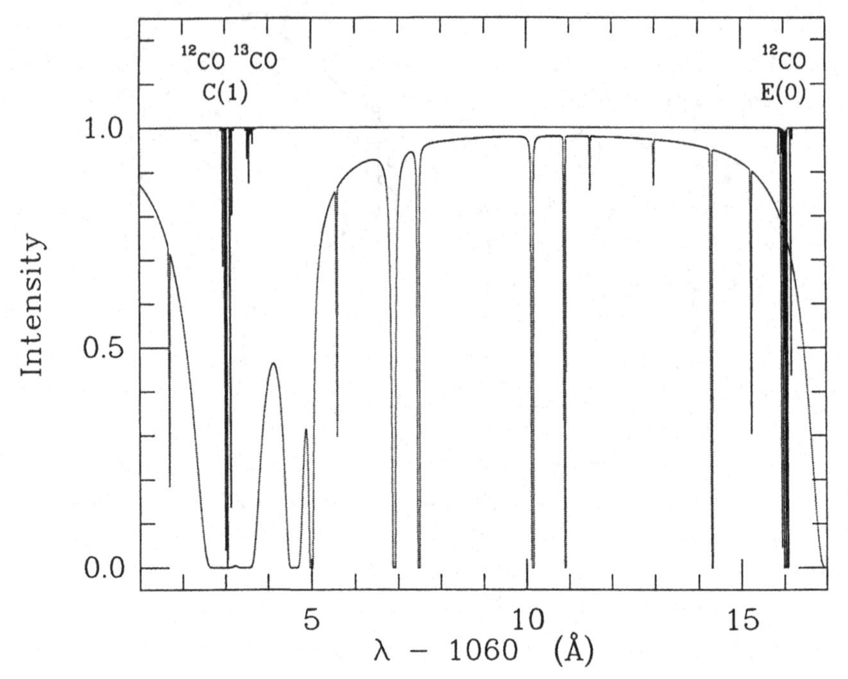

as that for the photodissociation rate of the more abundant H_2 molecule. The case of CO clearly illustrates the many details that need to be considered in a proper depth-dependent calculation of the abundance of a molecule whose photodissociation is dominated by line absorptions.

3.2.5. Nitrogen chemistry

The formation routes for nitrogen-bearing molecules in diffuse clouds are particularly uncertain. The ionization potential of atomic nitrogen is also higher than 13.6 eV, so that nitrogen will be mostly neutral. A small amount of N^+ can be maintained by cosmic ray ionization, but the reaction with H_2 to form NH^+ is now known to be endothermic by a small amount. Most models have assumed that the exothermic reaction

$$N + H_3^+ \rightarrow NH_2^+ + H \tag{3.10}$$

proceeds rapidly. However, this is a very unusual reaction in which two protons are transferred. Detailed theoretical calculations indicate indeed that large barriers exist along the reaction path, so that the reaction is likely to be very slow at low temperatures.

Figure 3.5. The most important reactions involving nitrogen-bearing molecules.

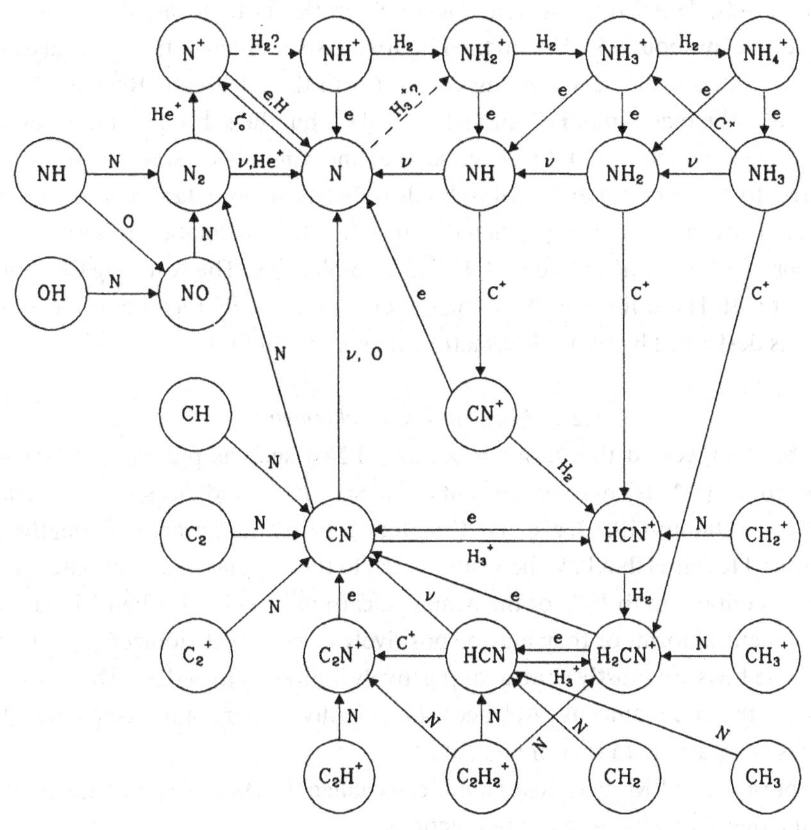

Alternative gas-phase reactions through which nitrogen-containing molecules can be formed are illustrated in Figure 3.5. They include neutral–neutral reactions of atomic nitrogen with small oxygen and carbon-containing molecules such as OH, CH and C_2 to form CN and NO, and ion–molecule reactions of atomic nitrogen with small hydrocarbon ions to form e.g. H_2CN^+, which can then recombine to form CN and HCN. The rate coefficients for the neutral–neutral reactions are usually small at low temperatures and may vary considerably from species to species. Few neutral–neutral rates have been measured at low temperatures in the laboratory. The photodissociation rate of CN is uncertain, but is likely to be larger than thought previously.

3.2.6. Sulfur chemistry

Although chemically, sulfur is usually considered to be similar to oxygen, the two species behave very differently under interstellar conditions (Oppenheimer and

Dalgarno 1974). This is due to the fact that the ionization potential of S is less than 13.6 eV, so that it is mostly ionized, whereas oxygen is predominantly neutral in diffuse clouds. However, the reaction of S^+ with H_2 to form SH^+ is highly endothermic by about 0.9 eV, and the radiative association reaction to form SH_2^+ is also predicted to be very slow, in contrast with the analogous Reaction (3.7) for carbon. The reaction of H_3^+ with S forms SH^+, but the subsequent reaction of SH^+ with H_2 is again endothermic, so that the amount of sulfur hydrides is expected to be very small in cold clouds. The most abundant sulfur-bearing molecule in diffuse clouds is predicted to be CS, whose formation is initiated by reactions of S^+ with the abundant CH and C_2 molecules. The resulting CS^+ ions can react with H_2 to form HCS^+, which then dissociatively recombines to form CS. CS is destroyed by photodissociation at an uncertain rate.

3.2.7. Large molecule chemistry

It has been suggested that large molecules (LMs) such as polycyclic aromatic hydrocarbons (PAHs) may be present in substantial abundances in interstellar clouds (e.g. Omont (1986); see also Chapter 8). From the observed strengths of the infrared features that have been attributed to them, it has been estimated that they may contain up to 10% of the available carbon in a cloud. The LMs can be either in neutral form, or they may be positively or negatively ionized. In diffuse clouds, the LMs are mostly neutral or positively ionized at the edges. Deeper into the cloud, the abundance of LM^+ decreases rapidly, and the abundance of LM^- becomes comparable to that of neutral LM.

The presence of LMs in diffuse clouds can change the chemistry both quantitatively and qualitatively through the reactions

$$\left.\begin{array}{l} \mathrm{LM} + \mathrm{X}^+ \rightarrow \mathrm{LM}^+ + \mathrm{X} \\ \mathrm{LM}^- + \mathrm{X}^+ \rightarrow \mathrm{LM} + \mathrm{X} \end{array}\right\} \qquad (3.11)$$

where X^+ can be any ion such as C^+, a metal ion or a molecular ion. Thus the main effect of the LMs will be to neutralize ions. Note that in this respect the specific identification of the LMs as PAHs is not crucial: the same chemical effects will occur for any other large molecule or small grain which is capable of neutralizing ions.

3.3. Steady-state models

3.3.1. Approaches

Since the early work of Bates and Spitzer (1951), many steady-state models have been developed to describe the chemistry in diffuse clouds. Two approaches can be distinguished. One is to compare molecular abundances for a large number of lines-of-sight with simple models that reproduce the observed trends (see e.g.

Federman (1982), Danks, Federman and Lambert (1984)). These models usually only provide the relative densities of the species at a given temperature and density, and it is assumed that they can be equated to relative column densities. The observations of CH, CO and CN are well bracketed by models with densities ranging from 200 to 2500 cm^{-3}. This suggests that the adopted chemical network is correct in first approximation, and that the physical conditions in the various diffuse clouds differ by no more than an order of magnitude.

In the opposite approach, detailed models are developed which reproduce the observed abundances of a wide variety of species in individual clouds. Such comprehensive studies provide the most rigorous tests of the chemical networks. The classic models in this category are those of Black and Dalgarno (1977; hereafter BD) for the cloud in front of ζ Oph and of Black, Hartquist and Dalgarno (1978) for the cloud toward ζ Per. Van Dishoeck and Black (1986; hereafter vDB) have recently updated the ζ Oph and ζ Per models, whereas Viala, Roueff and Abgrall (1988; hereafter VRA) have independently reinvestigated the ζ Oph cloud.

3.3.2. Physical conditions

Because the chemical reaction rates depend sensitively on the physical conditions, detailed information on the temperature T and density n_H in the cloud is a prerequisite for developing comprehensive models, as is the strength of the ultraviolet radiation field incident on the cloud surface. In this chapter, all intensities will be given by a scaling factor I_{UV} where $I_{UV} = 1$ refers to the intensity of the standard radiation field as given by Draine (1978). The physical parameters are usually constrained from observations of the excitation of atoms and molecules. Most modelers have adopted as the primary diagnostic the observed abundance of the H_2 molecule and its population distribution over the $v = 0, J = 0$ to $J = 7$ rotational levels of the ground state. As has been described in detail e.g. by BD and vDB (see also Section 3.2.1), the relative populations are determined by the competition between the specific formation process on dust, the destruction by photodissociation, the ultraviolet absorption and cascading process, and collisional excitation and deexcitation processes. The high rotational levels $J \geq 5$ are populated primarily by the ultraviolet pumping process illustrated in Figure 3.1, so that their populations are sensitive to the strength of the ultraviolet radiation field in the 912–1100 Å region. The population distribution over the lower levels $J = 0$ to $J = 2$ reflects the temperature structure in the cloud. The relative abundance of H_2 with respect to H depends on the combination of parameters $n_H y_f / I_{UV}$. If $y_f = 1$ is assumed, n_H, I_{UV} and T can be constrained in principle from observations of H_2 and H alone. The inferred parameters have to be consistent with those suggested by other observations. In particular, the higher rotational levels of the homonuclear molecule C_2 can also be populated by a pumping process, and its population distribution is sensitive to the parameters n/I_R, with $n = n(H) + n(H_2)$ and I_R the scaling factor for the radiation field in the near-infrared part of the spectrum (van Dishoeck and Black 1982). The CO rotational excitation and the C, C^+ and O fine structure excitation are controlled by collisional processes and are thus sensitive to the product $n_H T$. Finally, the ionization balance of the atomic species is determined by the ratio n_H/I_{UV}.

The analyses of the various diagnostic observations have often led to conflicting results. In the extreme case of the ζ Oph cloud, central temperatures between 20 and 60 K, central densities n_H between 200 and 2500 cm^{-3} and enhancement factors for the radiation field I_{UV} = 2–6 have been inferred. The original two component plane-parallel model of BD had about $\frac{2}{3}$ of the material at $n_H = 500$ cm^{-3}, $T = 110$ K with $I_{UV} = 2.5$, and the remainder at $n_H = 2500$ cm^{-3} with $T = 20$ K. The structure is illustrated in Figure 3.6. However, the self-shielding in the H_2 ultraviolet lines through which the dissociation and excitation occur was significantly underestimated in these models. Crutcher and Watson (1981) subsequently argued that the observed CO and HD rotational excitations imply a low density, $n_H \approx 200$ cm^{-3}, throughout most of the cloud, whereas the absence of significant ^{13}CO fractionation suggested a rather high temperature, $T \approx 60$ K. Roberge (1981) and vDB used a corrected treatment of the radiative transfer in the H_2 lines and showed that the original BD structure was inconsistent with observations. In particular, the intensity of the radiation field had to be increased and the density decreased to reproduce the H and H_2 measurements. vDB subsequently developed new models for the ζ Oph cloud and three other diffuse clouds, in which continuous gradients of temperature and density have replaced the unrealistic two component structure. The clouds are in hydrostatic equilibrium and are pressure bounded. The vDB models are not unique, but they typically have densities of a

Figure 3.6. Temperature and density structure in the Black and Dalgarno (1977; BD), van Dishoeck and Black (1986; vDB) and Viala *et al.* (1988; VRA) models of the ζ Oph cloud (from van Dishoeck and Black (1988a)).

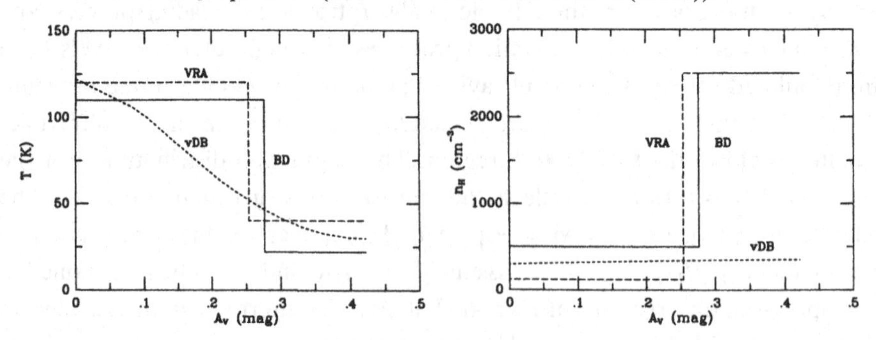

few hundred per cubic centimeter at the edge with temperatures of 100–200 K and I_{UV} = 2–6, depending on the adopted grain scattering properties (see Chapter 15). The densities increase to 300–800 cm^{-3} at the center and the temperatures decrease to 20–30 K. The relatively high temperature at the edge is needed to reproduce the observed H_2 populations in $J = 2$ and $J = 3$, whereas the low temperature in the core is suggested by the H_2 $J = 0$ to $J = 1$ population ratio. In addition, the observed $J = 0$ to $J = 2$ population ratios of C_2 suggest low temperatures in the centers of diffuse clouds, in particular the ζ Oph cloud. The temperature and density structure in the preferred model G of vDB of the ζ Oph cloud is included in Figure 3.6.

The VRA models do not have continuous variations of temperature and density but retain the BD 'core + envelope' structure. The temperature and density structure in the preferred model 6 of VRA is included in Figure 3.6 for comparison. VRA share the conclusions of vDB that a single component model cannot reproduce the observational data towards ζ Oph; that a significant part of the cloud must have a low density to reproduce the observed H/H_2 ratio; that an enhanced temperature of about 100–200 K is required at the edge by the H_2 $J = 2$ and $J = 3$ populations; and that an enhanced radiation field is necessary to reproduce the high J population of H_2. The VRA models differ, however, in a few details. In particular, the density in the core of the VRA models of the ζ Oph cloud is not well constrained by the H_2 and H observations alone, and can vary between 1000 and 5000 cm^{-3}. In VRA model 6, the central density is 2500 cm^{-3}. As discussed by vDB (see also van Dishoeck and Black 1988*a*), the available data on the C_2 and CO rotational excitation and the carbon fine structure excitation favor a lower density $n_H < 1000$ cm^{-3}. The enhancement factor of the incident radiation field was chosen in the VRA models such as to reproduce about half of the observed H_2 $J = 5$ column density, and is therefore lower than that in the vDB models. Table 3.2 compares the computed H_2 excitation in the two models. The slight discrepancies are probably due to the different treatments of the self-shielding in the H_2 lines and the continuum attenuation by dust. Compared with observations, all models generally reproduce well the observed populations in the various levels, with the exception of the $J = 3$ and $J = 4$ levels, for which the computed populations are too low by factors of a few.

In summary, the various efforts to model the ζ Oph cloud yield central densities that are uncertain by at least a factor of 2, although the evidence in favor of a rather low density throughout the cloud is compelling. The intensity of the incident ultraviolet radiation field is also not well determined, since it is constrained only by the rotational excitation of H_2. If other processes such as shocks (see Chapter 5) contribute to the population of the high J levels of H_2, the inferred

Table 3.2. *Comparison of steady-state models with observations*[a]

Species	ζ Oph vDB[b] model G	ζ Oph VRA[c] model 6	ζ Oph Observed[d]	ζ Per vDB[b] model F	ζ Per Observed[d]
δ_C	0.52	0.28		0.50	
δ_D	0.30	0.40		1.10	
ζ_0 (s^{-1})	8.0(−17)	5.1(−17)		6.0(−17)	
k_7 ($cm^3\ s^{-1}$)	7.0(−16)	2.6(−16)		7.0(−16)	
H	5.1(20)	4.6(20)	(5.2 ± 0.2)(20)	6.7(20)	(6.5 ± 0.7)(20)
H_2	4.2(20)	4.8(20)	(4.2 ± 0.3)(20)	4.8(20)	(4.8 ± 1.0)(20)
H_2 $J = 0$	2.9(20)	3.3(20)	(2.9 ± 0.3)(20)	3.3(20)	(3.2 ± 1.3)(20)
1	1.3(20)	1.5(20)	(1.3 ± 0.2)(20)	1.5(20)	(1.5 ± 0.6)(20)
2	8.3(17)	3.2(18)	(4.0 ± 1.6)(18)	1.0(18)	(1.1 ± 0.1)(18)
3	2.2(16)	8.9(16)	(1.7 ± 1.2)(17)	2.7(16)	(2.0 ± 0.7)(16)
4	1.5(15)	9.1(14)	(5.3 ± 2.2)(15)	1.2(15)	(1.0 ± 0.1)(15)
5	4.5(14)	1.7(14)	(4.3 ± 0.5)(14)	2.3(14)	(2.3 ± 0.2)(14)
6	5.3(13)	2.8(13)	(4.9 ± 0.5)(13)	2.5(13)	
7	3.8(13)	1.4(13)	(3.6 ± 0.4)(13)	5.4(12)	
C	2.8(15)	3.7(15)	(3.2 ± 0.6)(15)	2.9(15)	(3.3 ± 0.4)(15)
C^+	3.2(17)	1.8(17)	(1.0 ± 0.5)(17)	3.8(17)	(3.0 ± 1.0)(17)
CH	2.0(13)	2.1(13)	(2.5 ± 0.3)(13)	1.9(13)	(2.0 ± 0.3)(13)
C_2	2.2(13)	2.5(13)	(2.5 ± 0.2)(13)	2.0(13)	(1.9 ± 0.2)(13)
C_2H	2.6(12)	2.1(12)		2.4(12)	
CH_2	5.1(13)	1.0(13)		4.9(13)	
CH^+	2.8(11)	2.3(10)	(2.9 ± 0.3)(13)	2.4(11)	(3.5 ± 0.4)(12)
OH	4.4(13)	4.7(13)	(4.8 ± 0.5)(13)	4.7(13)	(4.2 ± 0.5)(13)
H_2O	4.9(12)	7.2(12)	≤2.2(13)	5.2(12)	
H_3^+	1.6(14)	2.6(13)		2.1(14)	<1.0(15)
HD	2.6(14)	6.0(14)	(2.1 ± 1.0)(14)	5.8(15)	(3.8 ± 1.4)(15)
CO	8.2(13)		(2.0 ± 0.3)(15)	8.9(13)	(6.1 ± 3.0)(14)
CN	4.1(11)	3.8(11)	(2.9 ± 0.3)(12)	4.4(11)	(3.0 ± 0.3)(12)
NH	2.2(10)	5.4(11)	<7.5(12)	2.3(10)	<6.3(11)
CS	6.0(10)[e]		(2 ± 1)(12)[e]		

[a] The table lists column densities in cm^{-2}.
[b] From van Dishoeck and Black (1986, 1988*a*) with the new CO photodissociation rate, $k_9 = 10^{-10}\ cm^3\ s^{-1}$, and $k_{10} = 10^{-12}\ cm^3\ s^{-1}$.
[c] Model 6 of VRA with $k_g = 2 \times 10^{-7} T^{-1/2}\ cm^3\ s^{-1}$.
[d] See van Dishoeck and Black (1986, 1988*a*) for references to the observations.
[e] Drdla, Knapp and van Dishoeck (1989).

scaling factors provide only upper limits to the strength of the radiation field. Further information on the strength of the radiation field may come from studies of the ionization balance of the atomic species, but the observational data are uncertain and the theoretical interpretation may be affected by the presence of LMs.

Table 3.2 also includes the results of vDB for the ζ Per cloud. This cloud has column densities of H and H_2 very similar to those observed toward ζ Oph, suggesting that $n_H y_f / I_{UV}$ has similar values in the two clouds. However, the excited rotational levels $J \geq 5$ are less populated in the ζ Per cloud, indicating that it is exposed to less intense radiation, and that it consequently has a lower density. In contrast with the H_2 excitation, the C_2 rotational population distribution is very similar for the two lines-of-sight. Since I_R is not expected to vary significantly from place to place, the C_2 data suggest that the densities in the two clouds are very similar. This inconsistency between the various diagnostics could be resolved if some other mechanism such as shock excitation contributes to the population of $J \geq 5$ in the ζ Oph cloud, but to a much smaller extent to that of $J = 5$ in the ζ Per cloud. Alternatively, the grain formation efficiency may be different in the two clouds, or the radiation field in the red may be enhanced in the ζ Oph cloud. The comparison between the ζ Per and ζ Oph clouds illustrates the importance of studying more than one line-of-sight. Differences in the C_2 and H_2 excitation for the various clouds may eventually provide significant clues to their structure. Note that the computed level populations in the ζ Per model agree well with observations, even for the $J = 3$ and $J = 4$ levels.

3.4. Chemistry results

Given the physical structure of a cloud, the chemical network can be solved at each depth, and column densities or line intensities can be calculated for direct comparison with observations. The uncertainties in the physical parameters discussed above directly affect the confidence with which the chemistry networks can be tested.

3.4.1. Carbon bearing species

In Table 3.2, the computed column densities of various species in the vDB models are compared with observational data for the ζ Oph and ζ Per clouds. At first sight, the agreement appears to be very good in most cases. However, this conclusion is deceptive since for each of these species there is at least one unknown reaction rate or parameter that can be adjusted *ad hoc* to obtain the best agreement with observations. For example, the CH abundance depends directly on the rate coefficient of Reaction (3.7) for the radiative association between C^+ and H_2. Since the main destruction route of CH by photodissociation is now fairly well understood, a value $k_7 \approx (5\text{–}7) + 10^{-16}$ cm^3 s^{-1} appears necessary to reproduce the observations. This value is consistent with the theoretical estimates mentioned in Section 3.2.3, but is quite close to the measured limit. Note, however, that the experimental limit pertains to the reaction of C^+ with normal hydrogen having an ortho:para ratio of 3:1, whereas the interstellar models require the rate for an ortho:para ratio of about 1:2. Since the radiative association with ortho-H_2 is expected to be slower than that with para-H_2, the upper limit under interstellar conditions may be higher. The inferred value of k_7 is lower in the VRA models bcause of the reduced radiation field and the higher density in that model.

The abundance of C_2 is directly related to that of CH and CH_2. In this case, a rather large branching ratio to CH_2 was adopted in the dissociative recombination of CH_3^+ to get the best agreement with observations.

The abundances of the carbon-bearing molecules are also sensitive to the carbon depletion factor δ_C. In the vDB models, δ_C was adjusted so as best to reproduce the neutral atomic carbon column density. Carbon appears to be only mildly depleted in diffuse clouds. The ionization balance of carbon, and thus the inferred carbon abundance, may be affected, however, by the possible presence of LMs (Lepp *et al.* 1988). If 1–7% of the carbon is in the form of 50-atom LMs, the inferred carbon abundances can be lowered by up to a factor of 3.

An important aspect of developing models of interstellar clouds is not only to reproduce existing observations, but also to predict abundances of other species that could further test the models. For example, the C_2H column density is quite large in the models, and the molecule could be observable through its millimeter emission lines. If the CH_3^+ branching ratio to CH_2 is indeed large, the predicted CH_2 column density is substantial in the models. Unfortunately, the oscillator strengths of electronic transitions in CH_2 are not known and the molecule lacks suitable transitions at millimeter wavelengths.

Although the steady-state models can account quite well for the observed CH and C_2 abundances, they fail to reproduce the observed column densities of CH^+ by at least one order of magnitude. The large observed abundance of CH^+ in diffuse clouds has been a puzzle since its detection, and excellent reviews of the problem have been given by Dalgarno (1976*b*), and most recently by Black (1988). Since the reaction between C^+ and H_2 to form CH^+ cannot proceed at low temperatures, the formation of CH^+ occurs mostly through photoionization of CH and reactions of C with H_3^+. CH^+ is destroyed efficiently by photodissociation, by reactions with H_2 to form CH_2^+, by reactions with H to form C^+ and H_2, and possibly by dissociative recombination. As long as at least one of these removal reactions is rapid, alternative formation channels for CH^+ have to be found. Stecher and Williams (1974) suggested that the exothermic reaction between C^+ and vibrationally excited molecular hydrogen, H_2^*, could produce significant amounts of CH^+. This process has recently been included in the vDB models, which calculate in detail the amount of vibrationally excited H_2 due to ultraviolet pumping. However, the resulting CH^+ column densities are increased by only 20% or less, because the concentration of vibrationally excited H_2 is large only in a narrow boundary layer of the cloud. Other theories within the steady-state framework include sublimation of CH_4 from dust grains followed by photoionization and photodissociation (Bates and Spitzer 1951), but none of these mechanisms can account quantitatively for the observed CH^+ abundances. The only reasonable alternative solution appears to be formation of CH^+ in shock-heated regions where Reaction (3.6) can proceed rapidly (see Chapter 5).

3.4.2. OH

The abundances of all oxygen-bearing molecules are directly proportional to the cosmic ray ionization rate ζ_0. In the models of vDB, a fairly high ionization rate, $\zeta_0 \approx (7 \pm 3) \times 10^{-17}$ s^{-1} is needed to reproduce the observed OH abundances. With this rate, the computed abundances of H_2O^+ and H_2O are consistent with the measured upper limits, even if the H_3O^+ dissociative recombination favors H_2O. The vDB models adopt a dissociative recombination rate of H_3^+ of $k_9 = 10^{-10}$ cm^3 s^{-1}. With this low rate, the H_3^+ abundance is large and the main formation route of OH is through the O + H_3^+ reaction. If k_9 were as large as 10^{-8} cm^3 s^{-1}, however, the H_3^+ abundance would be significantly lower, and the value of ζ_0 would have to be increased even further. On the other hand, the analysis assumes that all the observed OH is formed by steady-state gas-phase reactions. As is discussed in Chapter 5, OH can also be produced in shock-heated gas, in which case the inferred values of ζ_0 should be regarded as upper limits. However, at least for the ζ Oph cloud, the amount of shock-produced OH seems to be at most 20% of the total OH abundance (Crutcher 1979). A direct test of the inferred cosmic ray ionization rate would be provided by observations of the H_3^+ ion, since its abundance is also directly proportional to ζ_0. If the dissociative recombination of H_3^+ is indeed as slow as 10^{-9}–10^{-10} cm^3 s^{-1}, the predicted H_3^+ column densities are large enough that its absorption lines at near-infrared wavelengths would become detectable. The inferred value of ζ_0 is smaller in the VRA model due to the lower radiation field and the higher temperature throughout most of their model.

3.4.3. HD

The observed HD abundances could provide another measure of the cosmic ray ionization rate, if the overall deuterium abundance [D]/[H] in interstellar clouds were known. Alternatively, the values of ζ_0 derived from the oxygen chemistry can be used to infer [D]/[H] from the measured HD column densities. The derived deuterium abundances in the models of vDB are in the range $(0.5–2.0) \times 10^{-5}$ for the various clouds, consistent with other estimates of the deuterium abundance in the interstellar medium, $[D]/[H] = (1.5 \pm 1.0) \times 10^{-5}$. The models of the ζ Per cloud favor the upper part of this range, whereas those of the ζ Oph cloud give somewhat lower values, due to the order of magnitude lower HD column density.

Note that the possible presence of LMs also affects the inferred deuterium abundance, because of the large reduction of both the H^+ and D^+ concentrations through Reactions (3.11).

3.4.4. CO

In contrast with CH, C_2 and OH, the models produce CO column densities that are too low by an order of magnitude if the new photodissociation rate discussed in Section 3.2.4 is adopted. The main formation route of CO in diffuse clouds is through the reaction of C^+ and OH, and the vDB models presented in Table 3.2 take into account the fact that the rate for this reaction is enhanced at low temperatures due to the ion–dipole interaction. It is difficult to find a significantly more efficient route. Shock production of CO is excluded by the narrow width of its millimeter emission line, as discussed in Chapter 4. The discrepancy must therefore be due to the large destruction rate. The photodissociation rate of CO could be lowered if the intensity of the radiation field were smaller. As discussed in Section 3.3, the intensity of the radiation field in the models is constrained by the observed high-J population of H_2, under the assumption that ultraviolet pumping at 912–1100 Å is the dominant population mechanism of $J \geq 5$. The photodissociation of CO occurs in the same wavelength region, but the rate is sensitive to the shape of the radiation field at $\lambda < 1000$ Å, which is not well determined. If the radiation field at $\lambda < 1000$ Å were to decrease more rapidly than thought previously, the CO photodissociation rate could be lowered significantly without affecting the H_2 $J \geq 5$ excitation by more than about 50%. Such modifications of the shape of the ultraviolet radiation field have been explored by van Dishoeck and Black (1988*b*) and CO column densities up to a few times 10^{14} cm^{-2} may be obtained. They can be increased by another factor of 2 if the abnormally high ultraviolet extinction in the ζ Oph and ζ Per clouds is taken into account. Although these modifications may not be sufficient to remove the discrepancies for all lines-of-sight, they can significantly reduce them.

3.4.5. CN and NH

A similar problem is encountered for the CN molecule, for which the unshielded photodissociation rate is now thought to be a factor of 2–4 larger than that adopted by BD. If in addition Reaction (3.10) is taken to be negligibly small at interstellar temperatures, the models produce too little CN by nearly an order of magnitude. The computed NH abundances are well below the measured upper limits.

Additional formation of CN could result from grain surface formation of NH_3 followed by reactions with C^+. However, the rapid photodissociation of NH_3 also produces NH, and estimates of the efficiency of this process result in NH abundances that are close to the current upper limits for some clouds (Mann and

Williams 1985). More sensitive searches for NH in diffuse clouds would put significant constraints on any grain surface production of molecules. The fact that the observed OH and CH abundances are at least two orders of magnitude higher than that of NH in diffuse clouds is still one of the strongest arguments against significant grain surface formation of carbon- and oxygen-bearing molecules (Crutcher and Watson 1976).

The problem of the CN abundance in diffuse clouds could be alleviated by adopting a less intense radiation field. In particular, the CN photodissociation also occurs mostly at short wavelengths, $\lambda < 1000$ Å. If a modified radiation field which decreases more rapidly at $\lambda < 1000$ Å is used, the CN column densities can be increased to about 10^{12} cm^{-2}.

3.4.6. *CS*

Although CS has not been seen in diffuse clouds by its ultraviolet absorption lines, a tentative detection of its $J = 2 \rightarrow 1$ millimeter emission line has recently been reported by Drdla, Knapp and van Dishoeck (1989) toward ζ Oph. The inferred column density is uncertain, about 7×10^{11}–5×10^{12} cm^{-2}. If the sulfur chemistry outlined in Section 3.2.6 is included in the vDB models, the computed CS column density is close to the lower limit of the observed range. The predicted CS abundance is sensitive to the rate of the reaction of S^+ with CH at low temperatures, and to the adopted CS photodissociation rate.

3.4.7. *Discussion*

Compared with the original models of diffuse clouds of BD, many molecular abundances have decreased significantly. This is due mostly to the lower density in the new models, the larger scaling factor for the radiation field, and larger photodissociation cross sections, many of which were underestimated by up to an order of magnitude in the older work. As a result, it has become much more difficult to construct models that reproduce both the H_2 excitation and the observed molecular abundances. Although parameters such as the rate coefficient for Reaction (3.7) and the cosmic ray ionization rate can be adjusted to reproduce the CH and OH abundances, no such parameters appear available to increase the CO and CN abundances. If the photodissociation cross sections for these species are indeed as large as currently thought, the only remedy appears to be a reduction of the intensity of the radiation field in the models. Whether the intensity has to be lowered over the full wavelength region or only specifically at the shortest wavelengths $\lambda < 1000$ Å is still an open question. In the latter solution, the H_2 excitation is not affected too much, and ultraviolet pumping could still provide most of the population in $J \geq 5$. In the former case, some other mechanism such as shock excitation must provide at least half of the $J \geq 5$

populations. A shocked layer of the cloud appears necessary anyhow to explain the observed CH^+ abundances. However, as illustrated in Chapter 5, most shock models that are in harmony with both the measured H_2 high-J populations and the CH^+ column density, may produce too much CH and OH compared with observations. Note also that the amount of shock excitation appears smaller in the ζ Per cloud compared with the ζ Oph cloud. The relative contributions of ultraviolet pumping and shock excitation to the H_2 $J = 5$ to $J = 7$ populations could be determined by observations of lines out of vibrationally excited levels of H_2, which can be populated only by the radiative mechanism. Unfortunately, such observations have to await the launch of the Hubble Space Telescope with its high-resolution ultraviolet spectrograph. Other observational tests which could discriminate between the various models have been discussed by van Dishoeck and Black (1988*a*). Note that if the radiation field is lowered in the models, the inferred values of k_7 and ζ_0 could also be decreased, as found in the VRA models.

Can time-dependent effects account for some of the weaknesses of the steady-state models? The time scales for all chemical processes at the edges of diffuse clouds are so short, $t \approx 10^2$–10^3 yr, that chemical equilibrium is rapidly attained on typical dynamical time scales of 10^5–10^6 yr. Even at the centers of diffuse clouds, the time scales for most processes are less than 10^5 yr. The only exception is the H/H_2 chemistry, for which the time to reach chemical equilibrium in the center of a diffuse cloud like ζ Oph is about 5×10^6 yr. Wagenblast and Hartquist (1988, 1989) have treated the time dependence of the H/H_2 abundance and the H_2 rotational excitation in diffuse clouds recently in detail. The results depend on the history of the cloud, and in particular on the initial H/H_2 ratio. Time-dependent models of the ζ Oph cloud have very low number densities, $n_H \approx 50$ cm^{-3}, if all hydrogen was initially in molecular form, and if the cloud is younger than about 5×10^5 yr. Models in which hydrogen was initially atomic have densities greater than those found under the assumption of steady-state. The inferred ultraviolet radiation field is not a sensitive function of initial conditions or cloud age.

An additional complication in the steady-state modeling of diffuse clouds is the presence of small-scale structures or 'clumps', as discussed in Chapter 4 and illustrated for the case of the π Sco cloud by Jenkins *et al.* (1989). The extent to which clumps can shield each other will depend on the adopted geometry. Calculations of the radiative transfer in both lines and continuum through the various clumps will be a non-trivial exercise.

3.5. Translucent molecular clouds

All of the models and observations discussed so far have been concerned with the classical diffuse clouds that have $A_V \approx 1$ mag. In recent years, a growing body of data has become available for more reddened lines-of-sight with $A_V \approx 2$–5 mag.

These clouds are of interest because they provide the bridge between diffuse and dense interstellar clouds, and may give insight into the chemical processes occurring in both regions. They are denoted as 'translucent' clouds to indicate that photoprocesses play an important role in the chemistry throughout the cloud, even though the photo rates diminish rapidly toward the center. Although the translucent clouds are usually taken to refer to isolated small clouds, they may also represent the outer edges of dense molecular clouds. Translucent clouds have the virtue that they can be studied observationally not only by absorption line techniques, provided that a suitable background star is available, but also by millimeter emission lines. As discussed by Crutcher (1985), each of the two observational techniques has its merits. The main advantage of the millimeter observations are: (1) their routinely high spectral resolution ($\lambda/\Delta\lambda \approx 10^6$–$10^7$), so that the observed lines are resolved; (2) the fact that most lines are seen in emission, so that it is possible to map the spatial distribution of the clouds; and (3) the fact that complex molecules are more readily detected at millimeter than at optical wavelengths, so that the millimeter observations may provide more information for interstellar chemistry. On the other hand, the optical technique has the advantage that: (1) the angular resolution is about 10^5 times higher than that of radio techniques, so that the same area of the cloud is sampled in all measurements; (2) the derivation of column densities from optical data is more reliable and straightforward than for the millimeter data; (3) lines of atoms in various stages of ionization can be measured, so that direct information on the depletion of elements in the cloud can be obtained, as well as an estimate of the electron abundance; (4) important molecules, such as H_2 and C_2, can be observed only at visible and ultraviolet wavelengths; and finally (5) the extinction and polarization can be measured so that information on the properties of the dust in the clouds can be obtained.

A number of such translucent clouds, or small molecular clouds, have been detected by searches for absorption lines toward highly-reddened stars (see e.g. van Dishoeck and Black (1989) for a summary). Compared with the classical diffuse clouds, the column densities of CH, C_2 and CN are larger by up to an order of magnitude, whereas the CO column density is larger by several orders of magnitude. A good example of a translucent cloud is provided by the small cloud in front of the star HD 169454. The CO column density in this cloud lies in the interesting regime of parameter space where it starts to account for a large fraction of the available carbon. A detailed study of the rapid increase in CO abundance with increasing cloud thickness using the new spectroscopic data on the CO photodissociation has been made by van Dishoeck and Black (1988*b*). Specific models for the HD 169454 cloud are collected in Table 3.3., where they are compared with observations. The temperature and density in the models are fairly

Table 3.3. *Computed column densities (in cm^{-2}) for the HD 169454 cloud*[a]

Species	Model 1	Model 2	Observations
n_H (cm^{-3})	500	700	400–1000
T (K)	20	15	15^{+10}_{-5}
I_{UV}	2	1	
H_2	1.5(21)	2.2(21)	(1–2)(21)
H	6.3(20)	3.3(20)	>2(19)
C	1.1(16)	2.1(16)	
C^+	1.4(17)	9.8(16)	
CH	4.7(13)	7.6(13)	(4.6 ± 0.8)(13)
C_2	9.9(13)	1.6(14)	(7.3 ± 1.4)(13)
CN	1.4(13)	4.9(13)	(5.6 ± 0.9)(13)
CO	2.0(16)	1.0(17)	(1–9)(16)
OH	3.5(14)	2.1(15)	
C_2H	2.7(13)	5.2(13)	
C_3	5.7(9)	1.5(10)	
CS	2.3(13)[b]	5.2(12)[b]	(0.4–0.7)(13)[b]
A_V (mag)	2.25	2.95	3.5

[a] From Jannuzi *et al.* (1988).
[b] Drdla *et al.* (1989).

well constrained by the observed rotational excitation of C_2. The modeling is further simplified by the fact that no observations on the rotational excitation of H_2 are available, so that the scaling factor I_{UV} for the ultraviolet radiation field can be chosen to be close to unity. Another difference with the classical diffuse clouds is that no information is available on the column densities of atomic or ionized carbon, so that the carbon depletion factor δ_C can be treated as a free parameter as well. If $k_7 \approx (5\text{–}7) \times 10^{-16}$ cm^3 s^{-1} is adopted based on the diffuse cloud results, the depletion of carbon needs to be fairly large, $\delta_C \approx 0.1$, in order to reproduce the CH and C_2 data. The observed CN abundance is also well reproduced for this particular cloud. CS has recently been detected toward HD 169454 by its millimeter emission lines, and the observed amount can be well accounted for in the model.

In spite of the fact that the measured CN column density can be explained with reasonable parameters for the HD 169454 cloud, large variations in the abundance of CN with respect to those of CH and C_2 have been found for other translucent clouds. These differences may be caused by abnormal extinction curves, or by variations in the nitrogen and oxygen depletions from cloud to cloud, and may provide further clues to the nitrogen chemistry. They also emphasize the importance of a detailed treatment of the radiative transfer in translucent cloud models. More observations of chemically related species are needed.

3.6. High latitude clouds

Another class of diffuse or translucent clouds is formed by the high latitude molecular clouds detected by Magnani, Blitz and Mundy (1985) through CO millimeter observations. These clouds have visual extinctions $A_V \approx 1$ mag similar to those of the classical diffuse clouds such as the ζ Oph and ζ Per clouds; yet their CO column densities and CO/H_2 ratios are at least an order of magnitude higher. Such high CO column densities for low A_V can be produced within the steady-state framework only if the clouds are exposed to an ultraviolet radiation field that is smaller than the average background radiation field by a factor of 2–4 (van Dishoeck and Black 1988*b*). A reduction in the intensity by a factor of 2 is not implausible because the early type stars in the galactic plane may illuminate the high latitude clouds from one side only.

The molecules OH and H_2CO have also recently been detected in a number of high latitude clouds through observations at centimeter wavelengths. Within the large observational uncertainties, the OH column densities are in the range 10^{14}–10^{15} cm^{-2}, an order of magnitude larger than those found for the classical diffuse clouds, while the H_2CO column densities are in the range 10^{12}–10^{13} cm^{-2}. H_2CO has not yet been detected in any classical diffuse cloud. CS has been detected in one high latitude cloud by its millimeter emission lines.

Because of their low extinction, the high latitude clouds are amenable to absorption line observations. However, only few bright stars are available as background light sources. Interstellar CH and CH^+ have been detected in a few high-latitude clouds, but their column densities are not significantly larger than those found in the classical diffuse clouds.

Steady-state models appropriate to high latitude clouds have been developed by van Dishoeck and Black (1988*a,b*). As for the translucent clouds, the modeling is simplified by the fact that few constraints on the density, temperature and strength of the radiation field are available. Two examples of models with n_H = 500 cm^{-3}, T = 40 K and I_{UV} = 0.5, but with different total H_2 column densities, are presented in Table 3.4. Model 1 is appropriate for the high-latitude cloud toward HD 210121, whereas Model 2 represents a thicker cloud. Column densities of CO in excess of 10^{16} cm^{-2} are readily obtained in such models. The abundances of other species depend strongly on the adopted depletion factor of carbon. In particular, if the depletion of carbon is taken to be similar to that found in diffuse clouds, δ_C = 0.4, and if k_7 = (5–7) $\times$ 10^{-16} cm^3 s^{-1} is retained, the computed CH abundances are large compared with observations, whereas those of OH are somewhat low. On the other hand, if carbon is more depleted, the OH column density is increased significantly, because its removal rate through the reaction with C^+ is decreased. The depletion factor $\delta_C \approx 0.1$ adopted in the models in Table 3.4 appears to give CH and OH column densities that are

Table 3.4. *Computed column densities (in cm^{-2}) for high-latitude clouds*[a]

			Observations	
Species	Model 1	Model 2	HD 210121[b]	Average[c]
n_H (cm^{-3})	500	500		
T (K)	40	40		
I_{UV}	0.5	0.5		
H_2	8.0(20)	2.0(21)	(8 ± 2)(20)	
H	6.6(19)	7.8(19)	(1–5)(20)	
C	3.9(15)	1.1(16)		
C^+	6.6(16)	7.2(16)		
CO	7.4(15)	1.1(17)	(1 ± 0.5)(16)	6(16)
CH	4.1(13)	8.3(13)	(3.5 ± 0.5)(13)	
C_2	5.4(13)	1.3(14)		
CH^+	3.9(10)	5.1(10)	(6 ± 2)(12)	
CN	7.1(12)	2.8(13)		
CS	6.0(12)[d]		(0.4–0.8)(13)[d]	
OH	2.0(14)	2.1(15)		9(14)
H_2CO	8.7(8)	1.5(10)		5(12)
A_V (mag)	1.0	2.6	1.0	0.7:

[a] From van Dishoeck and Black (1988*a*,*b*).
[b] Observed column densities for the high-latitude cloud toward HD 210121 from de Vries and van Dishoeck (1988).
[c] Average properties from Magnani, Blitz and Wouterlout (1988).
[d] Drdla *et al.* (1989).

consistent with the observations of high latitude clouds. Observations of other carbon-bearing molecules such as C_2H and C_3H_2 may provide additional constraints on the gas-phase carbon abundance in these clouds. The observed amount of CS is somewhat larger than that found in the models. The computed H_2CO column densities are at least an order of magnitude below observations. The amount of H_2CO could be increased if an additional source of gas-phase CH_4 were added to the chemistry. Such a source may be provided by reactions of C and C^+ with LMs to form CH_4, or by the disruption or evaporation of grain mantles. If 5–10% of the carbon were in the form of LMs in high latitude clouds, the H_2CO column density could be increased to about 10^{12} cm^{-2}. Note that similar reactions of O and N with LMs would produce significant amounts of H_2O and NH_3 in high latitude clouds. Searches for species such as NH or metal hydrides would provide additional indications whether grain processes or LMs play an important role in the chemistry in these clouds. Further searches for CH^+ might assess the relative importance of shock-induced processes.

3.7. Concluding remarks

Although diffuse clouds may still provide the best environment in which to investigate the basic chemical networks, rigorous tests are currently hampered by a number of problems. The largest uncertainty lies in the physical structure of the cloud, in particular in the strength of the incident radiation field. If shocks contribute indeed significantly to the high-J populations of H_2, the intensity of the radiation field may have been overestimated in most steady-state models. However, shock models which reproduce the observed H_2 high-J population distribution and the CH^+ column density raise other problems, such as large CH and OH abundances, which have to be resolved before definite conclusions can be drawn. It is still not known how ubiquitous shocks are in diffuse clouds, and to what extent they contribute to the formation of molecules other than CH^+. An additional problem is that the reaction rate coefficients of the crucial reactions, such as those of Reactions (3.7) and (3.9), are still not well determined. As a result, there is almost no molecule in diffuse clouds for which we can be *fully* certain that the adopted gas-phase reaction scheme is correct or complete.

The study of diffuse clouds has suffered from a lack of new observational data over the last decade, especially from the fact that no sensitive high resolution spectrograph is available in space for searches at ultraviolet wavelengths, where most molecules have their strongest electronic transitions. However, new impetus for the modeling of diffuse clouds has come from recent high resolution millimeter observations, and from high quality optical data. In addition, further studies of the chemically related translucent and high latitude clouds through combined millimeter and optical observations are expected to provide new insights, not only into the chemistry of diffuse clouds, but also into the processes occurring in dense clouds. It is exciting to note that the studies of diffuse and dense clouds, which have evolved separately both observationally and theoretically over the last 25 years, may finally come together.

REFERENCES

Allison, A. C. and Dalgarno, A. 1970, *Atomic Data,* **1**, 289.
Bates, D. R. 1986, *Ap. J.* (*Letters*), **306**, L45.
Bates, D. R. and Spitzer, L. 1951, *Ap. J.,* **113**, 441.
Black, J. H. 1988, *Adv. Atom. Mol. Phys.,* **25**, 477.
Black, J. H. and Dalgarno, A. 1973*a*, *Astrophys. Letters,* **15**, 79.
Black, J. H. and Dalgarno, A. 1973*b*, *Ap. J.* (*Letters*), **184**, L101.
Black, J. H. and Dalgarno, A. 1977, *Ap. J. Suppl.,* **34**, 405 (BD).
Black, J. H., Hartquist, T. W., and Dalgarno, A. 1978, *Ap. J.,* **224**, 448.
Crutcher, R. M. 1979, *Ap. J.,* **231**, L151.
Crutcher, R. M. 1985, *Ap. J.,* **288**, 604.

Crutcher, R. M. and Watson, W. D. 1976, *Ap. J.*, **209**, 778.
Crutcher, R. M. and Watson, W. D. 1981, *Ap. J.*, **244**, 855.
Crutcher, R. M. and Watson, W. D. 1985, in *Molecular Astrophysics,* eds. G. H. F. Diercksen, W. F. Huebner and P. W. Langhoff, NATO ASI Series **157**, Reidel, Dordrecht, p. 255.
Dalgarno, A. 1976*a*, in *Frontiers of Astrophysics,* ed. E. H. Avrett, Harvard University, p. 352.
Dalgarno, A. 1976*b*, in *Atomic Processes and Applications,* eds. P. G. Burke and B. L. Moiseiwitsch, North Holland, Amsterdam, Chap. 5.
Dalgarno, A. 1988, *Astro. Lett. and Communications,* **26**, 153.
Dalgarno, A. and Black, J. H. 1976, *Rep. Prog. Phys.*, **39**, 573.
Dalgarno, A., Black, J. H. and Weisheit, J. C. 1973, *Astrophys. Letters,* **14**, 77.
Dalgarno, A. and Stephens, T. L. 1970, *Ap. J.* (*Letters*), **160**, L107.
Danks, A. C., Federman, S. R. and Lambert, D. L. 1984, *Astr. Ap.*, **130**, 62.
de Vries, C. P. and van Dishoeck, E. F. 1988, *Astr. Ap.*, **203**, L23.
Draine, B. T. 1978, *Ap. J. Suppl.*, **36**, 595.
Drdla, K., Knapp, G. R. and van Dishoeck, E. F. 1989, *Ap. J.* **345**, in press.
Eddington, A. S. 1926, *Proc. Roy. Soc. A,* **111**, 424.
Federman, S. R. 1982, *Ap. J.*, **257**, 125.
Herbst, E. 1982, *Ap. J.*, **252**, 810.
Herbst, E. and Klemperer, W. 1973, *Ap. J.*, **185**, 505.
Hollenbach, D. and Salpeter, E. E. 1971, *Ap. J.*, **163**, 155.
Jannuzi, B. T., Black, J. H., Lada, C. J. and van Dishoeck, E. F. 1988, *Ap. J.*, **332**, 995.
Jenkins, E. B., Lees, J. F., van Dishoeck, E. F. and Wilcots, E. M. 1989, *Ap. J.*, in press.
Kramers, H. A. and ter Haar, D. 1946, *Bull. Astr. Inst. Neth.*, **10**, nr. 371.
Lepp, S., Dalgarno, A., van Dishoeck, E. F. and Black, J. H. 1988, *Ap. J.*, **329**, 418.
Letzelter, C., Eidelsberg, M., Rostas, F., Breton, J. and Thieblemont, B. 1987, *Chem. Phys.*, **114**, 273.
Luine, J. A. and Dunn, G. H. 1985, *Ap. J.* (*Letters*), **299**, L67.
Magnani, L., Blitz, L. and Mundy, L. 1985, *Ap. J.*, **295**, 402.
Magnani, L., Blitz, L. and Wouterloot, J. G. A. 1988, *Ap. J.*, **326**, 909.
Mann, A. P. C. and Williams, D. A. 1985, *MNRAS,* **214**, 279.
O'Donnell, E. J. and Watson, W. D. 1974, *Ap. J.*, **191**, 89.
Omont, A. 1986, *Astr. Ap.*, **164**, 159.
Oppenheimer, M. and Dalgarno, A. 1974, *Ap. J.*, **187**, 231.
Roberge, W. G. 1981, Ph.D. thesis, Harvard University.
Solomon, P. M. 1965, as quoted by Field, G. B., Somerville, W. B. and Dressler, K. 1966, *Ann. Rev. Astr. Astrophys.*, **4**, 207.
Solomon, P. M. and Klemperer, W. 1972, *Ap. J.*, **178**, 389.
Stecher, T. P. and Williams, D. A. 1974, *MNRAS,* **168**, 51P.
Stephens, T. L. and Dalgarno, A. 1972, *J. Quant. Spectrosc. Rad. Transf.*, **12**, 569.
Swings, P. and Rosenfeld, L. 1937, *Ap. J.*, **86**, 483.
Turner, J., Kirby-Docken, K. and Dalgarno, A. 1977, *Ap. J. Suppl.*, **35**, 281.
van Dishoeck, E. F. 1988, in *Rate Coefficients in Astrochemistry,* eds. T. J. Millar and D. A. Williams, Kluwer, Dordrecht, p. 49.
van Dishoeck, E. F. and Black, J. H. 1982, *Ap. J.*, **258**, 533.
van Dishoeck, E. F. and Black, J. H. 1986, *Ap. J. Suppl.*, **62**, 109 (vDB).
van Dishoeck, E. F. and Black, J. H. 1988*a*, in *Rate Coefficients in Astrochemistry,* eds. T. J. Millar and D. A. Williams, Kluwer, Dordrecht, p. 209.

van Dishoeck, E. F. and Black, J. H. 1988*b*, *Ap. J.*, **334**, 771.
van Dishoeck, E. F. and Black, J. H. 1989, *Ap. J.*, **340**, 273.
Viala, Y. P., Roueff, E. and Abgrall, H. 1988, *Astra. Ap.*, **190**, 215 (VRA).
Wagenblast, R. and Hartquist, T. W. 1988, *MNRAS*, **230**, 363.
Wagenblast, R. and Hartquist, T. W. 1989, *MNRAS*, **237**, 1019.
Watson, W. D. 1973, *Ap. J. (Letters)*, **182**, L73.
Watson, W. D. 1978, *Ann. Rev. Astr. Astrophys.*, **16**, 585.
Yoshino, K., Stark, G., Smith, P. L., Parkinson, W. H. and Ito, K. 1988, in prep.

4

Observations of velocity and density structure in diffuse clouds

WILLIAM D. LANGER

Princeton University, Princeton, USA

4.1. Introduction

Our understanding of the structure of the interstellar medium has progressed enormously in the last 15 years largely as a result of the development of radio emission and ultraviolet absorption observations. From CO radio surveys we now know that about one-half of the interstellar gas is in the form of giant molecular clouds that are the formation sites of O and B stars, while measurements of highly excited gas, such as hot bubbles and expanding shells, indicate that a significant volume of the interstellar medium contains dilute, hot gas. The physical properties and dynamics of the interstellar gas are strongly influenced by the interaction of the massive O and B stars and supernovae. Supernova shells sweep up interstellar matter creating clouds and also impact them causing compression and fragmentation. The O and B stars produce large HII regions which expand and compress the interstellar gas. Thus on a theoretical basis we expect to find complex velocity fields, shock structures, and density inhomogeneities in the diffuse and translucent component of the interstellar clouds.

Today it is hard to imagine studying the diffuse interstellar clouds without relying on observations of molecular lines. In an excellent review article by Dalgarno and McCray (1972) on the heating, ionization and cooling processes of HI regions, these authors stated that 'The processes leading to interstellar molecule formation are only partially understood . . .'. The central role of molecules as both important components and physical probes was just becoming evident. Since that time Professor A. Dalgarno has made, and continues to make, major contributions to our understanding of the interstellar cloud chemistry. To interpret the diffuse cloud properties and test the chemical models, observations of molecular abundances have been important. In the past couple of years it has become evident that the velocity, density, and temperature structure of the clouds is also crucial to these interpretations.

In this review I will discuss recent measurements of molecular absorption and emission lines that provide important information on the velocity and density structure of diffuse interstellar clouds. These observations constrain the steady state and shock models of diffuse clouds which are reviewed in this volume by van Dishoeck and Hartquist, Flower and Pineau des Forêts, respectively (Chapters 3 and 5). The modeling of diffuse clouds has important implications for larger and denser molecular cloud structures because their envelopes resemble the 'classical' diffuse clouds in density, temperature, and radiation field. Indeed from CO emission maps it is now well appreciated that many of the classical diffuse clouds studied with optical and ultraviolet absorption along a single line of sight to a star are diffuse parts of much larger clouds or cloud complexes.

4.2. Observational probes of diffuse cloud regions

One of the earliest methods for studying the interstellar clouds in the solar neighborhood was the line of sight absorption measurements in the optical toward O and B stars. Optical studies have provided important information on the amount of gas and the number of absorbing components along the line of sight (e.g. Hobbs 1973). This approach is quite powerful because it can be carried out from the ground with high signal-to-noise and moderately high resolution (about 1–2 km s^{-1}). Very small column densities can be detected and, indeed, the first interstellar molecules were detected this way.

Most atoms and molecules, especially the important cloud component H_2, have their ground state transitions in the ultraviolet. Systematic studies at these wavelengths had to await rocket and satellite observations because ultraviolet cannot be observed from the Earth's surface. Indeed a great wealth of data was provided by the *Copernicus* satellite. Spitzer and Jenkins (1975) and Morton (1975) review many of the results from these ultraviolet absorption line measurements, though there have been important revisions of the original analyses (cf. Chapter 3). While the ultraviolet data have provided a rich source of information regarding atomic, molecular hydrogen, and ionic column densities, temperatures, and H_2 rotational populations, the velocity resolution, typically about 10 km s^{-1}, is inadequate to answer questions regarding the velocity fields of the diffuse clouds. Recognizing the need for better observations Jenkins *et al.* (1988) have developed an echelle instrument for rocket flight and observed H_2 in one source, π Sco, with a resolution of 2.4 km s^{-1}.

Even the ground based optical absorption line studies have important limitations. First, the spectral resolution is generally insufficient to probe the narrowest components of interstellar clouds, which have velocity dispersions of 0.3 km s^{-1} or less. The instrumental resolution is usually about 1 km s^{-1} and many of the useful lines, like Na D, have hyperfine or other internal splittings of the same order of magnitude. (The hyperfine structure of the Na D lines has been resolved by Blades, Wynne-Jones, and Wayte (1980) using a Michelson interferometer with a resolution of 0.5 km s^{-1}.) Second, the lines of sight are limited to relatively bright stars and probe only one spatial dimension of the absorbing cloud; hence, the global aspects of the clouds cannot be understood with this technique. Radio observations of molecules in diffuse clouds have a number of advantages in this regard, as has been discussed by Liszt (1979) and Langer, Glassgold, and Wilson (1987; hereafter LGW). The radio observations can be made with very high spectral resolution, better than 0.1 km s^{-1}, and can map a source. They have disadvantages too, since the molecules are observed in emission there must be sufficient excitation (density and temperature) to detect them readily and the telescope beam size is often very large at centimeter wavelengths so that the spatial structure may be poorly resolved. However, there are many diffuse regions of the interstellar medium lacking bright background stars, such as the envelopes of dark clouds and the recently detected high latitude clouds (Blitz, Magnani and Mundy 1984).

Finally, the infrared emission from clouds is useful for investigating the physical conditions within the clouds. While infrared emission is broad band so that no velocity information is present, it does contain information about the nature of the grains and the properties of the interstellar radiation field. A number of investigators have found a correlation between the 100 μm emission (Boulanger and Perrault 1988) intensity and the 100 μm opacity (Langer *et al.* 1988) and the atomic and molecular gas distribution, respectively. The IRAS observations are particularly important because they provide a convenient, readily available, large scale survey of the emission throughout much of the Galaxy.

4.3. Velocity and density structure in diffuse clouds

To discuss the velocity, shock, and density structure in diffuse clouds I will highlight the optical, ultraviolet, and radio observational studies towards O and B stars (in particular towards ζ Oph, one of the best studied examples of this kind). Results will also be reviewed for other diffuse regions including those at high latitudes, where background O and B stars are not prevalent, and the diffuse envelopes of larger denser structures.

4.3.1. Observations towards O and B stars

The diffuse matter along the line of sight to ζ Oph has become a classic case in the study of the interstellar medium (Herbig 1968, Morton 1975). Extensive observations at optical, ultraviolet, and radio wavelengths show the atomic and molecular gas along the line of sight with different velocities and abundances. These observations have stimulated numerous attempts to model this cloud (see Chapters 3 and 5), but none of the existing theories provides a completely satisfactory explanation of the column densities and components. The strong optical absorption near zero velocity has been identified by Hobbs (1973) as having four velocity components. Two of these, at $V_{lsr} = -3.5$ and $+5.5$ km s^{-1}, have only atomic components, while the others have molecular components. The dominant molecular component at -0.4 km s^{-1} contains strong ultraviolet and optical absorption lines of H_2, CO, CH, and OH, while the other component containing CH^+ is at $+0.6$ km s^{-1} (Hobbs (1973) corrected for a change in wavelength by Hawkins, Jura and Meyer (1985)).

At first sight this velocity structure is suggestive of a shock environment in the neighborhood of ζ Oph. When a shock passes through a cloud, the postshock gas is blue-shifted with respect to the preshock gas for the case of the shock moving toward the sun. Due to the varying physical conditions, each part of the shock has a characteristic chemical composition and a particular sequence should be observed: (1) cool, compressed postshock gas with neutral atoms and molecules; (2) warm, postshock gas with high temperature chemical reaction products such as CH^+; and (3) preshock gas with abundances characteristic of low density diffuse clouds. Elitzur and Watson (1978, 1980) identified this sequence with three of the four velocity components towards ζ Oph, the -0.4 km s^{-1} component with cool postshock gas, the $V \doteq +0.6$ km s^{-1} with the warm, postshock gas, and the component near $+5$ km s^{-1} with preshock gas. Particular significance is attached to the CH^+ because all attempts to explain its abundance in ζ Oph, other than shock chemistry, have failed (see Chapters 3 and 5). Chemical models predict that other molecules will also be present in the shocks.

Radio observations of velocity structure

One of the early efforts to look at the velocity structure of ζ Oph was that of Liszt (1979) who observed emission from CO at 2.6 mm and CH at 9 cm. Liszt found that both these species had spectral shapes consisting of a narrow component at -0.7 km s^{-1}, linewidth 0.5 km s^{-1}, sitting atop a broad pedestal with emission extending out to 0.5 km s^{-1}. This lineshape was consistent with the shock velocity structure for ζ Oph outlined above. Crutcher (1975) also detected an OH 18 cm emission with a feature at $+5.5$ km s^{-1}, close to the $+4.5$ km s^{-1} NaI and CaII cloud detected in absorption (Hobbs 1973, Crutcher 1975). Unfortunately, the CH and OH measurements were made with very poor spatial resolution, 9′ and 18′ respectively. The optical absorption measurements, and even more so the ultraviolet ones, did not have sufficient resolution to detect the structure observed in the radio emission spectra, as can be seen in the CH and CH^+ observations of Palazzi, Mandolesim, and Crane (1988) and Hawkins, *et al.* (1985).

To improve on these velocity studies LGW mapped the 2.6 mm line of CO in the region surrounding the line of sight to ζ Oph obtaining very high signal-to-noise spectra with velocity resolution as small as 0.07 km s^{-1}. Their high resolution spectra (Figure 4.1) showed that the CO emission toward ζ Oph was coming from at least four narrow (FWHM about 0.5 km s^{-1}) components partially blended together, so that under poorer resolution they appear as one broad line or under lower signal-to-noise the result resembles the spectrum seen by Liszt. Thus, at the level of the sensitivity of LGW's observations, there is no evidence for a shock feature, even though the CO emission overlaps in velocity space the velocity where CH^+ is detected. Just as significant is the map, shown in Figure 4.2, which gives evidence that many of the velocity features vary significantly in strength over small distances.

LGW concluded on the basis of the spectral map that there are at least five components responsible for the CO emission. The region consists of a relatively

Figure 4.1. The spectrum of ^{12}CO $J = 1 \rightarrow 0$ emission at 2.6 mm along the line of sight to the O star ζ Oph taken with 50 kHz resolution (corresponding to a velocity resolution of 0.13 km s^{-1}). The emission intensity is expressed in antenna temperature and the rms noise per channel is 32 mK. Under this high signal-to-noise and small velocity resolution the line profile is observed to be composed of 4–5 narrow features, rather than the one broad feature seen in the ultraviolet absorption data or the two features previously deduced from the earlier CO work. (Adapted from Langer *et al.* (1987).)

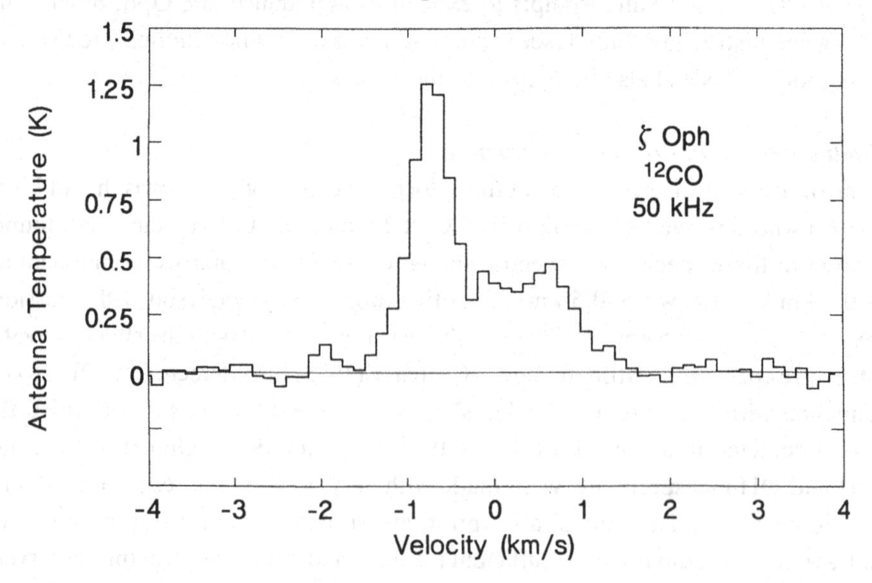

uniform component plus several clumps. One of the strong CO components is at $+0.6$ km s^{-1}, which is the nominal center of the broad (4 km s^{-1} FWHM) CH^+ absorption line measured by Hobbs (1973). However, it would be incorrect to identify these two components too closely with one another. First, the absolute velocity scale of the optical measurements is uncertain by at least ±0.3 km s^{-1}. Indeed, Lambert and Danks (1986) have recently reported the CH^+ velocity as -0.7 km s^{-1}, barely compatible with Hobb's result. Second, Hobb's lineshape appears Gaussian, but the broad CH^+ feature could be the result of several close components mimicking a Gaussian at low resolution.

OH is another molecule observed in emission in the radio and in absorption at ultraviolet wavelengths towards ζ Oph (Crutcher 1979). The line of sight shows two components one centered at a velocity of -0.5 km s^{-1} and the other at $+5.5$ km s^{-1} and the column density from the radio data agrees with that obtained from the ultraviolet absorption measurements. Crutcher identifies the $+5$ km s^{-1} component with the preshock gas toward ζ Oph, as also indicated by optical NaI and CaII data, and the -0.5 km s^{-1} component with the cool compressed postshock gas. He suggests that these OH data fit in well with the shock model of ζ

Figure 4.2. A map of the ^{12}CO emission around the line of sight toward ζ Oph. The emission of the main peak is nearly uniform in intensity whereas the other components vary across the map. The region shown is about 0.2 pc × 0.2 pc, so that the inhomogeneities are on a scale the order of 0.1 pc or less. (Taken from Langer *et al.* (1987).)

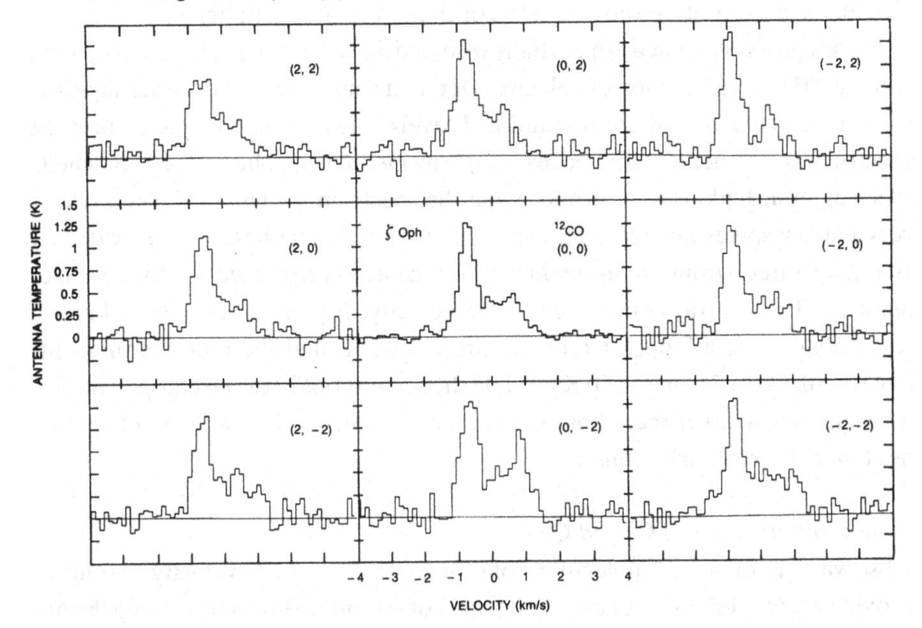

Oph, but require that the preshock gas be located in high density patchy dust clouds, rather than a very low density gas. The overlap in velocity between OH and CH^+ also suggests that some OH is made in shocks.

The early shock models of Elitzur and Watson (1980) tended to produce too much OH, but that does not appear to be a problem in the magnetic shock models (see Chapter 5). It is, however, puzzling that OH should be detected in the cool preshock gas, but not CO (LGW). The widely spaced OH map of ζ Oph perhaps sheds some light on this puzzle since Crutcher finds that this component's velocity varies with position by up to 2 km s^{-1} and that it is weakest at ζ Oph. In contrast the -0.5 km s^{-1} component shows little variation in intensity and velocity. The OH observations were taken with an 18′ half-power beam diameter, and it is quite possible that there is no detectable OH within the smaller region around ζ Oph mapped in CO. While this possibility would argue against OH in the preshock gas the -0.5 km s^{-1} component is still consistent with a shock.

High resolution ultraviolet observations of H_2

Jenkins *et al.* (1988) have used an objective-grating echelle spectrograph with ten times the resolving power of the instruments available on the *Copernicus* satellite to observe the H_2 profiles in π Sco. The actual instrumental resolution during the rocket flight was 2.4 km s^{-1} (FWHM). Many absorption features were observed for rotational levels up to $J = 5$, but the features for $J = 0$ and 1 are too broad to analyze for velocity structure. Jenkins *et al.* find no convincing evidence for a difference in radial velocity of H_2 among the J levels. They conclude that this provides a constraint on shock models of the rotational excitation.

The H_2 profiles in π Sco are partially resolved and allow a calculation of column density $N(H_2)$ as a function of velocity. Detailed comparison of the lineshape and column densities for different rotational levels leads to the conclusion that the spectrum is not being resolved even at the medium resolution of the echelle spectrograph. Jenkins *et al.* propose that the spectrum seen in H_2 is composed of very narrow spikes not resolved even at 2.4 km s^{-1}. The best fit is provided by assuming three strong components at line center surrounded by four weaker features. The central components have velocity dispersions $b = 0.91$ km s^{-1} corresponding to a Doppler temperature of 100 K, and the outer components have $b = 0.7$ km s^{-1} (or $T = 60$ K). While the separations between components for π Sco are much larger than those seen in the CO data for ζ Oph the multicomponent model is strikingly similar.

Optical observations of C_2 *and* CN

Observations of many molecular rotational states at high velocity resolution provide a powerful probe of cloud physical conditions and structure along the line of sight to bright stars. While H_2 is the most pervasive and abundant molecule in the diffuse clouds (and also the precursor for forming most other molecules in the interstellar chemistry) it is presently difficult to observe in absorption at the required sensitivity and resolution. An alternative molecule that is useful for probing the cloud conditions is C_2. This molecule has been observed in absorption toward a number of reddened early-type stars (Chaffee and Lutz 1978, Hobbs 1979, Danks and Lambert 1983, van Dishoeck and de Zeeuw 1984). The rotational structure and radiative properties of the C_2 molecule make it suitable for determining the kinetic temperature and density along the line of sight (van Dishoeck and Black 1986). However, it probably only traces the central region of the cloud according to current models of molecule production.

In the case of ζ Oph many different bands have been observed and values of the kinetic temperature derived from the rotational populations have ranged from 20 K to 60 K (Snow, 1978, Hobbs and Campbell 1982, Danks and Lambert 1983, van Dishoeck and Black 1986). The determination of the kinetic temperature from the rotational temperature requires an excitation model for C_2 including collisions, radiative absorption, and infrared cascade (cf. van Dishoeck and Black (1986), Le Bourlot, Roueff and Viala (1987)). The absorption lines of the $A^1\Pi_u - X^1\Sigma_g^+$ Phillips system of $(v', v'') = (2,0)$ and $(3,0)$ bands at 8750 Å and 7720 Å, respectively, are particularly convenient to such studies. Van Dishoeck and Black (1986) have observed the (3,0) Phillips band toward ζ Oph, which offers some advantages over other bands, and conclude that the central temperature is $\approx$30 K and the density $\approx$200 cm^{-3}. Other well studied clouds whose kinetic temperatures have been determined are ζ Per and χ Oph (van Dishoeck and de Zeeuw 1984); in the latter the temperature and density are $\approx$40 K and 160 cm^{-3}, respectively. Finally, as was pointed out by van Dishoeck in Chapter 3, the central line velocities for the various rotational states agree to within $\approx$1–2 km s^{-1} up to as much as $J = 14$, which implies that very little if any of the C_2 is produced in shocks in ζ Oph or χ Oph.

Absorption line studies of CN have been made to investigate the microwave background temperature, isotopic ratios (cf. Crane and Hegyi (1988)), and the carbon chemistry in diffuse clouds. Recently Federman and Lambert (1988) have made such carbon chemistry studies toward moderately reddened stars behind diffuse regions and combined their observations with data from lower extinction clouds. Their studies compare absorption data from lines of CN and C_2 versus CH. They find that there are two regimes indicated by a change in slope of the column density relationships. They attribute this change to a transition from a photochemical regime for directions with small CH column densities to a chemical regime for directions with larger CH.

Optical data of CH^+ *and* CH

Both CH and CH^+, two of the three molecules first detected in space, have been observed in absorption towards over 100 sources in an attempt to understand the origins of these molecules in diffuse clouds (e.g. look for evidence of shock chemistry in the velocity differences). While in broad terms the CH chemistry appears to be well understood (see Chapter 3) that of CH^+ is not. Comparisons of abundances and velocities of CH and CH^+ have been made by a number of authors to study the different models of molecule production. One thorough study of this problem is that of Lambert and Danks (1986) who surveyed 65 lines of sight towards bright O and B stars for CH^+ at 4232 Å in the southern Milky Way. They made a careful determination of its column density and radial velocities and compared these to column densities of the rotational states of H_2. They found an excellent correlation between the CH^+ and the column densities of rotationally excited H_2. They conclude that this result is consistent with CH^+ production in a warm gas behind a shock front.

In shock models of the diffuse cloud gas and chemistry one must identify three components: the preshock gas, the warm postshock gas, and the cooled compressed postshock gas. The shock models require a shock velocity $V_s \sim 10$ km s^{-1} to explain the CH^+ abundance (see Chapter 5). The preshock gas should be at a velocity about $+3V_s/4$ relative to the CH^+ in the warm postshock gas, and the cool component at $-V_s/4$ km s^{-1}. Evidence of a velocity difference of $V_s/4$ or about 2.5 km s^{-1} would support the shock models of CH^+ production. Lambert and Danks made a comparison of CH and CH^+ along 12 lines of sight for which they had good detections of both molecules and found a mean absolute difference of 0.7 ± 0.2 km s^{-1}. They conclude that their comparison provides only weak evidence for the velocity structure expected for a shock.

On the other hand they suggest that comparisons of CH^+ velocities with CaI and KI lines at 4226 and 7699 Å respectively might also be used to study hot shock and cool postshock velocity differences, because these neutral atoms should reside in the cool gas component. Here they find a mean difference of 1.2 ± 0.2 km s^{-1} which does indicate $V_s > 7.6$ km s^{-1}, and is evidence of a shocked gas environment associated with CH^+.

An alternative explanation for CH^+ production in diffuse regions, not necessarily involving shocks, has been given by White (1984) and involves the heating of a cloud surface by ultraviolet. In this case the ultraviolet pumps the H_2 into excited states and heats the gas so that the endothermic reaction forming CH^+ takes place. Given the variety of conditions found in diffuse regions it would not be surprising that both, or even other, mechanisms operate to produce CH^+ in different diffuse regions.

4.3.2. Observations of diffuse regions around molecular clouds

Maps of CO emission have helped to identify the diffuse envelopes surrounding molecular clouds. While, in general, the emission intensities are weak the recent development of very sensitive millimeter receivers has made it possible to study these regions routinely. Figure 4.3 shows an example of such a region around the dark cloud B335 taken from the survey of Frerking, Langer and Wilson (1987). It is apparent that the opaque portion of the cloud is only a small fraction of the total cloud's area and that much of the region has very low column density corresponding to visual extinctions typically less than 1–2 mag. It is possible to estimate the density, temperature, and chemical abundances from excitation models of CO emission by combining observations of the $J = 1 \rightarrow 0$ and $J = 2 \rightarrow 1$ transitions. To date very few studies of this type have been made because of the time consuming

Figure 4.3. The optical image of the small dark globule B335 on the blue POSS print. Superimposed in it are the extinction contours in visual magnitudes; the derivation of A_V (or likewise $N(H_2)$) from CO is discussed by Frerking *et al.* (1987) from whose article this figure is taken. Note that the opaque portion of the cloud seen on the optical plate is only a small fraction of the total area observed in CO emission. Much of the cloud is contained in a diffuse envelope at a visual extinction less than 2 mag. The envelope is similar to the 'classical' diffuse clouds.

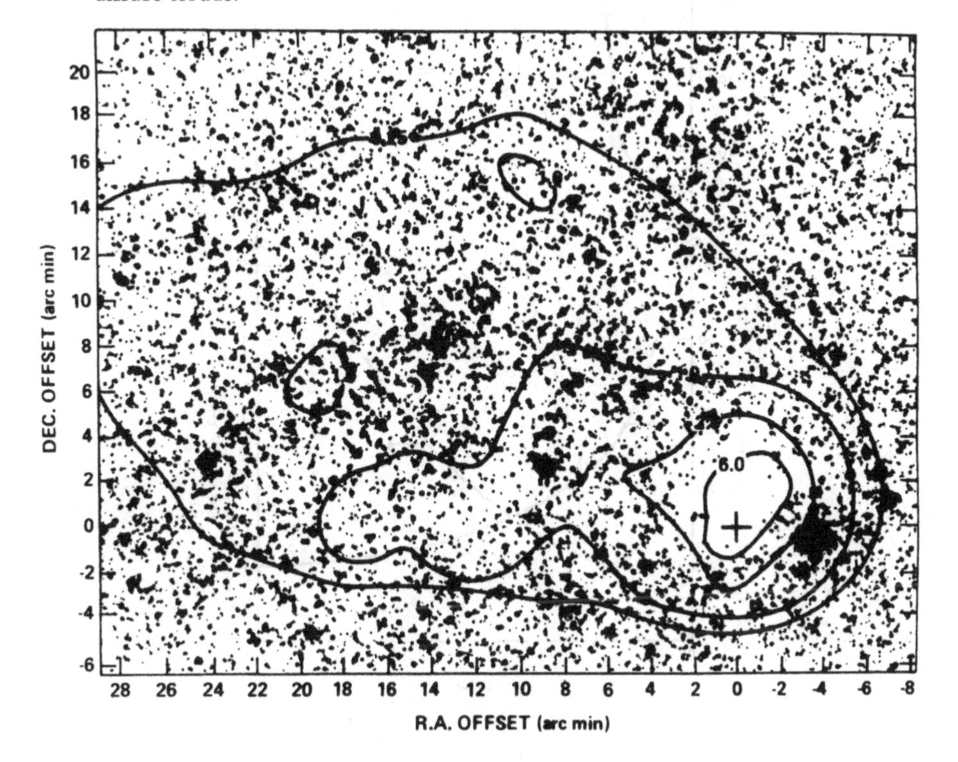

effort needed to collect the data at two transitions. Young *et al.* (1982) made one of the earlier efforts of this kind in their study of the dark cloud B5 in the Perseus complex. An example of their results can be seen in Figure 4.4, which shows the abundance profile of CO in the cloud B5 across a cut going from the edge through the core. One can readily see the sharp decrease in abundance at the cloud's edge, corresponding to diffuse regions with visual extinction less than 1–2 mag. Results like these are very important for modeling the photochemistry of regions such as those found in diffuse clouds or cloud envelopes. In general these studies of cloud structure find conditions in the envelope similar to those determined for the 'classical' diffuse clouds (i.e. $T_{\rm kin} > 30$ K, density ~200 cm^{-3}, and ultraviolet radiation accompanied by active photochemistry).

Figure 4.4. The physical properties across a strip through the cloud B5 as determined from the $J = 2 \rightarrow 1$ and $1 \rightarrow 0$ emissions of three CO isotopes. Shown are the density, kinetic temperature, and CO fractional abundance from the center out to the edge. The properties of the edge are similar to those determined towards diffuse gas. This figure is adapted from Young *et al.* (1982).

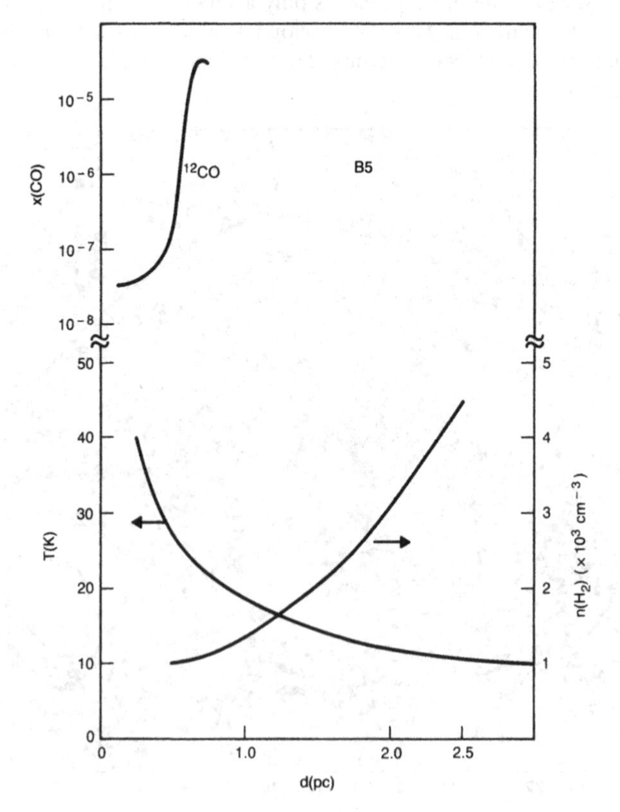

4.3.3. Observations toward high-latitude diffuse clouds

One of the recent developments in the study of diffuse clouds has been the discovery of diffuse, high latitude molecular clouds (HLCs) by Blitz, Magnani, and Mundy (1984) by means of their CO emission. Their relatively late discovery in the history of the study of the diffuse medium is due to the lack of bright O and B stars located conveniently behind these clouds.

One of their interesting features is the strong intensity of the CO 2.6 mm emission from regions whose mean visual extinction is 0.6 mag (Magnani and de Vries 1986). Lada and Blitz (1988) observed eight HLCs with the 1.3 mm transition of CO and compared them to similar observations of 16 diffuse clouds in the plane. They conclude that there are two populations of diffuse molecular clouds characterized by their being either CO-rich or CO-poor. The differences between these groups is suggested to be the result of different ultraviolet radiation fields, gas densities, and carbon abundance because models of CO abundance are sensitive to the self-shielding of CO to ultraviolet photodissociation (Glassgold, Huggins, and Langer 1985).

To correlate the radio observations of HLCs with absorption experiments one must rely on background sources other than O and B stars. These weaker sources, of course, make optical and ultraviolet absorption experiments more difficult. Hobbs *et al.* (1988) have made one of the first studies of this type by searching for the D lines of NaI and the K lines of CaII toward seven stars (A–F type) in the vicinity of the two nearest HLCs detected so far, MBM 12 and 16, and by searching for the 4300 Å line of CH toward three background stars. It is intriguing to note that CH could not be detected in these HLCs, and hence, unlike CO, is not 'overabundant'. These differences should provide useful data for modeling the chemistry of diffuse regions.

4.4. Future prospects

A volume dedicated to Professor A. Dalgarno will certainly touch on subjects that directly or indirectly have been influenced by his research. Yet, I know him well enough to expect that he will be as productive, creative, and influential in the years to come so I will conclude by pointing out the future prospects for observations on the chemical abundances, density and velocity structure in diffuse clouds.

The diffuse interstellar matter observed along lines of sight toward O and B stars has been studied intensively for over 50 years. With new measurement techniques, it remains a strong focus of research and provides valuable insights into the nature of interstellar clouds. Advances in the sensitivity of optical detectors promise to expand these earlier studies toward less bright background A and F stars. At the same time the development of low noise receivers at 1 mm where the CO $J = 2 \rightarrow 1$ transition can be observed will permit the mapping of this molecular component of diffuse interstellar regions. Combined with the $J = 1 \rightarrow 0$ observations the excitation conditions may be determined more readily in the diffuse envelopes, as has already been done in B5.

Recent developments at sub-millimeter wavelengths offer additional promise of being able to study atoms and ions in diffuse clouds. The first detection of neutral carbon emission at 345 GHz in a diffuse cloud has been reported by Keene, Blake, and Phillips (1987). In the future this type of work will allow important comparisons of abundance and velocity structure of the atomic and molecular gas. At even higher frequencies observations can now be routinely made of C^+ in hot sources (Stutzki *et al.* 1988) and with further development of more sensitive receivers, observations of diffuse clouds may become possible. Other high frequency possibilities include the direct observation of emission from the rotational levels of hydrides such as OH, CH, and CH^+. Given the high velocity and spatial resolution available with millimeter and sub-millimeter systems the structure and chemistry of diffuse clouds will be open to a new stage of study.

Looking to future satellite observations the possibility for high resolution ultraviolet absorption studies of H_2 and other species will be very significant. Finally, the infrared wavelengths, may find increased utility in interpreting the nature of large and small grains and their interaction with ultraviolet in the diffuse regions. I expect that these observational developments will keep this field fresh and active and give the theoreticians and modelers much to think about in the years to come.

REFERENCES

Beichman, C. A., Wilson, R. W., Langer, W. D., and Goldsmith, P. F. (1988). Infrared Limb Brightening in the Barnard 5 Cloud, *Astrophysical Journal Letters,* in press.

Black, J. H., and van Dishoeck, E. F. (1988). Unresolved Velocity Structure In Diffuse Clouds, *Astrophysical Journal,* in press.

Blades, J. C., Wynne-Jones, I., and Wayte, R. C. (1980). Very-high-resolution Spectroscopy of Interstellar NaI, *Monthly Notices Royal Astronomical Society,* **193**, 849–66.

Blitz, L., Magnani, L., and Mundy, L. (1984). High-Latitude Molecular Clouds, *Astrophysical Journal Letters,* **282**, L9–12.

Blitz, L., Magnani, L., and Wandel, A. (1988). Broad Wing Molecular Lines Without Internal Energy Sources, *Astrophysical Journal,* in press.

Boulanger, F., and Perault, M. (1988). Diffuse Infrared Emission From the Galaxy I – Solar Neighborhood, *Astrophysical Journal,* **330**, in press.

Chaffee, F. H., and Lutz, B. L. (1978). The Detection of Interstellar Diatomic Carbon Toward Zeta Ophiuchi, *Astrophysical Journal Letters,* **221**, L91–3.

Crane, P., and Hegyi, D. J. (1988). Detection of Interstellar ^{13}CN Toward Zeta Ophiuchi, *Astrophysical Journal Letters,* **326**, L35–8.
Crutcher, R. M. (1975). Velocity Component Structure of the Zeta Ophiuchi Interstellar Lines, *Astrophysical Journal,* **200**, 625–33.
Crutcher, R. M. (1979). Radio Observations of OH Toward Zeta Ophiuchi: Evidence for an Interstellar Shock, *Astrophysical Journal Letters,* **231**, L151–3.
Dalgarno, A., and McCray, R. A. (1972). Heating and Ionization of HI Regions, *Annual Reviews of Astronomy and Astrophysics,* **10**, 375–426.
Danks, A. C., and Lambert, D. L. (1983). Interstellar C_2 in the Ophiuchi Clouds, *Astronomy & Astrophysics,* **124**, 188–96.
Elitzur, M., and Watson, W. D. (1978). Formation of Molecular CH^+ in Interstellar Shocks, *Astrophysical Journal Letters,* **222**, L141–4 (*erratum*: **226**, L157).
Elitzur, M., and Watson, W. D. (1980). Interstellar Shocks and Molecular CH^+ in Diffuse Clouds, *Astrophysical Journal,* **236**, 172–81.
Federman, S. R., and Lambert, D. L. (1988). The Carbon Chemistry in Interstellar Clouds Toward Moderately Reddened Stars, *Astrophysical Journal,* **328**, 777–84.
Frerking, M. A., Langer, W. D., and Wilson, R. W. (1987). The Structure and Dynamics of Bok Globule B335, *Astrophysical Journal,* **313**, 320–45.
Glassgold, A. E., Huggins, P. J., and Langer, W. D. (1985). Shielding of CO From Dissociating Radiation in Interstellar Clouds, *Astrophysical Journal,* **290**, 615–26.
Hawkins, I., Jura, M., and Meyer, D. M. (1985). The $^{12}C/^{13}C$ Isotopic Ratio Toward Zeta Ophiuchi, *Astrophysical Journal Letters,* **294**, L131–5.
Herbig, G. H. (1968). The Interstellar Line Spectrum of Zeta Ophiuchi, *Zeitschrift fur Astrophysik,* **68**, 243–77.
Hobbs, L. M. (1973). Interstellar NaI, KI, CaII, and CH^+ Line Profiles Toward Zeta Ophiuchi, *Astrophysical Journal Letters,* **180**, L79–82.
Hobbs, L. M. (1979). Interstellar C_2 Molecules toward ζ Perzei, *Astrophysical Journal Letters,* **232**, L175–8.
Hobbs, L. M., and Campbell, B. (1982). Interstellar C_2 Molecules Towards Zeta Ophiuchi, *Astrophysical Journal,* **254**, 108–10.
Hobbs, L. M., Blitz, L., Penprase, B. E., Magnani, L., and Welty, D. E. (1988). On the Nearest Molecular Clouds, *Astrophysical Journal,* **327**, 356–63.
Jenkins, E. B., Lees, J., van Dishoeck, E., and Wilcox, E. (1988). High Resolution Observations of Interstellar H_2 in π Sco, *Astrophysical Journal,* in press.
Keene, J., Blake, G. A., and Phillips, T. G. (1987). Comparison of Submillimeter and Ultraviolet Observations of Neutral Carbon Toward Zeta Ophiuchi, *Astrophysical Journal,* **313**, 396–9.
Lada, E. A., and Blitz, L. (1988). Two Populations of Diffuse Molecular Clouds, *Astrophysical Journal Letters,* **326**, L69–73.
Lambert, D. L., and Danks, A. C. (1986). On the CH^+ Ion in Diffuse Interstellar Clouds, *Astrophysical Journal,* **303**, 401–15.
Langer, W. D., Glassgold, A. E., and Wilson, R. W. (1987). Radio Observations of Carbon Monoxide Toward Zeta Ophiuchi, *Astrophysical Journal,* **322**, 450–62 (denoted LGW in text).
Langer, W. D., Wilson, R. W., Goldsmith, P. F., and Beichman, C. A. (1988). Dust and Gas Emission In Barnard 5, *Astrophysical Journal,* in press.
Le Bourlot, J., Roueff, E., and Viala, Y. (1987). Rotational Equilibrium of C_2 in Diffuse Interstellar Clouds: (I) Static Model, the case of Zeta Ophiuchi, *Astronomy and Astrophysics,* **188**, 137–45.
Liszt, H. S. (1979). Radiofrequency Molecular Emission Spectra Observed Toward Zeta Ophiuchi, *Astrophysical Journal Letters,* **233**, L147–50.

Magnani, L., and de Vries, C. P. (1986). High Latitude Molecular Clouds: Distances and Extinctions, *Astronomy & Astrophysics,* **168**, 271–83.
Morton, D. C. (1975). Interstellar Absorption Lines in the Spectrum of Zeta Ophiuchi, *Astrophysical Journal,* **197**, 85–115.
Palazzi, E., Mandolesim, N., and Crane, P. (1988). Interstellar CH Toward Zeta Ophiuchi, *Astrophysical Journal,* in press.
Snow, T. P. (1978). The Kinetic Temperature in the Interior of the Zeta Ophiuchi Cloud from *Copernicus* Observations of Interstellar C_2, *Astrophysical Journal Letters,* **220**, L93–6.
Spitzer, L. Jr, and Jenkins, E. B. (1975). Ultraviolet Studies of the Interstellar Gas, *Annual Reviews of Astronomy and Astrophysics,* **13**, 133–64.
Stutzki, J., Stacey, G. J., Genzel, R., Harris, A. I., Jaffe, D. T., and Lugten, J. B. (1988). Submillimeter and Far-Infrared Line Observations of M17 SW: A Clumpy Molecular Cloud Penetrated by UV Radiation, *Astrophysical Journal,* in press.
van Dishoeck, E. F., and de Zeeuw, T. (1984). Observations of Interstellar C_2 toward Chi Oph, HD 154368, 147889, and 149404, *Monthly Notices Royal Astronomical Society,* **206**, 383–406.
van Dishoeck, E. F., and Black, J. H. (1986). Detection of the (3, 0) Phillips Band of Interstellar C_2 Toward Zeta Ophiuchi, *Astrophysical Journal,* **307**, 332–6.
White, R. E. (1984). Interstellar Matter Near the Pleiades. II. CH^+ Formation, *Astrophysical Journal,* **284**, 695–704.
Wayte, R. C., Wynne-Jones, I. and Blades, J. C. (1978). Detection of Hyperfine Structure of Interstellar NaI in the Alpha Cygni Sight-line, *Monthly Notices Royal Astronomical Society,* **182**, *Short Communications,* 5p–10p.
Young, J. S., Goldsmith, P. F., Langer, W. D., Wilson, R. W., and Carlson, E. R. (1982). Physical Conditions and Carbon Monoxide Abundance in the Dark Cloud B5, *Astrophysical Journal,* **261**, 513–31.

5

Shock chemistry in diffuse clouds

T. W. HARTQUIST

Max Planck Institute for Physics and Astrophysics, Institute for Extraterrestrial Physics, Garching, FRG

D. R. FLOWER

Physics Department, The University of Durham, Durham, England

G. PINEAU DES FORÊTS

DAMAP Observatoire de Paris, Meudon, France

5.1. Introduction

The high observed column densities of CH^+, one of the first identified (Douglas and Herzberg 1941) interstellar molecules, and of CO apparently indicate that existing static, equilibrium models do not provide adequate descriptions of the natures of diffuse molecular interstellar clouds. (See Chapter 3.) It has been argued that velocity structures in lines formed in such clouds provide evidence for the existence of shocks in them (e.g. Crutcher (1979), but see the detailed assessment by Langer in Chapter 4). If such shocks do exist, they will drive the production of detectable column densities of a number of chemical species.

The chemistry in shocked gas can be exceptionally rich since many reactions which, because they are endothermic or have activation barriers, are unimportant in cool, static gas, can proceed in shocked gas. For instance, the endothermic reactions $C^+ + H_2 \rightarrow CH^+ + H$ (Elitzur and Watson 1978a) and $S^+ + H_2 \rightarrow SH^+ + H$ (Millar *et al.* 1986) can initiate hydrogen abstraction sequences in shocked gas but are unimportant in static, cool diffuse clouds. A neutral–neutral sequence (Aannestad 1973) which is of no relevance to low temperature chemistry but which plays a major role in shock chemistry is $O + H_2 \rightarrow OH + H$; $OH + H_2 \rightarrow H_2O + H$. The fractional abundances of CH^+ and OH are high in some diffuse cloud shocks, and SH^+ may serve as a diagnostic of shocks.

Collisionally induced rotational excitation of molecular hydrogen can also occur in diffuse cloud shocks. Aannestad and Field (1973) speculated that all of the rotationally excited H_2 observed with the *Copernicus* satellite is produced by collisional excitation behind shocks rather than by the ultraviolet pumping–cascade mechanism (refer to Chapter 3) assumed to be responsible in static equilibrium diffuse cloud models. The contention that the H_2 excitation is due to collisions in shocked gas receives some support from the conclusion that the observed CO abundances can be explained only if the ultraviolet radiation field is much weaker than is inferred with the standard ultraviolet pumping–cascade modelling procedure or the spectral shape is different in the ultraviolet than was assumed. (See Chapter 3.) One of our objectives in writing this chapter is to elucidate to what extent the outstanding discrepancies between model and observed chemical abundances can be removed if we assume that most of the more highly rotationally excited molecular hydrogen observed with Copernicus is in shocked gas.

In the the following we give the steady multifluid magnetohydrodynamic (MHD) equations used to describe a shock which propagates perpendicularly to the upstream magnetic field in a molecular cloud. The chemistry affecting the structures of shocks in diffuse molecular clouds is discussed. The question of whether CH^+ can be formed in shocks at the column densities observed is addressed, and molecular diagnostics of shocks are considered. Results for chemical column densities in shocked gas exposed to a weak ultraviolet radiation field and a reduced cosmic ray flux are presented; we conclude that if the radiation field is weak, both shock and cold cloud contributions to observed column densities can be important.

We note that Professor A. Dalgarno has had a long standing interest in the contributions of dynamic regions to observed molecular abundances. In 1976, he suggested to one of us (TWH) that he should consider the possibility that CH^+ in diffuse clouds is produced primarily in evaporative flows from molecular clouds. Only simple estimates could feasibly be made at that time, but now, with the use of multifluid codes, steady models of the flows in such regions can be built. With the current state of development of codes to treat the problem of H_2 photodissociation and photoexcitation in moving media, the quantitative study of the chemistry in the flows is possible. The work of all three of us on interstellar shock chemistry has been influenced by Alex's contributions.

5.2. The structures of shocks in molecular clouds

Because molecular clouds are weakly ionized, the coupling between ionized species, which are subject to electromagnetic forces, and neutral species is weak, and multifluid descriptions of the flow must often be adopted.

Consider steady, plane parallel flow in the $\hat{z}$-direction which is perpendicular to the magnetic field in the $\hat{x}$-direction. The mass, momentum, and energy conservation equations can be derived in a manner analogous to that used by Boyd and Sanderson (1969) in Section 3.3 of their book to obtain the equations governing flow in a two-fluid system. For the jth fluid the equations are

$$\frac{\mathrm{d}}{\mathrm{d}z}(\rho_j v_{jz}) = S_j^{\mathrm{r}} \tag{5.1a}$$

$$\frac{\mathrm{d}}{\mathrm{d}z}\left(\frac{\rho_j v_{jz}}{m_j}\right) = D_j^{\mathrm{r}} \tag{5.1b}$$

$$\frac{\mathrm{d}}{\mathrm{d}z}\left(\rho_j v_{jz}^2 + \frac{\rho_j}{m_j} k_{\mathrm{B}} T_j\right) = F_{jz}^{\mathrm{e}} + F_{jz}^{\mathrm{r}} + \frac{\rho_j q_j e}{m_j}\left(E_z - \frac{v_{jy}}{c} B_x\right) \tag{5.1c}$$

$$\frac{\mathrm{d}}{\mathrm{d}z}(\rho_j v_{jz} v_{jy}) = F_{jy}^{\mathrm{e}} + F_{jy}^{\mathrm{r}} + \frac{\rho_j q_j e}{m_j}\left(E_y + \frac{v_{jz}}{c} B_x\right) \tag{5.1d}$$

$$\frac{\mathrm{d}}{\mathrm{d}z}\left(\rho_j v_{jz}\frac{5k_{\mathrm{B}}T_j}{2m_j}\right) + \frac{d}{dz}\left(\frac{\rho_j v_{jz}\bar{u}_j}{m_j}\right) - v_{jz}\frac{\mathrm{d}}{\mathrm{d}z}\left(\frac{\rho_j k_{\mathrm{B}} T_j}{m_j}\right) = G_j^{\mathrm{e}} + G_j^{\mathrm{r}} - C_j - v_{jz}(F_{jz}^{\mathrm{e}} + F_{jz}^{\mathrm{r}}) \tag{5.1e}$$

S_j^{r} and D_j^{r} describe the effects of chemical exchange between fluids. The $\mathbf{F}^{\mathrm{e}}$s and $\mathbf{F}^{\mathrm{r}}$s describe momentum transfer due to elastic and reactive scattering between particles in different fluids. The G_j^{e}s and G_j^{r}s describe interfluid energy transfer due to elastic and reactive scattering, and the C_js represent the radiative losses; m_j, $\bar{u}_j$, and q_j are the mean mass, average internal energy and charge per particle in the jth fluid. In Equation (5.1e), small terms containing $v_{jy}F_{jy}$ have been neglected. From Equation (5.1d) for electrons and by neglecting inertial and drag terms we see that the magnetic field strength is given by

$$v_{ez}B_x = v_s B_0 = -cE_y \tag{5.1f}$$

where v_{s} is the shock speed and B_0 is the magnetic field strength in the ambient medium. E_z can be eliminated by assuming quasi-neutrality.

$$\sum_j \frac{\rho_j q_j}{m_j} = 0 \tag{5.1g}$$

Current conservation is also assumed

$$\Sigma_j \rho_j q_j v_{jz}/m_j = 0 \tag{5.1h}$$

This formulation differs somewhat from that given by Draine (1980) and by Draine, Roberge, and Dalgarno (1983) but is appropriate when a more rigorous treatment (which may be required when the preshock number density is about 10^7 cm^{-3} or greater or when the shock speed is less than v_{Ai} but greater than $B_0/(4\pi\rho_{\mathrm{g}})^{1/2}$ where the subscript g indicates grain) of the grain dynamics is desired (Havnes, Hartquist, and Pilipp 1987). When the jth fluid is composed of grains, Equation (5.1e) is replaced by $T_j = 0$.

The evaluation of the **F**s and Gs is discussed in detail in the papers by Draine

(1986a) and by Flower and Pineau des Forêts (1986) and in the references therein. Collisions between particles in the jth and kth fluids lead to

$$\mathbf{F}_j^{\mathrm{e}} = \sum_{k \neq j} \frac{\rho_j \rho_k}{m_j + m_k} \int \sigma_{jk} |\mathbf{v}_k - \mathbf{v}_j| (\mathbf{v}_k - \mathbf{v}_j) \phi_j(\mathbf{v}_j) \phi_k(\mathbf{v}_k)\, \mathrm{d}^3 v_j\, \mathrm{d}^3 v_k \tag{5.2a}$$

and

$$G_j^{\mathrm{e}} = \sum_{k \neq j} \frac{\rho_j \rho_k}{(m_j + m_k)^2} \int \sigma_{jk} |\mathbf{v}_k - \mathbf{v}_j| (\mathbf{v}_k - \mathbf{v}_j) \cdot (m_j \mathbf{v}_j + m_k \mathbf{v}_k) \cdot \phi_j(\mathbf{v}_j) \phi_k(\mathbf{v}_k)\, \mathrm{d}^3 v_j\, \mathrm{d}^3 v_k \tag{5.2b}$$

where σ_{jk} is the momentum transfer cross section and $\phi_j(\mathbf{v}_j)$ and $\phi_k(\mathbf{v}_k)$ are the particle distribution functions which are generally taken to be velocity-shifted Maxwellians.

In modelling diffuse cloud shocks, only three fluids need be considered, those consisting of neutrals, ions, and electrons. Quasi-neutrality then requires that $v_{\mathrm{iz}} = v_{\mathrm{ez}}$, and the coupling between the charged particles and the neutrals is dominated by ion-neutral coupling. Then to a good approximation

$$F_{\mathrm{nz}}^{\mathrm{e}} = -F_{\mathrm{iz}}^{\mathrm{e}} = \frac{\rho_{\mathrm{n}} \rho_{\mathrm{i}}}{m_{\mathrm{n}} + m_{\mathrm{i}}} \alpha_{\mathrm{in}} (v_{\mathrm{iz}} - v_{\mathrm{nz}}) \tag{5.3a}$$

$$G_{\mathrm{n}}^{\mathrm{e}} = \frac{\rho_{\mathrm{n}} \rho_{\mathrm{i}}}{(m_{\mathrm{n}} + m_{\mathrm{i}})^2} \alpha_{\mathrm{in}} [m_{\mathrm{i}} (\mathbf{v}_{\mathrm{i}} - \mathbf{v}_{\mathrm{n}})^2 + 3 k_{\mathrm{B}} (T_{\mathrm{i}} - T_{\mathrm{n}})] + F_{\mathrm{nz}}^{\mathrm{e}} v_{\mathrm{nz}}.$$

The subscripts i, e, and n stand for ion, electron, and neutral and α_{in}, the ion–neutral scattering rate coefficient, is the product $\sigma_{\mathrm{in}} |\mathbf{v}_{\mathrm{i}} - \mathbf{v}_{\mathrm{n}}|$ and has a constant value of $1.9 \times 10^{-9}\ \mathrm{cm^3\ s^{-1}}$ for a heavy singly-ionized species colliding with H_2 at the velocities attained in the diffuse cloud shocks which we consider.

We now describe briefly the qualitative natures of the solutions to Equations (5.1). Consider a moving piston which drives motions of the ion fluid only. An electric field generated by a small charge separation will cause the electrons to move with the ions, and the neutrals will be accelerated by the ion–neutral drag. Even if the piston affects other fluids, in an MHD shock with a speed sufficiently less than $v_{\mathrm{Ai}} \equiv B_0/(4\pi\rho_{\mathrm{i}})^{1/2}$ in weakly ionized plasma, the ions will initially move at nearly the shock speed while the neutrals are accelerated slowly. The length-scale over which the neutrals are accelerated and the shock energy is dissipated in such a shock is of the order of

$$\Delta = B_0^2 / 4\pi \alpha_{\mathrm{in}} \rho_{\mathrm{n}} n_{\mathrm{i}} v_{\mathrm{s}}, \tag{5.4}$$

a result which one obtains by equating the tension ($\approx B_0^2/4\pi\Delta$) in the magnetic field to the frictional drag between the ions, which are assumed to be moving relative to the neutrals at a speed v_{s}, and the neutrals.

The behavior described above obtains when v_{s} is sufficiently less than v_{Ai} and not too much greater than $v_A \equiv B_0/(4\pi\rho)^{1/2}$ with $\rho = \rho_n + \rho_i + \rho_e$. When v_s is less than v_{Ai} but is sufficiently larger than v_A, the neutral flow will become discontinuous. The former type of shock is referred to as C-type, and the latter type of shock is referred to as J-type where C and J denote continuous and jump. If $v_s > v_{Ai}$, the flow will be discontinuous in all fluids. If $v_s < v_{Ai}$, C-type shocks exist for a wide range of v_s when the neutral fluid cools efficiently.

In general the shock velocity will not be perpendicular to the upstream magnetic field. Wardle and Draine (1987) have studied oblique multifluid shocks in diffuse interstellar clouds. If the angle between the shock velocity and the upstream field is θ and the upstream field strength is B_0, the structure of an oblique C-type shock is similar to that of a perpendicular ($\mathbf{B} \cdot \mathbf{v} = 0$) C-type shock with the same velocity propagating into a medium with a field strength of $B_0(\sin\theta)^{1/2}$. Clearly, if θ is small, J-type shocks occur for very low velocities. A strong oblique J-type shock has a structure similar to that of a strong perpendicular J-type shock propagating into a medium with a field of strength $B_0 \sin\theta$.

Figures 5.1 and 5.2 show the neutral, ion, and electron speeds and temperatures as functions of distance for the C-type shock models (a) and (b) defined by the parameters given in Table 5.1. As will be shown in Section 5.3, the shock structure is affected by the chemistry which leads to a rapid initial drop in the ion flux. The length of the zone over which the neutrals are accelerated is given approximately

Figure 5.1. Neutral and ion velocities for the shock models (a) and (b).

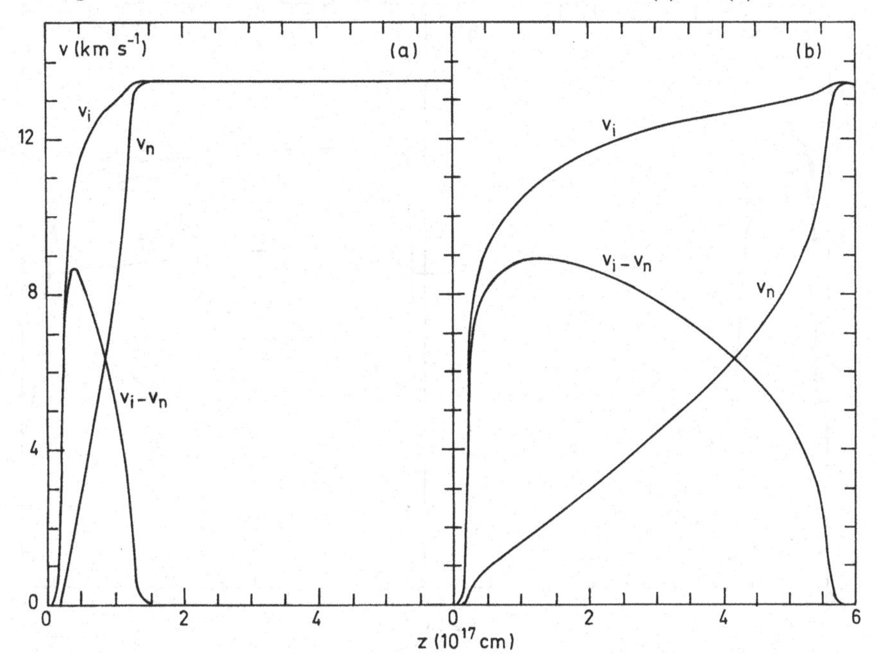

Table 5.1. *Parameters describing shock models*

	(a)	(b)
ζ	$4\,(-17)$ s^{-1}	$4\,(-18)$ s^{-1}
v_s	15 km s^{-1}	15 km s^{-1}
B_0	5 μG	5 μG
χ	1	0.1
$n(H^0)$	4 cm^{-3}	4 cm^{-3}
$n(H_2)$	8 cm^{-3}	8 cm^{-3}
$n(He)$	2 cm^{-3}	2 cm^{-3}
$n(C^0)$	4.12 (-6) cm^{-3}	6.29 (-5) cm^{-3}
$n(O^0)$	8.50 (-3) cm^{-3}	8.50 (-3) cm^{-3}
$n(S^0)$	9.53 (-8) cm^{-3}	1.49 (-6) cm^{-3}
$n(H^+)$	5.35 (-4) cm^{-3}	3.41 (-5) cm^{-3}
$n(H_2^+)$	1.82 (-8) cm^{-3}	1.76 (-9) cm^{-3}
$n(H_3^+)$	3.72 (-5) cm^{-3}	3.82 (-6) cm^{-3}
$n(He^+)$	2.63 (-5) cm^{-3}	2.39 (-6) cm^{-3}
$n(C^+)$	6.59 (-3) cm^{-3}	6.52 (-3) cm^{-3}
$n(S^+)$	3.17 (-4) cm^{-3}	3.15 (-4) cm^{-3}

The number densities given are for the distant upstream gas. $\chi = 1$ indicates that the standard interstellar radiation field is assumed. ζ is the cosmic ray ionization rate.

Figure 5.2. Neutral, ion, and electron temperatures for the shock models (a) and (b).

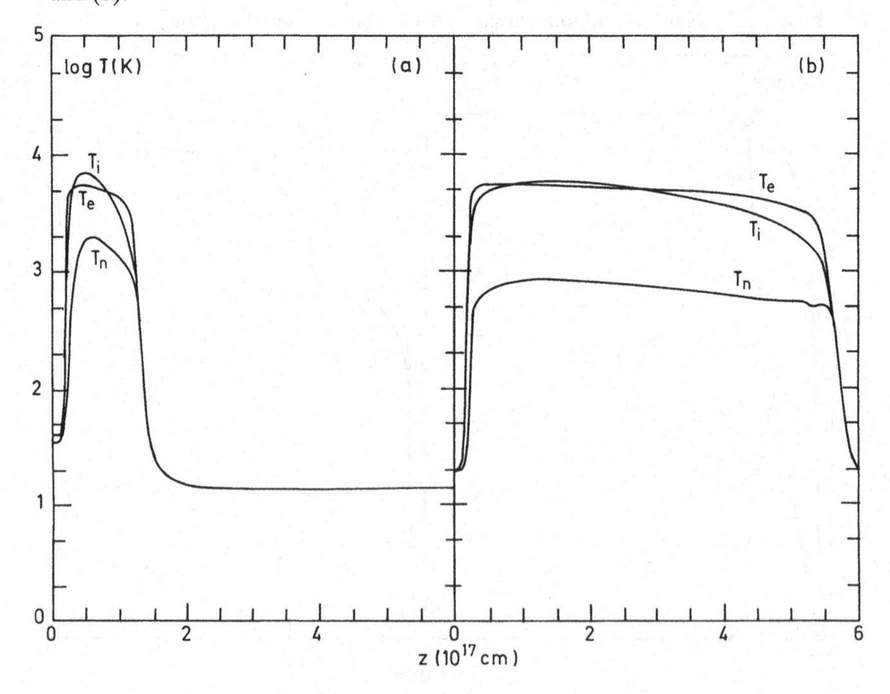

by inserting into Equation (5.4) the ion number density which obtains in the shocked gas after the ion flux has dropped to a nearly constant value.

5.3. The effect of chemistry on shock structure

As we stated in the introduction, the outstanding discrepancies between computed and observed column densities in diffuse molecular clouds constitute a prime concern. However, as Alex frequently argues, the chemistry is not only interesting in itself but also affects the macroscopic structures of the sources in which it occurs and possesses utility in the diagnosis of conditions in those sources. In this section, we consider the ways in which chemistry affects the structure of a shock in a diffuse molecular cloud; in later sections, we will address the discrepancies with observations and consider possible diagnostic studies.

In low velocity shocks in which the neutral temperature remains much less than 10^3 K, radiative cooling of the neutral gas is dominated by emission from molecules containing heavy elements following excitation in collisions primarily with H, H_2, and He. The chemical processes which determine the abundances of OH and H_2O are particularly important; they are the reactions

$$O + H_2 \rightarrow OH + H \tag{5.5a}$$

and

$$OH + H_2 \rightarrow H_2O + H \tag{5.5b}$$

their inverses, and the photodissociation of OH and H_2O. Reaction (5.5a) is endothermic and has a barrier, and Reaction (5.5b) has a barrier; the terms exp $(-2980/T)$ and exp $(-1490/T)$ appear in the expressions for their rate coefficients. In diffuse molecular cloud shocks in which $T_n \gtrsim 10^3$ K, the cooling is dominated by the radiative decay of the H_2 $(v = 0, J = 4)$ level after its collisional excitation (Elitzur and Watson 1978b; Flower, Pineau des Forêts, and Hartquist 1986).

The chemistry affects the shock structure more dramatically by determining the ionization structure which, as shown in Equation (5.4), affects the length over which the neutral gas is accelerated and the shock energy is dissipated. In the preshock gas, nearly all gas phase carbon is in the form of C^+; H^+ produced by cosmic ray induced ionization is also abundant. However, when heating occurs and/or the ion–neutral streaming speed in the neutral acceleration zone becomes large, the endothermic reaction

$$0.4\,\text{eV} + C^+ + H_2 \rightarrow CH^+ + H \tag{5.6}$$

can occur. Subsequent hydrogen abstraction reactions ($CH_n^+ + H_2 \rightarrow CH_{n+1}^+ + H$) can take place, but, in general, the recombination of the molecular ions containing carbon is five orders of magnitude faster than is the radiative recombination of C^+. The formation of carbon-bearing molecular ions in the shocked gas leads to rapid recombination, a significant drop in the ion flux, and an increase in the length of the acceleration zone (Flower, Pineau des Forêts, and Hartquist 1985a). The corresponding decrease in the ion–neutral collisionally induced heating of the neutrals suppresses jumps in the neutral flow (Flower, Pineau des Forêts, and Hartquist 1985b).

Ultimately, the reduction of the C^+ density is limited because photodissociation and photoionization reactions break down the larger carbon bearing species and lead to reionization. The shock structure is highly dependent on the photo-reaction rates which are assumed. Many of these rates (e.g. $CH_2 + h\nu \rightarrow CH_2^+ + e$; $CH_2^+ + h\nu \rightarrow CH^+ + H$) are highly uncertain. The combination of hydrogen abstraction reactions, dissociative recombination, photodissociation, and photoionization can lead to a drop in the carbon ion flux of about an order of magnitude in many diffuse cloud shocks.

The H^+ density in the shocked gas depends primarily on the cosmic ray ionization rate (which determines the preshock H^+ abundance), its removal rate in charge transfer reactions with O, and production by the photodissociation of CH^+ to form C and H^+ (Kirby 1980). For many shock models, the timescale for H^+ to be formed or removed is longer than the timescale during which gas is in the dissipation zone or precursor. Hence, initially the number density of H^+ increases proportionally to B_x or inversely proportionally with v_{iz} as measured in the shock frame. However, when equilibrium obtains, the H^+ number density in a cold diffuse cloud is insensitive to the cloud number density, and, hence, the H^+ number density eventually drops in the cooled postshock region to its preshock value. The H^+ number density is comparable to and can exceed the C^+ number density in the warm regions. The H^+ number density must be followed carefully in order to calculate the shock structure correctly.

5.4. The production of CH^+ in shocks

Dalgarno (1976) reviewed low temperature chemical models for the production of CH^+ in diffuse clouds. He concluded that the model column densities are sometimes 1–2 orders of magnitude lower than those observed.

Elitzur and Watson (1978a) investigated the possibility that column densities comparable to those observed could be formed in shocked gas. At the time of their work, three-fluid treatments of shock structure were not prevalent, and they employed a one-fluid nonmagnetic shock model. We argue below that the CH^+ column densities in C-type shocks will be substantially greater than in one-fluid models.

The chemistry which dominates the CH^+ formation is the same as the carbon chemistry which was discussed in the previous section, where we considered the ionization structure in the shocked gas. CH^+ is formed by Reaction (5.6.) and removed by the hydrogen abstraction reaction

$$CH^+ + H_2 \rightarrow CH_2^+ + H \tag{5.7}$$

for which we will take the reaction rate coefficient to be α_7. If every C^+ ion were converted to CH^+ and CH^+ could in no way be reformed after it was removed by Reaction (5.7), its column density along a line of sight normal to a plane parallel shock would be about $n_0(C^+)v_s/[4\alpha_7 n_0(H_2)]$ where $n_0(C^+)$ and $n_0(H_2)$ are the distant upstream number densities of C^+ and H_2. In fact, photodissociation and photoionization and, in gas in which the atomic to molecular hydrogen ratio is of order one or greater, the reverse of the hydrogen abstraction reaction sequence lead to the formation of CH^+ from the larger molecules. The cycling timescale, τ_r, for the reformation of CH^+ from larger molecules depends on the photoabsorption and inverse hydrogen abstraction reaction rates. If τ_{cool} is the cooling timescale of the gas, the column density of CH^+ in a hydrodynamic shock is roughly $n_0(C^+)v_s\tau_{cool}/[4\alpha_7 n_0(H_2)\tau_r]$. The timescale over which CH^+ is in the magnetic precursor of a hydromagnetic shock is larger than the cooling timescale in a hydrodynamic shock containing the same column density of H_2 ($J = 4$); this flow time should be substituted for τ_{cool} in the above expression to obtain an estimate for the column density of CH^+ in a hydromagnetic C-type shock.

The observations of H_2 in rotational levels with $J \geq 4$ comprise a major constraint on the shocks which might exist in diffuse molecular clouds. If the radiative decay of H_2 ($J = 4$) is the major radiative loss mechanism in the shock, its column density is constrained by

$$N_{H_2}(J = 4) \lesssim \rho_0 v_s^3/2(\Delta E_{42})A_{42} \tag{5.8}$$

where ρ_0 is the distant upstream mass density, v_s is the shock speed, ΔE_{42} is the energy difference between the $J = 4$ and $J = 2$ levels of H_2 ($v = 0$), and A_{42} is the radiative decay probability of the $J = 4$ level. Hence, if v_s is sufficiently large N_{H_2} ($J = 4$) depends only on the value of $\rho_0 v_s^3$ and not on the assumed value of the $J = 2 \rightarrow 4$ collisional excitation rate coefficient. However, the model temperature in the dissipation zone of a C-type shock is sensitive to the value of the $J = 2 \rightarrow 4$ rate coefficient. The model results for N_{H_2} ($J = 5$) and N_{H_2} ($J = 6$) depend on the values taken for the $J = 3 \rightarrow 5$ and $J = 4 \rightarrow 6$ excitation rates which in turn depend on the temperature and hence on the $J = 2 \rightarrow 4$ rate coefficient. The uncertainties in the rotational excitation rates for H_2 are discussed in Chapter 12.

Draine (1986b), Draine and Katz (1986), Pineau des Forêts, Flower, Hartquist, and Dalgarno (1986), and Pineau des Forêts *et al.* (1987) have calculated the column densities of CH^+ and of rotationally excited H_2 for various shock models. Different reaction networks and rates were used. One specific point to note is that

Draine (1986b) and Draine and Katz (1986) took the products of CH^+ photodissociation to be C^+ and H, whereas they are principally C and H^+ (Kirby 1980); when Draine (1987, private communication) used the correct dissociation products in his best fit ζ Oph model and used the correct endothermicity in the H^+ charge transfer reaction with O to determine the upstream H^+ abundance, the model CH^+ column density was decreased by a factor of almost 3. To this date, the special conditions necessary for the production in shocks of the observed CH^+ column densities and of column densities of H_2 in $J \geq 3$ which do not exceed those observed have not been found.

We now elaborate on the dependence of the column densities N_{CH^+} and N_{H_2} $(J = 4)$ on the adopted parameters. Attention is restricted to C-type shock models, because, as we argued above, the CH^+ column densities in them are much larger than in shocks (i.e. J-type shocks) for which one-fluid models are appropriate.

Pineau des Forêts, Flower, Hartquist, and Dalgarno (1986) considered the dependence of N_{CH^+} and of N_{H_2} $(J = 4)$ on shock speed, on preshock magnetic field strength, and on $n(H)/n(H_2)$. In the range of shock speeds 8–15 km s^{-1}, N_{H_2} $(J = 4)$ increases more rapidly than v_s^3, because H_2 is not the dominant coolant in this regime. Throughout a large part of the same speed range, the abundance of CH^+ increases even more rapidly with speed but less rapidly than the abundances in the $J = 5$ and 6 levels; because they increase most rapidly, the $J = 5$ and 6 level populations provide the strongest constraints on shock speed. When the ambient medium has a number density of 20 cm^{-3} and a magnetic field with strength less than 10 μG, the abundances are fairly insensitive to the field strength. If the shock speed is greater than about 8 km s^{-1}, the CH^+ abundance is fairly insensitive to the preshock $n(H_2):n(H)$ ratio.

However, Draine and Katz (1986) have shown that, in slower shocks, the CH^+ column density rises as the $n(H_2):n(H)$ ratio increases. They also found that the CH^+ column density increases by a factor of 4–5 in the shock speed range of 10–15 km s^{-1} if the radiation field is enhanced by a factor of 5 over the typical interstellar value. This higher CH^+ concentration is a consequence of decreasing τ_r, the cycling timescale for the reformation of CH^+ from larger molecules.

A detailed study of the dependence of model column densities on the adopted values of individual reaction rates has not been made. For some of the key photodissociation and photoionization rates the values assumed by Draine and Katz (1986) and Draine (1986b) and by Pineau des Forêts *et al.* (1986, 1987) differ substantially.

The model results are sensitive to the assumed H_2 collisional excitation rates. Danby, Flower, and Monteiro (1987) constructed shock models using two different sets of values for H_2–H_2 collisional excitation rates. One set derived

Table 5.2. *Results for shock models, and column densities observed towards ζ Oph*

	(a)	(b)	ζ Oph
N(OH)	6.6 (13) cm^{-2}	4.4 (13) cm^{-2}	5.1 (13) cm^{-2}
N(CH)	2.7 (13)	1.6 (14)	3.4 (13)
$N(CH^+)$	2.9 (13)	1.6 (13)	3.4 (13)
N(CO)	6.8 (12)	9.3 (13)	2.0 (15)
N_{H_2} ($J = 3$)	3.9 (17)	5.9 (17)	1.2 (17)
N_{H_2} ($J = 4$)	3.0 (16)	2.5 (16)	4.8 (15)
N_{H_2} ($J = 5$)	7.1 (15)	1.2 (15)	5.9 (14)
N_{H_2} ($J = 6$)	1.4 (14)	1.05 (13)	3.8 (13)
T_n(max)	1880 K	800 K	
$(v_i - v_n)_{max}$	8.65 km s^{-1}	8.84 km s^{-1}	
l	1.4 (17) cm	6.0 (17) cm	

l is the length of the precursor or dissipation zone.

from their quantum mechanical studies of H_2–H_2 collisions. The other set came from an independent quantum mechanical study of He–H_2 collisions. The models were of 12 km s^{-1} shocks propagating into a medium with $n(H_2) = 9.8$ cm^{-3}, n(H) $= 0.4$ cm^{-3}, and a transverse magnetic field strength of 5 μG. The column densities of CH^+ and H_2 ($J = 4$) were the same for both models. The CH, OH, H_2 ($J = 5$), and H_2 ($J = 6$) column densities differed by 10, 35, 7, and 20%, respectively.

Table 5.2 gives results for the CH^+, CH, OH, and CO column densities and the column densities of the various rotational levels of H_2 for the two models defined in Table 5.1. The observed column densities towards ζ Oph are given for comparison. Inspection of Table 5.2 gives an impression of the difficulties faced when one attempts to construct a shock model with a CH^+ column density comparable to that observed without violating other constraints.

The effects of the ultraviolet radiation field on the H_2 rotational population distribution (see Chapter 3) were not included in any of the calculations described above. Their inclusion probably will enhance the discrepancies between model H_2 ($J > 3$) column densities and those measured; at best the ultraviolet radiation may make little difference if the shocked gas is shielded from the ultraviolet radiation by sufficient cold cloud material. Some quantitative results exist. Monteiro *et al.* (1988) have solved the H_2 line transfer problem for a plane-parallel C-type shock and have calculated the ultraviolet induced pumping and dissociation rates. The radiation field was taken to be χ times the unattenuated interstellar background at the front of the shock. The problem was solved for 9, 11, and 16 km s^{-1} shocks propagating into media in which $n(H_2) = 9.8$ cm^{-3} and $n(H) = 0.4$ cm^{-3}. For $v_s = 11$ km s^{-1} and $\chi = 4$, the ratio, $n(H)/n(H_2)$, increases to roughly $\frac{1}{2}$ at the back of the precursors. The CH^+ column density in the models did not depend sensitively on χ in the range $1 \leq \chi \leq 8$, but the column densities of H_2 ($J = 5$, 6, and 7) each increased by roughly an order of magnitude as χ was increased in this range. As mentioned above, in many cases, the radiation field at the front of the shock may be attenuated by cold cloud material between the shock and the sources of the background radiation; such screening will reduce the rates at which the shocked H_2 in lower lying J levels absorbs radiation, but the absorption rates in the $J \geq 4$ levels will not be altered substantially by the presence of intervening cold molecular gas.

5.5. Diagnostics

As mentioned in the introduction, some endothermic reactions or reactions with barriers can proceed rapidly in shocked gas. Among those are Reactions (5.5). They, along with the photodissociation of H_2O and OH, determine the OH abundance in shock heated gas. Results for models (a) and (b) show that gas in a shock can be the site of a significant column density of OH. In Chapter 4 Langer discusses the contention that an OH velocity component originating in a shock has been detected (Crutcher 1979).

The reaction 0.9 eV + S^+ + $H_2 \rightarrow SH^+$ + H is endothermic just as the $C^+ + H_2 \rightarrow CH^+ + H$ reaction is. As shown in detail by Pineau des Forêts, Roueff, and Flower (1986) potentially observable amounts of SH^+ can be formed in diffuse cloud shocks. SH^+, like CH^+, should have much higher column densities in shocks with extended acceleration zones than in strong J-type shocks. The observationally determined upper bound on the SH^+ column density towards ζ Oph is comparable to the column density expected to exist in a C-type shock (Millar and Hobbs 1988). The C_2 abundances in shocks may be high because of the enhanced concentrations, in the acceleration and dissipation zone, of neutral molecules and molecular ions containing carbon and of atomic carbon; reactions occurring amongst such species can lead to the formation of molecular species containing more than one carbon atom. Langer has reviewed the work of van Dishoeck and de Zeeuw (1984) who showed that the bulk of the C_2 in clouds does not seem to be shifted significantly in velocity relative to most of the cloud material. Pineau des Forêts and Flower (1987) have discussed more completely the possibility that other species containing two or more carbon atoms may serve as diagnostics.

CO also contains more than one heavy atom and, like C_2, might be expected to have a large column density in the precursor zone. However, model results show that the CO column densities in shocks should not be sufficient to explain the observations.

5.6. Shocks as a means of resolving problems with the diffuse cloud chemistries

The OH and C_2 column densities in diffuse clouds have substantial contributions from quiescent gas. Whilst model CH^+ column densities are not as high as those observed, the shock origin of CH^+ remains a distinct possibility.

As discussed by van Dishoeck in Chapter 3, the high photodissociation rate for CO implies that existing low temperature chemical models (with radiation fields as intense as those inferred from models of the ultraviolet pumping of H_2) fail to reproduce the observed column densities of CO in diffuse molecular clouds. Model results indicate that the shocked gas itself does not contain sufficient CO to contribute substantially to the observed column densities. However, if all of the highly rotationally excited H_2 were formed in shocks, the ultraviolet field could be considerably weaker than inferred from ultraviolet pumping models, leading to an increase in the abundance of species such as CO in both the shocked and quiescent gas. A decrease in the radiation field would require some decrease in the cosmic ray ionization rate to ensure that the cold cloud gas did not contain too much OH. We have constructed one shock model specified by the same parameters as model (a), except that $\chi = 0.1$ and $n(H^+)$ and $n(He^+)$ are appropriate to a cosmic ray ionization rate which is a factor of 10 smaller.

Inspection of Table 5.2 shows that, for model (b), the discrepancies with the column densities observed towards ζ Oph are generally no more severe than for model (a). The CH and H_2 ($J = 3$) column densities in model (b) are too high, but the CH^+ column density is within a factor of 2 of the observed value. In addition the H_2 ($J = 5$) and H_2 ($J = 6$) column densities in model (b) do not conflict strongly with the data. Given that the discrepancies with observations are of the same degree of severity for models (a) and (b), the following possibilities suggest themselves.

(1) Shocks are the sites of much of the more highly excited H_2 in diffuse clouds.
(2) Consequently, the radiation fields have been greatly overestimated by the procedure of modelling cold clouds in equilibrium.
(3) As a result, the photodestruction rates of molecules like CO have been greatly overestimated. A number of discrepancies between models of cold clouds and observations can be removed.

A clear demonstration of the validity of these conclusions will require considerable improvement of the shock chemistry models. To some extent, the improvements will depend on the availability of improved data for a number of reactions. In particular, the ionization structure and thermal structure in a shock model depend on the photoionization and photodissociation rates adopted for species of the types CH_n and CH_n^+, but other rates are important, too. More systematic explorations of the parameter space of the models might also prove to be profitable.

REFERENCES

Aannestad, P. 1973, *Ap. J.* Suppl., **25**, 223.
Aannestad, P. A. and Field, G. B. 1973, *Ap. J.*, **186**, L29.
Boyd, T. J. M. and Sanderson, J. J. 1969, *Plasma Dynamics* (New York: Barnes & Noble, Inc.).
Crutcher, R. M. 1979, *Ap. J.*, **231**, L151.
Dalgarno, A. 1976, in *Atomic Processes and Applications*, p. 109, eds. Burke, P. G. and Moiseiwitsch, B. L. (Amsterdam: North-Holland).
Danby, G., Flower, D. R. and Monteiro, T. S. 1987, *MNRAS*, **226**, 739.
Douglas, A. E. and Herzberg, G. 1941, *Ap. J.*, **94**, 381.
Draine, B. T. 1980, *Ap. J.*, **241**, 1021; **246**, 1045.
Draine, B. T. 1986a, *MNRAS*, **220**, 133.
Draine, B. T. 1986b, *Ap. J.*, **310**, 392.
Draine, B. T. 1987, private communication.
Draine, B. T. and Katz, N. S. 1986, *Ap. J.*, **310**, 392.
Draine, B. T., Roberge, W. G. and Dalgarno, A. 1983, *Ap. J.*, **264**, 485.
Elitzur, M. and Watson, W. D. 1978a, *Ap. J.*, **222**, L141; **226**, L157.
Elitzur, M. and Watson, W. D. 1978b, *Astr. Ap.*, **70**, 443.
Flower, D. R. and Pineau des Forêts, G. 1986, *MNRAS*, **220**, 149.
Flower, D. R., Pineau des Forêts, G. and Hartquist, T. W. 1985a, *MNRAS*, **216**, 775.
Flower, D. R., Pineau des Forêts, G. and Hartquist, T. W. 1985b, in *Cosmical Gas Dynamics*, p. 163, ed. Kahn, F. D. (Utrecht: VNU Scientific Press).
Flower, D. R., Pineau des Forêts, G. and Hartquist, T. W. 1986, *MNRAS*, **218**, 729.
Havnes, O., Hartquist, T. W. and Pilipp, W. 1987, in *Physical Processes in Interstellar Clouds*, p. 389, ed. Morfill, G. E. and Scholer, M. (Dordrecht: Reidel).
Kirby, K. 1980, in *Interstellar Molecules*, p. 283, ed. Andrew, B. H. (Dordrecht: Reidel).
Millar, T. J., Adams, N. G., Smith, D., Lindinger, W. and Villinger, H. 1986, *MNRAS*, **221**, 673.
Millar, T. J. and Hobbs, L. M. 1988, *MNRAS*, **231**, 953.
Monteiro, T. S., Flower, D. R., Pineau des Forêts, G. and Roueff, E. 1988, *MNRS*, **234**, 863.
Pineau des Forêts, G. and Flower, D. R. 1987, *MNRS*, **228**, 1p.
Pineau des Forêts, G., Flower, D. R., Hartquist, T. W. and Dalgarno, A. 1986, *MNRAS*, **220**, 801.
Pineau des Forêts, G., Flower, D. R., Hartquist, T. W. and Millar, T. J. 1987, *MNRAS*, **227**, 993.
Pineau des Forêts, G., Roueff, E. and Flower, D. R. 1986, *MNRAS*, **223**, 743.
van Dishoeck, E. F. and de Zeeuw, T. 1984, *MNRAS*, **206**, 383.
Wardle, M. and Draine, B. T. 1987, *Ap. J.*, **321**, 321.

III

Quiescent dense clouds

6

Chemical modelling of quiescent dense interstellar clouds

T. J. MILLAR

Department of Mathematics, UMIST, Manchester, UK

6.1. Introduction

Giant interstellar clouds are the most massive chemical 'factories' in our Galaxy containing around 80 molecules presently identified (neglecting isotopic variants) and ranging in complexity from H_2 and CO to large saturated molecules such as ethanol, CH_3CH_2OH, and highly unsaturated cyanopolyynes including the 13-atom chain $HC_{11}N$. Millimetre and sub-millimetre observations of interstellar molecules allow one to probe the densities, temperatures, and dynamics of interstellar clouds and can give information on the initial conditions for star formation. It is also of interest to understand the chemistry of these molecules since chemical kinetic modelling can be used together with observational data to constrain uncertain parameters such as elemental abundances and the cosmic ray ionisation rate. An understanding of deuterium fractionation in interstellar molecules can be used to determine the D/H ratio and thus has a bearing on cosmological models of the origin of the universe.

In recent years models of increasing chemical, physical, and computational complexity have been developed to study molecular formation and destruction in various astronomical regions. Models of interstellar cloud chemistry can be divided roughly into two classes: (1) *steady-state models*, in which chemical abundances are calculated through solving a coupled system of non-linear algebraic equations; and (2) *time-dependent models*, in which the variations of abundances as a function of time are followed through solving a coupled system of stiff, non-linear, first-order, ordinary differential equations. Traditionally, steady-state models have been the dominant tool for studying chemistry in diffuse clouds which, in the absence of shocks, reach steady-state well within their lifetimes. Such regions are discussed elsewhere in this volume. Some steady-state models have been applied to dense clouds, although at the present time time-dependent models are more widely studied.

In the following section we review briefly some of the processes by which chemistry is initiated in dense clouds. Section 6.3 outlines some of the strengths and weaknesses inherent in both steady-state models, discussed in Section 6.4, and time-dependent models, discussed in Section 6.5. We conclude by listing some areas for future research.

6.2. Initiation of the chemistry

It has been established since the pioneering work of Dalgarno and co-workers that molecule formation in interstellar clouds is not efficient unless an appreciable fraction of hydrogen is in molecular form. The H_2 is known to form on dust grains but all other important chemical processes are thought to occur in the gas phase although there may be localised regions, such as hot cores, in which molecular abundances reflect a surface chemistry. In dense interstellar clouds, H_2 is the most abundant molecule and reactions of neutral atoms with H_3^+, produced from H_2 by cosmic ray ionisation followed by

$$H_2^+ + H_2 \rightarrow H_3^+ + H \qquad (6.1)$$

are important in driving the chemistry. Simple hydride ions form through series of reactions with H_2, such as

$$C + H_3^+ \rightarrow CH^+ \rightarrow CH_2^+ \rightarrow CH_3^+ \rightarrow CH_5^+ \qquad (6.2)$$

and

$$O + H_3^+ \rightarrow OH^+ \rightarrow H_2O^+ \rightarrow H_3O^+ \qquad (6.3)$$

The corresponding reaction of N with H_3^+ is endothermic while the exothermic product NH_2^+ does not form because of a large activation energy barrier. Atomic sulphur reacts with H_3^+

$$S + H_3^+ \rightarrow SH^+ + H_2 \qquad (6.4)$$

but the subsequent hydrogen abstraction reactions of H_nS^+ ($n = 1,2$) with H_2 are very endothermic – a case opposite to that of nitrogen for which the initial proton transfer does not occur while the subsequent abstraction reactions are all exothermic.

The nitrogen hydride sequence is initiated through reactions of N^+ with H_2 but excited N^+ is required as the reaction is slightly endothermic. Excited N^+ results from reactions of He^+, which reacts very slowly with H_2, with nitrogen-bearing molecules such as CN and N_2 which form primarily in neutral–neutral reactions, as discussed below. In this case the sequence

$$N^+ \rightarrow NH^+ \rightarrow NH_2^+ \rightarrow NH_3^+ \rightarrow NH_4^+ \qquad (6.5)$$

occurs. The ions which terminate the chains (6.2), (6.3), (6.4) and (6.5) do not react with H_2 and are therefore able to do so with less abundant species, including electrons to give CH_4, H_2O and NH_3 among other products, and atoms, for example,

$$CH_5^+ + O \rightarrow H_3CO^+ + H \quad (6.6)$$

which leads to formaldehyde, H_2CO, via dissociative recombination

$$H_3CO^+ + e \rightarrow H_2CO + H \quad (6.7)$$

In reaction sequence (6.2) the final step in the chain is a radiative association reaction

$$CH_3^+ + H_2 \rightarrow CH_5^+ + h\nu \quad (6.8)$$

whose rate coefficient at low temperatures is uncertain but is much less than that typical for exothermic ion–neutral reactions, $\sim 10^{-9}$ cm^{-3} s^{-1}. Reactions of CH_3^+ which are important include

$$CH_3^+ + O \rightarrow HCO^+ + H_2 \quad (6.9)$$

$$CH_3^+ + HCN \rightarrow CH_3CNH^+ + h\nu \quad (6.10)$$

$$CH_3^+ + H_2O \rightarrow CH_3OH_2^+ + h\nu \quad (6.11)$$

$$CH_3^+ + CO \rightarrow CH_3CO^+ + h\nu \quad (6.12)$$

and lead to CO, methyl cyanide, CH_3CN, methanol, CH_3OH, and ketene, CH_2CO, respectively, upon recombining with electrons. The HCN molecule is one of a number of molecules which form in neutral–neutral reactions, as discussed below.

Although H_3^+ is probably the most abundant ion in dense clouds, proton transfer reactions of H_3^+ with stable neutrals often lead to the reformation of the neutral upon dissociative recombination. Helium ions are of particular importance because they can break down stable neutrals. They provide a source of carbon ions through their reaction with the abundant CO molecule

$$He^+ + CO \rightarrow C^+ + O + He \quad (6.13)$$

and this C^+ can react with methane, for example, to make the precursor ions of acetylene, C_2H_2, and the ethynyl radical, C_2H,

$$C^+ + CH_4 \rightarrow C_2H_3^+ + H \quad (6.14)$$

$$\rightarrow C_2H_2^+ + H_2 \quad (6.15)$$

Subsequent reactions of C^+ with the C_2-hydrocarbons and condensation reactions of acetylene ions with neutral hydrocarbons can build even larger molecules

$$C^+ + C_2H_2 \rightarrow C_3H^+ + H \quad (6.16)$$

$$C_2H_2^+ + C_2H_2 \rightarrow C_4H_3^+ + H \quad (6.17)$$

$$\rightarrow C_4H_2^+ + H_2 \quad (6.18)$$

The H_3^+ ion produced as a result of cosmic ray ionisation of H_2 has a low proton affinity and proton transfers with many species to form complex molecular ions; for example

$$H_3^+ + CO \rightarrow HCO^+ + H_2 \tag{6.19}$$

These molecular ions are neutralised through dissociative recombination with electrons, e.g.,

$$HCO^+ + e \rightarrow CO + H \tag{6.20}$$

and through charge transfer with metal atoms, M,

$$M + HCO^+ \rightarrow M^+ + HCO \tag{6.21}$$

The fractional ionisation, $x(\mathrm{e})$, is thus determined by the cosmic ray ionisation rate, ς, and the abundances of metal atoms. If the metal abundance is large, then M^+ ions carry the positive charge since they are destroyed in slow radiative recombination reactions

$$M^+ + e \rightarrow M + h\nu \tag{6.22}$$

Oppenheimer and Dalgarno showed that, in this case, $x(\mathrm{e}) \propto (\varsigma/n)^{1/3}$, where n is the total particle density. At large densities, the fractional electron abundance is small and recombination of metal ions and molecular ions with negatively charged grains dominates. In this case, $x(\mathrm{e}) \propto (\varsigma/n)^{1/2}$. Detailed chemical models of cold dense clouds, such as those described in Sections 6.4 and 6.5, indicate that the metal abundance is low so that molecular ions such as H_3^+, HCO^+ and H_3O^+ provide the major source of electrons and give $x(\mathrm{e}) \propto (\varsigma/n)^{1/2}$ irrespective of whether free electrons or negatively charged grains dominate the neutralisation.

Deuterium can be enhanced, or fractionated, in molecules through processes primarily involving HD, H_2D^+ and D atoms. The major channel at low temperatures is via H_2D^+ produced in the reaction

$$H_3^+ + HD \leftrightarrow H_2D^+ + H_2 + \Delta E_1 \tag{6.23}$$

where the reaction enthalpy $\Delta E_1/k \sim -227$ K. At low temperatures the reverse reaction in (6.23) is inhibited and the $[H_2D^+]/[H_3^+]$ ratio can be much greater than the $[HD]/[H_2]$ ratio. Fractionation into other species occurs in reactions such as

$$XH^+ + HD \leftrightarrow XD^+ + H_2 + \Delta E_2 \tag{6.24}$$

and

$$XH + H_2D^+ \leftrightarrow XHD^+ + H_2 + \Delta E_3 \tag{6.25}$$

followed by dissociative recombination. Such reactions form molecules such as DCO^+, N_2D^+, DCN and DNC. Since the extent of fractionation is very sensitive to the small changes in free energy occurring in these reactions, observations of deuterated molecules can give information on cloud temperatures and on the abundance of species which compete with D, HD and H_2D^+ for the reactant molecule. Some of these species, such as O_2, are unobservable.

An understanding of deuterium fractionation is extremely important since it allows one to study the D/H ratio on a galactic (and, ultimately, extragalactic) scale as well as giving information on the fractional ionisation, $x(e)$, which determines the coupling of the magnetic field to the cloud. If $x(e)$ is small enough, the field can no longer couple via the ions to the predominantly neutral gas and magnetic pressure will no longer be a support against self-gravity.

Dalgarno and Lepp have discussed the chemistry of DCO^+ which forms via

$$H_2D^+ + CO \rightarrow DCO^+ + H_2 \tag{6.26}$$

and

$$D + HCO^+ \rightarrow DCO^+ + H \tag{6.27}$$

with destruction through dissociative recombination with electrons.

Since H_2D^+ is deuterated by HD (Reaction (6.23)) and by D atoms

$$D + H_3^+ \rightarrow H_2D^+ + H \tag{6.28}$$

the observed DCO^+/HCO^+ abundance ratio together with a theoretical analysis of that expected, can be used to set limits on $x(e)$ and D/H if some assumptions, for example on the cosmic ray ionisation rate, are made. The lower limits for D/H derived by Dalgarno and Lepp imply that baryonic matter cannot close the universe.

Neutral–neutral reactions also play a rôle in the production of many molecules, particularly in heterogeneous bond formation. For example, OH, which may form in the dissociative recombination of H_3O^+ with electrons (see Section 6.5.2), leads to molecules such as O_2, SO, NO and N_2 in the reactions

$$O + OH \rightarrow O_2 + H \tag{6.29}$$

$$S + OH \rightarrow SO + H \tag{6.30}$$

$$N + OH \rightarrow NO + H \tag{6.31}$$

$$N + NO \rightarrow N_2 + O \tag{6.32}$$

and can convert carbon atoms to CO,

$$C + OH \rightarrow CO + H \tag{6.33}$$

while reactions between atomic nitrogen and neutral hydrocarbons, formed in the dissociative recombination of the ions generated in reaction sequence (6.2), can produce CN and HCN

$$N + CH \rightarrow CN + H \tag{6.34}$$

$$N + CH_2 \rightarrow HCN + H \tag{6.35}$$

6.3. Chemical models – general considerations

Before discussing the results of steady-state and time-dependent models of dense clouds we shall address some of the problems and shortcomings inherent in these two approaches.

The solution of time-dependent models generally requires a considerable amount of CPU time and has meant that, in the main, 'one point' models of interstellar clouds have been investigated. That is, during the chemical evolution of the cloud, which requires typically 10^6–10^8 yr, the physical conditions are assumed to remain constant and are represented by a single density, temperature, visual extinction and so on. Such models are generally called 'pseudo-time-dependent'. Calculated abundances are compared with column densities averaged over the beam size used for a particular molecular observation which implies that, particularly for distant sources such as in the Galactic Centre, column densities are averaged over a large physical area in which density, temperature, and extinction may vary. In addition, many regions show evidence of dynamic effects, such as turbulence, collapse, rotation, star formation, and shocks, which can change physical conditions on time-scales of $\lesssim 10^6$ yr. Recent time-dependent models on the chemistry in a hydrodynamically collapsing cloud, and one including star formation, shocks, and accretion have been published (see also Chapter 17). On the other hand, many clouds appear to be quiescent, at least at the present time, the most famous example of which is Taurus Molecular Cloud 1 (TMC-1). In this case, dynamics may be unimportant but the requirement of hydrostatic equilibrium enforces the presence of a pressure gradient, or equivalently temperature and density gradients, within the cloud. Such effects have not been included, as yet, in any time-dependent calculation.

Steady-state models are generally 'many point' models and so often include a pressure gradient but, of necessity, must neglect the evolution of physical parameters and, most importantly for quiescent dense clouds, the accretion of gas phase species onto dust grains. The most detailed steady-state models have been developed to describe chemistry in diffuse clouds for which accretion is unimportant (Chapter 3). Hydrostatic models of dense clouds such as TMC-1 and L134N have been explored by de Jong, Dalgarno, and Boland, by Boland and de Jong and by Hasegawa. Although the first two studies included accretion in an *ad hoc* fashion, the neglect of this process, which occurs on a time-scale of $\sim 3 \times 10^9/n$ yr, where n is the total hydrogen density per cubic centimetre, in a quantitative manner compromises their results. On the other hand, such studies can be used to investigate chemistry in cloud edges where the density is small enough to make accretion unimportant.

Finally we should mention that while accretion has been considered in some time-dependent models, we do not, as yet, observe clouds which are free of molecules other than H_2 and H_3^+. It appears that some process must prevent efficient accretion and/or return grain mantle material to the gas phase. The precise mechanisms by which these could occur are unknown, as is the chemical processing which might result. Some calculations which include surface processing and the release of mantle material have been performed, but cannot be regarded as anything more than speculative exercises.

In principle, detailed chemical models of dense clouds should contain several thousand reactions to describe the chemistry of a few hundred species. The largest model calculation of which I am aware is by Brown and Rice, who included 3167 reactions to describe deuterium chemistry in dense clouds. Many of the reactions included in any chemical model have unknown rate coefficients or product branching ratios. However this is not necessarily a problem for exothermic ion–molecule reactions whose rate coefficients can be estimated fairly accurately based on extensive laboratory measurements. Indeed these experimental studies have provided a firm base onto which chemistries, sometimes fairly fragile, may be built. In this regard it is worth noting that, with the exception of the atom–H_3^+ reactions, all of the ion–molecule reactions in (6.1)–(6.19) have been measured at or below room temperature.

Large chemical schemes are not always necessary, however, since many reactions of ions which react rapidly with H_2 are included but could be neglected safely because the abundance of such ions must always be small. Millar, Leung and Herbst have shown that a system of around 620 reactions, based on the dense cloud model of Millar and Nejad, gives results in excellent agreement with the 2000 reaction model presented by Herbst and Leung. In general, therefore, we can conclude that model calculations are not sensitive to many of the (unknown) rate coefficients although specific choices for important reactions can cause substantial differences as we shall discuss further in Section 6.5.

6.4. Steady-state models of dense clouds

The earliest detailed models which described dense cloud chemistry were steady-state models, in particular the pioneering studies of Herbst and Klemperer, and Mitchell, Ginsburg and Kuntz. These models were mainly concerned with the formation of small molecules, in part because of the relatively small number of molecules observed and in part because of the lack of relevant laboratory data. Subsequently Herbst, Suzuki, Freeman, and Millar extended the chemistry to include certain hydrocarbons and the simplest cyanopolyynes. They were able to find reasonable agreement with observations of TMC-1 and L183 (L134N) particularly if a large abundance of C atoms exists in these clouds. Despite their success all these models contain one major inconsistency in that accretion of the gas onto grains is ignored. In a cloud such as TMC-1, the accretion time-scale is

$\sim 3 \times 10^5$ yr whereas the time needed to reach steady-state is $\sim 10^7$ yr. Possible ways of avoiding the accretion catastrophe include adopting a smaller dust-to-gas ratio, which is usually taken from diffuse cloud observations, or to argue for some efficient, but unknown, desorption mechanism. Such a mechanism must return material essentially in the form in which it was accreted if the gas phase chemistry is not to be perturbed to any significant extent, a conclusion based on the agreement between the observations and theoretical calculations which ignore surface effects.

Most recent steady-state models of dense clouds haved focused on hydrostatic cloud models. These models, which determine self-consistently the temperature and density gradients in plane-parallel and spherical clouds, enable spatial effects to be considered, for example, the relative contributions of core and envelope to the total column densities of species. It has been shown recently that the millimetre wavelength absorption lines of HCN and HNC detected in spiral arm gas clouds arise in the envelopes of large clouds rather than in small dense, unresolved clouds as had been thought previously. Hasegawa has discussed the chemistry of fairly simple species in hydrostatic spherical cloud models in an attempt to reproduce column densities observed in the Bok Globules L134 and L183. His best fit models to the data for L183 uses a cloud mass of 80 $M_\odot$ and an external pressure $P/k = 2.9 \times 10^5$ (K cm^{-3}). This rather large pressure results in an outer boundary density of 3000 cm^{-3} and a central density of 1.8×10^4 cm^{-3}. His calculated column densities for species such as C^0, OH and CS agree reasonably well with those observed but he finds poor agreement for SO, CCH and CN; the latter species is overabundant, as it is in other models. It would be of interest to compare this type of model with the results of the recent observational study of L183 by Swade, who finds that molecules in L183 can have dissimilar spatial distributions despite being closely related chemically.

6.5. Time-dependent models of dense clouds

In this section we shall review pseudo-time-dependent models of dense interstellar clouds with particular emphasis on the sensitivity of calculated abundances to 'global' uncertainties in the models. These uncertainties include the number of reactions needed to describe the chemistry in an adequate fashion (discussed in Section 6.3), the effects of enhanced rate coefficients caused by ion–dipolar molecule collisions, the choice of neutral products in dissociative recombination reactions, and photodissociation caused by cosmic ray induced ultraviolet photons; however, we shall not discuss any models which include molecule formation on grain surfaces.

The earliest, detailed time-dependent studies of dense cloud chemistry were by Iglesias, and Prasad and Huntress. These latter articles were a major advance and now act as a 'test-bed' against which more recent calculations can be compared. This chemistry contained around 1300 reactions and several groups have used it as the basis for even larger reaction sets. The basic result of all these studies has been to show that molecular abundances peak at 'early times', typically $\sim 3 \times 10^5$ yr, before decreasing, often by several orders of magnitude, to steady-state values at $\sim 10^7$ yr, although this behaviour does not occur for all species. For example, the abundances of CO, NH_3 and O_2 generally increase monotonically to their steady-state values. The peak in abundances at early times is especially noticeable for carbon-bearing species and is related to the large abundance of C^0 which builds hydrocarbons efficiently. At steady-state, the C^0 abundance is much lower as it is processed into CO and the rate at which hydrocarbons form decreases. At early times, the abundances are generally in better agreement with the observed values but this is not an argument that clouds are young since the adopted initial conditions (hydrogen molecular, all other elements atomic) and the simplistic physical model are rather artificial. However, a general conclusion which can be made is that gas phase processes do appear capable of explaining the chemistry of many molecules in a fairly straightforward manner; in particular, the detection of molecular ions, the abundances of deuterated molecules and protonated ion/parent molecule abundance ratios. This conclusion is independent of uncertainties in rate coefficients since, for simple molecules, many of the important reactions, involving primarily H_2, have been measured in the laboratory, although certain problems, relating mostly to nitrogen-bearing molecules such as CN, do remain.

6.5.1. Ion–dipolar molecule reactions

In most chemical models of dense interstellar clouds, for reactions known to be fast at room temperature, the room temperature rate coefficients have been adopted at low temperatures, while for systems having no available measurement, the Langevin or average dipole orientation (ADO) rate coefficients have been used. However, for ion–dipolar molecule collisions, long-range attractive forces lead to rate coefficients which possess an inverse temperature dependence and which can be as large as 10^{-7} cm^3 s^{-1} at 10 K. Clary has reviewed theoretical models for calculating ion–molecule rate coefficients at low temperatures while Rowe has summarised the low-temperature experimental data. Herbst and Leung have studied the effects of enhanced rate coefficients and shown that while the abundances of neutral molecules are decreased by faster destruction involving H^+, He^+, H_3^+ and C^+ ions, the reduction is less than an order-of-magnitude. More significantly, however, very rapid proton transfer reactions between H_3^+ and polar neutrals lead to enhanced abundances of protonated molecules and appear to be essential in order to account for the observed HCS^+/CS and $HCNH^+$/HCN ratios. Other protonated species should be abundant in cold dense clouds.

6.5.2. Dissociative recombination branching ratios

To date, almost all model calculations have used neutral product branching ratios for dissociative recombination reactions estimated according to the ideas outlined by Green and Herbst which were supported, for some simple molecular ions, by the statistical calculations of Herbst. Green and Herbst argued that hydride ions would lose one or more hydrogen atoms during recombination, for example,

$$H_3O^+ + e \rightarrow H_2O + H \quad (6.36)$$

$$\rightarrow OH + 2H \quad (6.37)$$

Molecules containing a number of heavy atoms, such as a carbon chain backbone, preserve these, in general, during recombination, for example,

$$C_3H_2^+ + e \rightarrow C_3H + H \quad (6.38)$$

$$\rightarrow C_3 + 2H \quad (6.39)$$

A more recent theory due to Bates, and described by him in Chapter 11, envisages molecular ions as having a localised valence bond structure so that the incident electron enters an anti-bonding orbital and breaks one specific bond, although the neutral products may subsequently dissociate in some cases. Thus Bates argues that the H_3O^+ ion can recombine only to H_2O and not to OH thereby increasing the water abundance in dense clouds while reducing the abundance of OH. In his theory the carbon backbone in complex hydrocarbon ions can break and hence possibly reduce the abundance of certain neutral molecules.

In order to quantify the predictions of this theory, Millar and co-workers have investigated dense cloud chemistry with both the 'standard' approach to recombination and the new ideas due to Bates. The charge distribution of the molecular ions, necessary in the estimation of possible neutral products, was determined by quantal calculations. Millar *et al.* found that the charge was more delocalised than in the classical valence bond picture, thereby increasing the number of available product channels. As was expected, the abundance of complex molecules decreased in the new model of recombination although the decrease was significant only at steady-state, which is arguably not the phase in which molecular clouds exist. At early times, when molecular abundances peak, the decrease was small indicating that, at this time, molecule formation is robust enough to overcome the additional destruction caused in the new picture of recombination. While this result applies to many species of interest, it is important to note that several species, such as OH, O_2, N_2 and N_2H^+, are reduced considerably at early times.

The most important dissociative recombination is that of H_3O^+ since its mode of recombination can affect the abundances of several species which form from OH. If H_3O^+ recombines to H_2O only, then the water abundance becomes similar to that of CO and very little O and OH remain in the gas. Since these two species are reactive, especially in neutral–neutral reactions, the abundances of species such as O_2, NO and N_2, which forms from NO, decrease. On the other hand, the NH_3 abundance increases because NH_4^+ is assumed to recombine only to NH_3. The ion H_3O^+ becomes the most abundant molecular ion in the new model of recombination. A single interstellar line has been identified as arising from this ion but confirmation, as well as searches in several sources, is required.

6.5.3. Cosmic ray induced photodissociation

Cosmic rays ionise molecular hydrogen and release energetic electrons which collide with, and excite to higher electronic states, other H_2 molecules. The excited molecules relax through photon emission, principally in the Lyman and Werner bands, and hence provide an internal source of radiation in dense clouds (see Chapter 16). The spectrum of this radiation and its effects on molecular abundances have been investigated by Sternberg, Dalgarno, and Lepp, with particular reference to the formation of complex molecules, and by Gredel, Lepp, and Dalgarno with reference to the photodissociation of CO. The former authors have estimated photodissociation rates for several simple species including H_2O, NH_3 and CH_4. Typical rates are $\sim 10^2$–10^3 times the cosmic ray ionisation rate, ζ, i.e. $\sim 10^{-15}$–10^{-14} s^{-1}. For clouds with $n(H_2) \sim 10^4$ cm^{-3}, the rates are comparable with molecular loss rates in ion–molecule reactions but can be less than loss rates in certain fast neutral–neutral reactions. Since many radicals, such as OH, NH and CCH, are destroyed in fast neutral–neutral reactions while stable molecules are not, cosmic ray induced photodissociation is important only for these latter species. However, since complex molecules are thought to form as the end products of a sequence of chemical reactions and since photodissociation can occur at each stage in the sequence, the abundances of the complex hydrocarbons, in particular, may be much reduced by these photons although the increased C^0 abundance results in more efficient formation of certain species. This process must take place over a long time-scale, however, since the photons are generated by cosmic ray ionisation; hence order-of-magnitude effects are not expected for times less than 10^5–10^6 yr.

Gredel *et al.* have noted that the wavelengths of many of the H_2 emission lines coincide with CO photoabsorption lines and have calculated the resultant CO photodissociation rate. Although the rate is slow, typically 10^{-16}–10^{-15} s^{-1}, photodissociation of CO is important because it provides the major source of C atoms in dense clouds – destruction of CO by ions, principally He^+, leads to C^+ and not to C^0. The observed C^0/CO ratio in dense interstellar clouds is difficult to interpret; Gredel *et al.* show that the inclusion of CO photodissociation by cosmic ray induced photons can increase this ratio by several orders of magnitude although it is still somewhat less than the observational ratio.

6.5.4. Quantitative results

In order to quantify the processes outlined in Sections 6.5.1–6.5.3, we present in Tables 6.1 and 6.2 the results of some pseudo-time-dependent calculations of the chemistry in a cold ($T = 10$ K), dense ($n(H_2) = 10^4\ cm^{-3}$) cloud. The chemistry is similar to that described in the dense cloud models by Millar *et al.*, but includes the recently measured reaction

$$C_4H^+ + H_2 \rightarrow C_4H_2^+ + H \qquad (6.40)$$

which enhances C_4H formation over previous models. Model 1 assumes standard recombination and neglects both internal ultraviolet photons and ion–polar rate coefficients. In Model 2 we have included ion–polar rate coefficients. Model 3 is similar to Model 1 but with the new dissociative recombination branching ratios as determined by Millar *et al.* Finally, Model 4 is the same as Model 1 but with internal ultraviolet photons included. Since some effects occur over a long time-scale, early time (3.2×10^5 yr) and steady-state (5×10^7 yr) results are given in Tables 6.1 and 6.2 respectively.

Table 6.1. *Fractional abundances at early time* (3.16×10^5 *yr*) *for a number of dense cloud models. For details see text*

	Model number			
Species	1	2	3	4
C	2.4 (−5)	2.6 (−5)	4.7 (−5)	2.1 (−5)
O_2	1.7 (−6)	1.7 (−6)	4.2 (−8)	1.9 (−6)
N_2	1.7 (−6)	1.8 (−6)	7.7 (−8)	1.9 (−6)
CH	1.9 (−9)	1.1 (−9)	4.6 (−9)	3.6 (−9)
OH	1.6 (−7)	1.6 (−7)	7.3 (−9)	1.7 (−7)
CO	9.1 (−5)	9.3 (−5)	7.6 (−5)	1.0 (−4)
CN	3.6 (−8)	3.5 (−8)	1.3 (−8)	4.1 (−8)
NO	2.5 (−8)	2.5 (−8)	8.5 (−10)	2.9 (−8)
H_2O	1.4 (−5)	5.1 (−6)	4.3 (−5)	1.1 (−5)
CCH	1.9 (−8)	1.0 (−8)	4.1 (−8)	3.0 (−8)
NH_3	7.9 (−9)	4.6 (−9)	2.2 (−8)	6.9 (−9)
CH_4	1.9 (−5)	1.9 (−5)	1.5 (−5)	1.3 (−5)
C_2H_2	7.3 (−7)	5.5 (−7)	8.0 (−7)	6.2 (−7)
C_3H	3.7 (−7)	9.4 (−9)	5.5 (−7)	3.5 (−7)
C_4H	8.0 (−7)	7.6 (−7)	4.8 (−7)	6.7 (−7)
HC_3N	7.1 (−9)	6.8 (−10)	5.1 (−9)	6.3 (−9)
CH_3OH	7.1 (−8)	3.7 (−8)	1.6 (−7)	4.9 (−8)
e	4.8 (−8)	4.6 (−8)	4.6 (−8)	4.9 (−8)
H_3^+	2.2 (−9)	2.0 (−9)	1.7 (−9)	2.3 (−9)
HCO^+	2.0 (−9)	1.7 (−9)	1.1 (−9)	2.4 (−9)
N_2H^+	2.0 (−11)	1.9 (−11)	7.6 (−13)	2.2 (−11)
H_3O^+	2.3 (−9)	2.8 (−9)	3.2 (−9)	2.2 (−9)
$H_2C_3N^+$	8.6 (−12)	7.0 (−12)	8.5 (−12)	8.4 (−12)

The notation $a(b) \equiv a \times 10^b$.

The tabulated data hide a multitude of sins. For example, ion–dipolar rate coefficients have been calculated according to the approach given by Herbst and Leung, but it is unclear how accurately theory gives the actual rate coefficients at 10 K. Likewise the dissociative recombination branching ratios have been determined theoretically. There have been, at the time of writing, no experimental determinations of the neutral products of the recombination of ground state polyatomic ions; the only experiments have been on vibrationally excited ions which may behave differently from those in the ground state. The photodissociation cross-sections of many species important in interstellar chemistry are unknown. Where possible, cosmic ray-induced photodissociation rates have been taken from Sternberg *et al.* and Gredel *et al.*; otherwise we have assumed rather large values, $\sim 10^{-14}$ s^{-1}, in order to maximise the destructive effects of the radiation field. Thus the results in Tables 6.1 and 6.2 are not presented in any definitive sense but are useful in indicating general trends and substantiate the comments made in the preceding subsections.

Table 6.2. *Fractional abundances at steady-state for a number of dense cloud models. For details see text*

	Model number			
Species	1	2	3	4
C	3.2 (−10)	2.4 (−10)	6.3 (−8)	3.0 (−9)
O_2	8.3 (−5)	8.6 (−5)	3.9 (−7)	6.2 (−5)
N_2	2.1 (−5)	2.1 (−5)	2.7 (−7)	1.9 (−5)
CH	1.5 (−10)	8.0 (−11)	1.8 (−11)	5.8 (−11)
OH	6.8 (−7)	5.1 (−7)	7.2 (−10)	4.4 (−7)
CO	1.5 (−4)	1.5 (−4)	1.5 (−4)	1.5 (−4)
CN	8.1 (−10)	1.2 (−9)	4.3 (−10)	4.6 (−10)
NO	1.7 (−7)	1.3 (−7)	1.9 (−10)	1.1 (−7)
H_2O	3.8 (−6)	1.8 (−6)	1.1 (−4)	5.3 (−6)
CCH	1.7 (−9)	1.1 (−9)	1.6 (−10)	4.7 (−10)
NH_3	5.6 (−8)	1.6 (−8)	4.2 (−7)	5.1 (−8)
CH_4	1.1 (−7)	1.0 (−7)	1.5 (−7)	6.0 (−8)
C_2H_2	8.2 (−9)	5.8 (−9)	1.8 (−8)	3.2 (−9)
C_3H	3.0 (−9)	5.5 (−11)	1.6 (−10)	5.8 (−10)
C_4H	4.4 (−10)	2.6 (−10)	1.8 (−11)	7.8 (−11)
HC_3N	1.8 (−12)	3.3 (−13)	1.6 (−11)	1.3 (−12)
CH_3OH	7.7 (−11)	3.1 (−11)	1.7 (−9)	1.1 (−10)
e	5.2 (−8)	5.0 (−8)	4.2 (−8)	5.1 (−8)
H_3^+	3.6 (−9)	3.2 (−9)	1.2 (−9)	3.2 (−9)
HCO^+	5.5 (−9)	4.5 (−9)	8.4 (−10)	4.9 (−9)
N_2H^+	3.1 (−10)	2.8 (−10)	1.7 (−12)	2.6 (−10)
H_3O^+	9.6 (−10)	1.7 (−9)	4.3 (−9)	1.3 (−9)
$H_2C_3N^+$	2.9 (−15)	5.4 (−15)	2.5 (−14)	2.1 (−15)

The notation $a(b) \equiv a \times 10^b$.

At early times, which are thought to be most appropriate, the abundances in all four models are very similar in most instances implying that the chemistry is difficult to perturb. However, significant differences can occur for *specific* molecules such as H_2O, which is enhanced in abundance under the new recombination rules and protonated ion/parent neutral ratios such as H_3O^+/H_2O and $H_2C_3N^+/HC_3N$, which are enhanced through ion–dipolar molecule collisions. Although cosmic ray induced photons do not affect abundances appreciably, with the exception of C^0, from their values in the standard model, this conclusion changes if such photons are added to a model which incorporates the new branching ratios. In this case the photodissociation of H_2O becomes the major source of OH, which also promotes formation of species such as O_2, NO and N_2, and results in steady-state abundances increased over those of Model 3 for several species.

6.6. Concluding remarks

Several difficulties still remain in understanding the observed abundances of certain species. The C^0/CO abundance ratio is a well-known example (Chapter 14). A number of explanations have been advanced (see Chapters 15, 16, and 17) but none are without difficulties, particularly for quiescent dense clouds, although there is only a detection of C^0 in one such source (L183).

A more severe problem, at least in the sense that there are no suggested solutions, occurs for NH_3. As mentioned in Section 6.2, the initiating reaction in sequence (6.5) is endothermic and can only proceed at low temperatures if excited N^+ is available. However, very recent laboratory measurements of the $N^+–H_2$ system, involving normal, para and deuterated hydrogen, have determined the endothermicity rather accurately and indicate that the energy contained in N^+ cannot drive reaction sequence (6.5) at a rate sufficient to account for the observed abundance of NH_3 in cold clouds. On the other hand, this analysis was not carried out for the recombination scenario envisaged by Bates. Model 3 in Table 6.2 shows that the NH_3 abundance is large at steady-state $\sim 4 \times 10^{-7}$, and therefore a certain decrease in the efficiency of reaction sequence (6.5) can be allowed, given that the abundance of NH_3 observed in TMC-1 is $\sim 2 \times 10^{-8}$. The weak point in this argument is that the majority of molecular abundances are best fit at early time, whereas the results of Model 3 in Table 6.1 indicate that the NH_3 abundance is then only just consistent with that observed in TMC-1 and would be much less using the new laboratory data.

Finally, as models in interstellar chemistry become more complex, there has been an effort to describe the chemistry of species such as ethanol, CH_3CH_2OH, and its isomer dimethyl ether, CH_3OCH_3. In general, schemes for producing large molecules have involved radiative association reactions in which large molecular ions are formed directly, rather than by sequential reactions. Recently, the rate coefficients of several of the important radiative association reactions have been shown to be much smaller than previously thought and it appears some other processes, perhaps involving grains, must be involved in the formation of these complex species.

Despite the difficulties noted above, chemical models of quiescent dense interstellar clouds have reached a certain degree of maturity. Although most modern studies can contain over 1000 reactions, the rate coefficients for almost all of the important reactions, mainly reactions between ions and H_2, have been determined in laboratory studies. A major uncertainty lies in the identification of the products of dissociative recombination reactions, although as shown in the previous section, the changes in the models are generally small except at steady-state. The photodissociation cross-sections of many molecules are unknown which precludes an accurate assessment of the influence of cosmic ray induced photodissociation. However, it is likely that such photons play a minor role in cloud chemistry. An important exception is that the photodissociation of CO is probably the most important chemical source of C^0. Again, abundances are not affected appreciably on a short time-scale, while at steady-state the abundances of complex molecules do not change significantly due to the increased production rates (via the increased C^0 abundance) which offsets any additional destruction due to cosmic ray photons.

While chemical abundances at early times may be insensitive to the uncertainties discussed in the previous section, it is clear that a more realistic *physical* description of dense interstellar clouds needs to be incorporated into the theoretical models. Some recent attempts to do so have been made (Chapter 17) but much more work needs to be done. While the basic chemical networks, such as those outlined in Section 6.2, are fairly solid, the inclusion of more realistic physics in the models may well alter the parameters, such as cosmic ray ionisation rate and elemental depletions, which are derived from comparisons between observation and theory.

REFERENCES

Section 6.1

Hjalmarson, A. and Friberg, P. 1988. In *Formation and Evolution of Low Mass Stars* eds. A. K. Dupree and M. V. T. Lago (Kluwer), in press.

Section 6.2

Black, J. H. and Dalgarno, A. 1973, *Ap. Letters*, **15**, 79.
Black, J. H. and Dalgarno, A. 1977, *Ap. J. Suppl.*, **34**, 405.
Black, J. H., Dalgarno, A. and Oppenheimer, M. 1975, *Ap. J.*, **199**, 633.
Dalgarno, A. and Lepp, S. 1984, *Ap. J.*, **287**, 47.

Elmegreen, B. G. 1979, *Ap. J.*, **232**, 739.
Millar, T. J., Bennett, A. and Herbst, E. 1989, *Ap. J.*, **340**, 906.
Oppenheimer, M. and Dalgarno, A. 1974, *Ap. J.*, **187**, 201.
Oppenheimer, M. and Dalgarno, A. 1974, *Ap. J.*, **192**, 29.
Turner, J. L. and Dalgarno, A. 1977, *Ap. J.*, **213**, 386.
Watson, W. D. 1976, *Rev. Mod. Phys.*, **48**, 513.

Section 6.3

Anicich, V. G. and Huntress, W. T., Jr. 1986, *Ap. J. Suppl.*, **62**, 553.
Boland, W. and de Jong, T. 1984, *Astron. Astrophys.*, **134**, 435.
Brown, R. D. and Rice, E. H. N. 1986, *MNRAS*, **223**, 429.
Charnley, S. B., Dyson, J. E., Hartquist, T. W. and Williams, D. A. 1988. *MNRAS*, **231**, 269.
Hasegawa, T. 1988, *Publ. Astr. Soc. Japan*, **40**, 219.
Herbst, E. and Leung, C. M. 1986, *MNRAS*, **222**, 689.
de Jong, T., Dalgarno, A. and Boland, W. 1980, *Astron. Astrophys.*, **91**, 68.
Millar, T. J., Leung, C. M. and Herbst, E. 1987, *Astron. Astrophys.*, **183**, 109.
Millar, T. J. and Nejad, L. A. M. 1985, *MNRAS*, **217**, 507.
Prasad, S. S. 1987. In *Astrochemistry* eds. M. S. Vardya and S. P. Tarafdar (Kluwer), p. 259.
Tarafdar, S. P., Prasad, S. S., Huntress, W. T. Jr., Villere, K. R. and Black, D. C. 1985, *Ap. J.*, **289**, 220.

Section 6.4

Freeman, A. and Millar, T. J. 1983, *Nature*, **301**, 402.
Herbst, E. 1983, *Ap. J. Suppl.*, **53**, 41.
Herbst, E. and Klemperer, W. K. 1973, *Ap. J.* **185**, 505.
Millar, T. J. and Freeman, A. 1984, *MNRAS*, **207**, 405.
Mitchell, G. F., Ginsburg, J. L. and Kuntz, P. J. 1978, *Ap. J. Suppl.*, **38**, 39.
Swade, D. 1987, Ph.D. Thesis, Univ. Mass. at Amherst.

Section 6.5

Iglesias, E. 1977, *Ap. J.*, **218**, 697.
Prasad, S. S. and Huntress, W. T. Jr. 1980, *Ap. J.*, **239**, 151.
Prasad, S. S. and Huntress, W. T. Jr. 1980, *Ap. J. Suppl.*, **43**, 1.

Section 6.5.1

Clary, D. C. 1988. In *Rate Coefficients in Astrochemistry* eds. T. J. Millar and D. A. Williams (Kluwer), p. 1.
Herbst, E. and Leung, C. M. 1986, *Ap. J.*, **310**, 378.
Millar, T. J., Adams, N. G., Smith, D. and Clary, D. C. 1986, *MNRAS*, **216**, 1025.
Rowe, B. R. 1988. In *Rate Coefficients in Astrochemistry* eds. T. J. Millar and D. A. Williams (Kluwer), p. 135.
Ziurys, L. M. and Turner, B. E. 1986, *Ap. J.* **302**, 131.

Section 6.5.2

Bates, D. R. 1986, *Ap. J.*, **306**, L45.
Green, S. and Herbst, E. 1979, *Ap. J.*, **229**, 121.
Herbst, E. 1978, *Ap. J.*, **222**, 508.
Hollis, J. M., Churchwell, E. B., Herbst, E. and De Lucia, F. C. 1986, *Nature*, **322**, 524.

Millar, T. J., DeFrees, D. J., McLean, A. D. and Herbst, E. 1988, *Astron. Astrophys.*, **194**, 250.
Wootten, A., Boulanger, F., Bogey, M., Combes, F., Encrenaz, P. J., Gerin, M. and Ziurys, L. M. 1986, *Astron. Astrophys.*, **166**, L15.

Section 6.5.3

Gredel, R., Lepp, S. and Dalgarno, A. 1987, *Ap. J.*, **323**, L137.
Prasad, S. S. and Tarafdar, S. P. 1983, *Ap. J.*, **267**, 603.
Sternberg, A., Dalgarno, A. and Lepp, S. 1987, *Ap. J.*, **320**, 676.

Section 6.5.4

Gredel, R., Lepp, S. and Dalgarno, A. 1987, *Ap. J.*, **323**, L137.
Herbst, E. and Leung, C. M. 1986, *Ap. J.*, **310**, 378.
Millar, T. J., DeFrees, D. J., McLean, A. D. and Herbst, E. 1988, *Astron. Astrophys.*, **194**, 250.
Millar, T. J., Leung, C. M. and Herbst, E. 1987, *Astron. Astrophys.*, **183**, 109.
Sternberg, A., Dalgarno, A. and Lepp, S. 1987, *Ap. J.*, **320**, 676.

Section 6.6

Brown, P. D., Charnley, S. B. and Millar, T. J. 1988, *MNRAS*, **231**, 409.
Herbst, E. 1987, *Ap. J.*, **313**, 867.
Herbst, E., DeFrees, D. J. and McLean, A. D. 1987, *Ap. J.*, **321**, 898.
Marquette, J. B., Rebrion, C. and Rowe, B. R. 1988, *J. Chem. Phys.*, **89**, 2041.

7

Interstellar grain chemistry

V. BUCH

Department of Chemistry, University of Illinois, Chicago, USA

7.1. Introduction

The subject of this article is dust in dense interstellar clouds – its composition and its chemical evolution. The relevant astronomical data include extinction and polarization in the infrared, visible and ultraviolet spectral ranges. Identification of the carriers of the observed spectral features is a non-trivial task. The features appear to be notoriously non-unique; significant fractions of them have been assigned to two or more dissimilar materials. Also, it now appears that grains are made largely of highly disordered and/or composite materials. In disordered mixtures spectral features of molecules can be altered considerably with respect to known spectra of pure crystalline materials, which also complicates the identification.

Clearly, additional sources of information are needed. The sources which I chose to employ are the recently available data on the composition of Halley's comet, and on the structure, the composition and the spectral properties of the *interplanetary dust particles* (*IDPs*). I thus assume that cometary and IDP materials preserve many of the characteristics of the dust in the original interstellar cloud. As a rationale I quote a recent review of Geiss (1987) on the results of exploration of Halley's comet: (*a*)'. . . The abundance data show that a large fraction of material in Halley's nucleus condensed at very low temperature' (*b*) '. . . comets are regular members of the solar system which have preserved the original charactistics of the condensed and accreted matter better than other bodies in this system.'

IDPs collected in the stratosphere are apparently the most pristine material available for *direct* examination in terrestrial laboratories (Bradley, Sandford, and Walker 1988). They are believed to originate from comets and asteroids and are unequilibrated aggregates of smaller <0.1–3 μm grains. Their pristine origin is suggested by their high porosity, the presence of delicate microstructures, and the lack of evidence for post-accretional processing. While the grains in IDPs were certainly altered to *some* extent with respect to their interstellar cloud beginnings (e.g. by processes which 'glued' them together) they appear to be the best presently available source of information on the non-volatile component of dust in the presolar cloud.

The assumption that comets and IDPs are closely related to the 'typical' dense cloud dust is far from certain, but combined with astronomical observations, laboratory data, and current theoretical understanding, it yields a reasonably consistent picture of dust in dense clouds. In Sections 7.2 and 7.3 a summary will be presented of the composition of Halley's comet and of the IDPs. Section 7.4 will be devoted to the identification and the chemistry of the non-mineral grain component, including the 'dirty ices' and the organic or the carbonaceous material. In Section 7.5 the inferred composition of the dense cloud dust will be summarized.

I finally apologize for the incomplete coverage of the pertinent literature, which is mostly due to limited space. For other reviews on the subject see e.g. Duley and Williams (1984), Greenberg and Grim (1986), and Tielens and Allamandola (1987a,b).

7.2. Composition of the ejecta of Halley's comet

Recent exploration of Halley's comet yielded a wealth of data including *in situ* measurements in the inner coma by the two Vega missions (closest approach 8000/8900 km) and by the Giotto mission (c.a. 600 km). The data are consistent with the comet being formed by accretion of grains coated by 'dirty ice' mantles (Greenberg and Grim 1986).

The primary material ejected from the comet can be roughly divided into a volatile fraction, which evaporates upon ejection, and a refractory fraction, which is detected as solid dust. The mass ratio of the two components in Halley's comet was estimated to be ~3 (Whipple 1987).

The dominant (≥80%) component of the volatile fraction was H_2O, in accord with Whipple's 'icy conglomerate' model. Other identified parent molecules included NH_3: 1–2% of the water production rate; CH_4: ~2%; CO_2: ~1.5%; H_2CO: ~5% (Allen *et al.* (1987); Moroz *et al.* (1987); identification of H_2CO was tentative). The second most abundant volatile CO (5–20% of water) and the radicals C_2 and CN were proposed to originate mostly from evaporating/decomposing grains, rather than being ejected directly from the nucleus (A 'Hearn *et al.* 1986, Festou *et al.* 1986, Eberhardt *et al.* 1987b, Matthews and Ludicky 1986). Infrared emission in the 3.2–3.5 μm range indicated ~20% abundance of CH_2/CH_3 hydrocarbon groups in molecules or small grains.

The refractory grain component included striking amounts of tiny grains in the mass range from 10^{-14} down to the detection limit of 3×10^{-17} g (Vaisberg *et al.* 1987, McDonnell *et al.* 1987). This mass range matches the masses of *individual* interstellar grains estimated from the interstellar extinction curves (Greenberg and Grim 1986). There was also evidence for continuous decomposition of the larger dust particles ($m \geq 10^{-14}$ g), which yielded the tiny 'interstellar sized' ones (Simpson *et al.* 1987). Preliminary information on the elemental grain composition is available from mass spectrometric measurements on grains of typical mass $\leq 3 \times 10^{-15}$ g (Langevin *et al.* 1987, Clark, Mason, and Kissel 1987):

Approximately $\frac{1}{3}$ of the particles could be interpreted as minerals with a compositional range similar to that encountered in carbonaceous chondrites. Mg rich silicates comprised $\frac{2}{3}$ of these grains. The Fe/Ca fraction varied by more than a factor 100 in different grains. Up to $\frac{1}{4}$ of mineral grains had low Si content and could be interpreted as metal sulfides, carbides and oxides. In particular, 3–4% of material could be interpreted as iron sulfide.

$\frac{1}{3}$ of the particles were dominated by light elements (CHON). This category of particles could be further divided into several subcategories according to the dominant elements present: CHON (48% of the events), H,C (17%), H,C,N (15%), H,C,O (13%), H,O (4%), and C,O (3%). The last three categories were mostly detected in the inner coma, and thus may have corresponded to volatile ices (and/or fragile polymers).

The remaining $\frac{1}{3}$ of grains were mixtures in varying amounts of the above two major components. The largest particles examined were of this nature.

Both Giotto and Vega measurements suggested remarkably low density for the 'interstellar sized' grains $m \leq 10^{-14}$ g. Smirnov, Vaisberg, and Anisimov (1987) suggested a density of 0.1–0.3 g cm^{-3} in the outer coma, increasing towards 1 g cm^{-3} in the inner coma. The PIA experiment aboard Giotto indicated even lower densities of 0.01–0.05 g cm^{-3}; moreover the density appeared to decrease with the mass! The CHON grains seemed to have a particularly low ratio of mass to volume (Schwehm *et al.* 1986, Clark *et al.* 1987). Although the above results were presented as tentative, they do suggest very high porosity of grains even on a very small scale.

7.3. Evidence from IDPs and meteorites

Stratospheric particles of the most thoroughly established extraterrestrial origin are the so-called 'chondritic' IDPs, in which the relative abundances of Mg, Al, S, Ca, Fe, and Ni are similar to those of carbonaceous chondritic meteorites. Some particles dominated by Fe, S, and Ni may also be extraterrestrial. Although the particles are small, ~10 μm, recent advances in microanalytic techniques enabled studies of their mineralogical, petrographic, spectral and isotopic properties in considerable detail (Bradley, Sandford, and Walker (1988) and references therein).

Typical IDPs are aggregates of smaller grains of diameter in the range <0.1–3 μm. Two major classes of IDPs were identified – highly porous IDPs dominated by anhydrous minerals, and compact IDPs dominated by the layer-lattice silicate smectite. The anhydrous IDPs appear to consist of the most primitive and unaltered materials, while hydrated silicates may have undergone aqueous processing in the parent body.

Anhydrous IDPs appear to be composed of four basic building blocks: single (crystalline) mineral grains, 'tar balls', amorphous glass, and a carbonaceous phase. The relative abundances of these species vary widely. The most common single mineral species are olivine ($(Mg,Fe)_2SiO_4$) and pyroxene (e.g. $(Mg,Fe)SiO_3$, $CaMgSi_2O_6$). Other less common minerals include magnetite (Fe_3O_4 or $FeO \cdot Fe_2O_3$), kamacite (iron–nickel alloy), iron nickel carbides, and chromite ($FeCr_2OH_4$).

The 'tar balls' are microaggregates of extremely small (<0.01–0.05 μm) rounded crystals embedded in a carbonaceous matrix. The elemental abundances in an individual 'tar ball' are often close to cosmic. Olivine and kamacite are the dominant minerals, together with minor amounts of iron rich sulfides. The tar balls are highly suggestive of coagulation of small grains, proposed to take place in dense clouds (Draine 1985, Mathis 1986a).

Interestingly, materials similar to those found in IDPs were obtained in experiments designed to simulate dust formation in circumstellar shells (e.g. Nuth and Donn (1983), Reitmeijer, Nuth, and Mackinnon (1986), Hecht *et al.* (1986)). Specifically, vapor condensation in the system $Mg–SiO–H_2$ was shown to produce amorphous silicate materials, which upon annealing at $T \geq 1000$ K were transformed to either olivine or pyroxene, depending upon the temperature and the duration of heating.

Sandford and Walker (1985) compared the ~10 μm feature of different IDPs to that of comet Kohoutek, which, in turn, is similar to the interstellar feature (Ney 1977, Merrill 1979). While infrared spectra of individual IDPs differed considerably from each other and from the astronomical feature, good match was obtained with their mixtures.

Both non-crystalline and crystalline grains in IDPs often appear embedded in, or coated with, carbonaceous non-crystalline material; the coatings can reach a thickness of 0.1 μm. The coatings appear to obscure entirely the silicate features in the Raman spectra. The Raman spectra of the IDPs were shown to be dominated by two broad features at 7.38 and 6.25 μm characteristic of poorly crystallized carbonaceous material (Wopenka 1987, Allamandola, Sandford and Wopenka 1987). The degree of crystallinity appeared to range from microcrystalline material (estimated microcrystallite size ~3 nm) to a completely disordered (*non-aromatic*?) substance. In the carbonaceous material ion probe studies revealed spots with D/H enhanced up to two orders of magnitude with respect to mean protosolar ratio ($\sim 2 \times 10^{-5}$; Geiss and Reeves (1981)).

The carbon Raman feature of some of the IDPs was strikingly similar to a ubiquitous infrared emission feature, attributed to gaseous polycyclic aromatic hydrocarbons (PAH; Léger and Puget 1984, Allamandola *et al.* 1987), or to hydrogenated amorphous carbon grains (HAC; Duley 1985). Also, some samples displayed broad red luminescence similar to the red emission detected in several reflection nebulae and in the general diffuse medium, and also attributed to PAHs or HACs (Duley 1985, Léger, Olofsson and Schmidt 1986).

One may finally note that a non-crystalline carbonaceous polymer, occasionally deuterium enriched, is similarly a major form of carbon in primitive stony meteorites (Anders and Hayatsu 1981). For a discussion of interesting minor components of meteorites of likely extrasolar origin the reader is referred to Anders (1988).

7.4. The grain mantles

7.4.1. Infrared features

A number of infrared features seen towards obscured infrared sources has been attributed to molecular material absorbed on interstellar grains (Allamandola 1984, Tielens and Allamandola 1987a). These features are seen only in absorption through dense and presumably cold molecular clouds. Their width and shape indicate vibrational transitions in a solid rather than gaseous material.

The most prominent feature at 3.08 μm is usually attributed to O—H stretching vibration in water ice. It is sometimes not visible until $A_V = 25$ mag, but it can appear at 4–6 mag (Mathis 1986b). The ratio of water to silicate depths varies from ≤ 0.04 in diffuse ISM to ≥ 1 in some embedded sources (Merrill 1979).

Amorphous ice condensed slowly from water vapor provided a good fit to the observed lineshape (Léger *et al.* 1979, Hagen, Tielens and Greenberg 1983). The additional weaker feature at 6.0 μm, which is correlated with the one at 3.08 μm was assigned (mostly) to bending of water molecules (Tielens and Allamandola 1987a). Based on spectroscopic studies of dirty ice mixtures, water molecules were proposed to comprise at least half of the mantles in the clouds. A possible difficulty in the above interpretation is the non-detection of additional longer wavelength ice absorption due to vibrational modes. It was proposed that in a mixture this absorption is shifted outside a range which is easily accessible to observations.

A weak feature at 4.675 μm, identified as solid CO, was detected in some sources known to possess the 3.08 μm ice band. (Lacy *et al.* (1984), Geballe

Table 7.1. *Composition of interstellar grain mantles* [a] *and of Halley's volatiles*

Species	Mantle infrared band (μm)	W3 IRS 5	AFGL 961	W33A	AFGL 2136	NGC7538 IRS 9	Halley's comet
H_2O	3.08, 6.0	1	1	1	1	1	1
CH_3/CH_2 groups	3.4, 6.85	0.81[b]	0.87[b]	0.55[b]	0.66[b]	0.65[b]	0.2[c]–0.35[d]
CO	4.675	0.05	0.06	0.02	0.04	0.16	0.05–0.2[e]
Carbonyl group	5.8	0.05	0.04		0.04	0.01	0.05[f]
CH_4	7.7		<0.015	<0.005		<0.01	0.02[g]
NH_3	2.97, 6.15	<0.10	<0.10	<0.10	<0.10	<0.10	0.01–0.02[g]
CO_2							0.015[h]

[a] From Tielens and Allamandola (1987a); all abundances normalized to H_2O.
[b] Methanol absorption strength used for the 6.85 μm band.
[c] From 3.2–3.5 μm emission which may belong either to molecules or to small grains (Moroz *et al.* 1987).
[d] Upper bound; all carbon in refractory grains assumed in hydrocarbons; elemental C abundance taken from Geiss (1987).
[e] May originate from chemical decomposition of grains; e.g. organic grain materials with carbonyl groups (Festou *et al.* 1986, Eberhardt *et al.* 1987b).
[f] H_2CO; tentative identification (Moroz *et al.* 1987).
[g] Allen *et al.* (1987). [h] Moroz *et al.* (1987).

(1986), Tielens and Allamandola (1987a), and references therein). The large width of the 4.675 μm peak was suggested as evidence for a mixture with other molecules rather than pure solid CO. In W33A the CO feature was accompanied by a broad feature at 4.62 μm, which was assigned alternatively to a CN group in hydrocarbons (Lacy *et al.* 1984); to a trapped OCN^- ion (Grim and Greenberg 1987); and to SiH bonds (Nuth and Moore 1988).

Two features detected in many protostellar sources – a shoulder of the ice band at 3.4 μm, and a peak at 6.85 μm – were assigned to stretching and bending vibrations of saturated hydrocarbon groups CH_3/CH_2. The relative intensities of the two features matched hydrocarbons with strongly electronegative groups such as CH_3OH (Tielens *et al.* 1984, d'Hendecourt *et al.* 1986, Tielens and Allamandola 1987a).

In some protostellar sources the spectra in the 5–8 μm region are broadened and include substructures at 5.8, 6.7 and 7.1 μm (Tielens and Allamandola 1987a). The 5.8 feature was tentatively assigned to C=O stretch in ketones or aldehydes (e.g. H_2CO). Broadening of the 6.8 peak was explained by the presence of a more complex mixture of hydrocarbons, possibly formed as a result of mantle processing by ultraviolet and cosmic rays.

Mean composition of grain mantles towards various sources derived from IR spectra by Tielens and Allamandola (1987a) is given in Table 7.1. For comparison, I included the data for Halley's comet. Taking into account the uncertainties in abundance estimates, Halley's composition appears quite similar to that of the mantles. The hydrocarbons may be less abundant in the comet.

Alternative non-hydrocarbon assignments of the 6.85 μm feature were also suggested – to hydrated silicates by Hecht *et al.* (1986), and to carbonates by Knacke and Kratschmer (1980). The carbonate band is very strong and appears to require order of magnitude less carrier than the 'hydrocarbon only' assignment (Knacke and Kratschmer 1980, Tielens *et al.* 1984). The 6.85 μm band is, however, a *dense cloud* feature, correlated with the ice absorption but not with the silicate band (Willner *et al.* 1982). The proposed mineral carrier must thus have formed under cold cloud conditions, and the feasibility of such formation has as yet never been demonstrated.

7.4.2. Chemical processes on grains

Surface chemistry in cold clouds proceeds in parallel to gas phase chemistry. Because of the low temperatures, both chemistries proceed via exothermic (energy releasing) reactions only. Also, reactions requiring any appreciable activation energy (i.e. associated with crossing potential barriers between reactants and products) are excluded. Otherwise surface chemistry proceeds via very different routes from gas phase chemistry. When gas particles are absorbed on grains, new chemical routes become available, resulting in significant reprocessing of the absorbed material.

Interstellar gas phase chemistry is believed to be driven mostly by ion–molecule reactions (Duley and Williams 1984, Dalgarno 1987). On the other hand on the grain surfaces the most basic reaction is recombination of radicals (particles with free electrons available for chemical bonding). Recombination becomes efficient on the surface (as opposed to the gas phase) because a third body – the grain – is available to accommodate the excess energy.

The best documented case of interstellar recombination is the formation of H_2 from H atoms in *diffuse* clouds (Jura 1975, Duley and Williams 1984). Specifically, H_2 was observed in small diffuse clouds in which self-shielding from ultraviolet is ineffective. From the estimated photodissociation rate the necessary formation rate of H_2 was calculated. The only known mechanism consistent with the required high rate is surface recombination. Moreover, an atom hitting a grain must recombine with an efficiency of order unity per collision, to account for the observations. Several detailed mechanisms were proposed to justify such high reaction efficiency (e.g. Hollenbach and Salpeter (1971), Tabak (1978), Tielens and Hagen (1982), Duley and Williams (1986), Buch 1989). Although no measurements of recombination rates were published for astrophysically relevant systems, surface recombination of H_2 is a generally accepted for both diffuse and dense clouds.

Additional important recombinations include hydrogenation of O, C and N atoms to form H_2O, CH_4 and NH_3 in reactions of the type

$$XH_n + H \rightarrow XH_{n+1} \qquad X{=}O,C,N \tag{7.1}$$

Note that analogous reactions with molecular hydrogen

$$XH_n + H_2 \rightarrow XH_{n+1} + H \qquad X{=}O,C,N$$

are not included in mantle chemistry because of large activation energies (with the exception of $OH + H_2$; Tielens and Hagen (1982)).

Additional reactions which become feasible on the surface, while being inefficient in the gas phase, include:

(*a*) Activated exothermic reactions with hydrogen: Such reactions proceed via tunneling. The tunneling probability per collision is very low; however, on the surface, in contrast to the gas phase, the smallness of the tunneling probability is compensated by the high collision frequency of the hydrogen atom with neighboring reaction partners. The upper bound on a reaction barrier, which can be tunneled efficiently by an H atom before evaporating, can be estimated as $E_b \sim 8000$ K at $T_{grain} = 10$ K, and $E_b \sim 2000$ K at $T_{grain} = 20$ K (Buch 1989). An important surface reaction involving tunneling is

$$CO + H \rightarrow HCO \tag{7.2}$$

which has a barrier of only 1000 K; and which was proposed to be an efficient source of stored radicals in the grain mantles (Tielens and Hagen 1982, d'Hendecourt, Allamandola, and Greenberg 1985). H_2 formation on grains may also involve tunneling reactions of the type $XH + H \rightarrow X + H_2$ where XH denotes a surface species such as hydrocarbons (Tielens and Hagen 1982, Duley and Williams 1986, Buch 1989).

(*b*) Photoprocessing, condensation and polymerization: Because of the low density in the gas phase the important forms of photoprocessing are unimolecular, i.e. photodissociation and photoionization. On the other hand in the mantle photon absorption can induce reactions involving two or more neighboring molecules. In particular, carbon bearing molecules are capable of condensing and polymerizing to larger organic species. Surface reactions may be thus the major source of the larger organic molecules observed in space.

D'Hendecourt *et al.* (1986) studied ultraviolet induced formation of new molecules in 10 K 'dirty ice' mixtures ($H_2O/CO/CH_4/NH_3$ or $H_2O/CO/O_2/CH_4/NH_3/N_2$). They demonstrated photon induced formation of the 4.62 μm band carrier, and of the 6.8 μm band carrier (which belonged to hydrocarbons other than methane since CH_4 does not absorb at 6.8 μm). CO was shown to be partially converted to H_2CO and CO_2. In addition, this and other studies demonstrated the formation of 'residues' of less volatile or non-volatile molecules/polymers as a result of irradiation of dirty ices by ultraviolet or by fast protons (e.g. Sagan and Khare (1979), Moore and Donn (1982), Strazzulla (1989), d'Hendecourt *et al.* (1986), Matthews and Ludicky (1986).

One may note that ultraviolet photons are available for mantle processing even in dense clouds without internal sources. Cosmic ray and x-ray ionization generates a flux of secondary ultraviolet photons, which photodissociates heavier molecules at a rate roughly quivalent to A_V = 4–6 (Sternberg, Dalgarno, and Lepp 1987).

7.4.3. Formation and evolution of grain mantles

I shall now address the question to what extent one can account for the available data with the current understanding of interstellar surface chemistry. Several studies aiming at computational modeling of mantle composition have been published (Watson and Salpeter 1972, Allen and Robinson 1977, Boland and de Jong 1982, Tielens and Hagen 1982, d'Hendecourt *et al.* 1985). Due to various uncertainties, accurate calculation of mantle composition is not yet within easy reach. Qualitative understanding was, however, advanced considerably.

Gas phase particles colliding with grains are generally believed to stick with an efficiency of order unity. This belief is based mostly on theoretical studies (Burke and Hollenbach 1983, Leitch-Devlin and Williams 1985, Buch 1989) and on the deduced surface formation mechanism of H_2 (which involves sticking as a first step). The estimated grain temperature in cold clouds (7–17 K), is sufficient to evaporate physically bound H, H_2 and He but not heavier particles such as CO (e.g. Léger (1983)). In the absence of an efficient removal mechanism, virtually all oxygen, carbon and nitrogen would be depleted from the gas phase on a timescale $\sim 10^{10}/n$ years, which is short compared to the estimated cloud ages. Since there seems to be no convincing evidence for such depletion, a mechanism must exist to remove efficiently the absorbed volatiles (Léger 1983, Irvine, Goldsmith and Hjalmarson 1987). Mantle removal will be briefly discussed later.

The composition of the mantles is largely determined by the composition of the impinging gas. The major heavy particles arriving from the gas phase are expected to be O, O_2, N, N_2, C, and CO. The CO abundance derived from the observations is (6–8) $\times 10^{-5}$, while the observed abundance of C atoms is 0.01–0.5 that of CO (Dickman 1978, Irvine *et al.* 1987, Blake *et al.* 1987). O, O_2, N, and N_2 are not easily observable in cold clouds, but were predicted to be the dominant forms of oxygen and nitrogen by models of gas phase chemistry (e.g. Leung, Herbst, and Huebner (1984)). Comparable amounts of some unobservable gaseous *polyatomic* appear unlikely, in view of the small abundances of the observable polyatomics. Observations of N_2H^+ suggest that nitrogen is in predominantly molecular rather than atomic form, at least in some molecular clouds (Linke, Guélin and Langer 1983).

The elemental abundance of the heavier particles in the cold cloud gas phase (with respect to hydrogen) is at most solar: $X_O = 6.8 \times 10^{-4}$; $X_C = 3.7 \times 10^{-4}$; $X_N = 1.2 \times 10^{-4}$. Atomic hydrogen appears to be present in excess of the heavy particles at the level of 10^{-2}–10^{-3} (Mahoney, McCutcheon, and Shuter 1976, Wilson and Minn 1977, Krolik and Kallman 1983). The flux of H atoms on grains is further enhanced by their small mass (and therefore high velocity, Duley and Williams (1984)). The H atom is thus a primary candidate for a reaction partner for all the heavier particles. The O,C,N atoms are expected to be converted to H_2O, NH_3 and CH_4 in a hydrogenation sequence (7.1). Such recombination may be a major source of water molecules identified in the mantles. Tielens and Hagen (1982) propose that O_2 as well as O can be converted to water; invoking a reaction sequence which involves tunneling and an H_2O_2 intermediate. In the model of d'Hendecourt *et al.* (1985) conversion of O_2 to water was not included.

CO was predicted to be converted partially to CO_2 and partially to H_2CO by ultraviolet photons (d'Hendecourt *et al.* 1986). All three molecules – CO, CO_2 and H_2CO – were identified in Halley's comet; CO and (tentatively) H_2CO were identified in grain mantles (see Sections 7.2 and 7.4.1). The various carbon bearing molecules are further expected to undergo condensation and polymerization reactions initiated by photons, cosmic rays and radicals impinging from the gas phase (this part of the chemistry is the most difficult to quantify).

The derived ratio $H_2O:NH_3$ in Halley's comet is about one order of magnitude larger than the solar O:N (see Table 7.1). This fact may be traced back to nitrogen in interstellar clouds being locked mostly in the N_2. The very stable triple bonded N_2 molecule is likely to be completely unreactive on the surface. It is also much more volatile than water. Its high volatility was invoked as a reason for the factor 3–7 depletion of elemental nitrogen in Halley's comet, with respect to the solar values (Geiss 1987). It is not clear if N_2 *evaporated* from Halley's comet, or did not condense in the mantles in the first place.

There is, however, some evidence that highly volatile species such as N_2, CO or CH_4 indeed do not condense efficiently in the mantles. The mantle abundances of CO and CH_4 derived from the infrared spectra were at most on a level of 10% and 1% respectively (see Table 7.1; Tielens and Allamandola (1987a)). These molecules can bind to the mantles only via weak van der Waals forces. On the other hand the major observed constituents – H_2O molecules – can bind to each other by much stronger hydrogen bonds. The second highly abundant group of mantle species are hydrocarbons (that is if one accepts the hydrocarbon identification of the 6.8 μm band); moreover, infrared spectra indicate hydrocarbon compounds with strongly polar groups such as OH, C=O and/or CN (Tielens *et al.* 1984, d'Hendecourt *et al.* 1986, Tielens and Allamandola 1987a). Such molecules are also relatively non-volatile and are capable of forming relatively strong hydrogen bonds with H_2O.

Theoretical studies predict under some conditions formation of mantles bound by weak van der Waals forces (Tielens and Hagen 1982, d'Hendecourt *et al.* 1985). For example, at higher $n_h \sim 10^5\ cm^{-3}$ densities and high optical depths, the atomic hydrogen density was calculated to drop well below the density of the heavy particles. Then hydrogenation does not proceed efficiently and CO and O_2 are calculated to be the major mantle components. These models assumed that all heavy particles have equal retention probability in the mantles. As far as I know, no mantles containing more CO than H_2O were ever identified; this may be another indication that van der Waals forces are insufficient to hold the mantles together. (However according to Krolik and Kallman (1983) theoretical models badly underestimate the abundance of atomic hydrogen, in which case the theoretical result may not be valid.)

One may note that very slow gas condensation on cold surfaces was shown experimentally to produce peculiar solid forms which are thermodynamically unstable, amorphous, highly porous, and volatile (Kouchi 1987, Mayer and Pletzer 1986). The physical reason seems to be low surface mobility; apparently the molecules are insufficiently mobile to 'find' the lowest energy configuration on the surface. In the grain mantles such unstable condensates are exposed to various local heating events such as surface reactions and photon absorption. The local heating may be fairly intense because amorphous and porous materials are poor heat conductors (Klinger 1980); and may conceivably result in preferential ejection of the more weakly bound species from the mantles.

Embedded radicals are another source of mantle instability (Greenberg and Yencha 1973). The radicals in mantles can be generated by ultraviolet photons, cosmic rays, and chemical reactions, and may also 'freeze in' from the gas phase. The chemical instability may cause a so-called 'chemical explosion', i.e. explosive mantle removal due to collective recombination of the embedded radicals. Chemical explosions were shown to occur in the laboratory upon irradiation of 10 K dirty ices followed by heating to $T \geq 30$ K (d'Hendecourt *et al.* 1982). The presence of some carbon bearing species in the ice appears to be necessary for the explosion. In space, explosion triggering temperature jumps were proposed to be induced by grain–grain collisions, cosmic rays and x-ray photons (Léger, Jura and Omont 1985, d'Hendecourt *et al.* 1985).

Additional solutions to the mantle removal problem were proposed, including shock waves (Williams and Hartquist 1984); and turbulent convection to outer parts of the cloud, followed by photodesorption (Boland and de Jong 1982). The relative contributions of the various processes are at present unknown. The data support, however, a general notion of periodic recycling of the mantle material (Greenberg and Grim 1986). That is, a period of mantle accumulation on a grain surface was proposed to be followed by some form of 'violent' activity which removes most of the mantle but leaves a small amount of non-volatile organic residue coating the grain. This coating should thus contain remnants of the ancient material accumulated during the grain lifetime.

The above picture is supported by Raman studies of IDPs which revealed the presence of carbonaceous coatings on silicate grains (see Section 7.3). Moreover, the carbonaceous material is the carrier of the largest deuterium enrichment in the IDPs, which strongly supports its ancient 'interstellar cloud' origin.

Chemical deuterium enrichment in cold clouds is a well known phenomenon. The enrichment mechanism for mantles was recently proposed by Tielens (1983). The deuterium enrichment chain is initiated by gas phase ion–molecule chemistry (e.g. Dalgarno and Lepp (1984)). Gas phase ions are generated by cosmic ray ionization, resulting in the formation of H_3^+:

$$H_2 + \text{cosmic ray} \rightarrow H_2^+ + e^-$$
$$H_2^+ + H_2 \rightarrow H_3^+ + H$$

The deuterated H_2D^+ ion is formed by the reaction

$$H_3^+ + HD \rightarrow H_2D^+ + H_2$$

which is exothermic by 227 K due to the differences in zero point energies. At cold cloud temperatures the opposite (endothermic) reaction is inhibited. As a result

$$[H_2D^+]/[H_3^+] \gg [HD]/[H_2] \sim 10^{-5}$$

The D enrichment is further propagated by reactions such as

$$H_2D^+ + CO \rightarrow DCO^+ + H_2$$

The observed ratio DCO^+/HCO^+ in cold clouds is enhanced by factor 5×10^2–10^4 with respect to mean interstellar D/H (Wooten, Loren and Snell 1982). Dissociative recombination

$$DCO^+ + e^- \rightarrow D + CO$$

results in enrichment of *atomic* deuterium with respect to the *atomic* hydrogen, i.e. in $[D]/[H] \gg [HD]/[H_2]$. The deuterium enrichment of atomic hydrogen was proposed by Tielens (1983) as a major reason for enrichment of mantle molecules. That is because hydrogenated molecules (water, ammonia, hydrocarbons) are generated predominantly by reactions such as (7.1) with *atomic* hydrogen/deuterium arriving from the gas phase rather than with the unreactive molecules H_2 and HD.

Enriched hydrocarbons in the IDPs may have then formed in mantles, during hydrogenation of C atoms and of CO to methane and formaldehyde respectively (the latter process may require ultraviolet induced atom transfer from the enriched water; d'Hendecourt *et al.* (1986)). Some of the enriched molecules may have been subsequently incorporated in the non-volatile residues and thus survived in the IDPs.

The actual enrichments calculated by Tielens (1983) are much higher than the ones detected in IDPs, but the quantitative difference may be due to the serious uncertainties of the model. A possibly more meaningful comparison can be made with the Orion Hot Core (Walmsley *et al.* 1987). There the observed high abundance of HDO and ammonia was proposed to be due to a release of all mantle material into the gas phase (Blake *et al.* 1987), The observed ratio $[NH_2D]/[NH_3] \sim 0.003$ is enhanced by two orders of magnitude with respect to its mean interstellar value and is similar to the maximal enhancement detected in D-rich spots in IDPs.

Most of the IDP material is enriched by only one order of magnitude with respect to mean interstellar (or protosolar) value. Similar enhancement was found in several objects in the outer solar system, including terrestrial ocean water, Halley's water, and methane in the atmospheres of Titan and of Uranus (Owen, Lutz, and de Bergh 1986, Eberhardt *et al.* 1987a). Owen *et al.* (1986) proposed that the enriched materials originate from a separate reservoir of deuterium established before the formation of the solar system. Grain mantles are a natural candidate for such a reservoir. The reservoir apparently formed after a rise in temperature, with respect to a period when the more enriched spots in IDPs formed.

The question is where the Halley's CHON particles fit into the picture of the dense cloud dust. Although CHON material seems related to the carbonaceous component of the IDPs, the carbon content of Halley's dust appears to be significantly higher than that of IDPs (8–32% as compared to 2–10%) (Langevin *et al.* 1987, Allamandola *et al.* 1987). Moreover, evidence is available for continuous decomposition and evaporation of at least some of Halley's CHON material propagating away from the nucleus (see Section 7.1).

The decomposing CHON component appears to be a phase intermediate between 'volatile' and 'refractory'; and may contain an intimate mixture of large and small organic molecules and water. The extremely low density of the smallest CHON grains would then correspond to a tenuous polymer, remaining after the evaporation of the volatile species. The IDPs, after spending up to $\sim 10^4$ yr in the inner solar system are expected to retain only the most 'resilient' fraction of the CHON material.

7.5. Summary: composition of dust in dense clouds

Based on the above data, the following classes of materials appear to be likely dust constituents in dense clouds:

(*a*) minerals: predominantly magnesium rich silicates in a variety of forms (glasses, olivines, pyroxenes, possibly hydrated silicates); a minor fraction of metals and of metal sulfides, carbides and oxides;

(*b*) poorly crystallized carbonaceous/organic polymer, coating mineral grains; this material may be present in 'cores' and not only in coatings, judging from the 'pure' CHON grains detected in Halley's comet;

(*c*) dirty ice mantles: mostly water and hydrocarbons with electronegative groups; minor contributions from CO, CO_2, NH_3 and CH_4.

The polymeric material and the volatile ices ((*b*) and (*c*)) do not necessarily appear in separate phases; the evaporating CHON grains are proposed to be an intimate mixture of the two. At the present level of understanding, a plausible scenario can be provided for the formation of the non-mineral grain constituents (*b*) and (*c*). Understanding of the data is, however, very far from complete; in particular, there is perplexing variability in the data which is as yet unaccounted for. Many relevant properties vary strongly, including infrared spectra of protostellar sources, compositions of individual IDPs, and the gas-to-dust ratios in different comets. Grain composition thus appears to be very sensitive to something – what? There is still a lot to learn.

Acknowledgements

I should like to thank Drs L. J. Allamandola, B. Donn, J. S. Mathis, J. A. Nuth, and R. M. Walker for kindly supplying their recent reviews and articles. Many thanks to Dr C. N. Matthews for the materials on Halley's comet. This work was funded through NSF grant AST-8720255.

REFERENCES

A'Hearn *et al.* 1986, *20th ESLAB Symp., ESA SP-250,* Vol. I, 483.

Allamandola, L. 1984, in *Galactic and Extragalactic Infrared Spectroscopy,* eds. M. F. Kessler and J. P. Phillips (Dordrecht: Reidel), p. 5.

Allamandola, L. J., Sandford, S. A. and Wopenka, B. 1987, *Science,* **237**, 56.

Allen, M. and Robinson, G. W. 1977, *Ap. J.,* **212**, 396.

Allen, M. *et al.* 1987, *Astr. Ap.,* **187**, 502.

Anders, E. 1988, to appear in *Meteorites and the Early Solar System,* ed. J. F. Kerridge (Tucson: University of Arizona).

Anders, E. and Hayatsu, R. 1981, *Topics in Current Chemistry,* **99**, 1.

Blake, G. A., Sutton, E. C., Masson, C. R. and Phillips, T. G. 1987, *Ap. J.,* **315**, 621.

Boland, W. and de Jong, T. 1982, *Ap. J.,* **261**, 110.

Bradley, J. P., Sandford, S. A. and Walker, R. M. 1988, to appear in *Meteorites and the Early Solar System,* ed. J. F. Kerridge (Tucson: University of Arizona).

Buch, V. 1989, to appear in *Evolution of Interstellar Dust and Related Topics,* eds. A. Bonneti and S. Aiello.

Burke, J. R. and Hollenbach, D. J. 1983, *Ap. J.,* **265**, 223.

Clark, B. C., Mason, L. W. and Kissel, J. 1987, *Astr. Ap.,* **187**, 779.

Dalgarno, A. 1987, in *Physical Processes in Interstellar Clouds,* eds. G. E. Morfill and M. Scholer (Dordrecht: Reidel).

Dalgarno, A. and Lepp, S. 1984, *Ap. J.* (*Letters*), **287**, L47.

D'Hendecourt, L. B., Allamandola, L. J., Baas, F. and Greenberg, J. M. 1982, *Astr. Ap.*, **109**, L12.
D'Hendecourt, L. B., Allamandola, L. J. and Greenberg, J. M. 1985, *Astr. Ap.*, **152**, 130.
D'Hendecourt, L. B., Allamandola, L. J., Grim, R. J. A. and Greenberg, J. M. 1986, *Astr. Ap.*, **158**, 119.
Dickman, R. L. 1978, *Ap. J. Suppl.*, **37**, 407.
Draine, B. T. 1985, in *Protostars and Planets*, eds. D. C. Black and M. Shapley Matthews (Tucson: University of Arizona), p. 621.
Duley, W. W. 1985, *MNRAS*, **215**, 259.
Duley, W. W. and Williams, D. A. 1984, *Interstellar Chemistry* (London: Academic Press).
Duley, W. W. and Williams, D. A. 1986, *MNRAS*, **223**, 177.
Eberhardt, P. *et al.* 1987a, *Astr. Ap.*, **187**, 435.
Eberhardt, P. *et al.* 1987b, *Astr. Ap.*, **187**, 481.
Festou, M. C. *et al.* 1986, *Nature*, **321**, 361.
Geballe, T. R. 1986, *Astr. Ap.*, **162**, 248.
Geiss, J. 1987, *Astr. Ap.*, **187**, 859.
Geiss, J. and Reeves, H. 1981, *Astr. Ap.*, **93**, 189.
Greenberg, J. M. and Grim, R. 1986, *20th ESLAB Symp., ESA SP-250*, Vol. II, 255.
Greenberg, J. M. and Yencha, A. J. 1973, in IAU Symp. 52, *Interstellar Dust and Related Topics*, eds. J. M. Greenberg and H. C. van de Hulst (Dordrecht: Reidel) p. 309.
Grim, R. and Greenberg, J. M. 1987, *Ap. J.* (*Letters*), **321**, L91.
Hagen, W., Tielens, A. G. G. M. and Greenberg, J. M. 1983, *Astr. Ap.*, **117**, 132.
Hecht, J. H., Russell, R. W., Stephens, J. R. and Grieve, P. R. 1986, *Ap. J.*, **309**, 90.
Hollenbach, D. J. and Salpeter, E. E. 1971, *Ap. J.*, **163**, 155.
Irvine, W. M., Goldsmith, P. F. and Hjalmarson, Å. 1987, in *Interstellar Processes*, eds. D. J. Hollenbach and H. A. Thornson (Dordrecht: Reidel), p. 397.
Jura, M. 1975, *Ap. J.*, **197**, 575.
Klinger, J. 1980, *Science*, **209**, 271.
Knacke, R. F. and Kratschmer, W. 1980, *Astr. Ap.*, **92**, 281.
Kouchi, A. 1987, *Nature*, **330**, 550.
Krolik, J. H. and Kallman, T. R. 1983, *Ap. J.*, **267**, 610.
Lacy, J. H. *et al.* 1984, *Ap. J.*, **276**, 533.
Langevin, Y., Kissel, J., Bertaux, J. L. and Chassefiere, E. 1987, *Astr. Ap.*, **187**, 761.
Léger, A. 1983, *Astr. Ap.*, **123**, 271.
Léger, A., Jura, M. and Omont, A. 1985, *Astr. Ap.*, **144**, 147.
Léger, A., Klein, J., de Cheveigne, S., Guinet, C., Defourneau, D. and Berlin, M. 1979, *Astr. Ap.*, **79**, 256.
Léger, A., Olofsson, G. and Schmidt, W. 1986, *Astr. Ap.*, **170**, 91.
Léger, A. and Puget, J. L. 1984, *Astr. Ap.*, **137**, L5.
Leitch-Devlin, M. A. and Williams, D. A. 1985, *MNRAS*, **213**, 295.
Leung, C. M., Herbst, E. and Huebner, W. F. 1984, *Ap. J. Suppl.*, **56**, 231.
Linke, R. A., Guélin, M. and Langer, W. D. 1983, *Ap. J.* (*Letters*), **271**, L85.
Mahoney, M. J., McCutcheon, W. H. and Shuter, W. L. H. 1976, *Astr. J.* **81**, 508.
Mathis, J. S. 1986a, *Ap. J.*, **308**, 281.
Mathis, J. S. 1986b, *NASA Conf. Pub.*, 2403, 29.
Matthews, C. N. and Ludicky, R. 1986, *20th ESLAB Symp., ESA SP-250*, Vol. II, 273.
Mayer, E. and Pletzer, R. 1986, *Nature*, **319**, 298.

McDonnell *et al.* 1987, *Astr. Ap.*, **187**, 719.
Merrill, K. M. 1979, *Ap. Space. Sci.*, **65**, 199.
Moore, M. H. and Donn, B. 1982, *Ap. J.* (*Letters*), **257**, L47.
Moroz, V. I. *et al.* 1987, *Astr. Ap.*, **187**, 513.
Ney, E. P. 1977, *Science,* **195**, 541.
Nuth, J. A. and Donn, B. 1983, *J. Geophys. Res. Suppl.*, **88**, A847.
Nuth, J. A. and Moore, M. H. 1988, preprint.
Owen, T., Lutz, B. L. and de Bergh, C. 1986, *Nature,* **320**, 244.
Rietmeijer, F. J. M., Nuth, J. A. and Mackinnon, I. D. R. 1986, *Icarus*, **66**, 211.
Sagan, C. and Khare, B. N. 1979, *Nature,* **277**, 102.
Sandford, S. A. and Walker, R. M. 1985, *Ap. J.*, **291**, 838.
Schwehm, G. H., Kissel, J., Sagdeev, R. Z., Grun, E. and Massonne, L. 1986, *20th ESLAB Symp., ESA SP-250,* Abstracts, p. 282.
Simpson, J. A., Rabinowitz, D., Tuzzolino, A. J., Ksanfomality, L. V. and Sagdeev, R. Z. 1987, *Astr. Ap.* **187**, 742.
Smirnov, V. N., Vaisberg, O. L. and Anisimov, S. 1987, *Astr. Ap.*, **187**, 774.
Sternberg, A., Dalgarno, A. and Lepp, S. 1987, preprint.
Strazzulla, G. 1989, to appear in *Evolution of Interstellar Dust and Related Topics,* eds. A. Bonneti and S. Aiello.
Tabak, R. G. 1978, *Ap. Space Sci.*, **53**, 279.
Tielens, A. G. G. M. 1983, *Astr. Ap.*, **119**, 177.
Tielens, A. G. G. M. and Allamandola, L. J. 1987a, in *Physical Processes in Interstellar Clouds,* eds. G. E. Morfill and M. Scholer (Dordrecht: Reidel), p. 333.
Tielens, A. G. G. M. and Allamandola, L. J. 1987b, in *Interstellar Processes,* eds. D. J. Hollenbach and H. A. Thornson (Dordrecht: Reidel), p. 397.
Tielens, A. G. G. M., Allamandola, L. J., Bregman, J., Goebel, J., d'Hendecourt, L. B. and Witteborn, F. C. 1984, *Ap. J.*, **287**, 697.
Tielens, A. G. G. M. and Hagen, W. 1982, *Astr. Ap.*, **114**, 245.
Vaisberg, O. L., Smirnov, V., Omelchenko, A. and Iovlev, M., 1987, *Astr. Ap.*, **187**, 753.
Walmsley, C. M., Hermsen, W., Mauersberger, R. and Wilson, T. L. 1987, *Astr. Ap.*, **172**, 311.
Watson, W. D. and Salpeter, E. E. 1972, *Ap. J.*, **174**, 231.
Wells, C. H. J. 1972, *Introduction to Molecular Photochemistry* (London: Chapman and Hall), pp. 117, 123.
Whipple, F. L. 1987, *Astr. Ap.*, **187**, 852.
Williams, D. A. and Hartquist, T. W. 1984, *MNRAS,* **213**, 157.
Willner, S. P. *et al.* 1982, *Ap. J.*, **253**, 174.
Wilson, L. and Minn, Y. K. 1977, *Astr. Ap.*, **54**, 933.
Wootten, A., Loren, R. B. and Snell, R. L. 1982, *Ap. J.*, **255**, 160.
Wopenka, B. 1987, *Lunar Planet. Sci. Conf. XVIII,* 1102.

8

Large molecules and small grains in astrophysics

STEPHEN H. LEPP

Harvard-Smithsonian Center for Astrophysics, Cambridge, Massachusetts, USA

8.1. Introduction

There is an increasing body of evidence for a population of large molecules in the interstellar medium. Large molecules have been suggested as the source of near infrared continuum radiation and the near infrared emission bands observed in reflection nebula, planetary nebula, HII regions and active galaxies (Duley and Williams 1981, Léger and Puget 1984, Allamandola, Tielens and Barker 1985, d'Hendecourt *et al.* 1986, Barker, Allamandola and Tielens 1987). Large molecules have also been proposed as the carriers of the diffuse interstellar bands (van der Zwet and Allamandola 1985, Léger and d'Hendecourt 1985, Crawford, Tielens and Allamandola 1985). The large molecules proposed have between 30 and 100 atoms with suggested abundances in the range 10^{-7}–10^{-6} relative to hydrogen.

Mathis, Rumpl and Nordsieck (1977; MRN) determined a size distribution for grains that would fit the extinction measured to many sources. However, existing extinction measurements do not extend far enough into the ultraviolet to infer the population of the smallest grains. An extrapolation of the MRN distribution to very small grains having between 30 and 100 atoms gives an abundance roughly the same as that proposed for the large molecules, suggesting that the large molecules may be an extension of the interstellar grain distribution.

It appears that a population of small grains or large molecules exists in much of the interstellar medium. We will use the terms large molecules and small grains interchangeably to refer to clusters of between 30 and 100 atoms. In this chapter we will consider the evidence for the presence of large molecules or small grains and identify potential candidates for these molecules. We will also explore the influence large molecules may have on the ionization balance, thermal balance and chemistry of molecular clouds.

8.2. Evidence and identification

Infrared emission bands at 3.3, 6.2, 7.7, 8.6 and 11.3 μm are observed in a large number of objects such as reflection nebulae, planetary nebulae and some galaxies (Dwek *et al.* 1980). The bands are collectively known as the unidentified infrared bands. Emission in these bands is associated with gas which is illuminated by intense ultraviolet radiation. Dwek *et al.* (1980) concluded that emission from small grains heated by ultraviolet photons is the origin of the unidentified infrared bands. Duley and Williams (1981) identified the bands with spectral features characteristic of surface groups (CH_3, OH, CHO, . . .) on aromatic molecules. They proposed that the bands are formed by thermal emission from small grains of amorphous carbon. The form of amorphous carbon which they discussed consists of graphitic platelets randomly oriented to make small grains.

In addition to emission in the bands there is also an underlying continuum emission. Sellgren (1984) has observed the underlying continuum emission in a number of reflection nebulae and finds that it has a color temperature of approximately 1000 K. The ratio of continuum to band emission and the color temperature of the continuum emission appear to be independent of both the source of ultraviolet radiation and the position observed relative to the exciting star.

The surface brightness of the emission is too strong by factors of 2–20 to be explained by reflected light. The temperature of 1000 K is much higher then the expected equilibrium temperature of dust in the reflection nebulae and the equilibrium temperature of dust would fall off with distance from the exciting star. The emission may be explained by small grains which may be heated to high temperatures temporarily by absorption of a single ultraviolet photon (Sellgren 1984) or by thermally isolated regions on large grains (Duley and Williams 1988).

The equilibrium temperature for grains may be found by calculating the temperature at which the energy absorbed by the grain is equal to the energy emitted. This equilibrium temperature is low, typically less then 100 K and it depends on the intensity of the radiation field. If grains are heated by radiation of a nearby star and cool by radiating as black bodies then the equilibrium temperature varies as $T \propto 1/D^{1/2}$ where D is the distance to the star. However, small grains of less then 100 atoms may be heated to temperatures of greater than 1000 K by the absorption of a single ultraviolet photon. The small grains then cool by emitting infrared radiation and remain cool until excited by another photon. The intensity of the infrared radiation is proportional to the intensity of the exciting ultraviolet radiation but the characteristic temperature depends only on the properties of the small grains.

Draine and Anderson (1985) calculated the nonequilibrium emission from a grain population which included small grains. Comparing these calculations to

IRAS observations Draine and Anderson (1985) found evidence for a small grain population in four diffuse interstellar clouds and one dark cloud. They suggested that the MRN distribution should be extended to grains as small as 20 carbon atoms to explain the infrared emission from four of these clouds. In one of the diffuse clouds the IRAS observations suggested that small grains were even more abundant than suggested by the MRN distribution.

Léger and Puget (1984) have proposed that the small grains are polycyclic aromatic hydrocarbon (PAH) molecules containing about 50 carbon atoms. The PAH molecules they consider are a graphitic plane with hydrogen atoms bound to the periphery. Many of the unidentified infrared bands are characteristic of PAH molecules. Figure 8.1 shows a comparison of the infrared observations with the

Figure 8.1. Observed infrared emission compared to the calculated spectrum for coronene (from Léger and Puget (1984)).

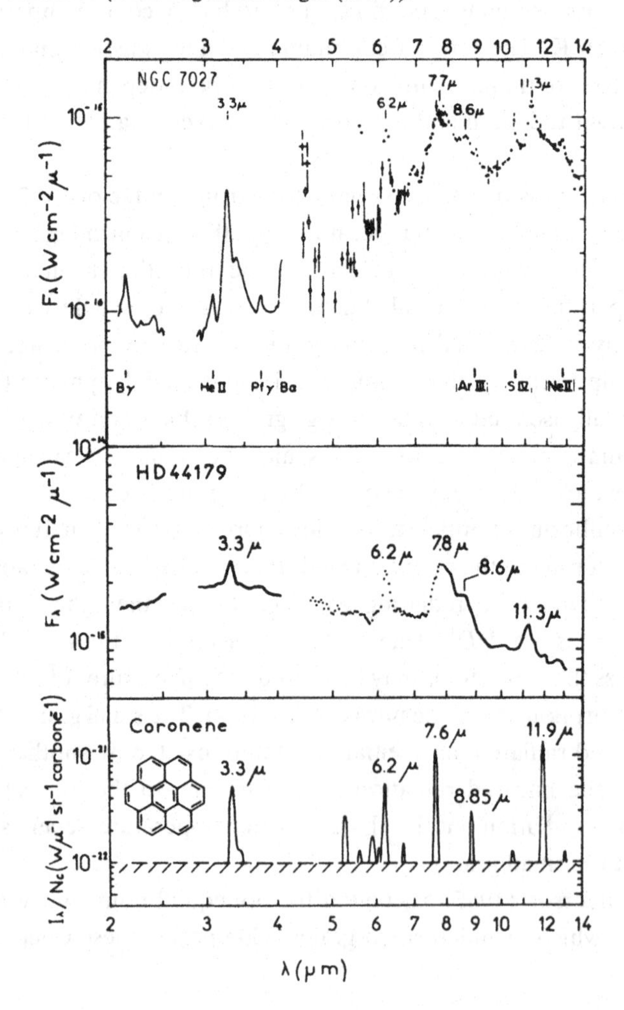

expected spectra of coronene, a PAH molecule. The 3.3 μm corresponds to C—H stretch, the 6.2 μm to C—C stretch in an aromatic ring and the 8.85 and 11.9 μm to C—H bending modes in and out of the plane respectively. PAH molecules are also robust to high temperatures and may be heated to temperatures of over 1000 K without subliming. The PAH molecule may convert ultraviolet radiation to infrared radiation by absorbing an ultraviolet photon to produce an electronically excited molecule followed by internal conversion to vibrationally excited states in a lower electronic state. Allamandola *et al.* (1985) suggested that this process may convert half of the energy of the ultraviolet photon into infrared radiation. The PAH molecules may also be attractive candidates for the absorbers producing diffuse interstellar bands in the visible (Léger and d'Hendecourt 1985).

8.3. Chemistry of large molecules

Small grains or large molecules such as PAHs appear to be ubiquitous in the interstellar medium. The abundances needed to explain the infrared observations are of the order of 10^{-7} large molecules per hydrogen atom, which requires that about 2% of the carbon is in large molecules. If large molecules are present in interstellar clouds at these abundances they play an important role in the chemistry.

A typical large molecule (LM) contains approximately 50 carbon atoms and may be in one of three ionization states: neutral (LM), positive ion (LM^+) or negative ion (LM^-). The relative abundance of these ionization states depends on interactions with other gas phase ions and neutrals and on the radiation field. A neutral LM may be photoionized to form a positive ion

$$LM + h\nu \rightarrow LM^+ + e \tag{8.1}$$

Photodetachment from a negative ion produces a neutral

$$LM^- + h\nu \rightarrow LM + e \tag{8.2}$$

The cross sections for these processes depend on the composition of the LMs. The ionization potential is about 6.5 eV for a 50 atom PAH and its ionization cross section lies between 10^{-16} and 5×10^{-16} cm^2 (Duley 1986, d'Hendecourt and Léger 1987). The electron affinities for PAHs are about 1 eV and the photodetachment cross sections are probably less than 10^{-16} cm^2 (Omont 1986).

The rates at which photoionization and photodetachment occur are calculated by integrating the cross sections over the photon flux. If we adopt a fit by Draine (1978) to the average interstellar ultraviolet radiation field and assume a diffuse blackbody spectrum with a temperature of 10^4 K for the visible we can use the above cross sections to derive rates of photoionization and photodetachment. We will scale the photon flux by F, a wavelength-independent multiplicative factor, and adopt 10^{-16} cm^2 for both cross sections. This gives a LM ionization rate of $1.6 \times 10^{-8} F$ s^{-1} and a LM^- photodetachment rate of $4.8 \times 10^{-8} F$ s^{-1}.

A positive ion may recombine to form a neutral

$$LM^+ + e \rightarrow LM \tag{8.3}$$

and a neutral may undergo electron attachment to form a negative ion

$$LM + e \rightarrow LM^- \tag{8.4}$$

For a 50 atom PAH the rate coefficients for recombination and electron attachment are 5.7×10^{-6} and 7.7×10^{-7} cm^3 s^{-1} (Omont 1986). Draine and Sutin (1987) found similar rate coefficients for small dielectric spheres, suggesting that the rates are characteristic of small particles and are not specific to PAHs.

The abundances of LM, LM^+ and LM^- may be determined in diffuse clouds by equating the photoionization rate with the recombination rate and the photodetachment rate with the electron attachment rate. The abundance of the ionization states depends only on the total LM abundance [LM], the electron density n_e and the ionization rate. As long as the density of large molecules is small compared to the electron density, the relative equilibrium abundances among the ion states of LMs are independent of [LM] and depend only on the ratio F/n_e. Figure 8.2 shows the equilibrium relative abundances of LM, LM^+ and LM^- as a function of F/n_e. For high n_e or low F, all the large molecules become negative ions. In diffuse molecular clouds, n_e is usually between 10^{-1} and 10^{-2} and a fraction between 0.1 and 0.5 of the large molecules are LM^- and less than 0.1 are LM^+.

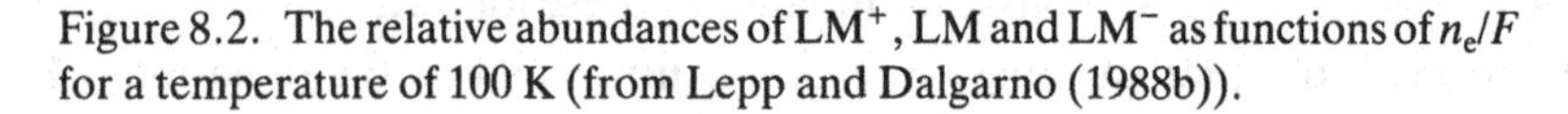

Figure 8.2. The relative abundances of LM^+, LM and LM^- as functions of n_e/F for a temperature of 100 K (from Lepp and Dalgarno (1988b)).

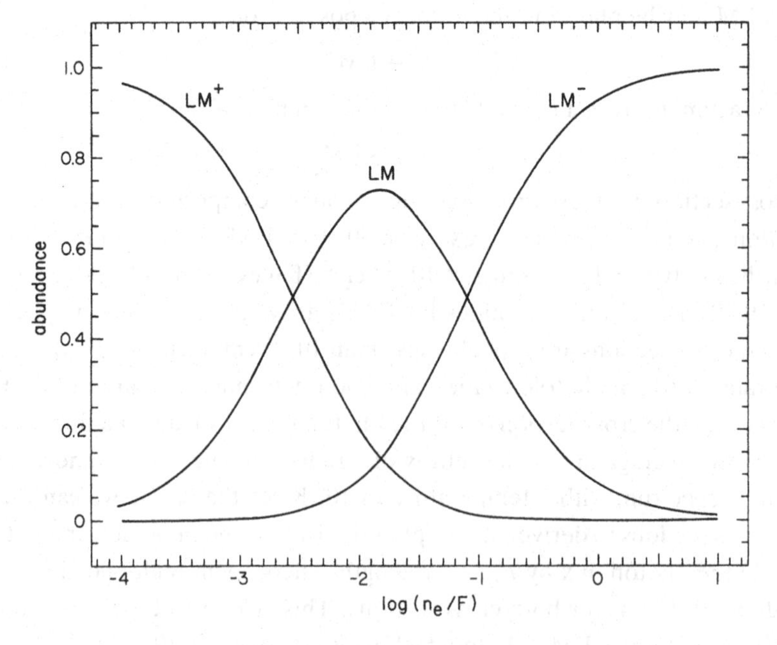

The chemistry of diffuse clouds is modified substantially by the presence of the negative ions LM^- and neutral LM (Lepp *et al.* 1988). The negative ions undergo mutual neutralization reactions

$$X^+ + LM^- \rightarrow X + LM \tag{8.5}$$

where X^+ represents any positive ion. Mutual neutralization occurs very rapidly because of the Coulomb attraction. For large molecules the rate coefficient is of order $1 \times 10^{-6} T^{-1/2}$ cm^3 s^{-1} (Omont 1986). The positive ions also charge exchange with the neutral LM

$$X^+ + LM \rightarrow X + LM^+ \tag{8.6}$$

which has a rate coefficient of order 4×10^{-9} cm^3 s^{-1} (Omont 1986). In diffuse clouds mutual neutralization with LM^- and charge exchange with LM reduce the ratio of positive ions to neutrals. The ratios between atoms and their ions are particularly sensitive to the presence of large molecules. Atomic ions recombine through a radiative process which is very slow so the effective recombination through Reactions (8.5) and (8.6) has a large effect on the ratio of atomic ions to neutral atoms. Fractional abundances of the order of 10^{-7} large molecules per hydrogen atom bring the ratio of atoms and atomic ions in models of diffuse clouds into best agreement with observations (Lepp *et al.* 1988).

In dark molecular clouds the photoprocesses are not important and the large molecules will attach an electron to form LM^- as in Reaction (8.3), and the LM^- will undergo mutual neutralization with positive ions to return to LM. The charge exchange Reaction (8.6) also occurs followed by recombination. If the total abundance of large molecules is greater then 10^{-8} relative to hydrogen then all the free electrons attach to LM to form LM^-. As in diffuse clouds the large molecule population decreases the ratio of atomic ions to atoms, but in addition the molecular ion abundance increases because all the electrons are attached to large molecules and fast dissociative recombinations do not occur. The chemistry is changed qualitatively as dissociative recombination is replaced as the dominant neutralization mechanism by nondissociative charge exchange reactions with LM and LM^- (Lepp and Dalgarno 1988a).

Table 8.1 shows the calculated abundances for a number of species in an equilibrium dark cloud model both with and without large molecules (Lepp and Dalgarno 1988a). The charge exchange reactions of C^+ with LM and LM^- substantially increase the neutral carbon abundance though the C^+ abundance changes little as it is primarily removed by reactions with O_2. Molecular ions such as H_3O^+ increase in abundance as the electron abundance is reduced and dissociative recombination becomes less important. Some molecular ions such as H_3^+ which react quickly with many molecules are unchanged by the addition of large molecules. In a reflection of the enhanced neutral carbon abundance, the

Table 8.1. *Abundances of various species relative to molecular hydrogen given for several abundance ratios of large molecules to total hydrogen.* $n_H = 2 \times 10^4$ cm^{-3} *and* $T = 25$ *K were assumed.*

n_{LM}/n_H	0	10^{-7}	10^{-6}
LM^-		6.8 (−8)	1.1 (−7)
LM		1.3 (−7)	1.8 (−6)
LM^+		1.2 (−9)	4.6 (−8)
e	4.4 (−8)	3.8 (−9)	6.4 (−10)
C^+	2.3 (−10)	2.3 (−10)	2.2 (−10)
C	6.2 (−10)	2.2 (−7)	6.7 (−7)
H_3^+	2.2 (−9)	2.2 (−9)	2.2 (−9)
HCO^+	5.7 (−9)	8.3 (−9)	7.6 (−9)
H_3O^+	6.1 (−9)	5.8 (−8)	5.2 (−8)
C_2H	1.0 (−10)	2.0 (−9)	5.9 (−9)
C_2H_2	1.8 (−10)	6.4 (−9)	2.6 (−8)
C_3H	8.6 (−12)	2.1 (−10)	1.3 (−9)
C_3H_2	4.5 (−12)	6.0 (−10)	1.1 (−8)

abundances of the hydrocarbon molecules increase. Also the hydrocarbons are more hydrogenated because of the diminishing importance of dissociative recombination during which a hydrogen atom is often lost.

In addition to charge exchange reactions, large molecules may also react chemically. Duley and Williams (1986) suggested that PAH molecules are destroyed rapidly by a sequence of chemical reactions in which a PAH molecule attaches an oxygen atom; when the PAH molecule is heated by a photon the oxygen is ejected along with a carbon atom which breaks the ring structure. The rate at which oxygen atoms are attached depends on the number of unoccupied radical sites and it is unclear whether the oxygen atom breaks the ring when it is ejected (Omont 1986). Further, large molecules may collide and stick together to form larger molecules. This coagulation of large molecules may continue until grains are formed or it may be balanced by the break up of clusters of large molecules by collisions or photons. Studies of the clustering of and chemical reactions of large molecules are needed to further understand the role of large molecules in the interstellar medium.

8.4. Heating by large molecules

A long standing problem in astrophysics is the source of heating in diffuse interstellar clouds (Dalgarno and McCray 1972). The major source of heating in these clouds is probably the photoelectric effect on grains (Dalgarno and McCray 1972, Draine 1978) but its adequacy is open to question. If large molecules are present their photoionization, Reaction (8.1), releases an energetic electron which heats the gas. The energy carried by the electron and available to heat the gas is approximately the energy of the photon minus the ionization potential. D'Hendecourt and Léger (1987) examined this process and concluded that photoionization of LM may be the dominant heat source in some regions. Heating also occurs through the photodetachment process, Reaction (8.2). Lepp and Dalgarno (1988b) found that this process may account for half of the heating by photoelectric effect on large molecules in diffuse clouds.

The heating is directly proportional to the large molecule abundance and is a function of the electron density divided by the photon flux, n_e/F. Figure 8.3 shows the heating by photoelectric effect on large molecules versus temperature for an abundance of large molecules of 2×10^{-7} relative to hydrogen and for various values of n_e/F. For comparison, the photoelectric heating rate for a standard grain model is also shown (Draine 1978). The heating due to large molecules is everywhere larger than or comparable to the standard grain heating. Lepp and Dalgarno (1988b) found that heating by photoelectric effect on large molecules is sufficient to balance the cooling in cores of diffuse clouds.

Heating by large molecule heating also extends to higher temperatures than standard grain heating (Figure 8.3). At some temperatures the average energy of the recombining electrons is equal to the average energy of the photoemitted

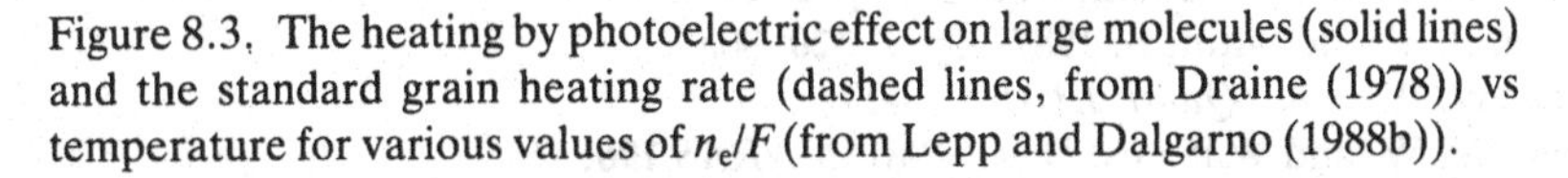
Figure 8.3. The heating by photoelectric effect on large molecules (solid lines) and the standard grain heating rate (dashed lines, from Draine (1978)) vs temperature for various values of n_e/F (from Lepp and Dalgarno (1988b)).

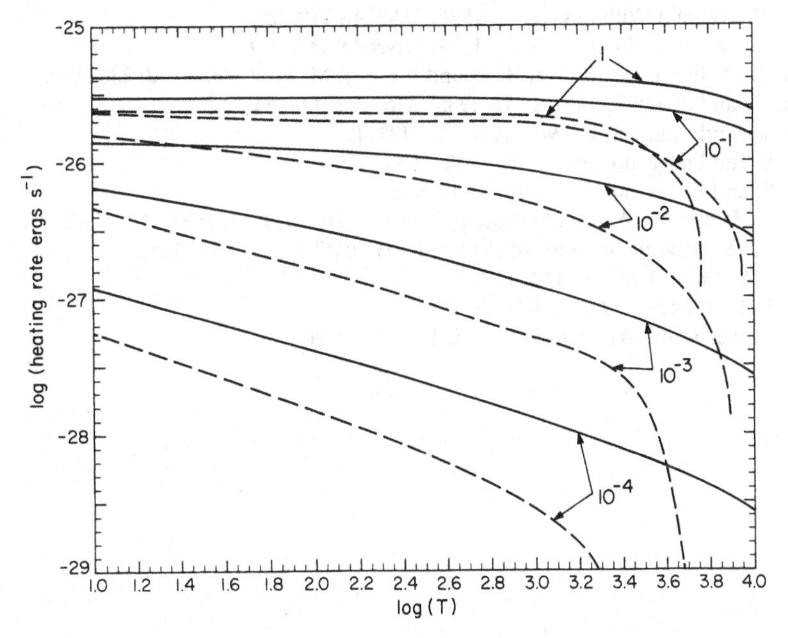

electrons and the photoelectric effect cannot heat the gas above this temperature. For large grains this maximum temperature is approximately 8000 K whereas for large molecules it is approximately 20 000 K. Thus heating by large molecules may be important in heating the intercloud medium and HII regions, if they exist in sufficient abundances in those regions (d'Hendecourt and Léger 1987, Lepp and Dalgarno 1988b).

Acknowledgements

I have benefitted from having A. Dalgarno as a colleague and friend and thank him for his help and encouragement. I would like to thank Drs T. Hartquist and K. Kirby for comments. This work has been supported by the National Science foundation, Division of Astronomical Sciences, under Grant AST-86-17675.

REFERENCES

Allamandola, L. J., Tielens, A. G. G. M. and Barker, J. R. 1985, *Ap. J. Lett.* **290**, L25.
Barker, J. R., Allamandola, L. J. and Tielens, A. G. G. M. 1987, *Ap. J. Lett.* **315**, L61.
Crawford, M. K., Tielens, A. G. G. M. and Allamandola, L. J. 1985, *Ap. J. Lett.* **293**, L45.
d'Hendecourt, L. B. and Léger, A. 1987, *Astr. Ap.* **180**, L9.
d'Hendecourt, L. B., Léger, A., Olofsson, G. and Schmidt, W. 1986, *Astr. Ap.* **170**, 91.
Dalgarno, A. and McCray, R. 1972, *Ann. Rev. Astr. Ap.* **10**, 375.
Draine, B. T. 1978, *Ap. J. Supp.* **36**, 595.
Draine, B. T. and Anderson, N. 1985, *Ap. J.* **292**, 494.
Draine, B. T. and Sutin, B. 1987, *Ap. J.* **320**, 803.
Duley, W. W. 1986, *Quart. JRAS* **27**, 403.
Duley, W. W. and Williams, D. A. 1981, *MNRAS* **196**, 269.
Duley, W. W. and Williams, D. A. 1986, *MNRAS* **219**, 859.
Duley, W. W. and Williams, D. A. 1988, *MNRAS* **231**, 969.
Dwek, E., Sellgren, K., Soifer, B. T. and Werner, M. W. 1980, *Ap. J.* **238**, 140.
Léger, A. and d'Hendecourt, L. B. 1985, *Astr. Ap.* **146**, 81.
Léger, A. and Puget, J. L. 1984, *Astr. Ap.* **137**, L5.
Lepp, S. and Dalgarno 1988a, *Ap. J.* **324**, 553.
Lepp, S. and Dalgarno 1988b, *Ap. J.*, in press.
Lepp, S., Dalgarno, A., van Dishoeck, E. F. and Black, J. H. 1988, *Ap. J.* **329**, 418.
Mathis, J. S., Rumpl, W. and Nordsieck, K. H. 1977, *Ap. J.* **217**, 425.
Omont, A. 1986, *Astr. Ap.* **164**, 159.
Sellgren, K. 1984, *Ap. J.* **277**, 623.
van der Zwet J. and Allamandola, L. J. 1985, *Astr. Ap.* **146**, 77.

IV

Studies of molecular processes

9

Molecular photoabsorption processes

KATE P. KIRBY

Harvard-Smithsonian Center for Astrophysics, Cambridge, Massachusetts, USA

9.1. Introduction

Molecules have been found to exist in a rich variety of astrophysical environments, including stellar atmospheres, comets, planetary atmospheres, and the dense and diffuse clouds in the interstellar medium. An understanding of molecular structure, spectroscopy, and photoabsorption processes is thus of critical importance in interpreting many of the current observations, in theoretically modeling these various astrophysical regions, and in judging the reliability of the available molecular data. Even in our own atmosphere, considerations of molecular photoabsorption determine 'windows' in the electromagnetic spectrum in which one can carry out ground-based observations. Accurate molecular spectroscopic data derived from laboratory experiments or theoretical calculations are necessary in order both to identify molecular species and to quantify abundances of these species from the observed absorption line wavelengths and intensities, respectively.

The vibrational and rotational nuclear degrees of freedom add significantly to the spectroscopic complexity of a molecule compared with an atom. As in an atom, transitions between different electronic energy levels in a molecule are generally observed at optical (4000–6000 Å) and ultraviolet (800–4000 Å) wavelengths, whereas transitions between vibrational energy levels and rotational energy levels generally take place in the infrared region (2–20 μm) and microwave region ($\lambda > 0.2$ mm) of the electromagnetic spectrum, respectively. This range of energies over which molecules are responsive to radiation makes them valuable as sensitive probes of the physical conditions of the astrophysical environments in which they are found.

Molecular photoabsorption may also lead to destruction of the molecule, through dissociation or ionization. Cross sections and/or rate constants for such processes are essential in interpreting or predicting observed abundances of various molecular species.

It is difficult in one chapter to do justice to the general subject of molecules and radiation, about which entire books have been written. Throughout this chapter most of the discussion will focus on diatomic molecules for several reasons. The first is the relative simplicity of treating molecular structure, symmetry, and selection rules pertaining to one rather than many internuclear coordinates. Secondly, only diatomics have been detected in the interstellar gas by optical absorption measurements (van Dishoeck and Black 1988). Finally, most diagnostic studies of diffuse molecular clouds (see Chapter 3) have focused on diatomic molecules. Explanations of their abundances present a continuing challenge to the astrochemist. Section 9.2 deals briefly with molecular structure as derived from the equations of quantum mechanics. Section 9.3 gives a cursory treatment of bound state molecular spectroscopy. Photodissociation and photoionization are treated in Sections 9.4 and 9.5. In the course of each section useful references which give more details are listed and the reader is encouraged to peruse these sources for further clarification.

9.2. Molecular structure

In order to understand what happens when molecules absorb photons, one must know some basic facts about molecular structure. Molecular structure and spectroscopy are intimately connected and the equations and ideas developed in this section will be used extensively in Section 9.3.

9.2.1. The Born–Oppenheimer approximation

The energies, E, and wave functions, Ψ, of a molecular system are solutions to the time-independent Schrödinger equation, $H\Psi = E\Psi$. The non-relativistic molecular Hamiltonian describing a system of electrons and nuclei can be written:

$$H = T_e + T_N + V_{NN} + V_{Ne} + V_{ee} \tag{9.1}$$

in which T_e and T_N are the electronic and nuclear kinetic energy operators, of the form $-(\hbar^2/2m_n)\,\Sigma_n \nabla_n^2$, where the index n refers either to electrons or nuclei; the potential terms refer to the nuclear–nuclear repulsion, the nuclear–electronic attraction, and electron–electron repulsion and are given as

$$V_{NN} + V_{Ne} + V_{ee} = \sum_{A<B} \frac{Z_A Z_B e^2}{|\mathbf{R}_A - \mathbf{R}_B|} - \sum_{A,i} \frac{Z_A e^2}{|\mathbf{R}_A - \mathbf{r}_i|} + \sum_{i<j} \frac{e^2}{|\mathbf{r}_i - \mathbf{r}_j|} \tag{9.2}$$

In this treatment, **R** and **r** will refer to coordinates of nuclei and electrons, respectively; Z_A is the nuclear charge, and the summations are over nuclei, A, or electrons, i.

The standard approach to solving the molecular Schrödinger equation is to assume that the wave function is separable into a product of a nuclear and an electronic wave function:

$$\Psi(\mathbf{r},\mathbf{R}) = \psi_{\mathrm{el}}(\mathbf{r}|R)\zeta_{\mathrm{nuc}}(\mathbf{R}), \tag{9.3}$$

in which $\psi_{\mathrm{el}}(\mathbf{r}|R)$ depends only on the electron coordinates $\mathbf{r}$ and parametrically on the internuclear coordinates R, and $\zeta_{\mathrm{nuc}}(\mathbf{R})$ describes the nuclear motion of vibration and rotation. The physical basis for this assumption is the fact that the nuclei are thousands of times more massive than the electrons and therefore the nuclei are effectively stationary, setting up a potential in which the electrons move. Substituting the product wave function into the Schrödinger equation for the molecule results in two equations,

$$H_{\mathrm{el}}\psi_{\mathrm{el}} = (T_{\mathrm{e}} + V_{\mathrm{Ne}} + V_{\mathrm{ee}})\psi_{\mathrm{el}} = E_{\mathrm{el}}\psi_{\mathrm{el}} \tag{9.4}$$

and

$$H_{\mathrm{nuc}}\zeta_{\mathrm{nuc}} = (T_{\mathrm{N}} + V_{\mathrm{NN}} + E_{\mathrm{el}})\zeta_{\mathrm{nuc}} = E_{\mathrm{total}}\zeta_{\mathrm{nuc}} \tag{9.5}$$

in which Equation (9.4) is solved for $\psi_{\mathrm{el}}(\mathbf{r}|R)$ and $E_{\mathrm{el}}(R)$ for fixed values of the nuclear coordinates. In Equation (9.5) the electronic energy, as a function of internuclear coordinates R, appears as part of the potential in which the nuclei move. The terms which have been neglected in making this separation of nuclear and electronic coordinates are generally of order 10^{-5}–10^{-7} times the electronic energies $E_{\mathrm{el}}(R)$.

9.2.2. Molecular electronic states

A number of theoretical *ab initio* techniques of varying degrees of approximation exist for solving the fixed-nuclei electronic energy eigenvalue equation (9.4). The Hartree–Fock approximation is one of the simplest standard approximations in which the wave function is written as an antisymmetrized product (a 'Slater determinant') of one-electron functions called molecular orbitals, $\varphi_i(\mathbf{r})$

$$\Phi(\mathbf{r}|R) = A|\varphi_1(\mathbf{r}_1)\ldots\varphi_n(\mathbf{r}_n)| \tag{9.6}$$

The molecular orbitals are usually expanded in a basis set of either Slater-type atomic orbitals or Gaussian-type functions. Substitution of Φ for ψ_{el} in Equation (9.4) and application of the variational principle results in a matrix eigenvalue equation for the electronic energy and expansion coefficients for the molecular orbitals in their basis set (see Roothaan (1951), Schaefer (1972)).

A more accurate approximation known as 'configuration interaction' (CI) can be made within the framework of the variational method by considering linear combinations of Slater determinants Φ:

$$\psi_{\mathrm{el}}(\mathbf{r}|R) = \sum_I C_I\Phi_I \tag{9.7}$$

Substitution of (9.7) into Equation (9.4) and application of the variational principle leads to the matrix equation

$$\mathbf{H}^{el}\mathbf{C} = E^{el}\mathbf{C} \tag{9.8}$$

in which the matrix elements of H are $\langle\Phi_J|H|\Phi_I\rangle$ and the vector $\mathbf{C}$ contains the coefficients C_I in (9.7). For an excellent review of this method see Shavitt (1977). Other approximations which go beyond Hartree–Fock make use of perturbation theory techniques (see e.g., Kutzelnigg (1977)).

In order to describe the variation of the electronic energy and wave function with internuclear coordinates the electronic Schrödinger equation, (9.4), is solved

Figure 9.1. Potential curves (energy versus internuclear separation) for molecule AB in three electronic states: U_1, U_2, and $U_3 \cdot U_1$ and U_3 are bound molecular states with vibrational levels labelled v'' and v', respectively. The dissociation energy of U_1 is indicated as D_e, and the equilibrium internuclear distance is R_e. U_2 is a repulsive (dissociating) molecular state. Electronic transitions between vibrational levels in U_1 and vibrational levels in U_3 are indicated by vertical arrows.

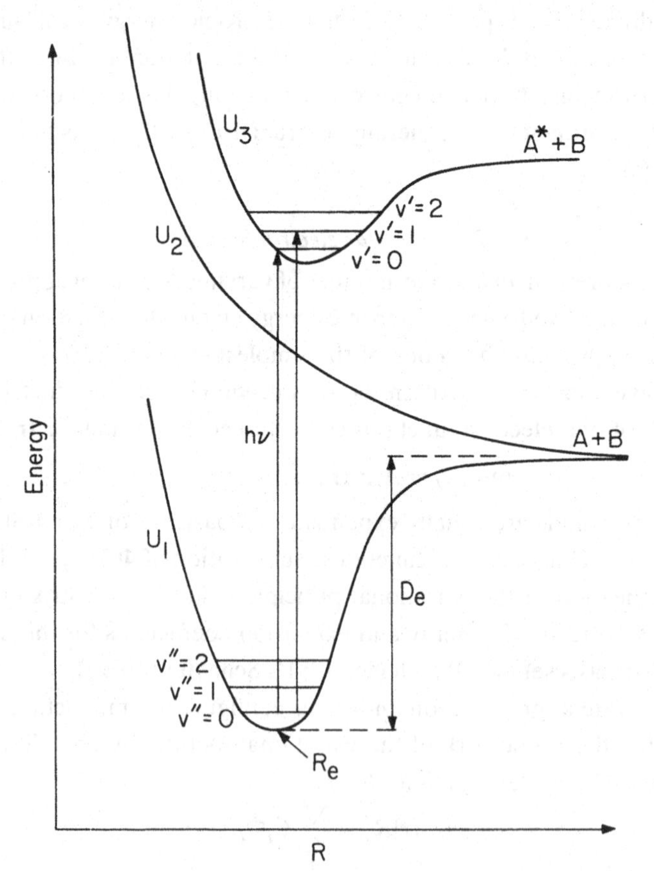

repeatedly at different values of **R**. Wave functions may include many hundreds of thousands of configurations, depending on the degree of accuracy desired in the energy and wave function. A great deal of computer time may be necessary, even on modern supercomputers, to obtain E_{el} and ψ_{el} at each **R**.

The potential energy in which the nuclei move, $U(R) = E_{el}(R) + V_{NN}$, can be plotted as a function of internuclear coordinates **R**. For a polyatomic system this gives rise to a multi-dimensional *potential energy surface* which is normally displayed in three dimensions by fixing all but two nuclear coordinate variables. For diatomic molecules the graph of $U(R)$ versus internuclear distance, R, is called a *potential energy curve*, illustrated in Figure 9.1, for a fictitious molecule, AB. The potential curve (surface) is an extremely useful concept in molecular physics, governing all molecular collision processes as well as molecular spectroscopy.

A stable, or bound, molecular electronic state (U_1 in Figure 9.1) exhibits a minimum with respect to the energy of its separated atoms, A + B, (at $R = \infty$). The minimum energy, $U(R_e)$, occurs at the equilibrium internuclear separation, R_e. The binding energy, D_e, is the well-depth, or the difference between $U(R_e)$ and the energy at the dissociation (separated atom) limit.

An unbound molecular state in which the nuclei do not have a stationary, bound configuration has no potential minimum and is often called a repulsive state ($U_2(R)$ in Figure 9.1). An excited molecular bound state is also shown in Figure 9.1 as $U_3(R)$, in which the dissociation limit involves an excited atomic state of one of the atoms, A*.

9.2.3. Symmetry and quantum numbers

Symmetry considerations are very useful and important in molecular structure and spectroscopy. The electronic wave functions and Hamiltonian must have the same spatial symmetry as the nuclear framework of the molecule. Many qualitative properties of an electronic state, such as degeneracies and selection rules for making transitions, may be deduced without complicated calculations. The symmetry groups of molecules are classified according to the behavior of the wave functions under the operations of rotation, reflection and inversion. Entire books have been dedicated to the explication of 'group theory' and the reader is referred to Cotton (1971) and Tinkham (1964). For simplicity, the following discussion pertains only to diatomic molecules (see Herzberg (1950)).

In an atom the electrons move in a spherically symmetric force field. In a diatomic molecule, the field is axially symmetric, and only the *component* of the total electronic orbital angular momentum **L** along the internuclear axis is a constant of the motion. This component is denoted Λ and diatomic molecular states are designated Σ, Π, Δ, Φ, . . . according to the value of $\Lambda = 0, 1, 2,$

3, ... For $\Lambda > 0$, the molecular states are doubly degenerate as the projected orbital angular momentum vector can have values $+\Lambda$ and $-\Lambda$. As in atoms, the electron spins couple together to form a total spin **S** which is integral or half integral depending on whether the number of electrons in the molecule is even or odd. When $\Lambda \neq 0$, the orbital motion of the electrons creates an internal magnetic field along the internuclear axis which causes **S** to precess about the field with component M_s or Σ (not to be confused with $\Lambda = 0$). There are $2S + 1$ values of Σ allowed, $\Sigma = S, S-1, \ldots, -S$, and $2S+1$ is called the multiplicity of the state.

The total electronic angular momentum about the internuclear axis is denoted Ω, where $\Omega = |\Lambda + \Sigma|$. The full designation of a diatomic molecular electronic state is ${}^{(2S+1)}\Lambda_\Omega$ but often the Ω quantum number is dropped. The non-degenerate Σ ($\Lambda = 0$) states are labelled Σ^+ or Σ^- according to whether their electronic wave functions remain unchanged or change sign on reflection at any plane passing through the nuclei. In the case of a homonuclear diatomic molecule, A_2, in which both nuclei have the same charge, the field in which the electrons move has a center of symmetry at the midpoint of the internuclear axis. The molecular states are labelled with subscript g (even) and u (odd) according to whether the electronic wave functions remain unchanged or change sign upon reflection through this center.

The ground state of a molecule has the label X before the quantum number designation, and excited electronic states which can be reached via dipole transitions from the ground state (see Section 9.3) are labelled with capital letters A, B, C, etc. Small letters, $a, b, c, \ldots$ are used to label states which are generally of different spin multiplicity than the ground state.

9.2.4. Nuclear vibration and rotation

The Born–Oppenheimer approximation assumes the separability of the total wave function (Equation (9.3)) and energy of the molecule: $E_{\text{total}} = E_{\text{el}} + E_{\text{nuc}}$. Translational motion of the center of mass through space will be ignored in the total energy of the system. For a diatomic molecule the relative motion of the two nuclei is described by $\zeta_{\text{nuc}}(R,\theta,\varphi)$ of Equation (9.5). To first approximation the nuclear coordinates are separable such that ζ_{nuc} can be written as a product of an angular function describing rotation, $\Theta_{\Lambda,J}(\theta,\varphi)$ and a vibrational part $\chi_{vJ}(\mathrm{R})/R$. Making the assumption that the molecule rotates rigidly with internuclear separation fixed at R_e, it can be shown (see e.g., Steinfeld (1974) pp. 93–104) that the rotational energy is quantized as

$$E_{\text{rot}} = hcBJ(J+1) \qquad J \geq 0, 1, 2, \ldots \tag{9.9}$$

where $B = h/8\pi^2 c\mu R_e^2$ (in units of cm^{-1}) is the rotational constant, J the nuclear angular momentum quantum number, and $\mu = m_A m_B/(m_A + m_B)$ is the reduced mass. The rotational energy is generally a very small fraction of the overall energy of the molecule. The spacings of the levels increase with J. The molecular hydrogen molecule has the largest rotational constant, ~60 cm^{-1} in its ground state.

With the angular variables of Equation (9.5) separated off, we are left with the radial Schrödinger equation for nuclear motion

$$-\frac{1}{2\mu}\frac{\delta^2}{\delta R^2}\chi_{vJ}(\mathrm{R}) + \left[U(\mathrm{R}) + \frac{1}{2\mu R^2}J(J+1) - E_{vJ}\right]\chi_{vJ}(\mathrm{R}) = 0 \qquad (9.10)$$

The rotational energy, parametrically dependent on J, appears as a centrifugal term in the potential energy governing the vibration of the nuclei. When $U(R)$ has a potential minimum, i.e., the molecule is bound, Equation (9.10) is an eigenvalue equation and v is the quantum number for nuclear vibration. In Figure 9.1 the vibrational energy levels for two bound state potentials, U_1 and U_3, are indicated. When $U(R)$ is repulsive, (see U_2 in Figure 9.1) the solutions to Equation (9.10) are continuum functions, χ_k, and the energies, E_k lie at positive energies with respect to the dissociation limit of the molecule. This case is discussed in Section 9.4.

Equation (9.10) is a relatively simple second order differential equation which can usually be solved by straight-forward numerical integration. It is convenient to note, however, that a molecular potential well can often be approximated by a harmonic oscillator potential $U(R) \approx \frac{1}{2}k(R - R_e)^2$ where k is the curvature of the potential around R_e. This leads to a simple analytic form for the quantized vibrational energy levels

$$E_{\mathrm{vib}} = hc\omega_e(v + \tfrac{1}{2}) \qquad v = 0, 1, 2 \ldots \qquad (9.11)$$

and ω_e is the vibrational frequency in cm^{-1}. It should be noted that for the lowest energy state, $v = 0$, there is still vibrational energy present, $E(0) = \frac{1}{2}hc\omega_e$. The vibrational energy level spacing is determined by the shape of the potential curve and the reduced mass of the nuclei, and is typically ~2000 cm^{-1} for light molecules.

In practice, of course, the molecule does not rotate as a *rigid* rotator, nor is the potential in which the nuclei vibrate a harmonic oscillator. Thus the energies given in Equations (9.9) and (9.11) have correction terms added such that the energy in cm^{-1} is given by

$$\frac{E_{\mathrm{nuc}}}{hc} = G(v) + F_v(J) \qquad (9.12)$$

where $G(v) = \omega_e(v + \frac{1}{2}) - \omega_e x_e(v + \frac{1}{2})^2 + \omega_e y_e(v + \frac{1}{2})^3 + \ldots$ and $F_v(J) = B_v J(J+1) - D_v J^2(J+1)^2 + \ldots$.† Accurate values for the constants ω_e, $\omega_e x_e$,

†$B_v = B_e - \alpha_e(v + \frac{1}{2}) + \ldots$ and $D_v = D_e + \beta_e(v + \frac{1}{2}) + \ldots$ where D_e is *not* to be confused with the binding energy given earlier in this chapter.

$\omega_e y_e$, B_v, and D_v can be obtained from detailed analysis of molecular spectra or may be calculated *ab initio* from solution of the electronic eigenvalue problem to obtain $U(R)$ and then solution of the equations of nuclear motion.

When molecules possess electronic orbital and/or spin angular momentum, interpretation of the rotational structure becomes somewhat more complex. As the electrons rotate in the molecule, the molecule rotates in space, and the constants of the motion usually involve the coupling of these angular momenta. The different coupling cases are set forth as *Hund's rules*. For lack of space, these will not be explained here, but the reader is referred to Herzberg (1950) for a complete description of these limiting cases.

9.3. Molecular absorption spectroscopy

The absorption of light by a molecule can produce electronic, vibrational, or rotational excitation and often a combination of these modes simultaneously. The energies involved are very different, however. Usually electronic excitation requires photons at optical and ultraviolet wavelengths, pure vibrational excitation requires infrared wavelengths, and pure rotational transitions occur in the microwave region of the spectrum. We shall first consider pure rotation and vibration spectra and then electronic transitions which involve simultaneously vibrational and rotational transitions.

9.3.1. *Rotational transitions*

Usually pure rotational transitions are observed for molecules in the interstellar medium in emission, not absorption, because background sources are weak in the microwave region. The frequencies at which radiation is emitted or absorbed can be calculated once the selection rule on the rotational quantum number J is known. The selection rule is obtained from the matrix element (see Herzberg (1950)) of the dipole moment, μ, evaluated between two rigid rotator eigenfunctions for levels J'' and J', $\langle \Theta_{J'M'} | \mu | \Theta_{J''M''} \rangle$. The matrix elements are different from zero only when the dipole moment is non-zero *and* $J' = J'' \pm 1$ or ($\Delta J = \pm 1$). The strength of the rotational lines is proportional to the square of this matrix element and thus to the square of the dipole moment. For homonuclear diatomics, such as H_2, which have no electric dipole moment, pure rotational electric dipole transitions are not allowed.

Using the expression for rotational energy $F_v(J)$ in Equation (9.12), the selection rule on J, and the convention that J' and J'' refer to the upper and lower levels respectively, the frequency of the emitted or absorbed radiation for a non-rigid rotator in vibrational level v is

$$F_v(J'' + 1) - F_v(J'') = 2B_v(J'' + 1) - 4D_v(J'' + 1)^3 \tag{9.13}$$

Since the constant D_v is often very small compared to B_v, the spacing of the lines in the rotational spectrum is essentially equidistant.

9.3.2. *Vibrational transitions*

The hypothetical case of a diatomic molecule vibrating as a harmonic oscillator will be examined first. In the case of vibration, the dependence of the electric dipole moment on the internuclear separation must be considered. Expanding the dipole moment in a Taylor series about R_e, keeping only the first derivative term, and examining the off-diagonal matrix elements of μ between harmonic oscillator function v'' and v', one finds non-vanishing terms for $\langle v''|(R - R_e)|v'\rangle$ only if $\Delta v = v' - v'' = \pm 1$ (see e.g., Steinfeld (1974), pp. 98–101). This is called a *fundamental* absorption and the strength of this transition depends on the magnitude of the *derivative* of the dipole moment with respect to R. The second derivative term of the dipole moment is usually very small, but matrix elements $\langle v''|(R - R_c)^2|v'\rangle$ are non-zero only for $\Delta v = \pm 2$. These weaker transitions are called *overtone* absorptions. A molecule such as CO may have a relatively small dipole moment, but à large dipole moment derivative and therefore absorb strongly in the infrared. Again, homonuclear diatomics which have no dipole moment, nor dipole moment derivative, show no pure vibrational absorption.

Of course a diatomic molecule is not really a harmonic oscillator. In that the vibrational wave functions depart from harmonic oscillator function, the selection rules for Δv are not rigorously obeyed. Usually, the fundamental transitions ($\Delta v = \pm 1$) are strongest, followed by the first overtone transitions ($\Delta v = \pm 2$) and the second overtone transitions ($\Delta v = \pm 3$).

The spacings of the vibrational levels can be obtained from Equation (9.12), keeping terms only through $\omega_e x_e$,

$$G(v + 1) - G(v) = \Delta G_{v+1/2} = \omega_e - 2(v + 1)\omega_e x_e \tag{9.14}$$

These are the frequencies at which the fundamental transitions are observed.

9.3.3. *Vibration-rotation spectra, spectroscopic constants and RKR curves*

The separation of vibration and rotation has been rather artificial, because, in fact, ‘pure’ vibrational transitions do not occur. ‘Pure’ rotational transitions do occur because the energies involved are too small to excite the vibrational modes. When a vibrational excitation occurs through photon absorption a rotational transition takes place simultaneously. The vibration-rotation transition manifests itself as a spectral band, a broad feature composed of lines arising from transitions between many different rotational levels J'' in vibrational level v'' and many different rotational levels J' in vibrational level v'. Low resolution observations may not resolve the individual lines in a band. In an absorption spectrum the usual selection rules $\Delta v = 1$, $\Delta J = \pm 1$ will be considered.

To a first approximation, the band spectrum appears centered around the frequency $\omega_0 = \Delta G_{v+1/2}$ of the pure vibrational transition given in Equation (9.14). In a transition when $J' = J'' + 1$, the resulting line is called an *R-branch* transition; when $J' = J'' - 1$, the line is called a *P-branch* transition. Neglecting terms higher than B in Equation (9.13), the frequencies of the line are

$$\left.\begin{aligned} \omega_P &= \omega_0 - (B_{v'} + B_{v''})J'' + (B_{v'} - B_{v''})J''^2 \\ \omega_R &= \omega_0 + (B_{v'} + B_{v''})(J'' + 1) + (B_{v'} - B_{v''})(J'' + 1)^2 \end{aligned}\right\} \qquad (9.15)$$

where B_v are rotational constants for the vibrational levels v' and v''.

Analyses of vibration-rotation band spectra lead to values for the spectroscopic constants ω_e, $\omega_e x_e$, B_e, etc. which are characteristic of a particular molecular electronic state. Once a number of line positions have been accurately determined, inversion procedures such as the Rydberg–Klein–Rees (RKR) method can be used to obtain the molecular potential curve as a function of R. The accuracy of such procedures is dependent on the number and accuracy of transitions measured. If only low-lying vibrational levels have been observed, the potential energy curve is not valid for higher energies.

9.3.4. Electronic transitions

Vibrational and rotational transitions occur simultaneously when an electronic transition is made. The manifestation of these transitions is a spectrum at visible or ultraviolet wavelengths consisting of a group of bands. The first identifications of molecules in the interstellar medium were made for CH, CN, and CH^+ through observations of several lines in the electronic band spectra of each (see e.g., Swings and Rosenfeld (1937), McKellar (1940)).

Selection rules for *electronic transitions* are obtained by examination of the dipole moment matrix elements between an upper (f) and lower (i) electronic state

$$\mu_{if}(R) = \langle \psi_{el,i}(R) | \mu | \psi_{el,f}(R) \rangle \qquad (9.16)$$

It can be shown that in order for Equation (9.16) to be non-zero $\Delta\Lambda = \Lambda_i - \Lambda_f = 0, \pm 1$ and $\Delta S = 0$. In addition $\Sigma^+ \leftrightarrow \Sigma^-$, and $u \leftrightarrow g$ *only* (for homonuclear diatomics).

We must now consider vibrational motion. The matrix element of the electronic dipole transition moment μ_{if} coupling vibrational functions v' and v'' of electronic states f and i, respectively, is

$$\mu_{if}^{v'v''} = \langle \chi_{v'}(R) | \mu_{if}(R) | \chi_{v''}(R) \rangle. \qquad (9.17)$$

If μ_{if} varies slowly with R, the *Franck–Condon approximation* can be made in which

$$\mu_{\text{if}}^{v'v''} \approx \bar{\mu}_{\text{if}} \langle \chi_{v'}(R) | \chi_{v''}(R) \rangle. \tag{9.18}$$

and the average value $\bar{\mu}_{\text{if}}$ is multiplied by the overlap of the two vibrational functions. The intensity of a transition between v' and v'' is proportional to the square of $\mu_{\text{if}}^{v'v''}$ in Equation (9.17), or $\bar{\mu}_{\text{if}}^2$ and $|\langle \chi_{v'}(R) | \chi_{v''}(R) \rangle|^2$, the latter quantity known as the *Franck–Condon factor*. Physically this approximation is based on the fact that the nuclei move much more slowly than the electrons, and in the time required for an electronic transition to occur, the nuclei have not changed their position. Thus only vibrational levels of the excited state potential curve which have significant, non-zero overlap with the lower state vibrational function are accessed. The molecule makes a 'vertical' transition as indicated by the arrows in Figure 9.1. Selection rules for rotation are similar to those that have been previously discussed: $\Delta J = \pm 1$, but also $\Delta J = 0$ if at least one of the electronic states has $\Lambda \neq 0$.

The energy of the absorbed photons can be resolved into an electronic, vibrational, and rotational contribution

$$E = E_{\text{e}} + E_{\text{vib}} + E_{\text{rot}}$$

where E_{e} is the difference between the energy minima of the two electronic potential curves involved in the transition. The vibrational and rotational frequencies can be expressed as differences in the spectroscopic term values

$$(E_{\text{vib}} + E_{\text{rot}})/hc = G_{v'} - G_{v''} + F_{v'}(J') - F_{v''}(J'') \tag{9.19}$$

In this case the (′) and (″) quantum numbers and spectroscopic constants denote the upper and lower electronic states, respectively. Looking at the vibrational structure of an electronic spectrum, ignoring for a moment the rotational structure and terms higher than quadratic in $(v + \frac{1}{2})$ we see

$$E_{\text{vib}}/hc = \omega_{\text{e}}'(v' + \tfrac{1}{2}) - \omega_{\text{e}}x_{\text{e}}'(v' + \tfrac{1}{2})^2 - [\omega_{\text{e}}''(v'' - \tfrac{1}{2}) - \omega_{\text{e}}x_{\text{e}}''(v'' + \tfrac{1}{2})^2] \tag{9.20}$$

This is the vibrational contribution to the frequency of the band (v', v''), of which there may be many in the complete electronic spectrum.

Under high resolution each of the vibrational bands is composed of large numbers of discrete lines due to rotational transitions with frequencies given by $(E_{\text{e}} + E_{\text{vib}})/hc + F_{v'}(J') - F_{v''}(J'')$. Three branches, R, P, and Q may occur for each band according to $\Delta J = \pm 1$, or 0. Formulae for the frequencies of the individual lines can be found in Herzberg (1950) p. 169.

A useful measure of the strength of an absorption band is the *oscillator strength* f. For an electronic transition from initial state i and vibrational level v'' to final state f and vibrational level v',

$$f_{v'v''} = \tfrac{2}{3} g \Delta E_{v'v''} |\langle \chi_{v'}(R) | \mu_{\text{if}}(R) | \chi_{v''}(R) \rangle|^2 \tag{9.21}$$

where g is a degeneracy factor equal to $(2 - \delta_{0,\Lambda'+\Lambda''})/(2 - \delta_{0,\Lambda''})$ and $\Delta E_{v'v''}$ is the transition energy. Historically, there has been considerable confusion as to the definitions and conventions used for intensity factors of rotational lines in diatomic molecular spectra. The reader is referred to Whiting *et al.* (1980) as well as Whiting and Nicholls (1974) for a more complete set of formulae than space permits here. The absorption oscillator strength is an essential quantity in the determination of molecular abundances from interstellar molecular absorption lines.

9.3.5. Perturbations

The assumptions we have made to simplify the molecular Hamiltonian are by and large justified by the fact that most molecular states are describable by a unique set of good quantum numbers. However, there are instances in which the presence of many nearby states affects the energy levels of a particular state under investigation and in which the small terms normally neglected in the molecular Hamiltonian become significant. Such cases manifest themselves as perturbations in the spectra of the affected levels, marked by irregularities in the pattern of a molecular band. Energy levels may be shifted, lines broadened, and intensities may be quite erratic. The origins of the perturbation, as electronic, vibrational, or rotational interactions, depend on the nature of the coupling terms involved. Due to lack of space, we will not discuss this subject further, but refer the reader to an excellent reference, Lefebvre-Brion and Field (1986). In the following section, a special case of perturbation which occurs between a bound and repulsive molecular state and causes predissociation will be treated.

9.4. Molecular photodissociation

Photodissociation ultimately involves a transition onto the repulsive part of a molecular potential curve at energies above the dissociation limit of that state. There are several different photodissociation processes which are distinguished by the way in which the dissociating transition occurs and these will be discussed in the following sections. A review of these processes with considerably more detail has been given by Kirby and van Dishoeck (1988). As is appropriate for molecules in interstellar environments, all transitions are assumed to originate from the lowest vibrational level ($v = 0$) of the ground state.

9.4.1. Direct photodissociation

The simplest dissociation pathway is through photon absorption into an excited electronic state of the molecule which is unbound with respect to nuclear motion as shown in Figure 9.2(a). The same effect is achieved if the transition occurs onto the repulsive wall above the dissociation limit of a bound molecular state. Rather than reradiating the photon, the molecule falls apart, creating atoms or fragment species which may be electronically excited or may have excess kinetic energy.

The direct photodissociation cross section, as shown in Figure 9.2(*a*), is continuous as a function of photon energy. The energy dependence of the cross section is governed to first approximation by the Franck–Condon principle, in that the maximum value lies near the vertical excitation energy indicated by the arrow in Figure 9.2(*a*). The shape of the cross section reflects that of the lower state vibrational wave function.

Figure 9.2. Potential curves for the molecular system AB illustrating three different photodissociation processes and the associated cross sections as a function of photon energy: (*a*) direct photodissociation, (*b*) predissociation, and (*c*) spontaneous radiative dissociation, from van Dishoeck (1988).

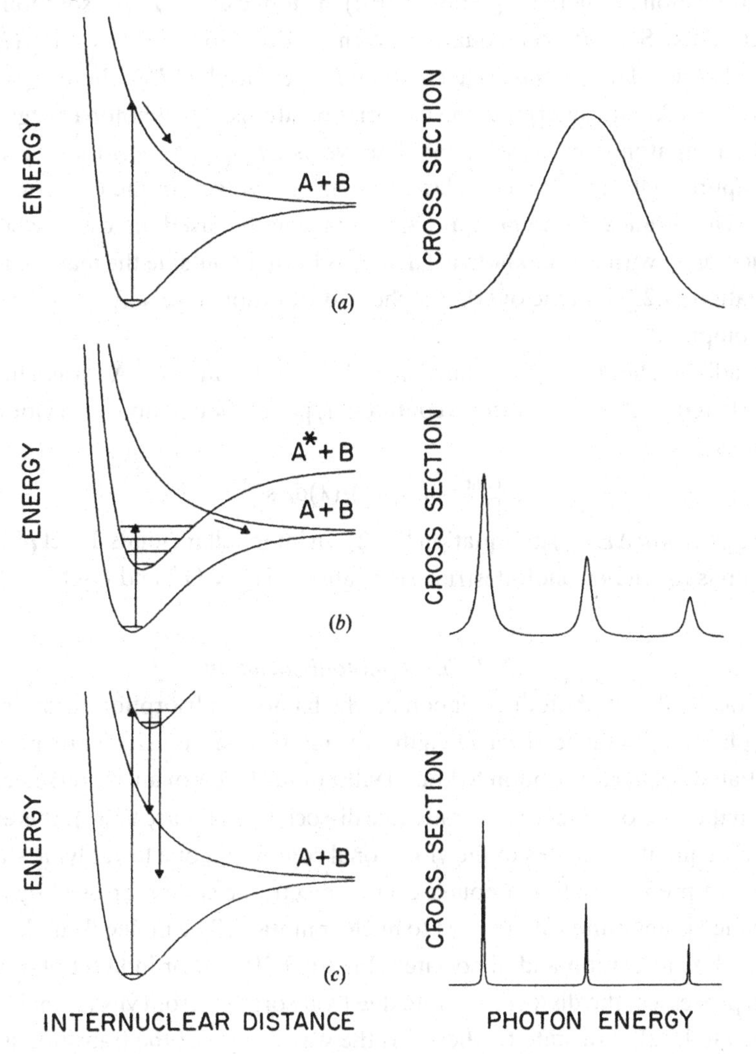

Experimental measurements of photodissociation cross sections and rates for molecules of astrophysical interest are often difficult to make. A number of these molecules are not easy to produce in the laboratory and their photodissociation continua may exist in spectral regions difficult to observe or may be obscured by other stronger absorption features. Most of the cross sections and rate constants used in modeling interstellar clouds have been calculated theoretically.

The cross section for direct photodissociation from vibrational level v'' of electronic state i into the vibrational continuum k' of electronic state f can be written

$$\sigma(\Delta E_{k'v''}) = 2.69 \times 10^{-18} g \Delta E_{k'v''} |\langle \chi_{k'}(R)|\mu_{\mathrm{fi}}(R)|\chi_{v''}(R)\rangle|^2 \ \mathrm{cm}^2 \qquad (9.22)$$

where $\chi_{v''}(R)$ is a bound vibrational wave function of state i, and μ_{fi} is the electric dipole transition moment (Equation (9.16)) in atomic units. $\chi_{k'}$ is a solution of the nuclear radial Schrödinger equation given in Equation (9.10) with $U(R)$ the potential curve of molecular state f and the E_{vJ} replaced by E_k, where $E_k = k^2/2\mu$ is the relative kinetic energy of the dissociating atoms. The photon energy of the transition (in atomic units) is denoted above as $\Delta E_{k'v''}$, and g is the degeneracy factor appropriate to the molecular symmetries involved in the electronic transition. The Franck–Condon approximation can be used in cases when the variation of μ_{fi} with R is unknown. Then μ_{fi}^2 is brought outside the matrix element in Equation (9.22) and the overlap of the two vibrational wave functions remains to be computed.

In a radiation field with mean intensity I (in photons $\mathrm{cm}^{-2}\,\mathrm{s}^{-1}\,\text{Å}^{-1}$) as a function of wavelength λ, the rate of direct (continuum) photodissociation from vibrational level v'' is

$$k_{\mathrm{pd}}^{\mathrm{cont}} = \int \sigma_{v''}(\lambda) I(\lambda) \mathrm{d}\lambda \ \mathrm{s}^{-1} \qquad (9.23)$$

where $\sigma_{v''}(\lambda)$ is $\sigma(\Delta E_{k'v''})$ in Equation (9.22). In interstellar clouds direct photodissociation is the major photodestruction route for H_2^+, CH^+ and H_2O.

9.4.2. Indirect photodissociation

In contrast to direct photodissociation in which photon absorption is continuous, taking place over a range of wavelengths, the indirect photodissociation processes are initiated by line absorption to bound vibrational levels of an excited electronic state. In the case of spontaneous radiative dissociation (Figure 9.2(*c*)) this excited state subsequently radiates to the vibrational continuum of a lower lying state. In the case of predissociation a bound excited electronic state is crossed by a third state which is unbound with respect to nuclear motion. The molecule undergoes a radiationless transition and dissociates (Figure 9.2(*b*)). For indirect photodissociation processes, the dissociation rate due to absorption from vibrational level v'' of state i to level v' of state f, where λ is the wavelength of the transition is

$$k_{\rm pd}^{\rm line} = 8.85 \times 10^{-21} \lambda^2 f_{v'v''} \eta_{v'} I(\lambda)\ {\rm s}^{-1} \quad (9.24)$$

$I(\lambda)$ has the same units as before, $f_{v'v''}$ is defined in Equation (9.21), and $\eta_{v'}$ is the dissociation efficiency of the vibrational state v', $0 \leq \eta_{v'} \geq 1$. By summation over all possible line absorptions which can give rise to dissociation, the total photodissociation rate of a molecule can be obtained.

Spontaneous radiative dissociation, is the primary photodestruction mechanism for molecular hydrogen in the interstellar medium. The absorption features, as illustrated schematically in Figure 9.2(*c*), are ordinary vibration-rotation band spectra. Thus this process can only be observed to occur through its emission spectra, appearing as a series of broadened peaks varying in appearance and extent depending on the number of upper state vibrational levels that have been populated. In H_2, the discrete absorptions contributing to dissociation take place into vibrational levels of the $B^1\Sigma_u^+$ and $C^1\Pi_u$ states (the Lyman and Werner systems). The total probability for radiative decay from each level v' to lower-lying bound and continuum levels can be calculated in order to obtain the $\eta_{v'}$ factor in Equation (9.24). Dalgarno and Stephens (1970) showed that for a uniform radiation field with a cut-off at the Lyman limit, 23% of the absorptions in the Lyman system lead to dissociation, whereas a negligible fraction contribute in the Werner system. In a more realistic radiation field the average fraction of absorptions leading to dissociation is about 11% (see Chapter 3). For most molecules spontaneous radiative dissociation is not a significant photodissociation channel. For strong absorptions with large oscillator strengths, the most probable decay route is radiation back to bound vibrational levels of the ground state.

Predissociation begins with line absorption into bound vibrational levels of an excited electronic state which is coupled to the vibrational continuum of a third state by small terms in the Hamiltonian which are often neglected. Such terms may include relativistic effects, such as spin–orbit, spin–spin and spin–rotation interactions, as well as the nuclear kinetic energy operators. The absorption feature characteristic of predissociation is spectral line broadening, as illustrated in Figure 9.2(*b*). The usual way of quantifying the predissociation is through measurement of the full-width at half maximum of the spectral lines, $\Gamma^{\rm pr}$. The predissociation lifetime $\tau^{\rm pr} = \hbar/\Gamma^{\rm pr} = 5.30 \times 10^{-12}/\Gamma^{\rm pr}$ if $\Gamma^{\rm pr}$ is in cm^{-1} and $\tau^{\rm pr}$ in s. A typical radiative lifetime for an electric dipole allowed transition is $\sim 10^{-8}$ s, corresponding to a line width of $\sim 5 \times 10^{-4}$ cm^{-1}. Only predissociations with significantly shorter lifetimes can be observed with standard spectroscopic techniques.

Usually the terms in H causing the predissociation are small and can be included within the framework of first-order perturbation theory. Then the predissociation width is given by the Golden Rule formula

$$\Gamma_{v'}^{\rm pr} = 2\pi |\langle \chi_{v'} | H_{\rm fd}(R) | \chi_k \rangle|^2 \quad (9.25)$$

in which H_{fd} is the coupling between electronic states f and d, the excited bound and dissociating states, respectively. The vibrational wave functions $\chi_{v'}$ and χ_k are solutions to the appropriate radial Schrödinger equation. The efficiency factor for predissociation, describing the competition between radiative and non-radiative decay is

$$\eta_{v'} = \frac{\Gamma_{v'}/\hbar}{\Gamma_{v'}/\hbar + 1/\tau_{v'}^{\mathrm{rad}}} \tag{9.26}$$

Predissociation is the mechanism responsible for the photodestruction of the abundant CO molecule. In special cases the interaction may be strong, particularly between states of the same symmetry, and a coupled states description of the process is necessary (see Kirby and van Dishoeck (1988)).

9.5. Molecular photoionization processes

Photon absorption by molecules can lead to ionization by several different routes:

$AB + h\nu \rightarrow$	$AB^+ + e^-$	direct photoionization
$\rightarrow$	$A^+ + B + e^-$	dissociative photoionization
$\rightarrow$	$AB^{**} \rightarrow AB^+ + e$	autoionization

These processes are illustrated in Figure 9.3. In direct photoionization energy is imparted to the electron directly, and the remaining molecular ion may also

Figure 9.3. Potential curves of molecules AB and AB^+ with vertical arrows illustrating three different photoionization processes. Direct photoionization is indicated by the dashed line, dissociative photoionization by the dotted line, and autoionization by the solid line.

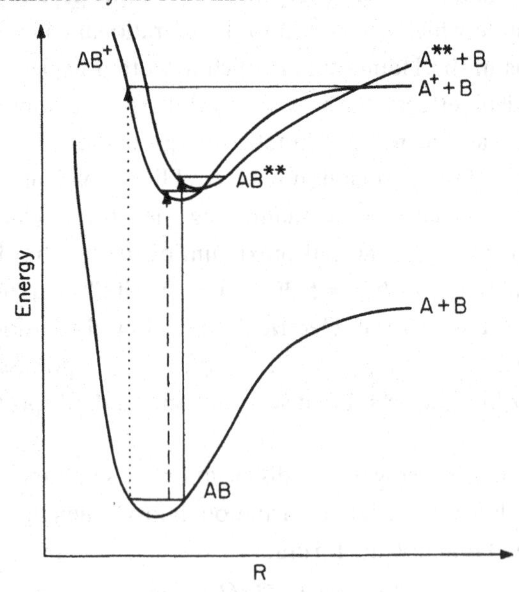

experience some vibrational and rotational excitation governed by the usual selection rules for electronic transitions. Dissociative photoionization takes place via a transition to the vibrational continuum of a molecular ion state. In autoionization, the electronic transition is first to an electronic state which by virtue of its electronic, vibrational, or rotational motion lies energetically above the molecular ion state. This excited electronic state, denoted AB^{**}, then decays to a molecular ion plus a free electron. In the first two processes, photon absorption is continuous, and the cross section is finite at threshold and generally a smooth function of photon energy. While not much information can be gathered from the photoionization continuum itself, if the ejected electron is energy analyzed, and the incoming photon wavelength is known, a number of details about the vibrational, rotational, and electronic excitation of the molecular ion are obtained. This kind of experiment is called *photoelectron spectroscopy*.

Autoionization is initiated by line absorption. The rapid decay to $AB^+ + e^-$ broadens the lines, which are known as *resonances*. Because these resonances lie above the photoionization threshold, the cross section has both a direct and a resonance contribution appearing as a smooth background superposed with resonance lines. Often, the contribution of resonances to the total photoionization rate near threshold may be much more significant than the background continuum contribution.

9.5.1. Direct and dissociative photoionization

Direct and dissociative photoionization will be discussed together, as they differ only in regard to the binding, or lack of binding, in the final molecular ion state of AB^+. Because photoionization is closely akin to a scattering process, many of the scattering theory methods developed for electron–molecule collisions have been applied to molecular photoionization. These methods cannot be reviewed here, but the reader is referred to reviews by McKoy and Lucchese (1983) and Langhoff (1985). We will give a very brief theoretical derivation of the photoionization cross section from the point of view of molecular structure.

Within the framework of the Born–Oppenheimer approximation, the electronic dipole photoionization matrix element is given by

$$\mu_{\mathrm{if}}(\varepsilon,R) = \langle \Psi_{\mathrm{i}}(\mathbf{r}|R)|\mu|\Psi_{\mathrm{f}}(\mathbf{r}|R,\varepsilon)\rangle \tag{9.27}$$

in which ε is the energy of the ejected electron. If $E_{v'J'}$ is the ionization threshold for absorption into a specific vibration-rotation level of AB^+, then the photon energy in the transition is $h\nu = E_{v'J'} + \varepsilon$; Ψ_{i} is the usual bound, electronic wave function, but Ψ_{f}, the final state wave function, must describe the molecular ion AB^+ plus the free electron. There are many different methods and approximations for describing Ψ_{f}. One of the simplest assumes an antisymmetrized product of a molecular ion wave function and a Coulomb function for the continuum electron (Ford, Docken, and Dalgarno (1975)).

The cross section for photoionization to a particular vibration–rotation level of the ion is

$$\sigma(\nu,v',J') = \frac{8\pi^3\nu}{3c}|\langle\chi_{v''J''}(R)|\mu_{\mathrm{if}}(\varepsilon,R)|\chi_{v'J'}(R)\rangle|^2 \tag{9.28}$$

where the radial vibration-rotation wave functions have been given previously (Equation (9.10)) and $S(J'',J')$ is an angular factor normalized so that $\Sigma_{J'}S(J'',J') = 1$. The total cross section at a particular photon energy $h\nu$ is obtained by summing the partial cross sections (Equation (9.28)) over all accessible vibrational and rotational levels of the ion. The photoionization rate expression is exactly equivalent to that of Equation (9.23) for photodissociation.

9.5.2. Rydberg series and autoionization

The process of photoionization can be viewed as any other molecular electronic transition, but in which the excitation imparted to the electron allows it to leave the systems. As the photoionization threshold is approached from the long wavelength side, a converging series of band systems, called a *Rydberg series*, can be seen. This series ultimately terminates as the continuous absorption of photoionization sets in. The Rydberg series arises from transitions into energetic *Rydberg states*. Such states are best described as an electron moving predominantly outside the ionic core AB^+ under the influence of a Coulomb-like central field. The successive states can be described by higher and higher principal quantum number n, and the orbital energies of this outer electron go approximately as $-(1/n^2)$ up to the ionization limit.

The Rydberg states of a molecule, converging to a particular electronic state of the molecular ion, lie nested below the molecular ion state, and have spectroscopic constants very similar to those of the molecular ion state. In certain wavelength regions there may be overlapping Rydberg series, in which Rydberg states converging to different ionic states lie at similar energies. Rydberg states play a critical role in the two types of autoionization which we will briefly mention, vibrational and electronic autoionization.

In vibrational autoionization, a vibrational level, $v' > 0$, of a Rydberg state converging to the first ionization threshold is excited. If the vibrational level lies above the $v = 0$ level of the molecular ion state the molecule may autoionize, giving up energy to the outgoing electron and ending up in a lower-lying vibrational level of the molecular ion.

In electronic autoionization, a Rydberg state in a series converging to an excited molecular ion state is excited. Because it lies at energies above the first photoionization threshold, this state may autoionize by coupling to the electronic continuum state through the electronic Hamiltonian.

The resonance contribution to the photoionization cross section is proportional to the following matrix element

$$\sigma_{\mathrm{r}}(\nu) \sim |\langle \chi_{v''}(R)|\mu_{\mathrm{ir}}(R)|F(R)\rangle|^2 \tag{9.29}$$

where μ_{ir} is the electronic dipole transition matrix element to the resonance state as in Equation (9.16) and $F(R)$ is a solution to the inhomogeneous radial Schrödinger equation (see e.g., Kirby *et al.* (1981)) which contains terms coupling the resonance state to the molecular ion plus electron continuum state.

9.6. Astrophysical considerations

Frequently, detailed and accurate data on molecular oscillator strengths, photodissociation cross sections and photoionization rates are lacking. Highly accurate theoretical *ab initio* calculations on even small molecular systems are still not routine. Various species prevalent in the interstellar medium may be difficult to produce and measure in the laboratory. Often educated guesses must be made based on known quantities for similar systems and good physical intuition. The reader can learn more about the way in which molecular photoabsorption processes affect the chemical composition of interstellar clouds by referring to a number of articles in this volume, including those by van Dishoeck, Hartquist, Flower and Pineau des Forêts, Gredel, and Roberge (Chapters 3, 5, 15 and 16).

9.6.1. Abundance determinations from optical and ultraviolet absorption observations

Molecular abundances derived from absorption line studies provide particularly important tests of current theories of interstellar molecule formation (Black 1985). Interstellar molecular absorption lines may be observed when a diffuse cloud of interstellar molecular gas lies in front of a bright background star. The integrated intensity of an absorption line, known as the equivalent width, depends on both the absorption oscillator strength (Equation (9.21)) and the column density of absorbing molecules in the initial state of the transition. This is only strictly true in the case of very weak, unsaturated absorptions. For stronger lines saturation effects must be taken into account. The need for accurate oscillator strengths in the determination of molecular column densities cannot be overemphasized.

In Table 9.1 seven interstellar molecules are listed, including one of the transitions by which each is observed, and the transition wavelength and oscillator strength. One of the best-studied diffuse interstellar clouds is the principal cloud

Table 9.1. *Selected transitions for diatomics studied in diffuse clouds*

Molecule	Transition $\Lambda' - \Lambda''$	$(v'v'')$	λ	$f_{v'v''}$
H_2	$B^1\Sigma_u^+ - X^1\Sigma_g^+$	(1,0)	1092	5.79 (−3)
CO	$A^1\Pi - X^1\Sigma^+$	(1,0)	1510	2.32 (−2)
OH	$A^2\Sigma^+ - X^2\Pi$	(0,0)	3078	1.10 (−3)
CN	$B^2\Sigma^+ - X^2\Sigma^+$	(0,0)	3876	3.38 (−2)
CH	$A^2\Delta - X^2\Pi$	(0,0)	4300	5.3 (−3)
CH^+	$A^1\Pi - X^1\Sigma^+$	(0,0)	4232	5.5 (−3)
C_2	$D^1\Sigma_u^+ - X^1\Sigma_g^+$	(0,0)	2312	5.4 (−2)

Table taken from Black (1985).

toward the star ζ Oph. All of the interstellar molecules listed in Table 9.1 have been observed in this cloud.

9.6.2. *Photodissociation*

Photodissociation may be a major destruction mechanism for both neutral and ionic molecules in any astrophysical environment in which there are ambient ultraviolet photons. Thus at the edges of dense clouds and in diffuse interstellar clouds, photodissociation plays a significant role in the chemistry. In dense clouds recent observations of neutral atomic species and the proposed Prasad–Tarafdar mechanism for producing ultraviolet photons suggest that photodissociation may be important there also.

Typical photodissociation rates in the unattenuated interstellar radiation field are $\sim 5 \times 10^{-10}\,s^{-1}$ for small interstellar species, but the range is about an order of magnitude on either side of this value. Van Dishoeck (1988) has compiled a table of interstellar photodissociation rates for over 50 different molecules. Whether photodissociation is caused by line or continuum absorption has profound implications for radiative transfer because line absorptions can be saturated much more readily than continuous absorptions. Molecules lying at greater column depths in an interstellar cloud are effectively shielded from dissociation due to reduced fluxes of line radiation.

The effectiveness of a photodissociation channel depends on the characteristics of the radiation field which change with depth in an interstellar cloud due to grain scattering and absorption. A channel with large cross section in a region of low photon flux in the interstellar radiation field may be much less significant for photodissociation than a channel with smaller cross section at the peak of the photon flux. This point is illustrated in Figure 9.4 for the case of CH^+. The photodissociation cross section through the $A^1\Pi$ state is several orders of magnitude smaller than the cross sections through the $3^1\Sigma^+$ and $2^1\Pi$ states. At the surface of the cloud, photodissociation is dominated by the channel with the largest cross sections, excited by high energy photons. As the higher energy photons are absorbed and scattered out of the field at greater optical depths, the $A^1\Pi$ channel becomes increasingly important. The $A^1\Pi$ channel produces C^+ and H, whereas the other channels produce C and H^+. As mentioned in Chapter 5, the CH^+ abundance in a diffuse cloud shock depends on the CH^+ photodissociation products.

9.6.3. Photoionization

Ionization energies for most of the small molecules found in interstellar clouds vary from ~6–16 eV. Several of the more ubiquitous molecular species have ionization potentials greater than 13.6 eV, most notably CN (14.1 eV), CO (14.0 eV) and H_2 (16.0 eV). For them photoionization is not a significant photodestruction process in the interstellar medium. Vibrationally excited ($v \geq 4$) molecular hydrogen may be photoionized at wavelengths longer than 912 Å (Ford *et al.* 1975) and Black and van Dishoeck (1987) have shown that this may be important in some astrophysical environments.

Ionization potentials and interstellar molecular photoionization rates have been tabulated by van Dishoeck (1988). There are uncertainties of at least a factor

Figure 9.4. Photodissociation rates in the interstellar radiation field arising from transitions to four excited electronic states of CH^+ as a function of optical depth. (From Kirby *et al.* (1980).)

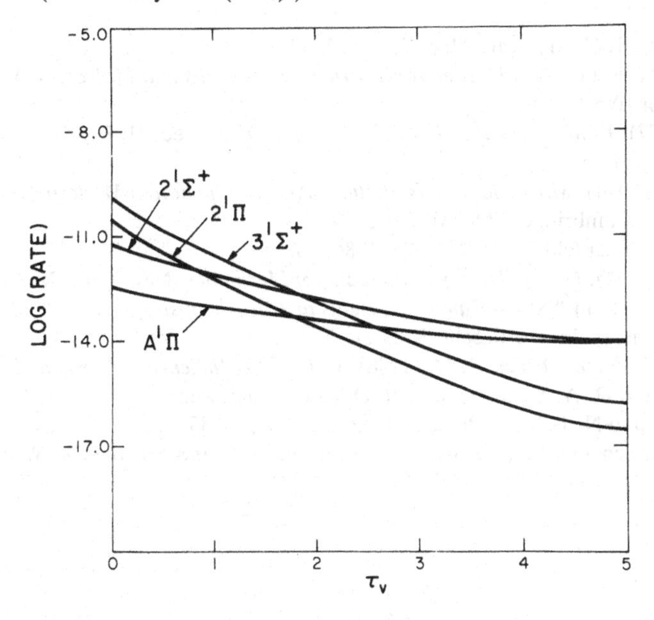

of 2 for most molecules. When photoionization rates are unknown, an order of magnitude estimate may be obtained by assuming a cross section of 10^{-17} cm^2 above the ionization threshold (van Dishoeck 1988).

REFERENCES

Black, J. H. (1985). In *Molecular Astrophysics,* eds. G. H. F. Diercksen *et al.*, p. 215. Dordrecht: Reidel.

Black, J. H. and van Dishoeck, E. F. (1987). *Ap. J.* **322**, 412.

Cotton, F. A. (1971). *Chemical Applications of Group Theory,* New York: Wiley-Interscience.

Dalgarno, A. and Stephens, T. L. (1970). *Ap. J. Lett.* **160**, L107.

Ford, A. L., Docken, K. K. and Dalgarno, A. (1975). *Ap. J.* **195**, 819.

Herzberg, G. (1950). *Molecular Spectra and Molecular Structure. I. Spectra of Diatomic Molecules,* Princeton: D. van Nostrand.

Hurley, A. C. (1976). *Introduction to the Electron Theory of Small Molecules,* New York: Acad. Press.

Kirby, K. and van Dishoeck, E. F. (1988). In *Advances in Atomic and Molecular Physics,* ed. D. R. Bates, in press.

Kirby, K., Roberge, W. G., Saxon, R. and Liu, B. (1980). *Ap. J.* **239**, 855.

Kirby, K., Uzer, T., Allison, A. C. and Dalgarno, A. (1981). *J. Chem. Phys.* **75**, 2820.

Kutzelnigg, W. (1977). In *Methods of Electronic Structure Theory,* ed. H. F. Schaefer, p. 129. New York: Plenum.

Langhoff, P. W. (1985). In *Molecular Astrophysics*, eds. G. H. F. Dierckson *et al.* p. 551. Dordrecht: Reidel.

Lefebvre-Brion, H. and Field, R. W. (1986). *Perturbations in the Spectra of Diatomic Molecules,* Orlando: Acad. Press.

McKellar, A. (1940). *Publ. Astron. Soc. Pac.* **52**, 307.

McKoy, V. and Lucchese, R. (1983). In *Electron–Molecule Collisions and Photoionization Processes,* eds. V. McKoy, *et al.*, p. 13. Dearfield Beach Fla: Verlag Chemie.

Roothaan, C. C. J. (1951). *Rev. Mod. Phys.* **23**, 69.

Schaefer, III, H. F. (1972). *The Electronic Structure of Atoms and Molecules.* Reading, MA: Addison-Wesley.

Shavitt, I. (1977). In *Methods of Electronic Structure Theory,* ed. H. F. Schaefer, p. 189. New York: Plenum.

Steinfeld, J. I. (1974). *Molecules and Radiation: An Introduction to Modern Molecular Spectroscopy.* Cambridge, MA: MIT Press.

Swings, P. and Rosenfeld, L. (1937). *Ap. J.* **86**, 483.

Tinkham, M. (1964). *Group Theory and Quantum Mechanics.* New York: McGraw-Hill.

van Dishoeck, E. F. (1988). In *Rate Coefficients in Astrochemistry,* eds. T. J. Millar and D. A. Williams, p. 49. Dordrecht: Kluwer.

van Dishoeck, E. F. and Black, J. H. (1988). In *Rate Coefficients in Astrochemistry,* eds. T. J. Millar and D. A. Williams, p. 209. Dordrecht: Kluwer.

Whiting, E. E. and Nicholls, R. W. (1974). *Ap. J. Suppl.* **235**, 1.

Whiting, E. E., Schadee, A., Tatum, J. B., Hougen, J. T. and Nicholls, R. W. (1980). *J. Mol. Spec.* **80**, 249.

10

Interstellar ion chemistry: laboratory studies

DAVID SMITH and NIGEL G. ADAMS
School of Physics and Space Research, University of Birmingham, Birmingham, UK

ELDON E. FERGUSON
Laboratoire de Physico-Chimie des Rayonnements, Université de Paris-Sud, France

10.1. Introduction

The wide variety of ionized and neutral molecules detected in diffuse and dense interstellar clouds are mainly synthesized in the gas phase by many sequential and parallel positive ion–neutral reactions. The product positive ions may then be neutralized by proton transfer or charge transfer reactions with ambient molecules and/or via dissociative positive ion–electron recombination reactions and/or (as has recently been proposed) via mutual neutralization (positive ion–negative ion recombination) reactions involving large negatively-charged molecules. The basic gas phase ion chemistry was outlined in the early 1970s (Solomon and Klemperer 1972, Herbst and Klemperer 1973, Dalgarno 1975, Dalgarno and Black 1976) and since that time more sophisticated quantitative ion–chemical models have been developed to describe the routes to, and relative abundances of, the increasing number of complex molecules detected in interstellar clouds (Leung, Herbst and Huebner 1984, Millar and Nejad 1985, Dalgarno 1986, van Dishoeck and Black 1986, Millar, Leung and Herbst 1987). These models require the input of a large amount of kinetic data which (ideally) has been obtained at the low temperatures pertaining to interstellar clouds. In particular, the rate coefficients and products of many positive ion–neutral reactions and recombination reactions are required. Hence, a good deal of effort has been made to design and exploit laboratory experiments to provide these data (Huntress 1977, Smith and Adams 1981a, Anicich and Huntress 1986, Adams and Smith 1987a, Rowe 1988, Adams and Smith 1988a). This has led to a better understanding of the molecular processes that can occur under the extreme conditions of low pressure and low temperature in quiescent interstellar clouds. It is with these processes that this chapter is largely concerned.

In addition, there is now a good deal of interest in the chemistry occurring in the shocked regions of interstellar gas (Dalgarno 1981, Adams, Smith and Millar 1984a, Draine and Katz 1986, Pineau des Forêts *et al.* 1987) and we will give this chemistry some attention, insofar as laboratory data exist, in order to quantify the proposed ion–chemical models for such regions. Also it appears that molecular synthesis on grain surfaces contributes to the production of some interstellar molecular species (notably the most abundant species H_2) (Williams (1987) and Chapter 7), but we will not be concerned with this here. Neither will we be concerned with reactions between neutral (radical) species which contribute to molecular synthesis in some interstellar regions (such as shocked regions; Dalgarno (1981), Mitchell (1987)).

This chapter is organized as follows. First we briefly consider the laboratory techniques which have been developed to study ion–neutral, ion–electron and ion–ion reactions and discuss the respective merits and weaknesses of these techniques in providing data relevant to interstellar chemistry. Then we discuss some basic ion–neutral reactions which are considered to be important in interstellar clouds. Laboratory studies of both binary and ternary (association) reactions are included and it is shown how the ternary data have been used to estimate radiative (binary) association rate coefficients. Then, the available laboratory data relating to dissociative recombination and mutual neutralization and their relevance to neutralization of ions in interstellar clouds are considered. We conclude with some general remarks regarding the current status of interstellar ion chemistry and mention the new experimental initiatives which are needed to further advance understanding of this subject.

10.2. Experimental techniques

It is not our intention to present a detailed description of the techniques used to study ionic reactions but rather briefly to describe the basic principles of the techniques and to discuss the applicability of the laboratory data obtained from their exploitation to the special conditions pertaining to interstellar clouds. The techniques will be categorized into those designed for the study of ion–neutral reactions and those designed for the study of recombination reactions. These techniques are in one of two categories: (i) The low pressure techniques, notably the ion cyclotron resonance technique, which are generally used to study only binary processes (although it is also possible with them to study very fast ternary processes). Ion and electron beam methods are included in this category. (ii) The collision-dominated higher pressure swarm methods, notably stationary and flowing afterglow, selected ion flow tube and drift tube techniques. Each of these techniques has particular merits for the study of ionic reactions.

10.2.1. Techniques for studying ion–neutral reactions

The ion cyclotron resonance (ICR) technique has been exploited (principally by W. T. Huntress Jr and his colleagues) for the study of a wide range of positive ion–neutral reactions of interstellar interest at ICR cell temperatures near room temperature (Huntress 1977). The low pressure ion trap (IT) technique was developed and exploited by G. H. Dunn and his colleagues to study ion–neutral reactions at very low temperatures (this IT technique has also been used to study positive ion–electron recombination, see below). The ICR and IT techniques both involve the trapping of ions under low pressure conditions ($<10^{-6}$ Torr) away from the surrounding cell walls by combinations of electric and magnetic fields. Details of the techniques are given in the papers by McIver (1978) (ICR) and Barlow, Luine and Dunn (1986) (IT). Trapping times can be up to a second in ICR cells, and very much longer (up to hours) in the IT cell. Thus, all but the longest-lived metastable ions will be radiatively relaxed in the IT and probably in the ICR. The addition of reactant gases to the cells allows binary reactions to be studied free from the influence of the higher pressure background (buffer) gases which are required in the collision-dominated methods.

A few important measurements of ternary association reaction rate coefficients have been carried out at relatively low pressures of third-body stabilizing gases in ICR cells, pressures much lower than those possible using the prolific flow tube techniques. The large amount of ICR data (Huntress 1977, Anicich and Huntress 1986) is of great value in suggesting synthetic routes to observed interstellar molecules. The recent ICR observations of binary (radiative) association reactions between polyatomic hydrocarbon ions and hydrocarbon molecules are particularly interesting (McEwan 1988). However, for some reactions the magnitudes of the rate coefficients (and the ion product distributions) determined using ICR experiments must be treated with caution, particularly those for slow reactions, the rate coefficients for which are most likely to be sensitive to the temperature or energy of the reactants. This is because there is evidence that the reactant ions in the ICR cells are kinetically excited (suprathermal) to a degree dependent on the particular ICR cell.

A good deal fewer data have been obtained from IT experiments; however, those available have been obtained at the extremely low temperatures characteristic of dense interstellar clouds and are thus all-the-more relevant. The variety of reactions that may be studied at very low temperatures is limited because of the condensation of ion source gases and reactant gases. Some of the important reactions studied using the IT and the ICR techniques will be referred to in later sections.

Of the collision-dominated, higher pressure experiments, the stationary afterglow (SA) was first used to determine the rate coefficients accurately for a few key reactions of ionospheric interest (Smith and Fouracre 1968) and, as it later turned out, one reaction of interstellar significance (the $He^+ + N_2$ reaction which is discussed later; Sayers and Smith (1964)). However, the development and exploitation of the flowing afterglow (FA) technique by E. E. Ferguson, F. C. Fehsenfeld and A. L. Schmeltekopf (1969) in the early 1960s represented a landmark in the study of ion–neutral reactions. It was the first of the now-familiar fast flow tube techniques which have been so prolific in providing ion–neutral reaction rate data (and latterly in providing ion–ion and ion–electron reaction rate data, see below).

In the FA, ions (either positively-charged or negatively-charged) together with electrons are created upstream in a fast-flowing inert carrier gas (usually helium) by some kind of ion source, such as a hot cathode discharge or a microwave cavity discharge, and this produces a thermalized afterglow plasma in the downstream region of the flow tube (the ions thermalize in collisions with the relatively high pressure carrier gas). The addition of reactant gases to this afterglow then allows the ion–neutral reaction rate coefficients and ion products to be determined under truly thermalized conditions. Reactant and product ions are monitored by a downstream mass spectrometer. A great advantage of the FA (and the other similar fast flow tube techniques) lies in its chemical versatility; the sequential addition of various gases along the length of the flow tube allowing a wide variety of ions to be synthesized and their reactions studied. Using FA apparatuses, both binary and ternary reactions can often be studied over appreciable ranges of temperature (usually from ~80 to 600 K, although one apparatus could be operated at 900 K!). A very large amount of ion–neutral reaction rate data has been obtained which is of a fundamental nature, is of great relevance to the ion chemistry of the terrestrial atmosphere and ionosphere, and is of some relevance to interstellar chemistry. Reviews of the FA technique and summaries of the many results obtained have been given previously (Ferguson *et al.* 1969, Fehsenfeld 1975, Adams and Smith 1983, 1988c).

The subsequent development of the selected ion flow tube (SIFT) technique by N. G. Adams and D. Smith (1976a,b) greatly expanded the potential of fast flow techniques. The principles of operation of the SIFT are closely similar to those of the FA but with the crucial difference that the reactant ions are not generated in the inert carrier gas but in a remote ion source and then injected into the carrier gas. By the use of various types of ion source, a great variety of ions (both positive and negative) can be produced, including weakly-bound (cluster) ions, doubly-charged ions, vibrationally and electronically-excited ions, etc., which survive the passage along the flow tube. Hence the reactions of these ions can be studied with any reactant gas or vapour that can be introduced in an adequate concentration into the flow tube. A remote ion source is particularly valuable for the study of the reactions of ions of interstellar interest since these ions are usually very reactive with their parent source gas (e.g. CH^+ reacts rapidly with any hydrocarbon source gas such as CH_4 and C_2H_2). Thus, it has been possible to study a very large number of reactions of interstellar interest, including those of ions in recognizable series, e.g. CH_n^+, NH_n^+, H_nS^+, etc., and including both binary and ternary reactions. The variable-temperature SIFT (VT-SIFT) allows such reactions to be studied over the approximate temperature range 80–600 K. This is particularly important for obtaining estimates of rate coefficients for reactions at the lower temperatures of dense interstellar clouds. Detailed reviews of the VT-SIFT technique have been published (Smith and Adams 1979a, 1988a) and many of the critical data obtained which are of interstellar interest have been described in review papers (Smith and Adams 1979b, 1981a, 1985, Adams and Smith, 1987a).

The FA and SIFT techniques as presently constructed can only be operated at pressures greater than ~0.2 Torr, and at temperatures above ~80 K. Thus, because of this pressure limitation, binary (radiative) association reactions cannot be studied directly and an indirect approach to the determination of binary (radiative) association reaction rate coefficients has been taken (see Section 10.3.3). The temperature restriction is quite inhibiting; indeed 80 K operation is only possible for reactant gases which have adequate vapour pressure at that temperature and this excludes many known interstellar species. This restriction has been overcome in the CRESU technique, developed by B. R. Rowe, J. B. Marquette and their colleagues, in which reactant gases are mixed with a carrier gas in a reservoir held at a temperature above the condensation temperature of the reactant gas or vapour and then the mixture is expanded via a de Laval nozzle to supersonic speeds to achieve very low temperatures (down to 8 K, Rowe and Marquette (1987)). Ions are created in the supersonic flow by a high energy electron beam and the reduction in density of these ions due to ion–neutral reactions is monitored using a mass spectrometer. In this way, rate coefficients have been determined at fixed temperatures within the range 8–160 K, a development of great importance to interstellar chemistry. The measurements of rate coefficients for the reactions of ions with polar molecules have been particularly valuable as a test of theoretical predictions of collisional rate coefficients (see Section 10.3.1). Soon, a 'SIFT-type' injector is to be added to the CRESU apparatus which will allow the study of the reactions of a very wide variety of ions, including molecular ions, and this will further enhance its value to studies of interstellar chemistry.

The flow drift tube (FDT) developed by the NOAA group in Boulder, Colorado (McFarland *et al.* 1973) combines the chemical versatility of the fast flow tube techniques with the ion energy variability of drift tubes. This allows the study of ion–neutral reactions within the chemically-interesting energy range from thermal to an electron volt or so. The subsequent development of the selected ion flow drift tube (SIFDT) and the variable-temperature SIFDT (i.e. the VT-SIFDT) has extended enormously the range of reactions that can be studied at elevated energies. Indeed with the VT-SIFDT it is possible to study reactions at temperatures for which the internal temperature of the reactant gas is low and at elevated ion–neutral interaction energies appropriate to the shocked regions of interstellar gas. The various forms of selected ion drift tubes have been described in the review by Lindinger and Smith (1983). Recently, these drift tubes have been used to study some reactions thought to be involved in the synthesis of molecular ions in interstellar shocks (e.g. the endothermic $C^+ + H_2$ reaction, see Section 10.3.1).

10.2.2. Techniques for studying dissociative recombination and mutual neutralization

Three major techniques have been developed for studying positive ion–electron dissociative recombination and positive ion–negative ion mutual neutralization. These are the stationary afterglow, microwave cavity (SA) technique, the merged beam (MB) technique and, more recently, the flowing afterglow/Langmuir probe (FALP) technique. All three techniques have been used to study dissociative recombination, whilst only the MB and the FALP have been seriously applied to the study of mutual neutralization.

In the SA technique developed by M. A. Biondi and his colleagues (Mehr and Biondi 1969), recombining afterglow plasmas are created in a microwave cavity and recombination coefficients are derived from observations of the decrease in the electron number density as monitored by the changing resonant frequency of the cavity. This pioneering work established the magnitudes of dissociative recombination coefficients for many positive ion species of varying complexity. The SA technique is limited in the type of ions that can be studied since the ions are created by discharging a mixture of a relatively-high pressure buffer gas with an ion source gas or gases. The high pressure buffer gas is necessary to inhibit diffusive loss of ions but unfortunately it also enhances ion clustering and thus may lead to unwanted and unrecognized complications. The details of the technique have been discussed by Bardsley and Biondi (1970) and by Johnsen (1987).

The MB technique has been used by two different groups, by J. R. Peterson and J. T. Moseley to study mutual neutralization (see Moseley, Olsen and Peterson (1975)) and by J. B. A. Mitchell and J. W. McGowan to study dissociative recombination (see Auerbach *et al.* (1977)). Mass selected beams of positive ions and electrons or positive ions and negative ions (as appropriate) are merged at high laboratory energies to be colinear so that the relative energies in the centre-of-mass frame can be varied down to near-zero energies. Cross sections for these neutralization processes are determined as a function of centre-of-mass energy by observing the attenuation of one or both of the beams. Rate coefficients for these processes are obtained by Boltzmann averaging the cross section data. The advantage of these techniques is that the neutralization reactions for a wide variety of ion species can be studied and lists of many rate coefficients derived from these experiments have been tabulated (Moseley *et al.* 1975, Mitchell and McGowan 1983). However, the disadvantage of these experiments lies in the uncertainties in the internal energy states of the reactant ions and neutrals and in the difficulties inherent in defining the centre-of-mass energies of the reactants at low energies. These can lead to great uncertainties in the derived rate coefficients at low centre-of-mass energies (temperatures) and thus the application of the data to the low temperature conditions of interstellar clouds is fraught with danger! Detailed descriptions of these techniques are given in several reviews (e.g. Moseley *et al.* (1975), Mitchell and McGowan (1983), Mitchell (1986)) and reference is made to some of the data obtained using these MB techniques in Sections 10.3.2 and 10.3.3.

The VT-FALP technique combines the versatility of the FA with the Langmuir probe diagnostic which is used to measure the electron and ion number densities with high spatial precision at any position along the axis of the afterglow plasma. Positive ion–electron or positive ion–negative ion plasmas can be established in which the major loss process is dissociative recombination or mutual neutralization and a measurement of the axial gradient of the density of one or both of the charged species (as appropriate) provides a value for the dissociative recombination or mutual neutralization coefficient under truly thermalized conditions. Experiments are performed at a carrier gas pressure (again, usually helium) of about 1 Torr (which is sufficient to inhibit diffusive loss of charged particles). This pressure is much smaller than in SA experiments and thus clustering of the ions to neutrals is less problematical. The temperature range accessible is 80–600 K. Whilst the number of different neutralization reactions that can be studied using the VT-FALP is not limitless (as, in principle, is the number accessible using MB techniques) the technique is nevertheless very versatile and productive and the rate coefficients for an appreciable number of dissociative recombination (many of interstellar importance) and mutual neutralization reactions have been determined to good accuracy. Some of these data are discussed in Sections 10.3.2 and 10.3.3. The details of the VT-FALP technique have been discussed previously (Smith and Adams 1983, 1984a).

10.3. Basic processes

A very large number of ionic reactions are involved in the chemistry of interstellar clouds. Many have been identified and included in extensive ion–chemical models which are used to explain the relative abundances of some of the observed interstellar molecules. It is not our purpose to describe the details of this chemistry and of these models since this has been done by others. Rather we discuss the nature of ion–chemical reactions which occur under the special conditions of interstellar clouds and, using key examples, show how the laboratory experiments referred to in Section 10.2 have enhanced the understanding of this subject.

The interstellar chemistry of diffuse and dense interstellar clouds differs because the major sources of ionization and the physical conditions (temperature and pressure) are different. However, the ionic processes that can occur are common to these two types of clouds although the relative importance of the various processes will again depend on the particular ambient physical conditions. What is certain is that only binary gas phase processes can occur since ternary processes are precluded at the very low pressures.

10.3.1. Binary ion–neutral reactions

Ionization is created in *diffuse* clouds by the action of ultraviolet starlight and galactic cosmic rays on the ambient gas. This produces C^+, H^+ and H_2^+ ions and electrons; H^+ can react via resonant charge transfer with O atoms to produce O^+ and H_2^+ reacts rapidly with H_2 to produce H_3^+. Thus C^+, O^+ and H_3^+ are important in the initial phases of diffuse cloud chemistry, a chemistry which leads to neutral molecules such as OH, CH, CN, HCN, C_2, CO and ionic species such as CH^+, all of which are observed in these regions. Only simple (diatomic and triatomic) molecular species have been detected in diffuse clouds because the ultraviolet radiation which penetrates these regions can readily dissociate large species even if they are formed (unlike in the dense clouds where shielding by dust protects polyatomic species from destruction). The chemistry of diffuse clouds is thus less complicated than dense clouds although problems still remain to be solved and these are under active consideration by theorists (modellers) and laboratory experimenters.

The reaction:

$$H^+ + O \rightleftharpoons O^+ + H \qquad \Delta E = 0.02\ \text{eV} \qquad (10.1)$$

is an example of a nearly-resonant charge transfer reaction. It has been of great interest ever since it was recognized that it was important in the ion chemistry of the upper terrestrial ionosphere and latterly because it is implicated in diffuse interstellar cloud chemistry. The rate coefficient, $k(1)$, for Reaction (10.1) has been measured using the FA technique (Fehsenfeld and Ferguson 1972) and $k(1)$ has been calculated as a function of temperature (Chambaud *et al.* 1980). The O^+ ions produced in this reaction initiate the production of the observed OH (and probably H_2O also) in diffuse clouds thus:

$$O^+ \xrightarrow{H_2} OH^+ \xrightarrow{H_2} H_2O^+ \xrightarrow{H_2} H_3O^+, \qquad (10.2a)$$

$$H_2O^+, H_3O^+ \xrightarrow{e} OH, H_2O. \qquad (10.2b)$$

The sequence of H-atom abstraction reactions (10.2a) is known from FA and SIFT studies to be rapid (Albritton 1978), each reaction in the sequence occurring essentially with unit reaction probability, i.e. the measured reaction rate coefficients are close to the collisional (or Langevin) rate coefficient, k_c (e.g. see Su and Bowers (1979); note that when the reactant molecule possesses a permanent dipole moment then k_c is increased above the Langevin value – see Section 10.3.2). Similar H-atom abstraction reaction sequences also occur in interstellar clouds and, for example, convert N^+ to NH_4^+ ions and CH^+ to CH_3^+ ions (see below). The H_2O^+ and H_3O^+ product ions of the reaction sequence (10.2a) can then dissociatively recombine with electrons in the clouds generating OH and/or H_2O (Rowe and Queffelec 1989, Herd *et al.* 1989). The products of reactions such as (10.2b) are not generally known and this is one of the great uncertainties in the qualitative and quantitative modelling of the ion chemistry of interstellar clouds (Rowe *et al.* (1988); this is referred to further in Section 10.3.5).

Uncertainties also exist in the magnitudes of the rate coefficients for the elementary reactions of C^+ with H and H_2 which, in *quiescent* low temperature clouds, can only proceed via binary (radiative) association producing CH^+ and CH_2^+ respectively. These reactions are so slow that they have not yet been observed in the laboratory. Attempts have been made to study the $C^+ + H_2$ binary association reaction using the IT technique (Barlow *et al.* 1986) and the guided beam (GB) technique (Gerlich and Kaefer 1987) but only upper limits to the rate coefficient have been obtained (1×10^{-15} cm^3 s^{-1} in the IT and 1×10^{-16} cm^3 s^{-1} in the GB). Fortunately, the rate coefficient for this critical reaction and for the $C^+ + H$ reaction have been obtained by calculation (Herbst, Schubert and Certain 1977, Abgrall, Giusti-Suzor and Roueff 1976). The reactions of both CH^+ and CH_2^+ with H_2 and with many other molecular species have been determined using ICR and SIFT methods (Albritton 1978, Adams and Smith 1983, Anicich and Huntress 1986) and are generally very rapid. In the reactions with H_2 the ultimate product ion is CH_3^+ which reacts only slowly with H_2 (by radiative association in interstellar clouds) and thus CH_3^+ is available to react with minority species in both diffuse and dense clouds. (We discuss the involvement of CH_3^+ in the chemistry of dense clouds below.) In diffuse clouds, CH_3^+ can react with N and O atoms producing molecular ions which include the CN and CO groupings, as has been shown by laboratory experiments (Fehsenfeld 1976, Federer *et al.* 1986).

H_3^+ ions play a major role in both diffuse and dense cloud chemistry. The small proton affinity of H_2 ensures that H_3^+ will donate its proton to most atomic and molecular species. Thus the presence of C atoms in interstellar clouds will mean that the proton transfer reaction

$$H_3^+ + C \rightarrow CH^+ + H_2 \qquad (10.3)$$

can occur as the first step in the production of small hydrocarbon ions and neutrals. As yet, Reaction (10.3) or indeed any reactions of ions with atomic carbon have not been studied at thermal energies and there is a need for such studies. However, no example is known of an exothermic proton transfer reaction which does not proceed rapidly (Bohme 1975). The analogous proton transfer reaction of H_3^+ with N atoms is endothermic and so cannot occur under the low temperature conditions of interstellar clouds, so it has been suggested that the reaction

$$H_3^+ + N \rightarrow NH_2^+ + H \qquad (10.4)$$

might proceed followed by the sequence of H-atom reactions leading to NH_4^+ and hence to NH_3 (reaction sequence (10.10)). However, SIFT experiments have indicated that Reaction (10.4) does not proceed at a significant rate ($k(4) < 10^{-11}$ cm^3 s^{-1} at 300 K) and this has recently been given substantial support by theory (DeFrees 1987). Thus there is a problem in explaining NH_3 synthesis in interstellar clouds quantitatively and this is referred to again below.

An outstanding problem in diffuse cloud chemistry is that the observed abundance of CH^+ ions in these regions is too large to be explained by any known ion chemistry in quiescent (thermal) regions. This led Elitzur and Watson (1978, 1980) to propose that production of CH^+ was occurring via the endothermic reaction of C^+ with H_2 in shocked regions thus:

$$C^+ + H_2 \rightarrow CH^+ + H - 0.43\ \text{eV} \qquad (10.5)$$

The rate coefficient for this reaction, $k(5)$, has been shown to increase rapidly with the C^+/H_2 centre-of-mass energy in SIFDT experiments (Adams *et al.* (1984a), see Draine and Katz (1986), Twiddy, Mohebati and Tichy (1986)) and in beam experiments (Ervin and Armentrout 1986), reaching 5×10^{-11} cm^3 s^{-1} at 0.43 eV, the threshold energy for this reaction. Subsequent detailed models of the ion chemistry in shocks have indicated that this proposed mechansim alone is not efficient enough to generate the observed amount of CH^+, and so other partial solutions to this problem have been sought. Flower *et al.* (1988) proposed that the loss rate of CH^+ via the reaction

$$CH^+ + H_2 \rightarrow CH_2^+ + H \qquad (10.6)$$

may be slower under shocked conditions than is usually assumed. This would clearly help to explain the greater-than-expected abundance of CH^+. Following this proposal, Smith and Adams (1988b) measured the rate coefficient for Reaction (10.6) as a function of the CH^+/H_2 centre-of-mass energy in their SIFDT apparatus and found that the reaction proceeds at the collisional rate from thermal energy to a centre-of-mass energy of the reactants of 0.35 eV. Thus this intriguing CH^+ problem remains.

The chemistry of *dense* interstellar clouds is much richer and more varied than that of diffuse clouds. The shielding from destructive radiations by grains in these regions allows the production and survival of very complex molecules. The primary ionization process in dense clouds is the production of H^+, H_2^+ and He^+ and free electrons by the action of galactic cosmic rays on the most abundant species H_2 and He. The H_3^+ produced in the $H_2^+ + H_2$ reaction does not react with H_2 and is thus available to react with C, N and O atoms (e.g. see Equations (10.3) and (10.4)) and with ambient molecules (see below). He^+ ions only react very slowly with H_2 (the rate coefficient is less than 10^{-13} $cm^3\ s^{-1}$ at 300 K) and thus He^+ can also react with ambient molecules such as the relatively abundant CO:

$$He^+ + CO \rightarrow C^+ + O + He \qquad (10.7)$$

This rapid dissociative charge transfer reaction ($k(7) = 1.7 \times 10^{-9}$ $cm^3\ s^{-1}$ and is sensibly independent of temperature) is the major source of C^+ ions in dense clouds and initiates (in parallel with Reaction (10.3)) the production of hydrocarbon molecules (e.g. see Dalgarno and Black (1976), Smith and Adams (1981a)).

He^+ ions also react with N_2 in dense clouds thus:

$$He^+ + N_2 \rightarrow N_2^+(C^2\Sigma_u^+, v = 4) \rightarrow N^+ + N + He \qquad (10.8a)$$

$$\rightarrow N_2^+ + He + h\nu \qquad (10.8b)$$

This reaction was one of the first ion–molecule reactions to be studied because of its importance in the terrestrial ionosphere (Ferguson 1975, Smith and Adams 1980a). $k(8)$ was measured to be 1.5×10^{-9} $cm^3\ s^{-1}$ in the early SA experiments (Sayers and Smith 1964), i.e. the reaction proceeds with unit collision efficiency. It has also been studied by many other techniques including the ICR, FA, SIFT, DT and CRESU techniques (Albritton 1978, Anicich and Huntress 1986, Ikezoe *et al.* 1987). It proceeds via resonant charge transfer which results in the production of the $N_2^+(C^2\Sigma_u^+, v = 4)$ a fraction (0.6) of which predissociates to N^+ and N and the remainder radiates to form $N_2^+(B^2\Sigma_u^+)$ (and ultimately $N_2^+X^2\Sigma_g^+$) (Schmeltekopf, Ferguson and Fehsenfeld 1968). N_2^+ rapidly reacts with H_2 in dense clouds producing N_2H^+ (an observed interstellar species).

More interesting is the behaviour of the N^+ which is released in Reaction (10.8a) with 0.14 eV of translational energy and these kinetically excited ions can promote the slightly endothermic reaction (Adams *et al.* 1984a).

$$N^+ + H_2 \rightarrow NH^+ + H \qquad (10.9)$$

This reaction was shown to have a small rate coefficient at very low temperatures for thermalized ions in IT experiments (Luine and Dunn 1985) but the rate coefficient is known to be large at 300 K as shown by SIFT experiments (Adams, Smith and Paulson 1980). From these IT data, subsequent SIFDT data (Adams *et al.* 1984a, Adams and Smith 1985a) and recent CRESU data (Marquette *et al.* 1985a), Reaction (10.9) has been shown to be endothermic by about 15 meV. Since Reaction (10.9) is probably a first stage of the reaction sequence via which NH_3 is synthesised in dense clouds:

$$N^+ \xrightarrow{H_2} NH^+ \xrightarrow{H_2} NH_2^+ \xrightarrow{H_2} NH_3^+ \xrightarrow{H_2} NH_4^+ \xrightarrow{e} NH_3 \qquad (10.10)$$

then it is important to know the effective rate coefficient of Reaction (10.9) under dense cloud conditions. This has recently been computed by Yee, Lepp and Dalgarno (1987) for several temperatures and, as predicted, is considerably enhanced above the true thermal values because of the kinetic excitation of the N^+ ions. Such kinetic excitation of the product ions (and neutrals) of other ion–molecule reactions, as yet unidentified, could be an important factor in influencing their subsequent reactivity. This phenomenon should be very carefully considered for those ions which react with H_2 with rate coefficients which are very dependent on temperature, for example, the fourth reaction in the sequence of reactions (10.10) (i.e. the $NH_3^+ + H_2$ reaction) (Fehsenfeld 1975, Smith and Adams 1981a, Luine and Dunn 1985) and the $C_2H_2^+ + H_2$ reaction (Herbst, Adams and Smith 1983, Adams *et al.* 1984a).

Identifying a convincing gas phase ion chemical route for the production of the H_2S observed in cold clouds has long been a problem. All the stages in the sequence of H-atom abstraction reactions

$$S^+ \xrightarrow{H_2} SH^+ \xrightarrow{H_2} SH_2^+ \xrightarrow{H_2} SH_3^+ \qquad (10.11)$$

are quite endothermic (0.85 eV, 0.55 eV and 0.25 eV from left to right in the sequence) and so it is unlikely that these reactions could be driven by the amount of residual kinetic energy which the ions might possess on formation. Thus the route to SH_3^+ is blocked energetically preventing the production of H_2S (by dissociative recombination of H_3S^+ ions). However, the reactions in sequence (10.11) could proceed in the shocked regions of interstellar gas and it was to provide kinetic data to investigate this quantitatively that a drift tube study of these reactions was carried out at Innsbruck. The results of this study were reported by Millar *et al.* (1986) who concluded that SH^+ production (and its detection) could be a tracer of C-type magnetohydrodynamic (MHD) shocks as opposed to shocks without magnetic precursors which, it was concluded, cannot efficiently produce SH^+ ions (via the $S^+ + H_2$ reaction).

Whilst discussing S^+ reactions, it is pertinent to mention the possible role that S^+ ions play in the synthesis of the molecules C_2S and C_3S which have recently been detected in quiescent interstellar clouds. A very recent SIFT study of the reactions of S^+ ions with several hydrocarbon molecules has revealed that the S^+ ions efficiently 'insert' into small hydrocarbon molecules, e.g.

$$S^+ + C_2H_2 \rightarrow HC_2S^+ + H \quad (10.12a)$$

$$S^+ + C_2H_4 \rightarrow H_3C_2S^+ + H \quad (10.12b)$$

$$S^+ + C_3H_4 \rightarrow H_3C_3S^+ + H \quad (10.12c)$$

and this implies that these 'sulphur insertion' reactions could be the first step in the synthesis of C_2S (Reactions (10.12a) and (10.12b)) and possibly also of C_3S (Reaction (10.12c)) (Smith *et al.* 1988). This study also showed that charge transfer (not sulphur insertion) was the dominant process when S^+ reacted with larger hydrocarbon molecules and so larger organo-sulphur molecules are not formed in this way.

We have indicated how sequences of exothermic reactions of ions with H_2 can rapidly synthesize hydrogen rich ions such as H_3O^+ (reaction sequence (10.2a)) and also NH_4^+ (reaction sequence (10.10)). The reaction sequence starting from C^+, i.e.

$$C^+ \xrightarrow{H_2} CH^+ \xrightarrow{H_2} CH_2^+ \xrightarrow{H_2} CH_3^+ \xrightarrow{H_2} CH_5^+ \quad (10.13)$$

can proceed rapidly through the second and third stages to CH_3^+, but the fourth (radiative association stage in interstellar clouds) is much slower. The heats of formation of most small (i.e. one-carbon and two-carbon) hydrocarbon ions are known sufficiently accurately to determine whether or not H-atom abstraction reactions with H_2 are exothermic and thus whether the reactions could proceed in cold interstellar gas. Nevertheless experimental checks are clearly desirable, and for reactions of larger hydrocarbon ions for which thermochemical data are sparse then experiments are essential. Thus considerable experimental effort, using ICR and SIFT techniques, has been and is continuing to be applied to the study of the reactions of ions in recognisable series such as the CH_n^+ and $C_2H_n^+$ (n = 0–4) series (Adams and Smith 1983) with the important interstellar neutral species especially H_2, CO, H_2O, NH_3 and hydrocarbons. This is providing a large data base comprising thousands of reaction rate coefficients and ion product distributions for modelling interstellar ion chemistry (Anicich and Huntress 1986).

A reaction mechanism which is worthy of special mention which such studies have identified is that of carbon insertion into hydrocarbon molecules (cf. Reaction (10.12)) which in effect lengthens the carbon chain according to the general mechanism

$$C^+ + C_nH_m \rightarrow C_{n+1}H_{m-x}^+ + xH. \quad (10.14)$$

For example, this process occurs rapidly for C_nH_m = CH_4, C_2H_2, C_2H_4, C_2H_6,

C_3H_8 and n-C_4H_{10} (Herbst *et al.* 1983). However, care must be exercised in assigning structures to the product hydrocarbon ions since different isomeric forms are possible. For example, the lowest energy form of $C_3H_3^+$ ions is cyclic and this particular isomer is generally (but not exclusively) formed in ion–molecule reactions. However, it is formed exclusively in the reaction of CH_3^+ with C_2H_2 (Adams and Smith 1987b). The SIFT technique is particularly valuable for such studies because the reactant ions are isolated from their parent gas with which they often react rapidly. For example, the reactions of ions in the series NH_n^+ (n = 0–4), H_nS^+ (n = 0–3) and H_nCO^+ (n = 0–3) have been studied with many neutral molecules, mostly at 300 K, and at lower temperatures in VT-SIFT experiments for some critical reactions (Adams and Smith 1983). These studies have indicated synthetic routes to many of the observed interstellar molecules which have been substantiated in many cases by quantitative ion–chemical models (Millar *et al.* 1987). Equally importantly, such experimental studies have revealed those ions which are relatively unreactive with H_2. Such ions can then survive for a long time and thus be candidates for detection in interstellar clouds. Examples are HCO^+, N_2H^+, HCS^+, H_3O^+ and H_2CN^+ which have all been detected in interstellar gas. Many such ions remain to be detected, such as CH_3^+, CH_5^+, $C_2H_3^+$, $C_2H_5^+$ and $C_3H_3^+$ to mention only a few. More polyatomic hydrocarbon molecular ions presumably exist in dense clouds; such ions are presumably the precursors to known interstellar molecules such as C_4H, C_5H, C_6H, etc.

As the complexity of the molecular ions and neutrals increases then a greater likelihood of isomeric forms arises. Recent SIFT studies have shown that both linear and cyclic forms exist of the relatively simple ions $C_3H_2^+$ and $C_3H_3^+$ which may be the precursors of the linear and cyclic forms of the C_3H and the cyclic C_3H_2 recently detected in interstellar clouds (this is discussed further in Section 10.3.3). More recent SIFT studies in our laboratory of the reactions of ions in the series $C_nH_m^+$ (n = 4–6, m = 0–6) have revealed two stable isomers of $C_6H_5^+$ which have quite different reactivities with H_2. These studies have also shown that only ions of low hydrogenation (i.e. those for which $m < 3$) react at a significant rate with H_2 via H-atom abstraction (producing $C_nH_{m+1}^+$ ions) and thus that the highly hydrogenated ions in this series cannot be produced by sequential hydrogenation (H-atom abstraction) reactions involving H_2. Thus, highly hydrogenated ions will presumably be formed in interstellar clouds in reactions between smaller hydrocarbon ions and hydrocarbon molecules e.g.

$$C_3H_3^+ + C_2H_2 \rightarrow C_5H_4^+ + H \qquad (10.15a)$$

$$\rightarrow C_5H_5^+ + h\nu \qquad (10.15b)$$

Note that Reaction (10.15b) is indicated to be a binary (radiative) association reaction; such are now being observed in low pressure ICR experiments. The collisional association analogue of Reaction (10.15b) and many other similar reactions are well known (Smyth, Lias and Ausloos 1982).

The majority of the laboratory measurements described above were carried out at room temperature using FA, SIFT and ICR apparatuses. A significant amount of data relevant to interstellar chemistry has also been obtained in VT-SIFT apparatuses at 80 K and 200 K (and indeed at temperatures above room temperature). Low temperature studies are only possible in a VT-SIFT for gases or vapours which do not condense onto the walls of the flow tube and this restricts the low temperature study of many reactions relevant to interstellar chemistry (fortunately this problem has been circumvented by the CRESU technique; see Section 10.2.1). This limits VT-SIFT studies of the reactions of ions with polar molecules for which recent theoretical work predicts the collisional rate coefficients to increase rapidly with decreasing temperature below room temperature, and especially rapidly below ~100 K for molecules possessing large dipole moments. However, the VT-SIFT can be used to study the reactions of ions with some polar neutrals down to 200 K and this has been done for the reactions of H_3^+ with HCN and HCl (Clary, Smith and Adams 1985) as a first critical test of the ACCSA theory (adiabatic capture and centrifugal sudden approximation, Clary (1985, 1987)) which predicts a rapid increase in the collisional rate coefficient with decreasing temperature. The close agreement between the experimentally measured and theoretically predicted rate coefficients was not only very satisfying scientifically but also clearly indicated that the collisional rate coefficients for reactions involving polar molecules will be much greater at the very low temperatures of interstellar clouds than the values which have hitherto been adopted in ion–chemical models of those regions. The ACCSA theory predicts that collisional rate coefficients will be largest when the polar molecules are in the lowest rotational state and this will usually be the case in low temperature interstellar clouds. For this situation, the rate coefficients for reactions involving molecules having large dipole moments can be as large as 10^{-7} cm^3 s^{-1} (Adams, Smith and Clary 1985). Subsequent theoretical work (Troe 1985) has provided a simple parameterization of the rate coefficient as a function of temperature at low temperatures for reactions involving polar molecules.

The realization that collisional rate coefficients may be very large (and very temperature sensitive at low temperatures) led directly to the solution of a long-standing problem in astrochemistry – i.e. the anomalously high abundance of HCS^+ relative to CS in dense clouds (Millar *et al.* 1985). Radio observations of interstellar clouds have indicated that the abundance ratio [HCS^+]/[CS] varies from cloud to cloud within the approximate range from 10^{-2} to 10^{-1}. This contrasts with the abundance ratio [HCO^+]/[CO] which is measured to be ~10^{-4} and hardly differs between clouds. Limited ion–chemical models were able to predict correctly the observed $[HCO^+]/[CO]$ ratio using a canonical rate coefficient of $\sim 10^{-9}\ cm^3\ s^{-1}$ for the proton transfer reaction of H_3^+ with CO which produces the HCO^+, but were quite unable to predict the much higher observed $[HCS^+]/[CS]$ ratio using the same rate coefficient for the proton transfer reaction of H_3^+ with CS. The fact that CS is very polar whereas CO possesses only a very small permanent dipole moment is the key to the difference. Hence, the $H_3^+ + CS$ reaction proceeds increasingly rapidly as the temperature decreases and the (collisional) rate coefficient reaches $\sim 5 \times 10^{-8}\ cm^3\ s^{-1}$ at 10 K as calculated using the ACCSA theory (see Millar *et al.* 1985). This large rate coefficient implies a more efficient production of HCS^+ and explains the large $[HCS^+]/[CS]$ ratio. The variability of this ratio from cloud to cloud is thus explained, at least in part, by different ambient temperatures. The development of the CRESU technique and the careful measurements of ion–molecule reaction rate coefficients carried out at very low temperatures (Marquette *et al.* 1985b) have verified that the collisional rate coefficients for reactions involving polar molecules (e.g. H_2O and NH_3) do indeed increase to large values at the low temperatures appropriate to interstellar clouds. Equally significant is the observation that the rate coefficients for reactions involving non-polar molecules (e.g. N_2 and O_2) are invariant with temperature down to 8 K as is always assumed (Rowe *et al.* 1985). Reviews of CRESU measurements have recently been published (Rowe and Marquette 1987, Rowe 1988).

One note of caution is pertinent here regarding the rate coefficients for ion–polar molecule reactions at low temperatures. It is sometimes observed that the measured values of rate coefficients are significantly smaller (by perhaps a factor of 2) than the predicted collisional values even though the temperature variation is as predicted by the theory. A good example of this is the study of the reactions of the structural isomers of $C_2H_2Cl_2$ with N^+ and H_3^+ (Rebrion *et al.* 1988). It was observed that the experimentally measured rate coefficient for the reactions of the non-polar *trans*-1,2-$C_2H_2Cl_2$ isomer was equal to the Langevin (collisional) value and invariant with temperature, whereas those for the polar 1,1-$C_2H_2Cl_2$ and *cis*-1,2-$C_2H_2Cl_2$ isomers increased with temperature in accordance with theory but with magnitudes equal to about half the predicted values. The reason for this is not yet clear (Rebrion *et al.* 1988) but whatever the true explanation of these results, it is a definite warning that more studies of the reactions of polar molecules are necessary at low temperatures and that caution must be exercised in estimating the rate coefficients for the reactions of polar molecules at the low temperatures of interstellar clouds.

10.3.2. Isotope exchange and isotope fractionation

Many interstellar molecular species are of sufficiently high column densities to be detected not only in those forms containing the common stable isotopes of the elements but also in forms containing rarer isotopes (e.g. D, ^{13}C, ^{15}N, ^{18}O, etc.). Hence isotopic ratios for these elements can be obtained from observations of these interstellar molecules and these observations indicate an abundance ratio of the rarer (heavier) isotope to the common isotope which often exceeds the ratio observed in the solar/terrestrial environment. The interesting astrophysical question is whether this is a reflection of an enhancement of the rare isotope in the interstellar cloud material as a whole or just in the particular molecular species. Laboratory studies of isotope exchange in ion–molecule reactions have shown that fractionation of the heavier isotope into molecules can occur at low temperatures due to zero-point-energy differences (Smith and Adams 1984b) and thus 'isotope fractionation' is almost certainly contributing to, if not entirely responsible for, the observed enrichment of the rare isotopes in some interstellar molecular species.

Detailed VT-SIFT studies have been made of isotope exchange in ion–molecule reactions of potential astrophysical importance. Special consideration has been given to D/H exchange. For example, $k_f(16)$ and $k_r(16)$, the forward and reverse rate coefficients for the reaction

$$D^+ + H_2 \rightleftharpoons H^+ + HD \qquad (10.16)$$

have been determined at different temperatures (Henchman, Adams and Smith 1981). Both the enthalpy (ΔH) and the entropy (ΔS) changes have been deduced from the kinetic data obtained and they closely agree with the ΔH calculated from consideration of bond strengths and ionization energies and with the ΔS calculated using partition functions. The kinetic measurements revealed that with decreasing temperature $k_f(16)$ increased reaching the collisional rate coefficient, k_c, for temperatures $T < |\Delta H|/R$, i.e. <500/2 K, and that the $k_r(16)$ decreased as expected for this endothermic reaction. Thus, at the temperatures of cold interstellar clouds, $k_f(16) = k_c(16)$ and $k_r(16)$ is negligibly small, and so under these conditions D is very efficiently fractionated into HD.

The importance of H_3^+ in interstellar clouds has been made clear previously. The importance of the H/D exchange reaction

$$H_3^+ + HD \rightleftharpoons H_2D^+ + H_2 \qquad (10.17)$$

is therefore obvious since it results in the fractionation of D into H_2D^+ ions which can then react with other interstellar species leading to the production of deuterated molecules. The large $|\Delta H|/R$ for Reaction (10.17) (= 220 K at 20 K; Adams and Smith (1981a), Smith, Adams and Alge (1982a)) implies that a substantial fraction of the H_3^+ in dense clouds will be deuterated and presently efforts are being made to detect H_2D^+ in such clouds. Again, the laboratory measurements have revealed that $k_f(17)$ approaches $k_c(17)$ at low temperatures and that $k_r(17)$ decreases dramatically. The reaction

$$CH_3^+ + HD \rightleftharpoons CH_2D^+ + H_2 \qquad (10.18)$$

similarly has a large $|\Delta H|/R$ of 335 K and so CH_2D^+ is also efficiently produced and the back reaction inhibited (Smith *et al.* 1982a,b). Thus, subsequent reactions of CH_2D^+ ions also lead to the production of deuterated molecules.

Many 'closed shell' ions such as HCO^+, N_2H^+, H_3O^+ and NH_4^+ do not undergo isotope exchange at a significant rate with HD and thus the deuterated analogues DCO^+, N_2D^+, etc. cannot be produced by this process in interstellar clouds (Smith *et al.* 1982a). However, reactions of HCO^+ and N_2H^+ with D atoms do lead to efficient isotope fractionation (Adams and Smith 1985b, Federer *et al.* 1985) e.g.

$$HCO^+ + D \rightleftharpoons DCO^+ + H \tag{10.19}$$

This process and the H_2D^+ + CO reaction both produce DCO^+ and they must both be considered when using observed abundance ratios of $[DCO^+]/[HCO^+]$ to estimate electron densities in dense clouds (Dalgarno and Lepp 1984). A fundamentally interesting question is why isotope exchange is facile in the HCO^+ + D reaction but not in the HCO^+ + HD reaction and this has been considered in a recent paper (Henchman *et al.* 1988). It is the fractionation of D into the ions by reactions such as (10.17), (10.18) and (10.19) that is responsible for the overabundance of D in many interstellar molecules. Fractionation of ^{13}C into CO can occur via the exchange reaction:

$$^{13}C^+ + {}^{12}CO \rightleftharpoons {}^{12}C^+ + {}^{13}CO \tag{10.20}$$

Detailed VT-SIFT studies of this reaction have also been carried out (Smith and Adams 1980b) and it has been shown that this fractionation can account for the observed overabundance of ^{13}C in interstellar CO. Similarly studies have also been carried out into the fractionation of ^{15}N into N_2H^+ (Adams and Smith 1981b).

A distinct feature which emerges from these kinetic studies of isotope exchange is that the k_f (for the exothermic direction) invariably approaches the k_c for the reaction at low temperatures. Indeed, at the temperature of cold interstellar clouds it is reasonable to equate k_f to k_c. At higher temperatures which do not greatly exceed $|\Delta H|/R$, then the experimental data reveal that $k_f + k_r \sim k_c$. These 'rules' are useful in quantitatively estimating the degree of isotope fractionation in interstellar molecules. A discussion of this and a detailed overview of the laboratory measurements are given by Smith and Adams (1984b). Studies of isotope exchange in ion–molecule reactions at very low temperatures using the CRESU apparatus would be very profitable and would provide a valuable check on the 'rules' outlined above. Studies of isotope exchange in reactions for which the ΔH is expected to be very small, such as those involving ^{32}S and ^{34}S exchange, could also be carried out using the CRESU apparatus.

10.3.3. Ternary (collisional) ion–molecule association reactions and radiative association

A common process in ionized gases at the pressures and temperatures typical of many laboratory experiments is ternary (or collisional or three-body) ion–molecule association exemplified by:

$$CH_3^+ + H_2 \rightarrow (CH_5^+)^* \xrightarrow{He} CH_5^+ + He \qquad (10.21)$$

In this reaction, the unstable $(CH_5^+)^*$ can be prevented from dissociating back to the reactants by a collision with an inert third body (a He atom in this example) which produces a stable, bound CH_5^+ ion. In this way, complex ions, including cluster ions consisting of a 'core ion' and many ligand molecules (e.g. $H_3O^+(H_2O)_n$), can be formed in ionized gases especially at high pressures and low temperatures. Such processes are especially important in the lower terrestrial atmosphere. A great deal of experimental work has been carried out to determine the rate coefficients, k_3, for many ternary association reactions, and their variation with temperature, and, in response, theories have been developed to describe the temperature variation of the k_3. Of note are the theories due to Bates (1979) and Herbst (1979) which predict that k_3 should vary as $\sim T^{-r/2}$, where r is the number of rotational degrees of freedom in the separated ion–molecule reactants. This prediction has been borne out by measurements of the k_3 for many reactions including Reaction (10.21) and several other association reactions of CH_3^+ ions which have been studied in VT-SIFT experiments.

Ternary association cannot be significant at the low pressures of interstellar clouds, but the analogous process of binary radiative association:

$$CH_3^+ + H_2 \rightarrow (CH_5^+)^* \rightarrow CH_5^+ + h\nu \qquad (10.22)$$

is very likely to occur at the low temperatures pertaining to these media. Such processes are very difficult to study under laboratory conditions but, nevertheless, some success has been obtained recently and the rate coefficient for Reaction (10.22) has been reported to be $1.8 \times 10^{-13}\,cm^3\,s^{-1}$ at 13 K using the low pressure, low temperature IT technique (Barlow, Dunn and Schauer (1984), see Section 10.2.1). However, Bates (1986a) has pointed out that the internal energies of the reactants in the IT experiment were not thermal and that the kinetic temperature of the ions was not 13 K but 53 K. Nevertheless these experimental measurements show that ion–molecule radiative association is a reality and recently binary association has also been observed between some complex hydrocarbon ions and hydrocarbon molecules in ICR experiments (McEwan 1988). Thus laboratory data on radiative association rate coefficients are slowly becoming available. Meanwhile estimates of radiative association rate coefficients are being obtained using theoretical models coupled with experimental values of ternary association reaction rate coefficients. This approach has been discussed in detail in several papers (see, e.g., Smith and Adams (1981a)) and, in brief, is as follows. An estimate of the lifetime of the intermediate excited ion, e.g. $(CH_5^+)^*$ in Reaction (10.21), can be obtained from a determination of the k_3 for the reaction and by adopting a value for the radiative lifetime (or the radiative rate coefficient) of the excited ion. The latter is commonly assumed to be 10^{-3}–10^{-2} s as is appropriate for pure vibrational state (infrared) transitions. However, it is very interesting to note that when the rate coefficient for Reactions (10.22) as *measured* using the IT experiment is combined with the VT-SIFT data for Reaction (10.21) in order to obtain a value for the radiative lifetime of the $(CH_5^+)^*$, a value of $\sim 3 \times 10^{-5}$ s is obtained. This has led Bates and Herbst (1988) to propose that stabilization of the $(CH_5^+)^*$ in Reaction (10.22) results from transitions between electronic states with small radiative lifetimes thus rendering the radiative association process much more efficient. This could be a common phenomenon, especially for polyatomic ion complexes. Even for the triatomic system

$$C^+ + H_2 \rightarrow CH_2^+(^2B_1) \rightarrow CH_2^+ + h\nu \qquad (10.23)$$

detailed calculations have shown that the reaction proceeds mainly via an electronic state which has a radiative lifetime as small as 10^{-5} s (Herbst *et al.* 1977). Reaction (10.23) is considered to be an important first step in the synthesis of hydrocarbon molecules in dense interstellar clouds (e.g. Dalgarno and Black (1976), Smith and Adams (1981a)). The ternary association analogue of Reaction (10.23) has been studied in a VT-SIFT (in helium carrier gas) and $k_3(23)$ varies as $T^{-1.3}$ (Adams and Smith 1981c). Again, studies of this type in the very low temperature CRESU experiment, perhaps also using para-H_2 as the reactant neutral, would be very instructive.

It is perhaps significant that as VT-SIFT studies are extended to include the reactions of polyatomic ions (with low recombination energies) it is observed that association reactions become more evident. Very often, the effective binary rate coefficients for these association reactions are independent of the carrier gas (usually helium) pressure down to the lowest pressures (~0.2 Torr) at which SIFT experiments can be carried out. This phenomenon is especially evident at low temperatures. It then becomes a challenge to determine whether the reaction is a 'saturated' ternary association reaction or a binary radiative association reaction. Evidence is mounting that some of these reactions do indeed proceed via radiative association (Babcock and Streit 1984). Several such reactions have been identified for which the reactant neutral is H_2 and this is then very interesting from the viewpoint of interstellar chemistry. A case in point is the reaction:

$$C_3H^+ + H_2 \rightarrow l\text{-}C_3H_3^+ \qquad (10.24a)$$

$$\rightarrow c\text{-}C_3H_3^+ \qquad (10.24b)$$

It has been shown that the products of this reaction are both the linear (l) and cyclic (c) isomers of $C_3H_3^+$ (Smith and Adams 1987), and it has been argued that the reaction could be proceeding via radiative association under the conditions of interstellar clouds. It has also been proposed that Reaction (10.24b) is the main source of c-$C_3H_3^+$ and hence of the abundant c-C_3H_2 in interstellar clouds (Adams and Smith 1987a).

Much more work needs to be done on association reactions which are likely to proceed via radiation emission and indeed efforts are needed to detect the emitted radiations. This is an exciting area of ion–molecule studies. Many questions are being asked that require answers, such as to what extent radiative association can occur in shocked interstellar gas (Herbst 1985). Recent VT-SIFDT studies of the association reactions of CH_3^+ ions in drift fields (Adams and Smith 1987c) may offer some help in answering this question.

10.3.4. Positive ion–electron dissociative recombination

Many of the molecular positive ions synthesised by ion–molecule reactions in interstellar clouds will be destroyed by reactions with ambient electrons thus producing neutral molecular fragments. This is the process of dissociative recombination, (exemplified by Reaction (10.25) below) which is the final step in the production of many of the observed interstellar molecules. Mutual neutralization of positive ions with negative ions (discussed briefly in Section 10.3.5), proton transfer and electron (charge) transfer involving species of low ionization energy (such as metal atoms) are other processes for production of neutral molecules from positive ions (although obviously the last two processes do not result in loss of ionization). Dissociative recombination is the most important deionization process in ionized gases in which electrons are dominant over negative ions, as is the situation, for example, in the terrestrial ionosphere (in the lower stratosphere and the troposphere negative ions dominate over electrons). Hence, much of the early laboratory work using the SA technique was directed towards the determination of the dissociative recombination coefficients, α_e, for atmospheric molecular positive ions such as O_2^+ and NO^+ (Kasner and Biondi 1968, Weller and Biondi 1968). The SA value of α_e for the reaction

$$O_2^+ + e \rightarrow O + O \qquad (10.25)$$

has been confirmed by the many checks subsequently made using other techniques and now it is established that $\alpha_e(O_2^+) = 2 \times 10^{-7}(T/300)^{0.7}$ over the thermal energy range. This has become the reaction which is used to check the validity of new techniques for determining α_e values. The strengths and weaknesses of the techniques used for determining α_e have been outlined in Section 10.2.2. Further discussions together with a list of the α_e, determined using the SA, MB and IT techniques are given in the review by Mitchell and McGowan (1983). Therefore, it is sufficient to note here that the SA is not suited to the study of many positive ion species involved in interstellar chemistry (although $\alpha_e(HCO^+)$ has been determined in the SA, Leu, Biondi and Johnsen (1973a)). The MB technique is more versatile but, as mentioned in Section 10.2.2, problems arise because the ions in the beams will often be internally excited, and also because it is difficult to translate the cross section data into α_e appropriate to the very low temperatures pertaining to interstellar clouds. Thus it is the most recent innovation, the VT-FALP technique, which is proving to be the most useful in the determination of the α_e for molecular ions of interstellar interest. This technique is particularly valuable because measurements of α_e can be made down to ~90 K for some ionic species. The determinations of $\alpha_e(O_2^+)$ and $\alpha_e(NO^+)$ over the temperature range 200–600 K using the VT-FALP are in excellent agreement with the SA data for $\alpha_e(O_2^+)$ and the IT data for $\alpha_e(NO^+)$ and this has established the accuracy of the VT-FALP method (Alge, Adams and Smith 1983). It was therefore surprising when VT-FALP studies clearly indicated that $\alpha_e(H_3^+)$ and $\alpha_e(D_3^+)$ (and by inference $\alpha_e(H_2D^+)$) were immeasurably small (i.e. $\alpha_e(H_3^+) < 2 \times 10^{-8}$ cm^3 s^{-1}) within the temperature range 95–550 K (Adams, Smith and Alge 1984b, Smith and Adams 1984c). This result is contrary to that obtained using both the SA and MB techniques which both indicated $\alpha_e(H_3^+)$ to be greater than 10^{-7} cm^3 s^{-1} (Leu, Biondi and Johnsen 1973b, Macdonald, Biondi and Johnsen 1984, Mitchell *et al.* 1984). However, about the same time as the VT-FALP data were obtained theoretical work by Michels and Hobbs (1984) indicated that $\alpha_e(H_3^+)$ should be immeasurably small for ground state H_3^+ ions but significant for vibrationally-excited H_3^+ ions (as was recognizable in the VT-FALP data!). This very important result has been discussed at length recently (Adams and Smith 1988a) and it now seems clear that: (i) $\alpha_e(H_3^+)$ is much less than 10^{-8} cm^3 s^{-1}, under the low temperature and low ionization density conditions of interstellar clouds, the latest estimate from recent VT-FALP work being $\lesssim 10^{-10}$ cm^3 s^{-1}; (ii) the large value obtained in the MB experiments was due to the presence of vibrationally excited H_3^+ ions in the ion beams. Recent MB work has shown a variation of $\alpha_e(H_3^+)$ with the vibrational state of the H_3^+ ions (Hus *et al.* 1988); (iii) the large value obtained in the SA experiments was due to impurity molecular ions in the afterglow plasmas (Johnsen 1987), probably H_5^+ and/or CH_5^+, the α_e for both these ions being large (Leu *et al.* 1973b, Adams *et al.* 1984b). A postscript to the studies of $\alpha_e(H_3^+)$ is the recent work of Amano (1988) who has studied spectroscopically the loss rate of ground vibronic state H_3^+ in a pulsed afterglow. From this work, a value for $\alpha_e(H_3^+)$ of 1.8×10^{-7} cm^3 s^{-1} was derived. It appears, however, that a major loss process for the H_3^+ ions in Amano's study was probably collisional-radiation recombination (Bates and Dalgarno 1962) and so this large value for $\alpha_e(H_3^+)$ should be treated with caution.

That $\alpha_e(H_3^+)$ and also $\alpha_e(H_2D^+)$ are small for vibrationally-relaxed ions under interstellar cloud conditions has important implications. It implies a greater number density of these ions in interstellar clouds (which improves the possibility of their detection) and also a more rapid chemistry. Also, the method due to Watson (1977, 1978) for estimating the electron number density, n_e, in dense clouds has to be abandoned since it expresses n_e in terms of $\alpha_e(H_2D^+)$ as well as other parameters, notably the observed $[DCO^+]/[HCO^+]$ number density ratio. This method has been replaced by other methods for estimating n_e (Smith and Adams 1984c, Dalgarno and Lepp 1984). A very small $\alpha_e(H_3^+)$ also has important implications to diffuse cloud chemistry (van Dishoeck 1988) and to the chemistry of the Jovian ionosphere (Dalgarno 1988). A more precise magnitude for $\alpha_e(H_3^+)$ rather than the upper limit value of 10^{-10} cm^3 s^{-1} (Adams and Smith 1988a) is required to satisfy the demands of current interstellar and Jovian atmospheric ion–chemical models. However, a new experimental approach will be necessary to achieve this.

From a practical viewpoint, the fact that H_3^+ does not recombine to a significant extent in the VT-FALP plasma offers an opportunity to determine the α_e for a wide range of ionic species of interstellar importance. Thus, an H_3^+/electron plasma is created in the upstream region of the VT-FALP and a molecular gas, X, to which H_3^+ can transfer a proton is added downstream *viz*:

$$H_3^+ + X \rightarrow XH^+ + H_2 \qquad (10.26)$$

Such proton transfer reactions are rapid if the proton affinity (*PA*) of X exceeds $PA(H_2)$ (see the discussion centred around Reaction (10.3)) which is the case for many X. Thus the α_e for HCO^+, N_2H^+ and CH_5^+ have been determined at 95 K and 300 K (Adams *et al.* 1984b) and the results indicate, in common with previous SA measurements for other ions that when α_e is large (as is the case of $\alpha_e(CH_5^+) = 1.1 \times 10^{-6}$ cm^3 s^{-1} at 300 K), only a gradual increase of α_e occurs with decreasing temperature, whereas for relatively small α_e (such as is the case for $\alpha_e(HCO^+)$) then a more rapid increase of α_e occurs with decreasing temperature (e.g. $\alpha_e(HCO^+) \sim T^{-1}$). Very recently, the α_e have been determined at 300 K using the VT-FALP for a variety of protonated positive ion species including the known interstellar ions HCO_2^+ and H_2CN^+ and several other species which are expected to be present in dense interstellar clouds. The measured values range from $\alpha_e(HCO_2^+) = 3.4 \times 10^{-7}$ cm^3 s^{-1} to $\alpha_e(C_2H_5OH_2^+) = 1.1 \times 10^{-6}$ cm^3 s^{-1}. At a temperature of dense clouds of say 20 K, the α_e for most of these 'normal' molecular ions (excluding H_3^+!), based on the data available on the temperature variation of α_e values, should lie within the limited range of $\sim(2–4) \times 10^{-6}$ cm^3 s^{-1} (Adams and Smith 1988b).

A major area of ignorance lies in the identity of the products of dissociative recombination reactions, yet a knowledge of this is clearly required if a proper understanding of plasma chemistry and especially interstellar chemistry is to be obtained. Some information is available regarding the states of the atomic products of Reaction (10.25) and a good deal of theoretical effort has been put into that reaction (Guberman 1987). Also, some data are available concerning the products of recombination of $N_2^+(v = 1)$ and NO^+, CO_2^+ and H_2O^+ in unknown states of excitation (see Adams and Smith (1988a) for a summary of the available data). However, from the viewpoint of interstellar chemistry, it is necessary to identify the products of recombination of many ground state polyatomic molecular ions. This is a difficult area for theory and thus experimental work is urgently required. To this end, both VUV absorption spectroscopy and laser induced fluorescence (LIF) spectroscopy are now being used in the Birmingham laboratory (N. G. Adams and D. Smith) and VUV absorption spectroscopy is being used in the Rennes laboratory (J. L. Queffelec and B. R. Rowe) to diagnose the products of dissociation recombination in FALP plasmas. A priority is to determine the products of the reaction:

$$H_3O^+ + e \rightarrow H_2O + H\,? \qquad (10.27a)$$

$$\rightarrow OH + H + H\,? \qquad (10.27b)$$

It is important to know if H_2O and/or OH radicals are generated in this reaction since theories are in conflict. Herbst (1978) argues for OH production whereas Bates (1986b) argues for H_2O production. Thus VUV absorption will be used to detect any H-atom product and LIF will be used to detect any OH product. Preliminary results indicate that OH is certainly produced and the fraction of each recombining H_3O^+ that leads to this product has been ascertained (Herd *et al.* 1989). Following this study similar questions will be asked about NH_4^+ (for example is there any NH or NH_2 product?) and H_2CN^+ (any CN product?). This promises to be an exciting and profitable study, both from the viewpoint of fundamental molecular physics and of interstellar chemistry.

10.3.5. Positive ion–negative ion mutual neutralization

When negative ions are present in significant concentrations in ionized gases or plasmas then positive ion–negative ion mutual neutralization becomes a potentially important ionization loss process. It is exemplified by the much-studied elementary reaction

$$NO^+ + NO_2^- \rightarrow NO + NO_2 \qquad (10.28)$$

It is known from emission spectroscopy that the energy released in this reaction appears largely as electronic excitation of the NO (Smith, Adams and Church 1978). Since it is now being seriously considered that negative ions, specifically negatively-charged polyaromatic hydrocarbons (PAH^-) (Omont 1986, Lepp and Dalgarno 1988) may exist in dense interstellar clouds, then mutual neutralization is a potentially important process for charge neutralization and hence for neutral molecule formation. Thus reactions such as

$$H_3^+ + PAH^- \rightarrow PAH + H_2 + H \quad (10.29)$$

could be effective in removing ionization from interstellar clouds. It is therefore important to be able to estimate the mutual neutralization (or binary ionic recombination) coefficients, α_i, for reactions such as (10.29) at appropriately low temperatures. Omont (1986) gives a theoretical expression for α_i in his paper. However, experimental α_i data are available for a variety of different reactions involving positive and negative ions of greatly differing complexities. These data were obtained using the VT-FALP technique and lists of the α_i values have been compiled (Smith and Adams 1983). Most of the data were obtained at room temperature and the somewhat surprising (but usefully simple) result is that α_i varies only within the limited range of (3–10) $\times 10^{-8}$ cm^3 s^{-1} at that temperature, there being no correlation with the complexity of the reactant molecular ions (note, however, that the α_i for reactions involving only atomic positive and atomic negative ions are usually very small for well understood reasons (Church and Smith 1978)). The VT-FALP work indicates that α_i varies with temperature as $\sim T^{-0.5}$ which is in accordance with the 'absorbing sphere theoretical model' for such reactions (Olsen 1972). On the basis of this T dependence and the accumulated VT-FALP data, the α_i appropriate to molecular ions in dense interstellar clouds at 20 K is predicted to be $\sim$(2–4) $\times 10^{-7}$ cm^3 s^{-1}, that is about an order-of-magnitude smaller than the α_e at the same temperature (see Section 10.34). Clearly, the relative importance of dissociative recombination and mutual neutralization will depend on the electron/negative ion number density ratio in the interstellar cloud as well as on α_e and α_i. Again little is known about the products of mutual neutralization reactions but further spectroscopic studies, such as those mentioned in relation to Reaction (10.28) and outlined in Section 10.3.4, for the products of dissociation recombination should begin to provide information at least for relatively elementary mutual neutralization reactions.

10.4. Concluding remarks

Great progress has been made in the understanding of interstellar chemistry over the last 15 years due to the combined efforts of radio-and-optical astronomers, theoretical modellers and laboratory kineticists. Many of the basic processes that initiate molecular synthesis in interstellar clouds have been identified and an overview of these processes has been presented in this chapter. However, problems remain concerning the observed abundances of some molecular species. For example, satisfactory quantitative gas phase routes to the production of NH_3 and H_2S have yet to be identified. Hence grain surface catalysis is invoked as a production process for these species. Whilst such surface catalytic processes cannot be ruled out, it is perhaps significant that, as a better understanding of ion–molecule reaction kinetics is obtained following developments in experiment and in theory, then some astronomical observations which could not initially be reconciled with gas phase ion–chemical models are being resolved. A good example is the problem of the apparently anomalous overabundance of HCS^+ in dense clouds which was resolved when it was shown that the rate coefficients for the reactions of ions with polar molecules (e.g. CS) increase dramatically with reducing temperature. So the optimistic view must be that future developments in the further understanding of reaction kinetics will resolve other problems. We await with anticipation the advancement in understanding that will surely take place with the exploitation of the ion-injected CRESU experiment, which will allow studies to be made of many molecule–ion reactions at very low temperatures, and the further exploitation of VUV and laser spectroscopic techniques for the identification of the products of recombination reactions in the VY-FALP plasmas. Such developments will accelerate the understanding of the fundamentals of ion–molecule and ion–electron reactions and consequently of interstellar chemistry. The future appears as bright as ever for laboratory ionic astrophysical and astrochemical research.

REFERENCES

Abgrall, H., Giusti-Suzor, A. and Roueff, E. (1976). *Ap. J.* (*Letters*), **207**, L69.
Adams, N. G. and Smith, D. (1976a). *Int. J. Mass Spectrom. Ion Phys.*, **21**, 349.
Adams, N. G. and Smith, D. (1976b). *J. Phys. B.*, **9**, 1439.
Adams, N. G. and Smith, D. (1981a). *Ap. J.*, **248**, 373.
Adams, N. G. and Smith, D. (1981b). *Ap. J.* (*Letters*), **247**, L123.
Adams, N. G. and Smith, D. (1981c). *Chem. Phys. Letts.*, **79**, 563.
Adams, N. G. and Smith, D. (1983). In *Reactions of Small Transient Species*, eds. A. Fontijn and M. A. A. Clyne (London, Academic Press), p. 311.
Adams, N. G. and Smith, D. (1985a). *Chem. Phys. Lett.*, **117**, 67.
Adams, N. G. and Smith, D. (1985b). *Ap. J.* (*Letters*), **294**, L63.
Adams, N. G. and Smith, D. (1987a). In *Astrochemistry*, eds. M. S. Vardya and S. P. Tarafdar (Dordrecht, Reidel), p. 1.
Adams, N. G. and Smith, D. (1987b). *Ap. J.* (*Letters*), **317**, L25.
Adams, N. G. and Smith, D. (1987c). *Int. J. Mass Spectrom. Ion Processes*, **81**, 273.
Adams, N. G. and Smith D. (1988a). In *Rate Coefficients in Astrochemistry*, eds. T. J. Millar and D. A. Williams (Dordrecht, Reidel), p. 173.
Adams, N. G. and Smith, D. (1988b). *Chem. Phys. Letts.*, **144**, 11.
Adams, N. G. and Smith D. (1988c). In *Techniques for the Study of Gas Phase Ion–Molecule Reactions*, eds. J. M. Farrar and W. H. Saunders Jr (New York, Wiley), p. 165.
Adams, N. G., Smith, D. and Paulson, J. F. (1980). *J. Chem. Phys.*, **72**, 288.

Adams, N. G., Smith, D. and Millar, T. J. (1984a). *Mon. Not. R. Astr. Soc.*, **211**, 857.
Adams, N. G., Smith, D. and Alge, E. (1984b). *J. Chem. Phys.*, **81**, 1778.
Adams, N. G., Smith, D. and Clary, D. C. (1985). *Ap. J.* (*Letters*), **296**, L31.
Albritton, D. L. (1978). *Atom. Data Nucl. Data Tables*, **22**, 1.
Alge, E., Adams, N. G. and Smith, D. (1983). *J. Phys. B.*, **16**, 1433.
Amano, T. (1988). *Ap. J.* (*Letters*), **329**, L121.
Anicich, V. G. and Huntress, W. J. Jr (1986). *Ap. J. Suppl. Series*, **62**, 553.
Auerbach, D., Cacak, R., Caudano, R., Gaily, T. D., Keyser, C. J., McGowan, J. W., Mitchell, J. B. A. and Wilk, S. F. J. (1977). *J. Phys. B.*, **10**, 3797.
Babcock, L. M. and Streit, G. E. (1984). *J. Phys. Chem.*, **88**, 5025.
Bardsley, J. N. and Biondi, M. A. (1970). *Adv. Atom. Mol. Phys.*, **6**, 1.
Barlow, S. E., Dunn, G. H. and Schauer, M. (1984). *Phys. Rev. Letts.*, **52**, 902; **53**, 1610.
Barlow, S. E., Luine, J. A. and Dunn, G. H. (1986). *Int. J. Mass Spectrom. Ion Processes*, **74**, 97.
Bates, D. R. (1979). *J. Phys. B.*, **12**, 4135.
Bates, D. R. (1986a). *Phys. Rev.*, **A34**, 1878.
Bates, D. R. (1986b). *Ap. J.* (*Letters*), **306**, L45.
Bates, D. R. and Dalgarno, A. (1962). In *Atomic Collision Processes*, ed. D. R. Bates (New York, Academic), p. 245.
Bates, D. R. and Herbst, E. (1988). In *Rate Coefficients in Astrochemistry*, eds. T. J. Millar and D. A. Williams (Dordrecht, Reidel), p. 17.
Bohme, D. K. (1975). In *Interactions between Ions and Molecules* ed. P. Ausloos (New York, Plenum), p. 489.
Chambaud, G., Launay, J. M., Levy, B., Millie, P., Roueff, E. and Tran Minh, F. (1980). *J. Phys. B.*, **13**, 4205.
Church, M. J. and Smith, D. (1978). *J. Phys. D.*, **11**, 2199.
Clary, D. C. (1985). *Mol. Phys.*, **54**, 605.
Clary, D. C. (1987). *J. Chem. Soc. Faraday Trans. II*, **83**, 139.
Clary, D. C., Smith, D. and Adams, N. G. (1985). *Chem. Phys. Lett.*, **119**, 320.
Dalgarno, A. (1975). In *Interactions between Ions and Molecules*, ed. P. Ausloos (New York, Plenum), p. 341.
Dalgarno, A. (1981). *Phil. Trans. R. Soc. Lond.*, **A303**, 513.
Dalgarno, A. (1986). *Quart. J.R.A.S.*, **27**, 83.
Dalgarno, A. (1988). In *Rate Coefficients in Astrochemistry*, eds. T. J. Millar and D. A. Williams (Dordrecht, Reidel), p. 321.
Dalgarno, A. and Black, J. H. (1976). *Rept. Prog. Phys.*, **39**, 573.
Dalgarno, A. and Lepp, S. (1984). *Ap. J.*, **287**, L47.
DeFrees, D. J. (1987). Private communications.
van Dishoeck, E. F. (1988). In *Rate Coefficients in Astrochemistry*, eds. T. J. Miller and D. A. Williams (Dordrecht, Reidel), p. 49.
van Dishoeck, E. F. and Black, J. H. (1986). *Ap. J. Suppl. Series*
Draine, B. T. and Katz, N. (1986). *Ap. J.*, **306**, 655.
Dunkin, D. B., Fehsenfeld, F. C., Schmeltekopf, A. L. and Ferguson, E. E. (1968). *J. Chem. Phys.*, **49**, 1365.
Elitzur, M. and Watson, W. D. (1978). *Ap. J.* (*Letters*), **222**, L141.
Elitzur, M. and Watson, W. D. (1980). *Ap. J.*, **236**, 172.
Ervin, K. M. and Armentrout, P. B. (1986). *J. Chem. Phys.*, **84**, 6738, 6750.
Federer, W., Villinger, H., Tosi, P., Bassi, D., Ferguson, E. E. and Lindinger, W. (1985). In *Molecular Astrophysics*, eds. G. H. F. Diercksen, W. F. Huebner and P. W. Langhoff (Dordrecht, Reidel), p. 649.

Federer, W., Villinger, H., Lindinger, W. and Ferguson, E. E. (1986). *Chem. Phys. Letts.,* **123**, 12.
Fehsenfeld, F. C. (1975). *Int. J. Mass Spectrom. Ion Phys.,* **16**, 151.
Fehsenfeld, F. C. (1976). *Ap. J.,* **209**, 638.
Fehsenfeld, F. C. and Ferguson, E. E. (1972). *J. Chem. Phys.,* **56**, 3066.
Fehsenfeld, F. C., Lindinger, W., Schmeltekopf, A. L., Albritton, D. L. and Ferguson, E. E. (1975). *J. Chem. Phys.,* **62**, 2001.
Ferguson, E. E. (1975). In *Atmospheres of Earth and the Planets,* ed. B. M. McCormac (Dordrecht, Reidel), p. 197.
Ferguson, E. E., Fehsenfeld, F. C. and Schmeltekopf, A. L. (1969). *Adv. Atom. Mol. Phys.,* **5**, 1.
Flower, D. R., Monteiro, T. S., Pineau des Forêts, G. and Roueff, E. (1988). In *Rate Coefficients in Astrochemistry,* eds. T. J. Millar and D. A. Williams (Dordrecht, Reidel), p. 271.
Gerlich, D. and Kaefer, G. (1987). *Proc. 5th Inter. Swarm Seminar,* Birmingham, England, 29–31 July 1987, p. 133.
Guberman, S. L. (1987). *Nature,* **327**, 408.
Henchman, M. J., Adams, N. G. and Smith, D. (1981). *J. Chem. Phys.,* **75**, 1201.
Henchman, M. J., Paulson, J. F., Smith, D., Adams, N. G. and Lindinger, W. (1988). In *Rate Coefficients in Astrochemistry,* eds. T. J. Millar and D. A. Williams (Dordrecht, Reidel), p. 201.
Herbst, E. (1978). *Ap. J.,* **222**, 508.
Herbst, E. (1979). *J. Chem. Phys.,* **70**, 2201.
Herbst, E. (1985). *Ap. J.,* **291**, 226.
Herbst, E. and Klemperer, W. (1973). *Ap. J.,* **185**, 505.
Herbst, E., Schubert, J. G. and Certain, P. R. (1977). *Ap. J.,* **213**, 696.
Herbst, E., Adams, N. G. and Smith, D. (1983). *Ap. J.,* **269**, 329.
Herd, C. R., Adams, N. G. and Smith, D. (1989). *Ap. J.,* in press.
Huntress, W. T. Jr (1977). *Ap. J. Suppl. Series,* **33**, 495.
Hus, H., Youssif, F., Noren, C., Sen, A. and Mitchell, J. B. A. (1988). *Phys. Rev. Letts.,* **60**, 1006.
Ikezoe, Y., Matsuoka, S., Takebe, M. and Viggiano, A. A. (1987). *Gas Phase Ion–Molecule Reaction Rate Constants through 1986* (Tokyo, Maruzen Co. Ltd.).
Johnsen, R. (1987). *Int. J. Mass Spectrom. Ion Processes,* **81**, 67.
Kasner, W. H. and Biondi, M. A. (1968). *Phys. Rev.,* **174**, 139.
Lepp, S. and Dalgarno, A. (1988). *Ap. J.,* **324**, 553.
Leu, M. T., Biondi, M. A. and Johnsen, R. (1973a). *Phys. Rev.,* **A8**, 420.
Leu, M. T., Biondi, M. A. and Johnsen, R. (1973b). *Phys. Rev.,* **A8**, 413.
Leung, C. M., Herbst, E. and Huebner, W. F. (1984). *Ap. J. Suppl. Series,* **56**, 231.
Lindinger, W. and Smith, D. (1983). In *Reactions of Small Transient Species,* eds. A. Fontijn and M. A. A. Clyne (London, Academic Press), p. 387.
Luine, J. A. and Dunn, G. H. (1985). *Ap. J. (Letters),* **299**, L67.
Macdonald, J. A., Biondi, M. A. and Johnsen, R. (1984). *Planet Space Sci.,* **32**, 651.
Marquette, J. B., Rowe, B. R., Dupeyrat, G. and Roueff, E. (1985a). *Astron. Astrophys.,* **147**, 115.
Marquette, J. B., Rowe, B. R., Dupeyrat, G., Poissant, G. and Rebrion, C. (1985b). *Chem. Phys. Lett.,* **122**, 431.
McEwan, M. J. (1988). Private communication.
McFarland, M., Albritton, D. L., Fehsenfeld, F. C., Ferguson, E. E. and Schmeltekopf, A. L. (1973). *J. Chem. Phys.,* **59**, 6610, 6620, 6629.
McIver, R. T. Jr (1978). *Rev. Sci. Instrum.,* **49**, 111.
Mehr, F. J. and Biondi, M. A. (1969). *Phys. Rev.,* **181**, 264.

Michels, H. H. and Hobbs, R. H. (1984). *Ap. J. (Letters)*, **286**, L27.
Millar, T. J. and Nejad, L. A. M. (1985). *Mon. Not. R. Astr. Soc.*, **217**, 507.
Millar, T. J., Adams, N. G., Smith, D. and Clary, D. C. (1985). *Mon. Not. R. Astr. Soc.*, **216**, 1025.
Millar, T. J., Adams, N. G., Smith, D., Lindinger, W. and Villinger, H. (1986). *Mon. Not. R. Astr. Soc.*, **221**, 673.
Millar, T. J., Leung, C. M. and Herbst, E. (1987). *Astron. Astrophys.*, **183**, 109.
Mitchell, G. F. (1987). In *Astrochemistry*, eds. M. S. Vardya and S. P. Tarafdar (Dordrecht, Reidel), p. 275.
Mitchell, J. B. A. (1986). In *Atomic Processes in Electron–Ion and Ion–Ion Collisions*, ed. F. Brouillard (New York, Plenum), p. 185.
Mitchell, J. B. A. and McGowan, J. W. (1983). In *Physics of Ion–Ion and Electron–Ion Collisions*, eds. F. Brouillard and J. W. McGowan (New York, Plenum), p. 279.
Mitchell, J. B. A., Ng, C. T., Forard, L., Janssen, R. and McGowan, J. W. (1984). *J. Phys. B.*, **17**, L909.
Moseley, J. T., Olsen, R. E. and Peterson, J. R. (1975). In *Case Studies in Atomic Physics*, Vol. 5, eds. M. R. C. McDowell and E. W. McDaniel (Amsterdam, North Holland), p. 1.
Olsen, R. E. (1972). *J. Chem. Phys.*, **56**, 2979.
Omont, A. (1986). *Astron. Astrophys.*, **164**, 159.
Pineau des Forêts, G., Flower, D. R., Hartquist, T. W. and Millar, T. J. (1987). *Mon. Not. R. astr. Soc.*, **227**, 993.
Rebrion, C., Marquette, J. B., Rowe, B. R., Chakravarty, C., Clary, D. C., Adams, N. G. and Smith, D. (1988). *J. Phys. Chem.*, **92**, 6572.
Rowe, B. R. (1988). In *Rate Coefficients in Astrochemistry*, eds. T. J. Millar and D. A. Williams (Dordrecht, Reidel), p. 135.
Rowe, B. R. and Marquette, J. B. (1987). *Int. J. Mass. Spectrom Ion Processes*, **80**, 239.
Rowe, B. R., Marquette, J. B., Dupeyrat, G. and Ferguson, E. E. (1985). *Chem. Phys. Lett.*, **113**, 403.
Rowe, B. R. and Queffelec, J. L. (1989). In *Dissociative Recombination: Theory, Experiment and Applications*, eds. J. B. A. Mitchell and S. L. Guberman (Singapore, World Scientific), p. 151.
Rowe, B. R., Vallee, F., Queffelec, J. L., Gomet, J. C. and Morlais, M. (1988). *J. Chem. Phys.*, **88**, 845.
Sayers, J. and Smith, D. (1964). *Disc. Farad. Soc.*, **37**, 167.
Schmeltekopf, A. L., Ferguson, E. E. and Fehsenfeld, F. C. (1968). *J. Chem. Phys.*, **48**, 2966.
Smith, D. and Adams, N. G. (1979a). In *Gas Phase Ion Chemistry*, Vol. 1, ed. M. T. Bowers (New York, Academic Press), p. 1.
Smith, D. and Adams, N. G. (1979b). In *Kinetics of Ion–Molecule Reactions*, ed. P. Ausloos (New York, Plenum), p. 345.
Smith, D. and Adams, N. G. (1980a). *Topics in Current Chemistry*, **89**, 1.
Smith, D. and Adams, N. G. (1980b). *Ap. J.*, **242**, 424.
Smith, D. and Adams, N. G. (1981a). *Int. Rev. Phys. Chem.*, **1**, 271.
Smith, D. and Adams, N. G. (1981b). *Mon. Not. R. Astr. Soc.*, **197**, 377.
Smith, D. and Adams, N. G. (1983). In *Physics of Ion–Ion and Electron–Ion Collisions*, eds. F. Brouillard and J. W. McGowan (New York, Plenum), p. 501.
Smith, D. and Adams, N. G. (1984a). In *Swarms of Ions and Electrons in Gases*, eds. W. Lindinger, T. D. Mark and F. Howorka (Vienna, Springer-Verlag), p. 284.
Smith, D. and Adams, N. G. (1984b). In *Ionic Processes in the Gas Phase*, ed. M. A. Almoster Ferreira (Dordrecht, Reidel), p. 41.
Smith, D. and Adams, N. G. (1984c). *Ap. J. (Letters)*, **284**, L13.

Smith, D. and Adams, N. G. (1985). In *Molecular Astrophysics,* eds. G. H. F. Dierksen, W. F. Huebner and P. W. Langhoff (Dordrecht, Reidel), p. 453.
Smith, D. and Adams, N. G. (1987). *Int. J. Mass Spectrom. Ion Processes,* **76**, 307.
Smith, D. and Adams, N. G. (1988a). *Adv. Atom. Mol. Phys.,* **24**, in press.
Smith, D. and Adams, N. G. (1988b). In *Rate Coefficients in Astrochemistry,* eds. T. J. Millar and D. A. Williams (Dordrecht, Reidel), p. 153.
Smith, D., Adams, N. G. and Alge, E. (1982a). *Ap. J.,* **263**, 123.
Smith, D., Adams, N. G. and Alge, E. (1982b). *J. Chem. Phys.,* **77**, 1261.
Smith, D., Adams, N. G. and Church, M. J. (1978). *J. Phys. B.,* **11**, 4041.
Smith, D., Adams, N. G., Giles, K. and Herbst, E. (1988). *Astron. Astrophys.,* in press.
Smith, D. and Fouracre, R. A. (1968). *Planet. Space Sci.,* **16**, 243.
Smyth, K. C., Lias, S. G. and Ausloos, P. (1982), *Comb. Sci. Technol.,* **28** 147.
Solomon, P. M. and Klemperer, W. (1972). *Ap. J.,* **178**, 389.
Su, T. and Bowers, M. T. (1979). In *Gas Phase Ion Chemistry,* Vol. 1, ed. M. T. Bowers (New York, Academic Press), p. 84.
Troe, J. (1985). *Chem. Phys. Letts.,* **122**, 425.
Twiddy, N. D., Mohebati, A. and Tichy, M. (1986). *Int. J. Mass Spectrom. Ion Processes,* **74**, 251.
Watson, W. D. (1977). In *CNO Processes in Astrophysics,* ed. J. Audouze (Dordrecht, Reidel), p. 105.
Watson, W. D. (1978). *Ann. Rev. Astr. Ap.,* **16**, 585.
Weller, C. S. and Biondi, M. A. (1968). *Phys. Rev.,* **172**, 198.
Williams, D. A. (1987). In *Astrochemistry,* eds. M. S. Vardya and S. P. Tarafdar (Dordrecht, Reidel), p. 275.
Yee, J. H., Lepp, S. and Dalgarno, A. (1987). *Mon. Not. R. Astr. Soc.,* **227**, 461.

11

Theoretical considerations on some collision processes

DAVID R. BATES

Department of Applied Mathematics and Theoretical Physics, The Queen's University of Belfast, UK

Several classes of reaction will not be discussed because of limitation on space.

Energy and momentum must be conserved in a collision process. This is only possible if there are at least two products of reaction or if a photon is emitted. Radiationless transitions control some processes. They are faster than radiative transitions by five or more powers of ten.

11.1. Recombination

11.1.1. Radiative recombination

Radiative recombination is conceptually the simplest type of recombination: it is the inverse of photoionization. The two linked processes may be written

$$X^+(i_0) + e \rightleftharpoons X(j) + h\nu \tag{11.1}$$

i_0 denoting the ground state of the ion and j any state of the neutral. Noting that the cross section $\sigma(ji_0,\nu)$ for photoionization (11.1) and the derivative $\mathrm{d}f(ji_0,\varepsilon)/\mathrm{d}\varepsilon$ of the absorption oscillator strength with respect to the energy of the photoelectron are related through

$$\sigma(ji_0,\nu) = (\pi e^2 h/mc)\,\frac{\mathrm{d}f(ji_0,\varepsilon)}{\mathrm{d}\varepsilon} \tag{11.2}$$

it may readily be proved that detailed balancing requires that the rate coefficient for recombination process (11.1) at temperature T is

$$\alpha(i_0 j,T) = \frac{(2\pi)^{1/2}e^2h^3}{c^3m^{5/2}(kT)^{3/2}}\,\frac{\omega(j)}{\omega(i_0)}\int_0^\infty \frac{\mathrm{d}f(ji_0,\varepsilon)}{\mathrm{d}\varepsilon}\,\nu^2 \exp(-\varepsilon/kT)\,\mathrm{d}\varepsilon \tag{11.3}$$

in which $\omega(j)$ and $\omega(i_0)$ are the electronic statistical weights indicated. If kT is small compared with the ionization potential of state j the variation of terms other than the exponential may normally be disregarded when evaluating the integral so that

$$\alpha(i_0j,T) = \frac{(2\pi)^{1/2}e^2h^3}{c^3m^{5/2}(kT)^{1/2}}\frac{\omega(j)}{\omega(i_0)}\left(\frac{\mathrm{d}f(ji_0,\varepsilon)}{\mathrm{d}\varepsilon}\nu^2\right)_{\varepsilon=0} \tag{11.4}$$

This $T^{-1/2}$ power law is followed quite accurately by the total recombination coefficient

$$\alpha(i_0,T) = \sum_j \alpha(i_0j,T) \tag{11.5}$$

unless T is extremely high. Quantal calculations show that because of the importance of approximately hydrogenic excited states $\alpha(i_0,T)$ does not vary much amongst the atomic ions present in the interstellar medium (cf. Bates and Dalgarno (1962)) and may be represented to within about 30% by

$$\alpha(i_0,T) = 4\times10^{-12}(250/T)^{1/2}\,\mathrm{cm}^3\,\mathrm{s}^{-1}. \tag{11.6}$$

Radiative recombination coefficients for molecular ions are probably around the same as for atomic ions. It is occasionally suggested that they are enhanced owing to transitions to the continua provided by the repulsive states of the neutral molecule. The suggestion is incorrect because the Franck–Condon principle ensures that only a part of the total oscillator strength of the transition can be utilized. The rate coefficients are seldom needed since dissociative recombination (see below) is usually the dominant recombination process. If dissociative recombination happens to be slow it may be necessary to consider the possibility of recombination occurring through the sequence of inverse pre-ionization followed by photon emission. The formula in Equation (11.3) is convenient for such an enquiry. In favourable circumstances the sequence may be faster than radiative recombination but is most unlikely to be much faster.

11.1.2. Dissociative recombination

Dissociative recombination is the name given to the process

$$XY^+ + e \rightarrow XY^* \rightarrow X + Y \tag{11.7}$$

It may be regarded as taking place in two stages: first the system comprising the ion and the free electron participates in a radiationless transition that results in a neutral molecule XY^* on a repulsive potential energy surface that crosses the XY^+ attractive potential energy surface near its minimum; second the two atoms X and Y rapidly move apart under the influence of their mutual repulsion and thus stabilize the recombination by preventing the occurrence of autoionization (the inverse of the initial radiationless transition). The rate coefficient for the process has been shown (Bates 1950) to be

$$\alpha_D = sh^3AF/2(2\pi mkT)^{3/2} \tag{11.8}$$

where s is the ratio of the electronic statistical weight of XY^* to that of XY^+, A is the reciprocal lifetime towards autoionization and

$$F = \int_0^\infty |x(r)|^2 (\mathrm{d}r/\mathrm{d}\varepsilon) \exp(-\varepsilon/kT) \mathrm{d}\varepsilon \qquad (11.9)$$

in which $x(r)$ is the initial vibrational wavefunction, ε is the energy supplied by the free electron in a vertical transition between the two surfaces. If the ion is in its zeroth vibrational level and if the attractive potential is crossed at its minimum then

$$F = kT(2\mu\nu/h)^{1/2}/(\mathrm{d}\varepsilon/\mathrm{d}r)_0 \qquad (11.10)$$

where μ is the reduced mass, ν is the vibrational frequency and $(\mathrm{d}\varepsilon/\mathrm{d}r)_0$ is the radial gradient of the repulsive potential at the crossing. Equations (11.8) and (11.10) together lead to a $T^{-1/2}$ variation. For examples of modern computations on the rate coefficient see Guberman (1988) and references there cited.

The process just described is called *direct* dissociative recombination to distinguish it from the *indirect* process introduced by Bardsley (1968). There is an extra stage in the indirect process the original radiationless transition being replaced by a succession of two radiationless transitions which bring about the same change but do so via an intermediate $XY(n,v)$ which is in a fairly high Rydberg level n and low vibrational level v, the combination making it energetically accessible: thus

$$\{XY^+ + e\} \rightarrow \{XY(n,v)\} \rightarrow \{XY^*\} \qquad (11.11)$$

In the case of diatomic ions the indirect process is responsible for structure in the dissociative recombination cross section in the low energy region but is not of major importance as regards the rate coefficient (cf. Compton and Bardsley (1984)). A quantitative treatment is difficult but calculations on H_2^+ have nevertheless been done (Giusti-Suzor, Bardsley and Derkits 1983, Nakashima, Hidekazu and Katamura 1987, Hickman 1987).

Measurements (cf. Bardsley and Biondi (1970), McGowan *et al.* (1979), Massey (1982), Compton and Bardsley (1984), Johnsen (1987)) show that the dissociative recombination coefficient α_D of most diatomic ions (excluding those of the inert gases) lies within a factor of 3 of

$$\alpha_D(\text{diatomic}) = 4 \times 10^{-7}(100/T)^{1/2}\ \mathrm{cm^3\,s^{-1}} \qquad (11.12)$$

The existence of a multiple valence bond does not hinder dissociative recombination the radiationless transition of which may involve the free electron entering an anti-bonding orbital and another electron switching to such an orbital from a bonding orbital (cf. Bardsley (1968), Lee (1977)). In a few instances the absence of a favourable crossing makes α_D much smaller than in Equation (11.12): thus Roberge and Dalgarno (1982) judge that dissociative recombination to HeH^+ is negligibly slow and Giusti-Suzor *et al.* (1983) have calculated α_D for H_2^+ to be only $1.7 \times 10^{-8}(100/T)^{1/2}\ \mathrm{cm^3\,s^{-1}}$ in satisfactory agreement with measurements by Hus *et al.* (1988).

Examination of data on the heats of formation of polyatomic ions has shown that the concept of valence bonds and their energies is as useful quantitatively as it is for polyatomic neutral molecules (Bates 1987). The ionized atom in the polyatomic ion has the valency of the isoelectronic atom (so C^+ and O^+ have a valency of 3 and N^+ has a valency of 4); and the neutral atoms have their usual valency. It is generally possible to write down the structure of the lowest electronic state by considering the valencies involved, the ionization potentials of the constituent heavy atoms and if need be the valence bond energies. Dissociative recombination again requires a crossing of potential energy surfaces and is caused by a radiationless transition (or pair of radiationless transitions) producing a switch from an attractive to a repulsive potential energy surface. Trustworthy calculations have not been carried out but because of the number of vibrational modes the indirect process is thought to be more important relative to the direct process than it is for diatomic ions (Bardsley 1968, Compton and Bardsley 1984). Consistent with the effect being quite marked the measured rate coefficients (McGowan *et al.* 1979, Adams and Smith 1988b) tend to be greater: thus most are within a factor of about 2 of

$$\alpha_D(\text{polyatomic}) = 1 \times 10^{-6}(100/T)^{1/2}\ \text{cm}^3\ \text{s}^{-1} \tag{11.13}$$

In the (rare) absence of a favourable crossing the ion is almost immune to dissociative recombination. The most notable example is H_3^+ (Michels and Hobbs 1984, Johnsen 1987) which is a simple species as are the known diatomic species without a favourable crossing.

Information on the products of dissociative recombination of polyatomic ions is needed for the development of interstellar cloud chemistry. Theoretical guidance is desirable because laboratory results are scarce. However, prediction is difficult because of a problem that has been almost completely ignored: the problem of knowing if there is a favourable crossing for a particular reaction path (Bates 1989a).

The prime reason for a favourable crossing being the norm in the diatomic case is that several states are accessible to one or both products and the number of potentials that are repulsive at the energy of the minimum of the attractive potential of the ion is quite large. For non-hydrides 'the vicinity' is a target of width δ typically around 0.15 Å. Here and elsewhere we arbitrarily take

$$\delta = 2^{1/2}d \tag{11.14}$$

where d is the range of the classically allowed motion in the zeroth vibrational state of the ion.

When a single bond of a saturated polyatomic ion is broken different considerations arise dependent on whether neutralization of the constituent ionized atom increases or decreases the number of valence bonds. Carbon and silicon are examples of the first type of atom; nitrogen, oxygen and sulphur are examples of the second type.

Consider the dissociative recombination channel

$$CH_3^+ + e \rightarrow CH_2 + H + 5.3\ \text{eV} \tag{11.15}$$

Because of the increase in the number of valencies both products are radicals. The methylene molecule has several accessible states: the X^3B_B ground state; the a^1A_0, b^1B_0, and c^1A_1 states the respective excitation energies of which are 0.39, 1.27 and less than 3.8 eV; and probably a low $^3\Pi_g$ state (Herzberg 1966, Calloman *et al.* 1987).

Use of the stretch vibrational frequencies computed by DeFrees and McLean (1985) gives that the target width δ is about 0.3 Å; and the effective width may be rather greater due to flexibility in the extent of the vibrational excitation of the methylene. Taking into account the number of repulsive potentials a favourable crossing is thus likely. Similar reasoning applies to other polyatomic ions of the C^+ family: for example a favourable crossing is likely in at least one, and perhaps both, of

$$CH_3C^+H_2 + e \rightarrow CH_3 + CH_2 + 4.1\ \text{eV} \tag{11.16a}$$

$$\rightarrow CH_3CH + H + (3.5\ \text{eV}) \tag{11.16b}$$

However, consider

$$H_4N^+ + e \rightarrow NH_3 + H + 4.8\ \text{eV} \tag{11.17}$$

Ammonia is a saturated molecule. Only the ground X^1A_1 state is accessible so that the products are here confined to a single potential. A favourable crossing thus seems rather unlikely. This is also the judgement regarding the corresponding channels for the dissociative recombination of H_3O^+, H_3S^+, $CH_3N^+H_3$, $CH_3O^+H_2$, $C_2H_5O^+H_2$ and $CH_3S^+H_2$. Yet these six species, like H_4N^+, are known (Adams and Smith 1988a,b) to have large dissociative recombination coefficients. We must dismiss the possibility of a favourable crossing in all seven cases where there is good cause for believing that in each case such a crossing is rather unlikely.

The problem is eased if two H atoms are shed, in that the products would again be radicals. But in many instances shedding two H atoms is an endothermic or only slightly exothermic process (so that only the ground state of the heavy product radical is accessible) unless they combine to form a H_2 molecule as in

$$H_4N^+ + e \rightarrow NH_2 + H_2 + 4.6\ \text{eV} \tag{11.18}$$

The formation of a H_2 molecule facilitates a favourable crossing in a manner that has only recently been recognized. Owing to the lowness of the bending frequencies of H_4N^+ (or other parent ion) the H—H distance is not closely determined. Consequently the H_2 molecule formed may, consistent with the

Franck–Condon principle, carry a fairly wide range of vibrational energies. Extracting the vibrational energy the energy of the relative translational motion of NH_2 and H_2 can be between about 1 and 2 eV. The conditions for a favourable crossing are met if the value of the repulsive potential at the initial separation lies anywhere in this interval. The two H atoms of the NH_2 radicle are located away from the H_2 molecule so that the repulsion is mainly that between the molecule in the broadside position and the N atom. The required repulsions in the other six cases listed above are much the same and it has been argued that H_2 formation is a common means of providing a favourable crossing (Bates 1989a). The radiationless transition probability for H_2 formation is doubtless rather less than the maximum possible because of the relatively low overlap between the initial and final electronic wave functions. Having a favourable crossing may be more important than this. Branching ratios are difficult to estimate.

There is no clear need for the H_2 channel in the carbon family but it is not excluded: thus an alternative to channel (11.15) is

$$CH_3^+ + e \rightarrow CH + H_2 + 5.1\ \text{eV} \qquad (11.19)$$

11.1.3. Mutual neutralization

$$M^+ + N^- \rightarrow M + N \qquad (11.20)$$

Mutual neutralization could in a certain circumstance provide the main sink for positive ions. The circumstance is that polycyclic aromatic molecules, PAHs, should be abundant in dense clouds. Infrared observations and the visible diffuse absorption bands provide evidence for the presence of molecules containing up to around 100 atoms and PAHs have been proposed (Léger and Puget 1984) because the strong binding of C atoms in their plane hexagonal lattice gives stability against photodissociation. An account of the properties of PAHs has been given by Omont (1986). The electron affinity is about 1 eV. In dense clouds the negative ions PAH^- are more abundant than free electrons (Omont 1986) and

$$M^+ + PAH^- \rightarrow M + PAH \qquad (11.21)$$

may proceed even more rapidly than dissociative recombination. Dalgarno (1987) and Lepp and Dalgarno (1988) have considered some of the chemical consequences.

There are a number of crossings between the ionic potential surface along which the approach begins and the potential surfaces of the neutral products. Let P be the probability that the transfer of the electron occurs at a particular one of these (assuming that it has not occurred earlier) and let R_X be the distance between M^+ and the closest PAX atom (which is the location of the excess electron). If I is the ionization potential of M in the product state and A is the electron affinity of PAX then to a sufficient approximation

$$R_X = 14\ \text{Å}/\Delta E \tag{11.22}$$

where

$$\Delta E = (I - A)\ \text{eV} \tag{11.23}$$

Clearly P increases sharply as R_X decreases because so does the relevant interaction matrix element. Calculations show that in the case of single electron transitions P is small compared to unity if ΔE is less than about 1 eV and is almost unity if ΔE is more than about 2 eV (Bates and Boyd 1956). It is hence probable that the electron transfer occurs at a crossing rather than at impact. The impact energy would be ΔE in this event. Insofar as a single crossing is of dominant importance the mutual neutralization rate coefficient is

$$\alpha_i = 1.1 \times 10^{-6}(10/\tilde{M}_i)^{1/2}(10/T)^{1/2}\Delta E^{-1}\ \text{cm}^3\ \text{s}^{-1} \tag{11.24}$$

where $\tilde{M}_i$ is in the mass of the positive ion in amu and ΔE is greater than about 2 eV (Bates and Boyd 1956).

The transfer of the electron may lead to dissociation if M^+ is a molecular ion. This would certainly happen in the case of an ion like $R^+ \cdot H_2$ that is bound by polarization forces (Bates and Herbst 1988a).

11.2. Ion–molecule reactions

The long range attraction between an ion and a neutral molecule can cause them to spiral inwards towards one another until they make a hard collision. The associated rate coefficient k_c is called the capture or close encounter rate coefficient. If exothermic reactions can occur it is usually (but not always) a satisfactory approximation to take k_c to be the reaction rate coefficient. Exothermic ion–molecule reactions form a large and important class. An invaluable compilation of data on them has been produced by Anicich and Huntress (1986).

Let $u(r)$ be the long range interaction potential of the ion and molecule, μ be their reduced mass, b be the impact parameter and ε be the incident energy of relative motion. Adding the centrifugal potential we have that the effective potential that controls the radial motion is

$$V(r) = u(r) + b^2\varepsilon/r^2 \tag{11.25}$$

Its maximum is at the value r_0 such that

$$V'(r_0) = u'(r_0) - 2b^2\varepsilon/r^3 = 0 \tag{11.26}$$

For passage over the maximum to be only just possible it is necessary that

$$\begin{aligned} \varepsilon = V(r_0) &\equiv V(r_0) + \tfrac{1}{2}r_0 V'(r_0) \\ &= u(r_0) + \tfrac{1}{2}r_0 u'(r_0) \end{aligned} \tag{11.27}$$

in which b does not appear. Equation (11.27) enables us to find r_0 for given ε and hence using Equation (11.26) the corresponding cross section

$$\pi b^2 = \pi r_0^3 u'(r_0)/2\varepsilon \tag{11.28}$$

If the molecule has an isotropic polarizability α and no permanent dipole or quadrupole moment the long range interaction is that between the ion and the induced dipole:

$$u(r) = -\alpha e^2/2r^4 \tag{11.29}$$

Equation (11.27) gives

$$r_0 = (\alpha e^2/2\varepsilon)^{1/4}$$

whence Equation (11.28) gives

$$\pi b^2 = \pi(2\alpha e^2/\varepsilon)^{1/2} = (2\pi/v)(\alpha e^2/\mu)^{1/2} \tag{11.30}$$

where μ is the reduced mass and v is incident velocity of relative motion. It follows immediately from Equation (11.30) that the rate coefficient, known as the Langevin rate coefficient is

$$\begin{aligned} k_L &= 2\pi(\alpha e^2/\mu)^{1/2} \\ &= 2.34 \times 10^{-9}\{\tilde{\alpha}/\tilde{\mu}\}^{1/2}\ \text{cm}^3\ \text{s}^{-1} \end{aligned} \tag{11.31}$$

the tildes over α and μ indicating that these quantities are here in the usual practical units (10^{-24} cm^3 and amu respectively).

The collision is more difficult to treat if the molecule has a permanent dipole or quadrupole moment because the long range interaction is then orientation dependent: for example if the molecule has a permanent dipole of moment D the interaction is

$$u(r) = -\alpha e^2/2r^4 + De\cos\Theta/r^2 \tag{11.32}$$

where Θ is the angle between the line of centres of the colliding pair and the axis of the dipole. The earliest accurate calculations were carried out by Takayanagi (1978) and by Sakimoto (1984) using the quantal perturbed rotating state (PRS) approximation. They established the main pattern. A number of other approximations have been developed but only three need be mentioned here: the adiabatic invariance approximation (Bates and Mendas 1985, Morgan and Bates 1987, Bates and Morgan 1987) the centrifugal sudden approximation (Clary 1985, 1988) and the statistical adiabatic channel approximation (Troe 1987a). Formally these three appear quite different from each other and from the PRS approximation. However, the results they yield are in such close accord that they must in fact be nearly equivalent. Attention will be confined to the adiabatic invariance approximation.

Let w be the sum of the potential energy of the molecule in the orientation dependent part of the field and its kinetic energy of rotation (or libration) and let τ

be the rotation (or libration) time period. If the radial distance varies very slowly in the sense that

$$\tau(\mathrm{d}r/\mathrm{d}t) \ll r \tag{11.33}$$

it is evident that

$$\mathrm{d}w/\mathrm{d}t = F(r)\,\mathrm{d}r/\mathrm{d}t \tag{11.34}$$

where $F(r)$ is some function of r only and hence that

$$w = \int^{r} F(r)\,\mathrm{d}r \tag{11.35}$$

There is therefore a combination of w and r that remains constant throughout the encounter. It may be proved from Hamilton's equations of motion that this adiabatic invariant is

$$I = \int p_{\Theta}\,\mathrm{d}\Theta \tag{11.36}$$

where p_{Θ} is the momentum conjugate to Θ and the integration is over a complete rotation or libration (Landau and Lifshitz 1960). Doing the integration of Equation (11.36) gives $w(r)$ which is the effective contribution of the orientation dependent part of the field to the interaction potential. After adding the contribution from the orientation independent part the capture cross section may be obtained from Equations (11.26) and (11.27) by numerical methods. Knowing the capture cross section as a function of energy of relative motion, it is, of course, straightforward to compute the capture rate coefficient as a function of temperature. When presenting results it is advantageous to tabulate not the capture rate coefficient but rather its ratio to the Langevin rate coefficient of Equation (11.31) because this ratio is independent of the reduced mass and thus of the positive ion. Table 11.1 gives the ratio for two polar molecules: NH_3, an oblate spherical top and CH_3F a prolate spherical top. The parameter ξ used is defined by

$$\xi \equiv D/(2\alpha kT)^{1/2} = 60.2\tilde{D}/(\tilde{\alpha}T)^{1/2} \tag{11.37}$$

where $\tilde{D}$ is D expressed in 10^{-18} esu. As may be seen the presence of a dipole moment makes the capture rate coefficient considerably greater than the Langevin rate coefficient – especially when T and the rotational quantum number J are low. For $J = 1$ the dependence on the component of angular momentum quantum number K is not marked but this dependence becomes greater as J is increased (cf. Bates and Morgan (1987)).

The molecules of a cool diffuse cloud are in the lowest (J,K) level of each nuclear spin state: for instance half of NH_3 is in the (0,0) level which is a nuclear spin quartet and half is in the (1,1) level which is a nuclear spin doublet.

Because of importance of the low levels adiabatic invariance results for them

Table 11.1. *Ratio* $k_c(J,K)/k_L$ *for* NH_3 *and* CH_3F

	ξ						
	16	14	12	10	8	6	4
NH_3: $\tilde{a} = 2.26$, $\tilde{D} = 1.47$; rotational constants (cm^{-1}) $A = 6.20$, $B = 9.44$							
T(K)	13.5	17.6	23.9	34.6	53.8	96.0	215
J,K				$k_c(J,K)/k_L$			
0,0	10.7	10.1	9.33	8.44	7.40	6.17	4.73
1,0	4.76	4.76	4.75	4.69	4.55	4.25	3.68
1,1	6.64	6.34	6.02	5.67	5.26	4.71	3.93
CH_3F: $\tilde{a} = 2.97$; $\tilde{D} = 1.85$; rotational constants (cm^{-1}) $A = 5.10$, $B = 0.85$							
T(K)	16.3	21.3	29.0	41.8	65.3	116	261
J,K				$k_c(J,K)/k_L$			
0,0	16.8	15.1	13.4	11.6	9.71	7.81	5.87
1,0	12.8	12.0	10.9	9.79	8.49	7.04	5.46
1,1	13.7	12.7	11.5	10.2	8.78	7.23	5.56

Table taken from Bates and Morgan (1987)

have been parameterized (Morgan and Bates 1987, Bates and Morgan 1987) so that $k_c(J,K)/k_L$ for linear and symmetrical top polar molecules can easily be obtained for any (D, a, B,T) combination. As already mentioned this ratio does not depend on the reduced mass μ; it also does not depend on the rotational constant A which is rather curious.

Should the thermal average over all levels $\overline{k_c/k_L}$ be required use may be made of a parametric formula that Su and Chesnavich (1982) found represents results they got by trajectory calculations. The formula is

$$k_c/k_L \begin{cases} = 0.4767\xi + 0.6200, & \xi \geq 2 \\ = 0.09500(\xi + 0.5090)^2 + 0.9754, & \xi \leq 2 \end{cases} \tag{11.38}$$

with ξ as in Equation (11.37). It was derived for linear polar molecules but it is a good fit to the abiabatic invariance results on NH_3 and CH_3F and is doubtless satisfactory for other symmetrical top polar molecules (Bates and Morgan 1987).

The effect of a permanent quadrupole moment is relatively minor.

11.2.1. Some exceptions

The k_c-rule (that the reaction rate coefficient may be taken to equal the capture rate coefficient) may fail badly. Measurements (Fehsenfeld *et al.* 1975, Smith and Adams 1981) show that at 400 K the exothermic process

$$NH_3^+ + H_2 \rightarrow NH_4^+ + H \tag{11.39}$$

has a reaction rate coefficient k_{39} of only about 1×10^{-12} cm^3 s^{-1} whereas the capture rate coefficient k_c is 1.6×10^{-9} cm^3 s^{-1}. As T is decreased, k_{39} at first also decreases but then passes through a minimum of about 2×10^{-13} cm^3 s^{-1} near 80 K and is again about 1×10^{-12} cm^3 s^{-1} near 10 K (Bohringer 1985, Luine and Dunn 1985, Barlow, Luine and Dunn 1986a). The behaviour in the high temperature indicates there is an activation barrier of 0.09 eV (or more). In order to account for the rise below 80 K Barlow *et al.* (1986a) have suggested that a complex is formed with a sufficiently long lifetime to allow tunnelling through the activation barrier to occur. Measurements on isotopic variants of Reaction (11.39) provided them with verification of the suggestion. The interpretation of recent detailed observations (Morrison *et al.* 1986, Kemper and Bowers 1986, Winniczek *et al.* 1987) on the effect of vibrational and translational energy have been discussed by Barlow and Dunn (1987).

A remarkable collaborative research effort (Rowe *et al.* 1984, Barlow *et al.* 1986b) has been devoted to the reactions of O_2^+ with CH_4 and (as a check on the model) with the deuterated methanes. At 300 K the $O_2^+ + CH_4$ reaction rate coefficient is only about 5×10^{-12} cm^3 s^{-1} – far below k_c which is 1.1×10^{-9} cm^3 s^{-1}. The reaction course is

$$O_2^+ + CH_4 \rightarrow CH_3O_2^+ + H \tag{11.40}$$

where $CH_3O_2^+$ is the methylene hydroperoxide ion:

$$H_2C^+\text{—}O\text{—}OH \rightleftharpoons H_2C\text{=}O^+\text{—}OH \tag{11.41}$$

The measured values of the reaction rate coefficient between 20 K and 200 K may be represented by

$$k_{40} = 2.8 \times 10^{-11}(100/T)^{1.8}\ \text{cm}^3\ \text{s}^{-1} \tag{11.42}$$

and are consistent with an extrapolation to $k_{40} = k_c$ at absolute zero. Barlow *et al.* (1986b) proposed that the first step is the formation of a long lived complex

$$O_2^+ + CH_4 \rightleftharpoons CH_4 \cdot O_2^+ \tag{11.43}$$

which may decay back into the reactants or may participate in the rate limiting step of H^- transfer

$$CH_4 \cdot O_2^+ \rightarrow CH_3^+ \cdot O_2H \tag{11.44}$$

which is followed almost immediately by H evaporation

$$CH_3^+ \cdot O_2H \rightarrow CH_3O_2^+. + H \tag{11.45}$$

They succeeded in satisfactorily modelling their data on process (11.40) and the related processes involving deuterated methane. A more sophisticated treatment has been given by Troe (1987b). If T is raised above 400 K the reactive rate coefficient increases owing to the opening of the endothermic channel

$$O_2^+ + CH_4 \rightarrow CH_3^+ + HO_2 - 5.5\,\text{kcal mol}^{-1} \tag{11.46}$$

which is rapid when there is enough energy (Barlow *et al.* 1986b).

Note that the formation of a complex is the norm in ion–molecule reactions. The abnormal common feature of processes (11.39) and (11.40) is the *slowness* of the direct or indirect decay of the complex into the products. As T is lowered the lifetime of the complex becomes longer. This facilitates a slow decay channel and in the case of processes (11.39) and (11.40) the probability of the decay rises towards unity so that although the k_c-rule fails badly as far as some of the laboratory measurements are concerned it does not do so at the temperatures of cool interstellar clouds. There must be many such cases amongst the ion–molecule reactions that entail merely H atom or ion transfer in the complex. However, if the reaction envisaged requires a major rearrangement then an impenetrable activation barrier is likely to exist. As examples collision induced dissociation studies by Jarnold *et al.* (1986) show that barriers prevent the occurrence of the exothermic reactions:

$$CH_3^+ + CH_3 \cdot OH \nrightarrow H_3O^+ + C_2H_4 \tag{11.47}$$

$$\nrightarrow CH_3CHOH^+ + H_2 \tag{11.48}$$

just as consideration of the rearrangements entailed would lead one to expect.

11.3. Charge transfer

11.3.1. Atomic species

An electronic transition occurs in charge transfer so that the adiabatic criterion of Massey (1949) is relevant. According to this the probability P of the transition is small if

$$\lambda_M \equiv l\Delta U/\hbar v \gg 1 \tag{11.49}$$

where l is the range of the transition interaction, U is the separation of the potential energy surfaces and v is the velocity of relative motion. The last two entities are functions of position. The identity on the left of Equation (11.49) may be written

$$\lambda_M \equiv 11\tilde{l}\tilde{\Delta}\tilde{U}\tilde{\mu}^{1/2}/\tilde{\varepsilon}^{1/2} \tag{11.50}$$

in which μ is the reduced mass, ε is the energy of relative motion and the tilde as always indicates practical units (Å for length, eV for energy, amu for mass). Defining l to be the range over which the interaction is more than $1/\pi$ of its peak value it has been found (Bates 1978) that for representative interactions:

$$P \sim 1 \text{ for } \lambda_M \lesssim 1, \qquad P \lesssim 0.1 \text{ for } \lambda_M \gtrsim 4 \tag{11.51}$$

and that P decreases rapidly as λ_M is increased further. Because of Equations (11.50) and (11.51) the rate coefficient for charge transfer is high if the ionization potentials happen to be the same or very nearly the same as in

$$\mathrm{H(ls) + O^+(^4S) \rightleftharpoons H^+ + O(^3P_1)}, \qquad \mathrm{H(ls) + O^+(^4S) \rightarrow H^+ + O(^3P_2) + 0.02\ eV} \tag{11.52}$$

(Stebbings, Smith and Erhardt 1964, Fehsenfeld and Ferguson 1972).

At an avoided crossing of the potential energy surface ΔU may be small enough for P to be appreciable. The crossing must be traversed twice during an encounter so that the resulting probability of charge transfer is

$$\mathcal{P} = 2P(1 - P) \tag{11.53}$$

which is small not only when P is small but also when P is close to unity. As was pointed out by Bates and Moiseiwitsch (1954) there is a class of charge transfer collision processes

$$\mathrm{X}^{n+} + \mathrm{Y} \rightarrow \mathrm{X}^{(n-m)+} + \mathrm{Y}^{m+} + \Delta E \tag{11.54}$$

having avoided crossings at separations R_X that may be determined approximately from the asymptotes to the potentials. The rate coefficient is low if R_X is either small or large. Butler and Dalgarno (1980a) have applied the Landau–Zener approximation to cases of astrophysical interest. They found that many are fast: thus the rate coefficients for Ne^{+4}, Mg^{+3}, Mg^{+4}, Si^{+4}, S^{+3}, S^{+4}, Ar^{+3} and Ar^{+4} reacting in H and N^{+4}, O^{+3}, O^{+4}, Ne^{+4}, Mg^{+3}, Mg^{+4}, Si^{+4} and Ar^{+4} reacting in He exceed 1×10^{-9} cm^3 s^{-1} at 1000 K.

However owing to the absence of suitable crossings the corresponding rate coefficients for Ar^{+2} in H and C^{+4}, Ne^{+2}, Ne^{+3} and Ar^{+3} in He are in contrast extremely small – less than the rate coefficient (of order 1×10^{-14} cm^3 s^{-1}) for radiative charge transfer (Allison and Dalgarno 1965, Butler, Guberman and Dalgarno 1977). Although the Landau–Zener approximation provides a good qualitative description of charge transfer an accurate description requires a quantal close coupling treatment using diabatic states (Heil, Butler and Dalgarno 1981, Gargaud, McCarroll and Valiron 1982, Butler, Heil and Dalgarno 1984, Gargaud and McCarroll 1988). The theory of charge transfer by multiply-charged ions has been reviewed by Janev and Presnyakov (1981) and by McCarroll (1986).

For singly charged ions any avoided crossings occur at rather short nuclear distances where the interaction may be so strong that Equations (11.50) and (11.51) make $\mathcal{P}$ very small in thermal collisions. This was once widely thought to be the norm (cf. Rapp and Francis (1962)). However measurements by Turner-Smith, Green and Webb (1973) on charge transfer into particular excited states in thermal He^+–Cd and Ne^+–Mg collisions show that it is not uncommon for $\mathcal{P}$ to be close to its maximum value. Aeronomy provides some instructive examples (cf. Bates (1989b)).

In many instances radiative charge transfer is probably faster than radiationless charge transfer. Cooper, Kirby and Dalgarno (1984) have calculated that the rate coefficient for He^+–Ne radiative charge transfer is 5×10^{-16} cm^3 s^{-1} at 300 K, a result consistent with measurements by Johnsen and Biondi (1979). Johnsen (1983) has observed the radiation emitted.

Even in the absence of a favourable avoided crossing, charge transfer may ensue due to the spin–orbit interaction causing a transition between molecular states of different species. Calculations by Butler and Dalgarno (1979) show that such a transition leads to the rate coefficient for N^+–H charge transfer being about 1.0×10^{-12} cm^3 s^{-1} at 10^4 K. The corresponding rate coefficient for C^+–H charge transfer is 1.4×10^{-17} cm^3 s^{-1}; that for S^+–H charge transfer is estimated to be at most 3×10^{-15} cm^3 s^{-1} (Butler and Dalgarno 1980b). The stronger spin–orbit interaction in heavy systems should give higher rate coefficients. The fastest non-resonance charge transfer process between a neutral atom and an atomic ion to have been reported is He^+–Hg charge transfer. A measurement at 300 K by Johnsen and Biondi (1980) gives that its rate coefficient is 2.5×10^{-9} cm^3 s^{-1}.

11.3.2. Molecular species

If at least one of the pair is molecular, charge transfer becomes a type of ion–molecule reaction. The effect of the strength of the interaction at rather short nuclear distances may be alleviated by small Franck–Condon factors: and the complex that is formed may be sufficiently long lived at low temepratures to allow many traversals of any avoided crossing. The principal limitation on the usefulness of the k_c-rule of Section 11.2 arises because charge transfer is in many instances in competition with fast exothermic channels like H^+ ion transfer or H atom evaporation when the charge transfer branching ratio may be low as in H_2^+–C_2H_6 collisions where it is 0.06 or high as in H_2^+–C_2H_2 collisions where it is 0.91 (Kim and Huntress 1975). In considering whether or not charge transfer is exothermic it is necessary to take into account that some vibrational excitation would occur unless the equilibrium configurations of the ion and neutral are almost the same: thus the products of all close CH_4^+–CH_4 collisions are CH_5^+ + CH_3 (Huntress, Laudenslager and Pinizzotto 1975) the charge transfer channel being endothermic. Should the ion's parent molecule be unstable as in the case of HeH^+, H_3^+, N_2H^+, H_3O^+, H_3S^+, H_2CN^+, NH_4^+ and CH_5^+ it would be expected that the chemically active fragment released by charge transfer would become bound to the collision partner. This indeed happens: for instance

$$CH_5^+ + CO \rightarrow HCO^+ + CH_4 \tag{11.55}$$

(Bohme, Mackay and Schiff 1980). The course is equivalent to proton transfer and is also generally the preferred course for HCO^+, HNO^+, HO_2^+, NH_3^+, H_2CO^+ and CH_3O^+.

The k_c-rule fails completely if the potential energy surfaces have not a suitable avoided crossing. For example measurements by Bohringer and Arnold (1986) show that the rate coefficient for reactive He^+–H_2 collisions is extremely low (not exceeding 2×10^{-13} cm^3 s^{-1} even at 18 K) in agreement with theoretical work by Mahan (1971).

11.4. Neutral chemistry

Chemical processes generally have enough activation energy to make them negligibly slow at the temperatures of cool interstellar clouds. This is immediately evident from the rule of Hirschfelder (1941) that the activation energy of the exothermic atom exchange process

$$A + BC \rightarrow AB + C \tag{11.56}$$

is 5% of the energy of the bond broken while that of the exothermic atoms interchange process

$$AB + CD \rightarrow AC + BD \tag{11.57}$$

is 28% of the sum of the energies of the two bonds broken.

However in the case of the atom exchange process (11.56) the activation energy may be very small or zero if both reactants are free radicles, that is, contain unpaired electrons (so N, O_2 and CH_3 are free radicles but He, N_2 and CH_4 are not). At least for some orientations a pair of free radicles can usually approach one another along the attractive energy surface belonging to the bound state made by pairing the originally unpaired electrons in each radicle. The complex thus formed may dissociate back into the reactants. Alternatively it may dissociate along the channel corresponding to atom exchange. Absence of activation energy in this channel is favoured (but not ensured) by the products being radicles and by the reaction being highly exothermic. Because of the unpaired electrons a pair of radicles may give rise to many states. Should a large fraction of these correlate with the products a fast reaction may ensue: thus of the 36 states arising from O + CH 30 correlate to energetically accessible states of H + CO and the rate coefficient is as great as 9.5×10^{-11} cm^3 s^{-1} while of the 16 states arising from N + CH only 6 correlate to energetically accessible states of H + CN which may explain the smaller rate coefficient of 2.1×10^{-11} cm^3 s^{-1} (Howard and Smith 1983). The role of complex formation has been reviewed by Quack and Troe (1977) and by Troe (1979).

The rate coefficients for atom exchange in collisions between radicles (simple or polyatomic) generally lie between 10^{-10} cm^3 s^{-1} and 10^{-11} cm^3 s^{-1} at 300 K and may have a slight negative temperature dependence (cf. Baulch *et al.* (1981, 1982), Howard and Smith (1983)). The temperature dependence is due in part to the long range interaction (a dipole–dipole interaction giving a $T^{-1/6}$ dependence and a dipole–quadrupole interaction a T^0 dependence), in part to the electronic partition function (Clary 1984, 1988). Clary and Werner (1984) have carried out quantal scattering calculations on

$$O + OH \rightarrow O_2 + H \qquad (11.58)$$

assuming that the reaction is controlled by the long range dipole–quadrupole interaction. They found that the thermal average and OH ($J = 0$) rate coefficients at 300 K are about 4.5×10^{-11} cm^3 s^{-1} and 5.8×10^{-11} cm^3 s^{-1} and that the OH ($J = 0$) rate coefficient at 25 K is about $8 + 10^{-11}$ cm^3 s^{-1}. The satisfactory agreement between their thermal values and the experimental results of Howard and Smith (1981) confirms the correctness of the assumption they made.

11.5. Radiative association

11.5.1. Atomic systems

If the colliding pair approach along an excited potential energy surface a radiative transition to a lower attractive potential energy surface may occur during the brief ($\sim 10^{-13}$ s) interval during which the energy lost by the emission of the photon is enough to prevent the pair moving apart again. A classical treatment of the relative motion of the nuclei is usually adequate. The important process

$$C^+ + H(A^1\Pi) \rightarrow CH^+(X^1\Sigma^+) + h\nu \qquad (11.59)$$

has been thoroughly investigated (Giusti-Suzor, Roueff and van Regemorter 1976, Abgrall, Giusti-Suzor and Roueff 1976). At 20 K a classical treatment of the relative motion leads to a rate coefficient of 1.1×10^{-17} cm^3 s^{-1}; making allowance for the quantal tunnel effect raises the rate coefficient slightly to 1.2×10^{-17} cm^3 s^{-1}; and taking into account the shape resonances resulting from the quasi-bound states in the well of the excited state effective potential behind the centrifugal barrier brings about a modest further increase to 1.6×10^{-17} cm^3 s^{-1}. There is little variation with temperature: at 1000 K all three treatments give the rate coefficient to be 1.3×10^{-17} cm^3 s^{-1}. The characteristics of

$$H^+ + H(^2\Sigma_u^+, 2p\sigma_u) \rightarrow H_2^+(^2\Sigma_g^+, 1s\sigma_g) \qquad (11.60)$$

are different because the excited potential energy surface is here repulsive. At 20 K the rate coefficient got, neglecting tunnelling, is 5.3×10^{-21} cm^3 s^{-1} (Bates 1951) and that allowing for tunnelling is 1.7×10^{-20} cm^3 s^{-1} (Ramaker and Peek 1976); while at 1000 K tunnelling is unimportant and the rate coefficient is 5.3×10^{-18} cm^3 s^{-1}. Shape resonances naturally do not arise with a repulsive potential.

Radiative association and radiative charge transfer are rather similar processes. The relatively very large rate coefficients that Allison and Dalgarno (1965) and Butler *et al.* (1977) calculated for the latter in the multiply charged ion case (Section 11.3.1) are due mainly to the photon energies involved being high. Radiative charge transfer rate coefficients in the singly charged ion case (Cooper *et al.* 1984) are less different from radiative association rate coefficients.

11.5.2. Molecular species

When molecules X^+ and Y collide a complex XY^{+*} is formed. The radiative association rate coefficient is

$$k_r = \frac{h^3}{(2\pi\mu kT)^{3/2}} \frac{q(XY^{+*})A}{q(X^+)q(Y)} \tag{11.61}$$

in which the first factor is the ratio of the translational partition function of XY^{+*} to the product of the translational partition functions of X^+ and Y, and q's are the internal partition functions indicated and A is the rate of the radiative transition that stabilizes the complex. A detailed discussion of the determination of $q(XY^{+*})$ and A will not be presented because this would duplicate material in a recent article by Bates and Herbst (1988b). Instead Formula (11.61) will be recast in a way that is physically revealing.

For simplicity attention will be confined to cases where the long range interaction is that between an ion and an induced dipole and use will be made of the modified thermal model (Bates 1980) in which the effect of the internal angular momenta of the reactants is ignored and a summation over quantum numbers is replaced by an integration. The dependence of the density of rovibrational states on the energy E and rotational quantum number J will also be neglected apart from, of course, the statistical weight factor $\{2J + 1\}\{2j\delta(3,r) + 1\}$ in which r is the number of external rotational degrees of freedom of the complex. Let ρ_{vib} be the energy density of vibrational states at threshold, σ be the symmetry number and

$$J_L \equiv (8\mu^2\alpha e^2E)^{1/4}/\hbar \tag{11.62}$$

be the Langevin orbital quantum number, that is the greatest value of J for which passage over the centrifugal barrier at the given E is possible. With the approximations specified it is apparent that

$$q(XY^{+*}) = (\rho_{vib}/\sigma) \int_0^\infty \int_0^{J_L} \{2J + 1\}\{2J\delta(3,r) + 1\} \exp(-E/kT)\, dJ\, dE \tag{11.63}$$

Evaluation of the double integral and substitution in Equation (11.61) yields

$$k_r = \tau_r A k_L \tag{11.64}$$

where k_L is the Langevin rate coefficient of Equation (11.31) and τ_r, which is evidently the effective lifetime of the complex, is simply related to the time

$$\tau_0 \equiv h\rho_{vib}; \tag{11.65}$$

thus for a linear ($r = 2$) complex

$$\tau_2 = \tau_0/\sigma q(X^+)q(Y) \tag{11.66}$$

and for a non-linear ($r = 3$) complex

$$\tau_3 = \lambda\tau_2 \tag{11.67}$$

where

$$\lambda = [32\Gamma(7/4)/3](\pi\mu)^{1/2}(ae^2kT/2)^{1/4}/h$$
$$= 6.75a^{1/4}\mu^{1/2}T^{1/4}. \quad (11.68)$$

Denoting the association energy of XY^+ (from $X^+ + Y$) by E_0, the zero-point energy by E_Z, the number of atoms by N and, supposing there is no internal rotation, the frequencies of the

$$s \equiv 3N - 3 - r \quad (11.69)$$

vibrational modes by ν_i, it may be shown that

$$\rho_{vib} = (E_0 + aE_Z)^{s-1}/(s-1)! \prod_i (h\nu_i) \quad (11.70)$$

a being an empirically determined factor somewhat less than unity (Whitten and Rabinovitch 1963). In order to illustrate qualitatively the dependence of the lifetime of a complex on E_0 and N we assume that a is 0.95 and that the ν_is cluster around one of the following three frequencies (with the number in the cluster as in the brackets): 100 cm^{-1} (1), 1500 cm^{-1} ($2N - 4$); 3000 cm^{-1} ($N - 3$). Table 11.2, which out of curiosity is extended to rather large N, shows the values of τ_0 obtained. Because of the appearance of the internal partition functions $q(X^+)$ and $q(Y)$ in the denominator on the right of Equation (11.66) the lifetme of the complex increases as T decreases. It is evident that the lifetime of a large strongly bound complex is extremely long at low temperatures.

For a particular complex τ_r may be calculated if the association energy E_0 is known and if the frequencies ν_i have been determined by *ab initio* quantal computations. It may also be calculated if the low ambient gas density limit k_0 to the rate coefficient for the related termolecular association process

$$X^+ + Y + He \rightarrow XY^+ + He \quad (11.71)$$

has been found from experimental data. The relation between τ_r and k_0 is

$$\tau_r = k_0/\beta_c k_L' k_L \quad (11.72)$$

where k_L' is the Langevin rate coefficient for XY^+–He collisions and β_c is the fraction of these collisions that stabilize the complex. Experiments by Cates and Bowers (1980) suggest that β_c is about 0.3 at 300 K. Smith (1989) has pointed out that if one of the reactants is of low mass, as in the important case of H_2 a modest enhancement of k_r arises from tunnelling through the centrifugal potential barrier.

Radiative stabilization may be due to vibrational relaxation or to an electronic transition. The rate A of the former is commonly taken to be around 10^3 s^{-1} for vibrational energies of a few electron volts; the effective rate of the latter may in favourable cases be greater (cf. Bates and Herbst (1988b)).

Table 11.2. *Time τ_0 defined by Equation (11.65) and appearing in Equation (11.66) as function of number N of atoms in, and association energy E_0 of, the non-linear complex.*

N	5	6	7	8	10	15	20
E_0 (eV)				τ_0 (s)			
0.5	2.0 (−11)	5.0 (−11)	1.2 (−10)	3.2 (−10	1.9 (−9)	1.8 (−7)	1.7 (−5)
1.0	2.3 (−10)	8.3 (−10)	2.6 (−9)	7.3 (−9)	5.7 (−8)	7.1 (−6)	7.9 (−4)
1.5	1.5 (−9)	7.7 (−9)	3.0 (−8)	1.0 (−7)	1.1 (−6)	2.0 (−4)	2.8 (−2)
2.0	7.0 (−9)	4.7 (−8)	2.4 (−7)	1.0 (−6)	1.5 (−5)	4.4 (−3)	7.9 (−1)
2.5	2.5 (−8)	2.3 (−7)	1.5 (−6)	8.0 (−6)	1.5 (−4)	7.7 (−2)	1.8 (1)
3.0	7.7 (−8)	9.0 (−7)	7.7 (−6)	5.0 (−5)	1.3 (−3)	1.0 (0)	3.5 (2)

Note 2.0 (−11) ≡ 2.0×10^{-11}.

REFERENCES

Abgrall, H., Giusti-Suzor, A. and Roueff, E., 1976, *Astrophys. J.* **207**, L69.

Adams, N. G. and Smith, D., 1988a, in *Rate Coefficients in Astrochemistry,* eds. T. J. Millar and D. A. Williams, p. 173, Kluwer, Dordrecht.

Adams, N. G. and Smith, D., 1988b, *Chem. Phys. Lett.* **144**, 11.

Allison, D. C. S. and Dalgarno, A., 1965, *Proc. Phys. Soc.* **85**, 845.

Anicich, V. G. and Huntress, W. T., 1986, *Ap. J. Suppl.*, **62**, 553.

Bardsley, J. N., 1968, *J. Phys. B: Atom. Molec. Phys.* **1**, 365.

Bardsley, J. N. and Biondi, M. A., 1970, *Adv. Atomic Molec. Phys.* **6**, 1.

Barlow, S. E. and Dunn, G. H., 1987, *Int. J. Mass Spect. Ion Processes* **80**, 227.

Barlow, S. E., Luine, J. A. and Dunn, G. H., 1986a, *Int. J. Mass Spect. Ion Processes* **74**, 97.

Barlow, S. E., Van Doren, J. M., De Puy, C. H., Bierbaum, V. M., Dotan, I., Ferguson, E. E., Adams, N. G., Smith, D., Rowe, B. R., Marquette, J. B., Dupeyrat, G. and Durup-Ferguson, M., 1986b, *J. Chem. Phys.* **85**, 3851.

Bates, D. R., 1950, *Phys. Rev.* **78**, 492.

Bates, D. R., 1951, *Mon. Not. R. Astron. Soc.* **111**, 303.

Bates, D. R., 1978, *Phys. Reports* **35**, 306.

Bates, D. R., 1980, *J. Chem. Phys.* **73**, 1000.

Bates, D. R., 1987, *Int. J. Mass Spect. Ion. Processes* **80**, 1.

Bates, D. R., 1989a, *Astrophys. J.* **344**, in press.

Bates, D. R., 1989b, *Planet. Space Sci.* **37**, 363.

Bates, D. R. and Boyd, T. J. M., 1956, *Proc. Phys. Soc.* **A69**, 910.

Bates, D. R. and Dalgarno, A., 1962, in *Atomic and Molecular Processes,* ed. D. R. Bates, New York: Academic, p. 245.

Bates, D. R. and Herbst, E., 1988a, in *Rate Coefficients in Astrochemistry,* eds. T. J. Millar and D. A. Williams, Dordrecht: Reidel, p. 41.

Bates, D. R. and Herbst, E., 1988b, *ibid.*, p. 17.

Bates, D. R. and Mendas, I., 1985, *Proc. Roy. Soc. A. Lond.* **402**, 245.

Bates, D. R. and Moiseiwitsch, B. L., 1954, *Proc. Phys. Soc.* **A67**, 805.

Bates, D. R. and Morgan, W. L., 1987, *J. Chem. Phys.* **87**, 2611.
Baulch, D. L., Cox, R. A., Crutzen, P. J., Hampson, R. F., Kerr, J. A., Troe, J. and Watson, R. T., 1982, *J. Phys. Chem. Ref. Data* **11**, 327.
Baulch, D. L., Duxbury, J., Grant, S. J. and Montague, D. C., 1981, *J. Phys. Chem. Ref. Data* **10**, Supplt. 1.
Bohme, D. K., Mackay, G. I. and Schiff, H. I., 1980, *J. Chem. Phys.* **73**, 4976.
Bohringer, H., 1985, *Chem. Phys. Lett.* **122**, 185.
Bohringer, H. and Arnold, F., 1986, *J. Chem. Phys.* **84**, 1459.
Butler, S. E. and Dalgarno, A., 1979, *Astrophys. J.* **234**, 765.
Butler, S. E. and Dalgarno, A., 1980a, *Astrophys. J.* **241**, 838.
Butler S. E. and Dalgarno, A., 1980b, *Astron. Astrophys.* **85**, 144.
Butler, S. E., Guberman, S. L. and Dalgarno, A., 1977, *Phys. Rev. A.* **16**, 500.
Butler, S. E., Heil, T. G. and Dalgarno, A., 1984, *J. Chem. Phys.* **80**, 4986.
Calloman, J. H., Hirota, B., Iijima, T., Kichitsu, K. and Lafferty, W. J., 1987, *Landolt-Bernstein* II **15**.
Cates, R. D. and Bowers, M. T., 1980, *J. Am. Chem. Soc.* **102**, 3994.
Clary, D. C., 1984, *Molecular Physics* **53**, 3.
Clary, D. C., 1985, *Molecular Physics* **54**, 605.
Clary, D. C., 1988, in *Rate Coefficients in Astrochemistry,* eds. T. J. Millar and D. A. Williams, Dordrecht: Reidel, in press.
Clary, D. C. and Werner, H. J., 1984, *Chem. Phys. Lett.* **112**, 346.
Compton, R. N. and Bardsley, J. N., 1984, in *Electron Molecular Collisions,* eds. I. Shimamura and K. Takayanagi, New York: Plenum, p. 275.
Cooper, D. L., Kirby, K. and Dalgarno, A., 1984, *Can. J. Phys.* **62**, 1622.
Dalgarno, A., 1987, *Int. J. Mass. Spect. Ion Processes* **81**, 1.
DeFrees, D. J. and McLean, A. D., 1985, *J. Chem. Phys.* **82**, 333.
Fehsenfeld, F. C. and Ferguson, E. E., 1972, *J. Chem. Phys.* **56**, 3066.
Fehsenfeld, F. C., Lindinger, W., Schmeltekopf, A. L., Albritton, D. L. and Ferguson, E. E., 1975, *J. Chem. Phys.* **62**, 2001.
Gargaud, M. and McCarroll, R., 1988, *J. Phys. B: Atom. Molec. Phys.* **21**, 513.
Gargaud, M., McCarroll, R. and Opradolce, L., 1988, *J. Phys. B: Atom. Molec. Phys.* **21**, 521.
Gargaud, M., McCarroll, R. and Valiron, P., 1982, *Astron. Astrophys.* **106**, 197.
Giusti-Suzor, A., Bardsley, J. N. and Derkits, C., 1983, *Phys. Rev. A.* **28**, 682.
Giusti-Suzor, A., Roueff, E. and van Regemorter, H., 1976, *J. Phys. B: Atom. Molec. Phys.* **9**, 1021.
Green, S. and Herbst, E., 1979, *Astrophys. J.* **229**, 121.
Guberman, S. L., 1988, *Planet. Space Sci.* **36**, 47.
Heil, T. G., Butler, S. E. and Dalgarno, A., 1981, *Phys. Rev. A.* **23**, 1100.
Herzberg, G., 1966, *Electronic Spectra and Electronic Structure of Polyatomic Molecules,* New York: van Nostrand.
Hickman, A. P., 1987, *J. Phys. B: Atom. Molec. Phys.* **20**, 2091.
Hirschfelder, J. D., 1941, *J. Chem. Phys.* **9**, 645.
Howard, M. J. and Smith, J. W. M., 1981, *J. Chem. Soc. Faraday Trans II* **77**, 997.
Howard, M. J. and Smith, J. W. M., 1983, *Prog. Reaction Kinetics* **12**, 55.
Huntress, W. T., Laudenslager, J. B. and Pinizzotto, R. J., 1975, *Int. J. Mass Spect. Ion Phys.* **13**, 33.
Hus, H., Yousif, F., Noren, C., Sen, A. and Mitchell, J. B. A., 1988, *Phys. Rev. Lett.* **60**, 1006.
Janev, R. K. and Presnyakov, L. P., 1981, *Phys. Reports* **70**, 1.
Jarnold, M. I., Kirchner, N. J., Liu, S. and Bowers, M. T., 1986, *J. Phys. Chem.* **90**, 78.

Johnsen, R., 1983, *Phys. Rev. A.* **28**, 1460.
Johnsen, R., 1987, *Int. J. Mass Spect. Ion Processes* **81**, 67.
Johnsen, R. and Biondi, M. A., 1979, *Phys. Rev.* **A20**, 87.
Johnsen, R. and Biondi, M. A., 1980, *J. Chem. Phys.* **73**, 5045.
Kemper, P. R. and Bowers, M. T., 1986, *J. Phys. Chem.* **90**, 477.
Kim, J. K. and Huntress, W. T., 1975, *J. Chem. Phys.* **62**, 2820.
Landau, L. A. and Lifshitz, E. M., 1960, *Mechanics,* Section 49, Oxford: Pergamon.
Lee, C. M., 1977, *Phys. Rev. A.* **16**, 109.
Lepp, S. and Dalgarno, A., 1988, *Astrophys. J.* **324**, 553.
Léger, A. and Puget, J. L., 1984, *Astron. Astrophys.* **137**, L5.
Luine, J. A. and Dunn, G. H., 1985, *Astrophys. J.* **299**, L67.
Mahan, B. H., 1971, *J. Chem. Phys.* **55**, 1436.
Massey, H. S. W., 1949, *Reports Prog. Phys.* **12**, 248.
Massey, H. S. W., 1982, in *Applied Atomic Collision Physics,* eds. H. S. W. Massey, E. W. McDaniel and B. Bederson, New York: Academic, Vol. 1, p. 22.
McCarroll, R., 1986, in *Recent Studies in Atomic and Molecular Processes,* ed. A. E. Kingston, New York: Plenum, p. 113.
McGowan, J. W., Mul, P. M., D'Angelo, V. S., Mitchell, J. B. A., Defrance, P. and Froelich, H. R., 1979, *Phys. Rev. Lett.* **42**, 81.
Michels, H. H. and Hobbs, R. H., 1984, *Astrophys. J.* **286**, L27.
Morgan, W. L. and Bates, D. R., 1987, *Astrophys. J.* **314** 817.
Morrison, R. J. S., Conaway, W. E., Ebata, T. and Zare, R. N., 1986, *J. Chem. Phys.* **84**, 5527.
Nakashima, K., Hidekazu, T. and Katamura, H., 1987, *J. Chem. Phys.* **88**, 726.
Omont, A., 1986, *Astron. Astrophys.* **164**, 159.
Quack, M. and Troe, J., 1977, in *Gas Kinetics and Energy Transfer,* eds. P. G. Ashmore and R. J. Donovan, Specialist Periodical Reports, Chem. Soc. London **2**, 175.
Ramaker, D. E. and Peek, J. M., 1976, *Phys. Rev. A.* **13**, 58.
Rapp, D. and Francis, W. E., 1962, *J. Chem. Phys.* **37**, 2631.
Roberge, W. and Dalgarno, A., 1982, *Astrophys. J.* **255**, 489.
Rowe, B. R., Dupeyrat, G., Marguette, J. B., Smith, D., Adams, N. G. and Ferguson, E. E., 1984, *J. Chem. Phys.* **1984**, 241.
Sakimoto, K., 1984, *Chem. Phys.* **85**, 273.
Smith, D. and Adams, N. G., 1981, *Mon. Not. R. Astr. Soc.* **197**, 377.
Smith, I. W. M., 1989, *Astrophys. J.,* in press.
Stebbings, R. F., Smith, A. C. H. and Ehrhardt, H., 1964, *J. Geophys. Res.* **63**, 2349.
Su, T. and Chesnavich, W. J., 1982, *J. Chem. Phys.* **76**, 5183.
Takayanagi, K., 1978, *J. Phys. Soc. Japan* **45**, 976.
Troe, J., 1979, *J. Phys. Chem.* **83**, 114.
Troe, J., 1987a, *Int. J. Mass Spect. Ion Processes* **80**, 17.
Troe, J., 1987b, *J. Chem. Phys.* **87**, 2773.
Turner-Smith, A. R., Green, J. M. and Webb, C. E., 1973, *J. Phys. B: Atom. Molec. Phys.* **6**, 114.
Whitten, G. Z. and Rabinovitch, B. S., 1963, *J. Chem. Phys.* **38**, 2466.
Winniczek, J. W., Braveman, A. L., Shen, M. H., Kelley, S. G. and Farrar, J. M., 1987, *J. Chem. Phys.* **86**, 2818.

12

Collisional excitation processes

EVELYNE ROUEFF

DAMAP and URA 812 CNRS, Observatoire de Meudon, Meudon, France

12.1. Introduction

A detailed knowledge of collisional excitation processes is important in various aspects of the study of interstellar clouds. Because local thermodynamic equilibrium rarely obtains in such environments, the diagnosis of physical conditions, such as temperature, particle density, and radiation density, requires a quantitative understanding of all microscopic processes (i.e. collisional excitation and de-excitation and radiative decay and absorption) which influence the excitation conditions. Quite often, only rotational excitation of simple molecules need be considered. For instance, dense, cold cloud gas is studied primarily by observing millimeter and submillimeter emission features arising from transitions between different rotational levels of molecules in their ground electronic and vibrational states. In diffuse clouds, simple diatomics such as H_2, CN and C_2 are observed through electronic absorption transitions involving different rotational levels of the ground vibrational and electronic state. Even some atomic species such as C and C^+ are observed in various fine structure states. This information is important for the diagnosis of diffuse clouds. Observable emission from vibrationally excited molecules arises in hotter gas or in regions exposed to a strong ultraviolet radiation field, but the collisional excitation of vibrational states is not well understood for the appropriate temperature range. However, the observed emissions due to the decay of collisionally excited fine structure levels of atomic species, such as C and C^+ can be used to investigate clouds.

Second, collisional excitation of atomic and molecular species is always followed by spontaneous radiative emission leading to a loss of energy from the medium which is an important cooling process of the interstellar gas. In galactic

HI regions, spin-change processes occurring in the collision of two hydrogen atoms determine the spin temperature (Purcell and Field 1956). In diffuse neutral clouds, rotational excitation of CO, the most abundant heavy molecule is usually structure line emission from C, C^+ and O dominate the cooling. In dense cold clouds, rotational excitation of CO, the most abundant heavy molecule is usually considered as the main coolant. Molecular hydrogen is important for energy loss from hotter regions such as those which have been shocked (Dalgarno and McCray 1972, Black 1987).

A. Dalgarno has made fundamental contributions to the understanding of the physics of atomic and molecular collisional excitation and its application to astrophysics.

I restrict myself to the study of collisional excitation due to helium and atomic and molecular hydrogen which are the main constituents of the interstellar material. Following the development in the next section of the basic theoretical formalism necessary to describe inelastic collisional processes and the simplifying approximations mainly used, I will discuss atomic and molecular excitation in order of increasing complexity of the species involved. I will discuss separately the important example of rotational excitation of molecular hydrogen $\Delta J = 1$ transitions which result from a purely reactive process.

12.2. Theoretical calculations

Two problems have to be solved when one tackles atom–atom, atom–molecule, or molecule–molecule collisions resulting in fine structure excitation or rotational and vibrational energy transfer. First, one needs to know the relevant potential energy surface(s) over a wide range of nuclear geometries and then the equations of motion have to be solved to obtain the different scattering cross sections. I do not consider here the various aspects bound to the computations of reliable potential energy surfaces which is the task of quantum chemistry. However, it should be mentioned that A. Dalgarno has also been involved in such studies, especially those concerning the use of long range expansions (Dalgarno 1968). Perturbation theory can then be applied and intermolecular potentials become analytic forms depending on R^{-n} where R is the interatomic or intermolecular distance.

We describe now the main features of the scattering problem in the case of low energy inelastic molecular collisions, for which many reviews are presently available (see for example Gianturco (1980), Bernstein (1979), Schinke and Bowman (1983)).

The range of energies considered determines whether classical, semi-classical or quantal theory can (or has to) be applied. One has first to compare the de Broglie wavelength $\lambda\!\!\!^{-} = \hbar/p$ where p is the momentum of the relative motion, with the mean range a of the interatomic forces which is typically taken as several Bohr radii (~5–10). If $\lambda \ll a$ classical and semi-classical theory can be used whereas in the opposite case quantal scattering theory has to be employed. A second important condition to be fulfilled in order to apply semi-classical formalism is that the inelasticity involved in the collision should be small compared to the relative kinetic energy since in semi-classical treatments, the relative motion is treated classically. Inelastic collisions involving one light species (such as He, H, H_2) at the low energies relevant to interstellar studies often have to be studied within the framework of the quantal theory. Additional complexity arises from the necessity to couple the relative angular momentum to the other internal angular momenta. However, at low energies, relatively few total angular momenta have to be taken into account.

I consider now the general quantal formalism describing the collision between two molecular systems. The total Hamiltonian of the colliding particles, in the center of mass frame is expressed as

$$H = \frac{\hbar^2}{2\mu}\nabla_R^2 + H_1 + H_2 + V \qquad (12.1)$$

where μ is the reduced mass of the colliding system, H_1 and H_2 are the Hamiltonians for the internal coordinates $\bar{r}_1$ and $\bar{r}_2$ of species 1 and 2 (electronic coordinate for a structured atom, rotational and vibrational degrees of freedom for molecule without electronic angular momentum, etc.). V is the interspecies molecular potential surface which is a function of $\bar{r}_1$, $\bar{r}_2$ and R, the interspecies distance, and vanishes at infinity in a non-reactive collision which preserves the identity of the two colliding species. Figure 12.1 shows the example of a collision between two diatomics.

The quantum mechanical treatment of the dynamics is performed by solving, in the coordinate representation, the time-independent Schrödinger equation containing the total Hamiltonian.

$$H\Psi(\bar{R}, \bar{r}_1, \bar{r}_2) = E_{\text{tot}}\Psi(\bar{R}, \bar{r}_1, \bar{r}_2) \qquad (12.2)$$

Several possible coupling schemes can be used when building the most suitable functional expansion for the unknown total wave function $\Psi(\bar{R}, \bar{r}_1, \bar{r}_2)$. They allow one to define approximate quantum numbers for which the equations are 'almost' diagonal. Irrespective of these choices, one can always describe the system in terms of the total angular momentum J including the internal and relative orbital angular momentum and its projection, M, onto a space fixed axis and the total parity ε corresponding to the inversion of all particles (electrons and nuclei). Inelastic processes arise from the exchange of angular momentum (torque) between internal and translational motions.

The procedures most generally adopted rely on the asymptotic expansion of the total wave function (Arthurs and Dalgarno 1960):

$$\Psi(\bar{R}, \bar{r}_1, \bar{r}_2) = \sum_{JM\varepsilon} |\psi^{JM\varepsilon}\rangle\langle\psi^{JM\varepsilon}|\Psi(\bar{R}, \bar{r}_1, \bar{r}_2) \tag{12.3}$$

Denote the total angular momentum of each colliding species by j_i ($i = 1, 2$). The eigenvectors for the free Hamiltonian H_i (r_i) of each species are $|\alpha_i j_i m_i\rangle$ and the corresponding eigenvalues are $E\alpha_i j_i$; j_1 and j_2 are coupled to yield j_{12} and the vectors

$$|\alpha_1 j_1 \alpha_2 j_2 j_{12} m_j\rangle \equiv |\gamma j_{12} m_j\rangle \tag{12.4}$$

are defined. They correspond to the energies $E_{\alpha_1 j_1} + E_{\alpha_2 j_2}$. The abbreviated notation $\gamma \equiv \alpha_1 j_1 \alpha_2 j_2$ has been adopted.

A set of functions, each of which is specified by J, M and ε and j_1, j_2, j_{12}, and l, the relative angular momentum, is chosen. J, M, and ε are constants of the motion, while j_1, j_2, j_{12}, and l are 'good' quantum numbers only when the interspecies separation is large. These basis functions are required to satisfy plane wave boundary conditions. They may be expanded in spherical harmonics, $Y_{lm_l}(\theta, \phi)$. Hence, the basis functions are given by:

$$|\gamma j_{12} l J M \varepsilon\rangle = \sum_{m_j m_l} |\gamma j_{12} m_j\rangle Y_{lm_l}(\theta, \phi)\langle j_{12} l m_j m_l|JM\rangle \tag{12.5}$$

The values of the angular momenta are restricted by the choice of the above expansion and the triangular relation between coupled angular momenta in the Clebsch–Gordan coefficients $\langle j_{12} l m_j m_l|JM\rangle$.

Figure 12.1. Collision between two diatomic molecules. G_1 and G_2 are the centers of gravity of the molecules, r_1 and r_2 are the internuclear distances, R is the intermolecular distance.

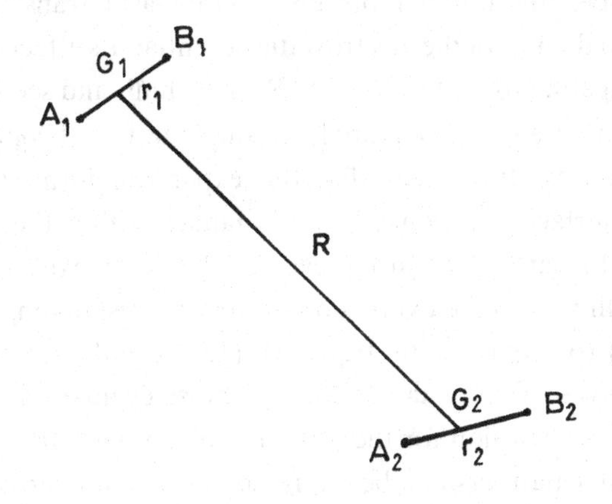

The required expansion functions of Equation (12.3) can then be written as

$$|\psi^{JM\varepsilon}\rangle = \sum_{m_j\, m_l} |\gamma j_{12} lJM\varepsilon\rangle(1/R)G^{J\varepsilon}_{\gamma j_{12}l}(R) \tag{12.6}$$

where the sum is restricted to γ, j_{12}, l values such that $\varepsilon = \varepsilon_{\alpha_1 j_1}\varepsilon_{\alpha_2 j_2}(-)^l$, where $\varepsilon_{\alpha_i j_i}$ is the parity of $|\alpha_i j_i m_i\rangle$, $i = 1, 2$. The superposition of the asymptotic states or basis functions is necessary because the anisotropic interspecies potential induces transitions from the initial asymptotic state $|(j_1 j_2)j_{12}l\rangle$ to the final asymptotic state $|(j_1' j_2')j_{12}' l'\rangle$.

Since the total Hamiltonian is invariant under rotation and inversion, the eigenfunctions (given by Equation (12.6)) of the total angular momentum and of definite parity are not coupled with each other via the potential.

For each $|j_1 j_2 j_{12} l\rangle$ initial state, one chooses the above expansion, (12.6), which is subsequently introduced in the Schrödinger equation. By left multiplying by the complex conjugate of $|\gamma j_{12} lJM\varepsilon\rangle$ and integrating over the angular variables, one obtains the familiar close-coupled (CC) equations:

$$\left(\frac{d^2}{dR^2} - \frac{l(l+1)}{R^2} + k^2_{\gamma j_{12}l}\right) G^J_{\gamma j_{12}l}(R) - \frac{2\mu}{\hbar^2}\sum V^J_{\gamma j_{12}l;\gamma' j_{12}' l'}(R)G^J_{\gamma' j_{12}' l'}(R) = 0 \tag{12.7}$$

or in matrix form

$$\left[\mathbb{1}\frac{d^2}{dR^2} - \frac{\mathbb{L}^2}{R^2} + \mathbb{K}^2 - \mathbb{V}^J(R)\right]\mathbb{G}^J(R) = 0 \tag{12.8}$$

where

$$(\mathbb{L}^2)_{\gamma j_{12}l;\gamma' j_{12}' l'} = l(l+1)\delta_{j_{12}j_{12}'}\delta_{ll'} \tag{12.8a}$$

$$(\mathbb{K}^2)_{\gamma j_{12}l;\gamma' j_{12}' l'} = \frac{2\mu}{\hbar^2}(E_{\text{tot}} - E_{\gamma j_{12}})\delta_{j_{12}j_{12}'}\delta_{ll'} \tag{12.8b}$$

$$(\mathbb{V}^J(R))_{\gamma j_{12}l;\gamma' j_{12}' l'} = (2\mu/\hbar^2)\int d\hat{R}\, d\hat{r}_1\, d\hat{r}_2 \langle \gamma j_{12} lJM|V|\gamma' j_{12}' l' JM\rangle \tag{12.8c}$$

One therefore sees that in this representation inelastic transitions take place because of the spatial range of the electrostatic potential. Its effect is first to distort the solution propagating to its asymptotic ($R \to \infty$) limit and second, because of the non-spherical nature of the potential, to couple different components of the angular eigenfunctions. It is clear that the expansion form, (12.8c), of the potential energy surface will control the dynamics within the representation described above. In particular, an increase in the complexity of the colliding partners appears directly in the expression of their corresponding interaction.

Equation (12.7) (or equivalently Equation (12.8)) is the scattering equation which must be solved. Before doing so, one must evaluate $\mathbb{V}^J(R)$. The most general form of the expansion for the potential electronic surface between two non-linear molecular partners can be expressed as a sum over the electrostatic interaction between multipolar components of two charge distributions (Gray 1968, Leavitt 1980)

$$V(R,\Omega_1,\Omega_2) = \sum_{j_1j_2lm_1m_2} V^{m_1m_2}_{j_1j_2l}(R)\left[\sum_{n_1n_2n} \langle j_1j_2n_1n_2|ln\rangle \mathscr{D}^{j_1}_{n_1m_1}(\Omega_1)\mathscr{D}^{j_2}_{n_2m_2}(\Omega_2)Y_{ln}(\hat{R})\right] \tag{12.9}$$

where the molecular orientations and the collision coordinates are all referred to a space-fixed frame. Equation (12.9) is of general validity. However, if one of the molecules possesses a point group symmetry, the individual rotation matrices, $\mathscr{D}^J_{nm}$, are replaced by linear combinations of rotation matrices. The linear combinations are chosen so that they block diagonalize the interaction potential; hence, only these irreducible representations which play a role in the particular collision event need be considered. However, intermolecular potential calculations are always performed in a body-fixed reference frame where the Z' axis is the intermolecular axis. This leads one to rewrite Equation (12.9) in a body-fixed reference frame with molecular orientations, Ω'_1, Ω'_2 which will be described below.

$$W(R,\Omega'_1,\Omega'_2) = \sum V^{m_1m_2}_{j_1j_2l}(R)\langle j_1j_2n_1n_2|l0\rangle \mathscr{D}^{j_1}_{n_1m_1}(\Omega'_1)\mathscr{D}^{j_2}_{n_2m_2}(\Omega'_2)((2l-1)/4\pi)^{1/2} \tag{12.10}$$

The properties of the Clebsch–Gordan coefficients imply that $n_1 = -n_2$ and only relative rotations of the two molecules about Z' modify the anisotropy of the potential. If one system has no angular momentum (an atom without electronic angular momentum or a molecule in a rotational state $J = 0$) and the other is an atom in a specific fine structure state J or a linear molecule without electronic angular momentum, the interaction reduces to

$$W(R,\Omega'_2) = \sum_{lm} V_{lm}(R)Y_{lm}(\theta'_2,\phi'_2) \tag{12.11}$$

Two different reference frames have been mentioned in the above discussion. One, which will be called the space-fixed (SF) frame, is appropriate for large interspecies separations and is defined, for example, relative to the initial velocity. It is in this frame that the close coupled equations, (12.7), are solved. The other will be more convenient for the description of the collision in the short and intermediate separation ranges where the two partners of the collision form a quasi-molecular system for which the intermolecular axis plays a major role and is defined as the Z' axis: this is the body-fixed (BF) reference frame. $W(R,\Omega'_1,\Omega'_2)$ is given for the BF frame by quantum chemistry calculations.

The two reference frames are evidently related through a frame transformation which is a rotation. Figure 12.2 shows the Euler angles defining the relation

Figure 12.2. SF and BF reference frames for the collision between two linear molecules. The primed coordinates define the *body* frame of reference while the unprimed coordinates are for the *laboratory* frame.

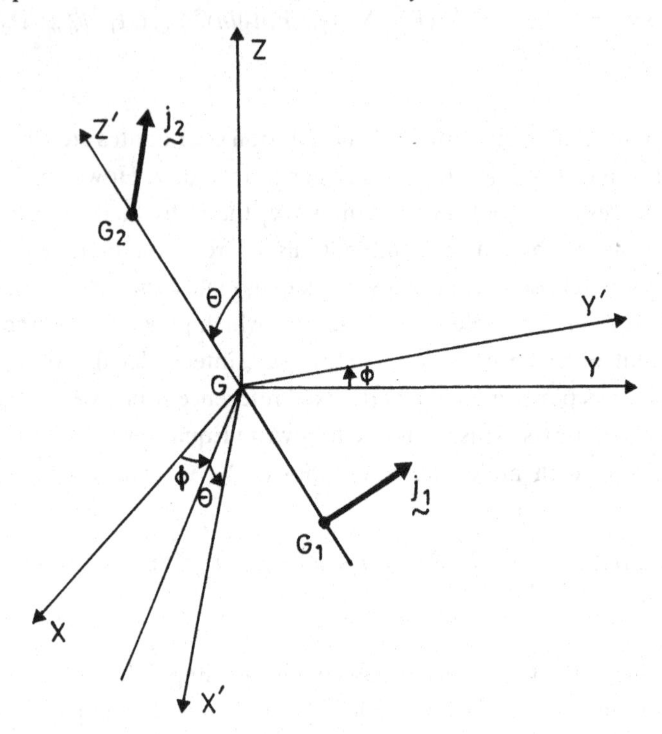

between the two frames. During the course of the collision the total wave function is better described in the BF reference frame. Indeed, the potential energy surface is diagonal in the quantum number Ω defined as the projection of the total angular momentum on the Z' axis.

The two sets of wave functions are related through the rotation transformation

$$|\gamma j_{12} lJM\varepsilon\rangle = \sum \mathcal{D}^{j_{12}^*}_{m_j\Omega}(\phi, \theta, 0)|\gamma j_{12}\Omega\rangle_{\mathrm{BF}} Y_{lm_l}(\theta, \phi)\langle j_{12} l m_j m_l | JM\rangle \quad (12.12)$$

where $\mathcal{D}^J_{m\Omega}(R)$ are the rotation matrices defined in the active point of view following the conventions of Rose (1957), Brink and Satchler (1961) and Messiah (1961).

The close coupled form of the scattering equation, (12.7), is thus obtained by first evaluating the potential interaction matrix elements W in the BF reference frame and then by transforming them in the SF reference frame.

$$V_{\mathrm{SF}} = RW_{\mathrm{BF}}R^{-1} \quad (12.13)$$

One can also choose to express all the operators involved in the Hamiltonian in the BF reference frame. This representation, sometimes called the helicity representation, allows the treatment of the incident and scattered waves in a symmetric manner. The Z' axis rotates during the collision so that the orbital angular momentum $\bar{L} = \bar{R} \times \bar{P}$ has a continuously vanishing projection along Z'.

The set of equations obtained by using wave functions defined in the BF reference frame, of course, has the same dimensionality as the set of SF equations but its structure will be different. The way in which approximately good quantum numbers are defined in order to reduce the dimensionality is therefore different according to which representation one chooses.

The BF expression of the wave function can be rewritten in terms of new eigenfunctions, namely of J^2, J_z, J_{12}, $|J_{12z'}|$ and parity π following Curtiss and Adler (1952), Mies (1973), Pack (1974) and Launay (1976)

$$|\gamma(j_1j_2)j_{12}\bar{\Omega}JM\rangle = (|\gamma j_{12}\Omega JM\rangle + \varepsilon p|\gamma j_{12} - \bar{\Omega}JM\rangle)/[2(1 + \varepsilon p\delta_{\Omega 0})]^{1/2} \tag{12.14}$$

where $p = \varepsilon_{\alpha_1 j_1}\varepsilon_{\alpha_2 j_2}(-)^{J-j_{12}}$ and $\bar{\Omega} = |\Omega|$

$$|\gamma j_{12}\bar{\Omega}JM\rangle = |\gamma j_{12}\Omega\rangle\left(\frac{2J+1}{4\pi}\right)^{1/2} \mathcal{D}^{J^*}_{M\Omega}(\phi, \theta, 0). \tag{12.15}$$

The angular momentum $\hat{l}$ operator involved in the Hamiltonian will now have non-diagonal matrix elements since it is not conserved in this representation. In the BF frame, the CC equations acquire the following structure

$$\frac{d^2}{dR^2} G^{J\varepsilon}_{\gamma j_{12}\bar{\Omega}}(R) + \left(\delta_{\gamma\gamma'}\delta_{j_{12}j'_{12}}k^2_\gamma - \frac{1}{R^2}\langle\gamma j_{12}\bar{\Omega}JM\varepsilon|\hat{l}^2|\gamma' j'_{12}\bar{\Omega}'JM\varepsilon\rangle \right.$$
$$\left. - \frac{2\mu}{\hbar^2}\langle\gamma j_{12}\bar{\Omega}JM\varepsilon|V|\gamma' j'_{12}\Omega' JM\varepsilon\rangle\right)$$
$$\times G^{J\varepsilon}_{\gamma' j'_{12}\bar{\Omega}'}(R) = 0 \tag{12.16}$$

where

$$\langle\gamma j_{12}\Omega JM|\hat{l}^2|\gamma j_{12}\Omega \pm 1JM\rangle = -\alpha \pm (j_{12}, \Omega)\alpha \pm (J, \Omega)$$
$$\langle\gamma j_{12}\Omega JM|\hat{l}^2|\gamma j_{12}\Omega JM\rangle = \beta(j_{12}, J, \Omega)$$
$$\alpha \pm (a, b) = [a(a+1) - b(b \pm 1)]^{1/2}$$
$$\beta(a, b, c) = a(a+1) - b(b+1) - 2c^2$$

and $\delta_{\gamma\gamma'}\delta_{j_{12}j'_{12}}k^2_\gamma$ are elements of a matrix which is similar to the one defined in (12.8b) and which is not diagonal in l, l'.

Now the couplings leading to inelastic transitions are due both to the non-spherically symmetric part of the electrostatic potential and to the rotation of the relative orbital angular momentum which couples states of different helicity. The matrix elements of the angular momentum operator are obtained through standard angular algebra; one should keep in mind that the total angular momentum J components in the BF reference frame satisfy reverse commutation laws (Mies 1973, Lefebvre-Brion and Field 1986).

The solution of the above coupled equations proceeds as in the SF representation and the computational problem is of the same order of complexity. However, in the BF reference frame, the off diagonal elements due to the $\hat{l}^2$ operator decrease very slowly (as R^{-2}) and are non-negligible even after the V coefficients of the potential energy surface have already died off. The main advantage of the BF coupled equations is that they can be used to introduce in a transparent way the various approximations discussed now. The $2j + 1$ degeneracy of the energy levels leads to a rapid proliferation of coupled quantum states and because in a standard (non-vectorial) computer the time needed to obtain a solution increases roughly as the cube of the number of coupled equations, intractable computational problems arise rapidly with increasing values of j. Moreover, the knowledge of the potential energy surface(s) is often incomplete for intermolecular interaction and approximations, which introduce uncertainties no greater than those originating from the inaccuracies of the potential surface(s), can be adopted. The approximations, which are based on physical arguments appropriate for the specific dynamics of the system being considered, lead to substantial reductions in the couplings between basis functions. We discuss briefly the centrifugal decoupling approximation and the Infinite Order Sudden approximation which are widely used for inelastic collisions.

Centrifugal decoupling approximation

This approximation, called also the coupled state (CS) approximation was introduced independently by Pack (1974) and McGuire and Kouri (1974). In the BF scattering equations, long range rotational coupling terms arise from the $\hat{l}^2$ operator. They can be eliminated by introducing an effective orbital angular momentum eigenvalue:

$$\hat{l}^2/2\mu R^2 \sim L(L+1)/2\mu R^2 \tag{12.17}$$

The validity of this approximation depends on the sensitivity on the effective potentials at the classical turning points and on the kinetic energy. Its accuracy improves with increasing energy and the approximation is better for cases when the collision is dominated by the repulsive part of the potential and only one turning point exists. In the early studies, the effective values of L were taken to be the total angular momentum J. More recently, the values of the diagonal matrix elements of the $\hat{l}^2$ operator have been adopted. Although this choice slightly complicates the long range form of the asymptotic boundary conditions, the coupling matrix elements have now a simpler structure because the value of Ω is conserved throughout the collision and the inelastic channels are only coupled through the electrostatic potential expressed in the BF reference frame.

The infinite order sudden (IOS) approximation

The infinite order sudden (IOS) approximation can be considered as a restricted form of the centrifugal decoupling approximation (Goldflam, Green and Kouri 1977, Pack 1972, Secrest 1975).

The replacement of the $\hat{l}^2$ operator by an effective eigenvalue can be further exploited by applying the closure property to the l, m_l quantum numbers of the complete set of the partial waves. A similar approximation may be used for the internal rotational energy levels. For molecule–molecule collisions the internal Hamiltonian $\hat{j}_2^2/2I$ is replaced by another effective eigenvalue form $j(j+1)/2I$ with the consequence that all k_j become equal to some value k. Thus, in the IOS approximation, the wave vector is the same in all the channels (energy sudden (ES) approximation). The differential equations are uncoupled and can be solved for each value L in the BF frame in which the interaction potentials and radial wave functions depend parametrically on the relative orientation between R and $\hat{r}_1$ and $\hat{r}_2$.

The inelastic transition probabilities are then obtained by calculating the elements of the orientation-dependent transition matrix connecting the initial and final states. Angular integration is then performed and usually accomplished by standard Gaussian quadrature procedures. The cross sections for the transition between two rotational states can be obtained by calculating only one row or column of the matrix of cross sections.

The simplicity of this approximation has led to a considerable interest in discussing its accuracy and range of validity. Because the rotational energy spacings are ignored compared to the kinetic energy, one expects the IOS approximation to be poor when only a few rotational levels are energetically accessible but in such case the more accurate CS approximation is often feasible. Because of the closure relationships, it also fails at energies close to threshold where closed channels play an important role. When many levels are energetically accessible the IOS method should be reliable but the comparison with more accurate results is often not available.

The IOS method has an ambiguity in the interpretation of the energy that defines the constant wave number. The recent suggestion that the IOS energy should be interpreted as the final kinetic energy appears to be a satisfactory procedure (Chapman and Green 1984), particularly for large ΔJ transitions.

12.3. Spin change and fine structure excitation

The relative populations of the hyperfine levels of atomic hydrogen and the spin temperature in galactic HI regions are controlled by spin-change processes occurring in collisions of two hydrogen atoms (Purcell and Field 1956). The two potential curves $X^1\Sigma_g$ and $^3\Sigma_u$ involved in the process are known to high precision

(Bishop and Shih 1976, Kolos and Wolniewicz 1965). The scattering formalism of spin-change processes has been studied by Dalgarno (1961), Glassgold (1963), and Dalgarno and Rudge (1965). The identity of the nuclei requires that the total wave function, including the nuclear and electronic spin and spatial wave functions, should be properly symmetrized with respect to the interchange of the two nuclei (which is equivalent to a rotation through an angle of π respective to an axis perpendicular to the internuclear axis). Initial and final states are specified only by their total angular momentum, F, so that one has to average over degenerate initial states and sum over all degenerate final states. Allison and Dalgarno (1969) have computed detailed quantal spin-change rate coefficients from 1 K to 1000 K and have shown the importance of quantal effects under 100 K.

Fine structure excitation studies relevant to interstellar space concern the most abundant heavy atomic and ionic species C, C^+ and O in collisions with H, He and H_2. Close coupling calculations have been performed for collisions between atomic hydrogen and the atoms of interest (Launay and Roueff 1977a,b), molecular hydrogen and C^+ (Flower and Launay 1977a,b), and He and O (Monteiro and Flower 1987). In the case of atom–atom collisions, the interatomic potentials are usually expressed as $v_{\Lambda S}$ where Λ is the absolute value of the projection on the internuclear axis of the total electronic angular momentum and S is the value of the total spin instead of the expansion given in Equation (12.11). The electrostatic interaction matrix elements in the basis adopted for the collision calculations take the form:

$$\langle (l_1 s_1) j_1 s_2 j_{12} lJ | V | (l_1 s_1) j_1' s_2 j_{12}' l' J \rangle = V^J_{\gamma l;\gamma' l'}(R) = \sum_{\Lambda S} g^{\Lambda SJ}_{\gamma l;\gamma' l'} v_{\Lambda S'} \quad (12.18)$$

where

$$g^{\Lambda SJ}_{\gamma l;\gamma' l'} = (-)^{2s_1+2s_2+2S}[S][j_1, j_{12}, l, j_1', j_{12}', l']^{1/2} \begin{Bmatrix} s_1 & s_2 & S \\ j_{12} & l_1 & j_1 \end{Bmatrix} \begin{Bmatrix} s_1 & s_2 & S \\ j_{12}' & l_1 & j_1' \end{Bmatrix}$$

$$\times \sum_{|m_{l_1}|=\Lambda;\Omega,m_s} \begin{pmatrix} l_1 & j_{12} & S \\ m_{l_1} & -\Omega & m_S \end{pmatrix} \begin{pmatrix} l_1 & j_{12}' & S \\ m_{l_1} & -\Omega & m_S \end{pmatrix} \begin{pmatrix} j_{12} & J & l \\ \Omega & -\Omega & 0 \end{pmatrix} \begin{pmatrix} j_{12}' & J & l' \\ \Omega & -\Omega & 0 \end{pmatrix} \quad (12.19)$$

where { } and ( ) are $6j$ and $3j$ coefficients, and $[a] = 2a + 1$. As already noted, the coupled equations are defined for each value of the total angular momentum J and parity π.

In the case of fine structure excitation the parity is defined as the product of the electronic parity of the structured atom and the nuclear parity $(-)^1$ of the relative angular momentum. Formula (12.18) implicity takes parity conservation into account and the electrostatic potential matrix splits into two separate submatrices. Table 12.1 gives the different channels arising in the cases of interest. In the case

Table 12.1. *Different channels included in fine structure excitation calculations and the relationship of the parity ε to the total angular momentum J*

Separated states	j	j_{12}	l_1	ε_1	l_2	ε_2
$^2P^o + {^1S}$	1/2	1/2	$J - 1/2$	$(-)^{J+1/2}$	$J + 1/2$	$(-)^{J-1/2}$
	3/2	3/2	$J - 1/2, J + 3/2$		$J + 1/2, J - 3/2$	
$^2P^o + {^2S}$	1/2	0	J			
		1	J	$(-)^{J+1}$	$J + 1$	$(-)^{J}$
	3/2	1	J		$J \pm 1$	
		2	$J, J \pm 2$		$J \pm 1$	
$^3P + {^1S}$	0	0			J	
	1	1	$J \pm 1$	$(-)^{J+1}$	J	$(-)^{J}$
	2	2	$J, J \pm 1$		$J, J \pm 2$	
$^3P + {^2S}$	0	1/2	$J + 1/2$		$J - 1/2$	
	1	1/2	$J + 1/2$		$J - 1/2$	
		3/2	$J + 1/2, J - 3/2$	$(-)^{J+1/2}$	$J - 1/2, J + 3/2$	$(-)^{J-1/2}$
	2	3/2	$J + 1/2, J - 3/2$		$J - 1/2, J + 3/2$	
		5/2	$J - 3/2, J + 1/2, J + 5/2$		$J - 5/2, J - 1/2, J + 3/2$	

of a ^3P atom perturbed by a structureless particle (such as He or H^+) one sees that only one parity is able to couple the 0 and 1 fine structure states of the ^3P atom. A detailed analysis of the electrostatic interaction shows moreover that the $\langle(l_1s_1)j_1s_2lJ|V|(l_1s_1)j_1's_2l'J\rangle$ interaction coupling matrix element with $\langle(11)00JJ|V|(11)10JJ\rangle$ values is vanishing from the triangular properties of the two $3j$ coefficients. The $0 \rightarrow 1$ transition can thus only occur from indirect coupling via the $J = 2$ state. Monteiro and Flower (1987) pointed out this property for O–He collisions for which they found that the $0 \rightarrow 1$ transition is about ten times smaller than the other transition cross sections. The extent to which this transition is forbidden depends essentially on the magnitude of the anisotropy in the interatomic potential i.e. the difference between $V^3\Sigma$ and $V^3\Pi$. The same considerations would apply for collisions with molecular hydrogen.

The fine structure excitation cross sections are calculated from the T matrix elements obtained from the resolution of the close coupled equations.

$$\sigma_{(l_1S_1)j_1,S_2 \rightarrow (l_1S_1)j_1',S_2}(E) = \frac{\pi}{k^2_{(l_1S_1)j_1S_2}} \sum_{J\varepsilon} \frac{2J+1}{(2j_1+1)(2s_2+1)} \sum_{ll'} |T^{J\varepsilon}_{\gamma l \rightarrow \gamma' l'}|^2 \quad (12.20)$$

Corresponding rate coefficients $k_{j \rightarrow j'}$ are then obtained by performing the subsequent Maxwellian average

$$k_{j \rightarrow j'}(T) = (8kT/\pi\mu)^{1/2} \int_0^\infty \sigma_{j \rightarrow j'}(E)\varepsilon \exp(-\varepsilon/kT)\, d\varepsilon \quad (12.21)$$

where ε denotes the kinetic energy on level j and E is the total (kinetic + internal) energy.

Table 12.2. *Rate coefficients* $k_{j \to j'}$ *in units of* $10^{-11}\ cm^3\ s^{-1}$

	T (K)						
	10	20	50	100	250	500	1000
$C^+_{1/2 \to 3/2} + H^a$	1.4 (−6)	1.41	24.5	64.1	116	147	177
$C^+_{1/2 \to 3/2} + H_2\ (J = 0)^c$	6.5 (−3)	0.70	12.6	33.8	64.6		
$C^+_{1/2 \to 3/2} + H_2\ (J = 1)^c$	9.45 (−3)	0.94	15.7	40.9	78.8		
$C_{0 \to 1} + H$	4.59	15.2	29.6	38.0	47.2	54.2	66.3
$C_{0 \to 2} + H^b$	0.10	2.14	13.6	25.1	42.1	59.1	80.6
$C_{1 \to 2} + H$	0.80	5.76	20.2	32.4	48.7	64.3	85.3
$O_{2 \to 1} + H$				0.56	4.37	10.69	19.9
$O_{2 \to 0} + H^b$					0.49	1.66	3.8
$O_{1 \to 0} + H$			0.39	1.33	3.39	5.69	9.22
$O_{2 \to 1} + He$				0.23	1.72	4.45	8.98
$O_{2 \to 0} + He^d$				0.04	0.63	1.92	4.00
$O_{1 \to 0} + He$				0.05	0.15	0.47	1.17

Numbers in parentheses are powers of ten.
[a] Launay and Roueff (1977a); for $T > 1000$ K, see Keenan *et al.* (1986)
[b] Launay and Roueff (1977b).
[c] Flower and Launay (1977).
[d] Monteiro and Flower (1987).

Fine structure excitation induced by molecular hydrogen requires the knowledge of the potential surface(s) for different geometries. Fine structure excitation of C^+ by molecular hydrogen has been studied by Chu and Dalgarno (1975) and Flower and Launay (1977). These authors have introduced five terms in the potential energy expansion (12.11) determined from the $^2A_1$, $^2B_1$, $^2B_2$ and $^2\Sigma$ and $^2\Pi$ adiabatic potential surfaces which are not known to high accuracy. As a consequence, the computed cross sections can be subject to some uncertainty at low energies. Table 12.2 gives the current best theoretical results for the excitation rate coefficients at some relevant temperatures.

Flower, Launay and Roueff (1977) made some predictions for the fine structure excitation of O and C by molecular hydrogen by considering only reduced mass effects and neglecting the possibly first order effect of parity conservation described above. The relevant rate coefficients will, however, depend on the real interaction potentials and in particular on the importance of anisotropic terms in the expansion.

12.4. Rotational excitation of diatomic molecules

The quantitative study of rotational excitation of astrophysically interesting diatomic molecules was undertaken in the seventies by Dalgarno, and by Green and their coworkers who also computed the relevant intermolecular potential energy surfaces.

Molecular hydrogen is the most abundant molecule in interstellar clouds but the observational data are restricted to ultraviolet electronic absorption transitions in diffuse clouds and to infrared emission through electric quadrupolar rovibrational transitions towards hot and dense cores. The availability of a recent accurate potential energy surface (Meyer, Schaefer and Liu 1980) has stimulated new work on complete close coupling calculations up to significant temperatures (Danby, Flower and Monteiro 1987b) with $J = 7$ as the maximum value of the rotational quantum number. Inelastic processes cannot change the ortho/para character of the molecule defined by its total nuclear spin so that only transitions with even ΔJ can occur. Care must be exercised in the definition of the collision cross sections between two identical molecules. Indeed, a factor of 2 appears between the definitions taken by Monchick and Schaeffer (1980) on the one hand and Green (1975a) on the other. However, the rates which determine the evolution of the physical system have to be the same and the above discrepancies have been elucidated by Danby *et al.* (1987b). Close coupled calculations have been compared with coupled states CS approximations and the agreement is seen to be good. All $\Delta J = 2$ transition rate coefficients for $J \leq 7$ induced by para-H_2 ($J = 0$) and ortho-H_2 ($J = 1$) are listed in Danby *et al.* (1987a) for kinetic temperatures 50 $\leq T \leq$ 600 K. A power law extrapolation to higher temperatures should not induce unacceptable errors. Their results for 600 K differ by about 20% from those obtained by using the power law deduced by Draine, Roberge and Dalgarno (1983). The rates of de-excitation by ortho-H_2 are shown to be somewhat larger than by para-H_2 which is probably due to supplementary terms in the potential energy surface which occur for $J > 0$. At the lower temperatures, the He rate coefficients agree remarkably well with ortho-H_2 rate coefficients while at the higher temperature the He rate coefficients are larger typically by a factor of 1.5. The inclusion of these accurate rate coefficients enables one to place more stringent constraints on interstellar shock models (Danby *et al.* 1987a, Monteiro *et al.* 1988; see Chapter 5).

Rate coefficients for the rotational excitation by H were obtained by Green and Truhlar (1979) for temperatures smaller than 100 K. The discrepancies with the former values obtained by Allison and Dalgarno (1967) are mainly due to the intermolecular potential surface chosen.

Rotational excitation of HD has been studied for collisions with He (Itakawa and Takayanagi 1972, Green 1974) and molecular H_2 (Chu 1975). The potential energy surface for He–H_2 and H_2–H_2 should also be appropriate when H_2 is replaced by HD, except that it must be expressed in terms of the coordinates of the center of mass of HD. The potential energy surface for H_2 perturbers involves only even terms in the expansion (12.11). When this is reexpanded in terms of the new coordinates, the expansion involves odd terms for the HD coordinates so that

$\Delta J = 1$ transition can occur. An additional difference appears in the long range expansion form (Buckingham 1968) since due to vibronic interaction HD possesses a (small) permanent dipole moment in addition to its permanent quadrupole moment.

Carbon monoxide has received particular attention and has been used in tests to check the various scattering approximations. Most recent calculations have been performed in the CS approximation for high temperatures (Schinke *et al.* 1985) and detailed close coupling calculations due to Flower and Launay (1985) have been performed with a different potential energy surface than in previously mentioned work. Para and ortho-hydrogen have there been separately considered at low temperatures. Viscuso and Chernoff (1988) have used these new data in their work on CO cooling rates at $T \lesssim 2000$ K which are relevant for interstellar shocks. Rotational excitation of CO by He was the subject of early study (Green and Thaddeus 1976); sometimes results for that system are used to deduce collisional rate coefficients for H_2 para ($J = 0$) collisions with CO. Schinke *et al.* (1985) have shown that the He–CO potential surface can be used to infer H_2–CO excitation rates if correct account is taken of the difference between the He and H_2 masses in the scattering calculations. Analytical fits of the various cross sections are given in Viscuso and Chernoff (1988). Collision rate coefficients for H collisions with CO are available at low temperatures (Green and Thaddeus 1976). The rotational excitation of the isovalent CS molecule in collisions with He and H_2 has been studied by Green and Chapman (1978).

Interstellar absorption features from C_2 rotational levels as high as $J = 14$ have been detected (Danks and Lambert 1983, van Dishoeck and de Zeeuw 1984); its excitation can only be understood from the knowledge of the various excitation processes. IOS calculations of the rotational excitation of C_2 due to H_2 ($J = 0$) collisions have been performed recently by using a preliminary potential energy surface obtained with the polarized atomic orbital technique (Chambaud *et al.* 1984) which enables one to reduce significantly the size of the configuration interaction calculations (Lavendy *et al.* 1988). The computed rate coefficients are given in Table 12.3.

The OH molecule was the first interstellar molecule to be detected in the centimeter wave region (Weinreb *et al.* 1963). The detected transitions between the hyperfine levels of the Λ doublet ground rotational state are at $\lambda = 18$ cm and the lines are affected by masing (Weaver *et al.* 1965). The first astronomical detections of pure rotational transitions of this radical have been made towards the Orion region (Storey, Watson and Townes 1981, Viscuso *et al.* 1985a,b, Watson *et al.* 1985), where collisions with H_2 molecules provide the most likely source of excitation. The theoretical treatment of its rotational excitation is subject to several difficulties. Firstly, the open-shell structure and the non-zero

Table 12.3. *C_2 deexcitation rate coefficients in units of $cm^3\,s^{-1}$ for collisions with H_2* (J = *0*)

T(K)	50	100	500	1000
$k(2 \rightarrow 0)$	1.54 (−12)	1.82 (−12)	1.04 (−11)	2.6 (−11)
$k(4 \rightarrow 2)$	1.93 (−12)	2.38 (−12)	1.56 (−11)	3.92 (−11)
$k(6 \rightarrow 4)$	1.48 (−12)	2.39 (−12)	1.79 (−11)	4.4 (−11)
$k(8 \rightarrow 6)$	1.47 (−12)	2.75 (−12)	2.0 (−11)	4.76 (−11)
$k(10 \rightarrow 8)$	1.54 (−12)	3.21 (−12)	2.22 (−11)	5.09 (−11)
$k(12 \rightarrow 10)$	2.26 (−12)	4.39 (−12)	2.47 (−11)	5.42 (−11)
$k(14 \rightarrow 12)$	3.39 (−12)	5.88 (−12)	2.75 (−11)	5.77 (−11)
$k(16 \rightarrow 14)$	4.36 (−12)	7.35 (−12)	3.04 (−11)	6.14 (−11)
$k(18 \rightarrow 16)$	5.45 (−12)	8.93 (−12)	3.35 (−11)	6.51 (−11)

Numbers in parentheses are powers of ten.

nuclear spin of OH lead to a relatively complex set of energy levels in which fine structure splitting, Λ-doubling and the hyperfine interaction manifest themselves. Secondly, in the interaction with molecular hydrogen, even in its $J = 0$ state, two adiabatic potential energy curves A' and A'', must be considered for a general geometry. The history of the derivation of the potential expansion, (12.11), from the adiabatic electronic potential energy surfaces A' and A'' has been tortuous due to phase convention misassignments. A factor of 2 discrepancy has remained for several years between different groups (Dewangan and Flower 1981, 1983, 1985) on the one hand and by Schinke and Andresen (1984) and Corey and Alexander (1988) on the other hand) for the value of the V_{22} coefficient in the expansion (12.11). Dewangan, Flower and Alexander (1987) have now clarified the origin of the discrepancy which arises from the erroneous assumption of factorization of the electronic rotational wavefunction describing the target OH molecule into electronic and rotation functions. It appears, however, that calculations using the incorrect V_{22} term (Dewangan, Flower and Danby 1986) give better comparison with experiments (Andresen, Haüsler and Lülf 1984). Consequently, Dewangan *et al.* (1987) have provided rate coefficients of rotational deexcitation of OH by para-H_2 between 15 and 500 K (which are the quantities required in astrophysical applications such as the calculation of infrared emission from shocks and the study of OH maser pumping models) by using this V_{22} value of the coefficient which is thus, in effect, adopted as an empirical factor. Λ-doublet and hyperfine structure transition cross sections at two different energies are given in the close coupling and coupled states formalism by Corey and Alexander (1988). These calculations predict a degree of selective excitation of the Λ-doublet levels in agreement with earlier work of Schinke and Andresen (1984) at a higher collision energy.

12.5. Rotational excitation of polyatomic linear molecules

The formal treatment of the electronic and rotational motion of a polyatomic linear molecule is very similar to the one applied to a diatomic molecule as long as vibrational excitation is neglected. The main difficulty arising in studies of the rotational excitation of polyatomic molecules is the calculation of the potential energy surface and comes from the small splitting of the rotational energy levels, due to the large reduced mass and leading to an increase of the number of channels involved in the calculations. Rotational excitation of various linear molecules has been studied mainly within the CS and IOS approximations (OCS, HC_3N, Green and Chapman (1978)). The treatment of molecular ions such as N_2H^+ (Green 1975b), HCO^+ and HCS^+ (Monteiro 1984a, 1985) does not differ from neutral systems except that long range forces are much more attractive so that collision rates are larger by a factor of approximately 5 than for neutral systems.

Anomalies observed in the intensities of the three hyperfine components of the $j = 1 \rightarrow 0$ transitions of interstellar HCN have led to detailed discussion of the corresponding excitation. Two independent sets of collisional rate coefficients obtained within the CS approximation (Monteiro 1984b, Stutzki and Winnewisser 1985) were in complete disagreement with each other. Monteiro and Stutzki (1986) have then performed full close coupling calculations of the temperature-dependent collisional rate coefficients for HCN–He between 10 and 30 K and discussed the origin of the previous discrepancy. It was due to the use of different ways of applying the centrifugal decoupling approximation. Monteiro (1984b) performed the CS approximation in the total angular momentum scheme $F = J + I$ where J and I are respectively the total (electronic + orbital) angular momentum and nuclear spin. Stutzki and Winnewisser (1985) first calculated nuclear spin free T matrices in the CS approximation; the T matrix in the F basis was then obtained by standard recoupling technique. Let us note that both treatments should be equivalent if exact scattering calculations are performed.

12.6. Rotational excitation of symmetric top molecules

12.6.1. The case of NH_3

The ammonia molecule is observed in dense interstellar clouds through rotation-inversion both at radio (see for example Guilloteau *et al.* (1983), Mauersberger *et al.* (1985)) and at infrared (see for example Townes *et al.* (1983), Keene, Blake and Phillips (1983)) wavelengths. The main exciting agent is molecular hydrogen (Ho and Townes 1983). The molecule has also been studied in the laboratory through microwave double resonance techniques (Daly and Oka 1970, Klaassen *et al.* 1982) and pressure broadening experiments (Klaassen, ter Meulen and

Dynamus 1983, Broquier and Picard-Bersellini 1985, Broquier *et al.* 1988) by various perturbers including He and molecular hydrogen. This has led to fruitful exchange and discussions, especially on the form of the intermolecular potential energy surface. The formalism for studying the collisional excitation of a symmetric top with states $|JKM\rangle$ where J is the rotational angular momentum and K and M are respectively the projection of J on the symmetry axis of the molecule and on a space fixed axis, by a structureless atom have been introduced by Green (1976, 1979, 1980) for the various quantal approximations.

Ammonia exists in two different (ortho and para) forms which correspond to different values of the total nuclear spin of the hydrogen atoms ($I = \frac{3}{2}$ and $\frac{1}{2}$ respectively). The requirement that the total wave function (nuclear × spatial) of ammonia should be antisymmetric under interchange of hydrogen nuclei implies that the value of K for ortho-ammonia is a multiple of three whereas para-ammonia has values of $K = 3n \pm 1$. Inelastic collisions cannot induce ortho–para transitions so that separate calculations have been performed for both species. Green (1980) has considered in detail the various effects of symmetry and of vibrational inversion motion on the scattering matrix elements because some (compensating) errors can be found in earlier literature. Rate constants obtained in the CS approximation are given by Green for para and ortho-NH_3 and He collisions for T between 15 and 300 K. The question of whether He and para-H_2 in its rotational ground state ($J = 0$) behave similarly as perturbers has been addressed recently in a large European collaborative work (Danby *et al.* 1986, 1987a) on close coupling quantal calculations. Billing and Diercksen (1985, 1986) have performed semi-classical calculations of NH_3–H_2 excitation rate coefficients with the same self consistent field (SCF) potential energy surface and a different treatment of the correlation energy. A detailed discussion of the differences occurring between the potential expansion terms is found in Danby *et al.* (1987a) and is shown to be the main origin of the discrepancies arising between the rate coefficients obtained by the two different scattering methods. On the other hand, Danby *et al.* (1987a) showed that rate coefficients for the rotational excitation of ammonia by H_2 can be deduced from scattering calculations using the NH_3–He potential and the real mass of H_2. This recommendation is in line with the suggestion of Green (1981) and Schinke *et al.* (1985). Comparison with experiments dealing with para-H_2 (Oka 1980, Broquier *et al.* 1988) shows some discrepancies which are most probably due to the uncertainties in the potential energy surface. However, the temperatures at which the experiments were performed ($T \sim 200$ K) are such that the $j = 2$ state of para-H_2 contains almost half of the populations; hence future theoretical work should include the rotational angular momentum of the hydrogen in the scattering calculations.

12.6.2. Methyl cyanide

Methyl cyanide, CH_3CN, is another symmetric top molecule which has been observed through various rotational transitions (Solomon *et al.* 1971, 1973, Cummins *et al.* 1983) which lie in the millimeter wavelength range. Radiative rotational transitions preserve the K angular momentum so that the total population of one K ladder relative to the other is determined only by collisions and is hence a function only of the kinetic temperature. Within a K ladder the populations radiatively relax and are thus a function of both kinetic temperature and density. The analysis of several rotational transitions within the context of a statistical equilibrium model allows a detailed picture of physical parameters in the molecular cloud if the state-to-state collisional excitation rates are known. Green (1985) performed calculations of the potential energy surface of the methyl cyanide–helium system by adopting the electron gas model and subsequently derived (Green 1986) the state to state collision rates relevant for H_2 within the IOS approximation.

12.7. Rotational excitation of asymmetric rotors

An accumulating body of observational data exists for several complex rotors, including H_2O, H_2CO, SO_2, SiC_2 and the recently discovered C_3H_2. Asymmetric rotors are potential probes of local conditions of molecular clouds insofar as their rotational levels and spectral lines are more densely spaced in energy and frequency than simple linear molecules, if the rates of the competing microscopic processes (i.e. collisional excitation and spontaneous radiative decay) can be or have been determined. All calculations of asymmetric rotor excitation made, to date, involve He as the perturber. The complexity of the determination of a potential energy surface for He rather than H_2 is considerably less and excitation by H_2 is not too different from excitation by He. Asymmetric top molecules are described by two quantum numbers: the rotational angular momentum J and its projection on a fixed axis M. The corresponding rotational wave function can be described as a linear combination of symmetric top wave functions $|JKM\rangle$

$$|J\tau M\rangle = \sum a^J_{\tau K}|JKM\rangle \tag{12.22}$$

K is the projection of J on the quasi-symmetry axis of the molecule and τ is an index that labels the asymmetric top functions. The expansion coefficients a_k are obtained by solving a standard secular equation

$$\sum a^J_{\tau K}[\langle JKM|H_{\text{rot}}|J'K'M'\rangle - E^J_\tau \delta_{KK'}] = 0 \tag{12.23}$$

where the rotational Hamiltonian is described by a rigid rotor where I_α and J_α are the moment of inertia and angular momentum operator about the α axis

$$H_{\text{rot}} = (2I_x)^{-1}J_x^2 + (2I_y)^{-1}J_y^2 + (2I_z)^{-1}J_z^2 \tag{12.24}$$

These axes are then used to define the interaction potential between the rotor and He. The matrix of the rotational Hamiltonian splits into two blocks defined by the parity of K since the functions should be either symmetric or anti-symmetric with respect to a 180° rotation about each principal axis. Moreover $a^J_{\tau K} = \varepsilon^J_\tau a^J_{\tau -K}$ where ε^J_τ takes values of ± 1. Nuclear statistics has to be taken into account when a pair of equivalent nuclei is present. If hydrogen nuclei are present as in H_2O, H_2CO and C_3H_2 two distinct forms of the molecule, ortho and para, are present and cannot be connected through radiative or inelastic transitions. When nuclei without nuclear spin are involved as in SiC_2 or SO_2, the wave functions vanish for odd values of K.

Collisional excitation of water, H_2O, was studied by Green (1980) in the CS approximation. Collision rate constants were given as a function of temperature between 20 and 250 K for para-H_2O and between 20 and 500 K for ortho-H_2O. Rates for the rotational excitation of water molecules colliding with the atoms have been recalculated for an extended temperature range by Palma *et al.* (1988) who used a new, accurate interaction potential. Green (1989) subsequently considered the rotational excitation of HDO.

The collisional excitation of ortho-formaldehyde at low energies has deserved particular attention (Garrison, Lester and Miller 1976, Garrison and Lester 1977, Bochetta, Gerratt and Guthrie 1988). Detailed close coupled and coupled states calculations were performed and show close agreement in most energy ranges. However, in two crucial regions at 32.2 and 47.7 K where Garrison *et al.* reported strong resonances, Bochetta *et al.* (1988), who used another more stable integration method with the same potential energy surface, found that the cross sections are smooth. However, strong resonances in the 20.2 K region were found in the latter calculations and characterized as being of Feshbach (compound state) type; others were predicted to exist in the 127 K region. Collision rate constants given in Garrison *et al.* (1976) should thus be used with some care. Green *et al.* (1978) using the CS approximation have extended the calculations of Garrison *et al.* (1976) to higher rotational levels and to kinetic temperatures up to 80 K. Rates for para-H_2CO have also been computed in the same paper and comparison between theoretical and experimental pressure broadening widths for several spectral lines are found to agree within the experimental uncertainties of 10%.

The IOS formalism for asymmetric top rotors has been discussed briefly by Green (1979) and applied for the first time in 1987 for the SiC_2 excitation by Palma and Green. They showed that the state-to-state cross sections $\sigma_{J\tau \to J'\tau'}$ can be computed in terms of 'generalized IOSA cross sections' $Q(L, M, M')$

$$\sigma_{J\tau \to J'\tau'} = (2J' + 1) \sum_{LMM'} C(J, \tau, J', \tau' | L, M, M') Q(L, M, M') \quad (12.25)$$

The $C(J, \tau, J', \tau' | L, M, M')$ are called 'spectroscopic coefficients' and contain all information about rotational wavefunctions.

$$C(J, \tau, J', \tau'|L, M, M') = \sum_{pp'qq'} a^J_{\tau p} a^J_{\tau q} a^{J'}_{\tau' p'} a^{J'}_{\tau' q'} \begin{pmatrix} J & L & J' \\ -p & M & p' \end{pmatrix} \begin{pmatrix} J & L & J' \\ -q & M & q' \end{pmatrix} \tag{12.26}$$

For linear and symmetric top rotors, each of the IOSA rates corresponds uniquely to a transition into or out of the lowest level and a prescription for interpreting the collision energy has been suggested for these cases which is based on the inelasticity of the transitions (Chapman and Green 1984). For asymmetric rotors however, one does not find this simple one-to-one correspondence between $Q(L, M, M')$ and transitions to the lowest level (see Equation (12.25) so that it is not entirely clear how to generalize the previous prescription. Future studies should document the reliability in general of IOS calculations for asymmetric top rotors and address the optimal interpretation of the IOSA energy parameter. The same approximation was used for SO_2 (Palma 1987) and cyclopropenylidene C_3H_2 (Green, De Frees and McLean 1987) excitations. The accuracy of the corresponding rate coefficients is difficult to assess. The authors claim a factor of 2 for the larger individual state-to-state rates whereas some of the smaller rates may be accurate only to an order of magnitude. It is, however, believed that they should be adequate for astrophysical applications which determine the temperature and density from relative line intensities.

12.8. Rotational excitation in a reactive collision: the example of ortho–para transitions in H_2

Non-reactive collisions cannot induce the ortho–para transitions in a molecule. However, even and odd values of the rotational quantum numbers (corresponding to para and ortho-H_2) are observed in absorption lines of H_2 towards diffuse interstellar clouds. Dalgarno, Black and Weisheit (1973) noticed that in cold gas, reactive collisions with protons may induce transitions for which ΔJ is odd. Black and Dalgarno (1976) suggested a rate coefficient of $1 \times 10^{-10}\,\mathrm{cm^3\,s^{-1}}$ for the $1 \rightarrow 0$ transition and obtain the downward rate coefficient for higher lying rotational levels by using the formula:

$$Q(J+2 \rightarrow J+1) = \frac{G(J+1)}{G(J+2)} Q(J+1 \rightarrow J) \tag{12.27}$$

where $G(J)$ is the (total) statistical weight of level J. Upward rate coefficients are obtained from the requirement of detailed balance and $\Delta J > 1$ transitions were ignored. These assumptions allowed them to conclude that the $J = 1$ level of hydrogen was thermalized in diffuse clouds and to deduce the temperature of the envelope (where $J = 1$ is mainly present) from the observed ratio of the column densities between $J = 1$ and $J = 0$ (Black and Dalgarno 1977).

The $H^+ + H_2$ system is the simplest one for which the corresponding electronic potential energy surface is accurately known (Giese and Gentry 1974). A most dynamically biased (MDB) statistical theory has been applied (Schlier 1980) to the $H^+ + H_2$ reaction and its isotopic variants $H^+ + D_2$ reaction and the results compare favorably to the available experimental data (Gerlich *et al.* 1980, Gerlich and Bohli 1981). The collision is seen to proceed, at low energies, through a strongly coupled intermediate complex which then dissociates into various channels.

In addition, Gerlich and Bohli (1981) showed the influence of the nuclear spin and symmetry on the $H^+ + H_2$ reaction. If one makes the assumption that the nuclear spin part of the total wave function of the (H^+, H_2) system is only weakly coupled to the other degrees of freedom, one finds additional good quantum numbers of the system i.e. I and M_I for the total internal nuclear ('spin') angular momentum and J and M_J for the total 'motional' (including electronic) angular momenta (Quack 1977). This leads to a decomposition of the collision S matrix in block diagonal form with each block corresponding to a value of the total nuclear spin I. A complex with $I = \frac{3}{2}$ (statistical weight 4) can only be formed from and decay to ortho-hydrogen whereas the decay of the $I = \frac{1}{2}$ complex (statistical weight 2) leads to the formation of either ortho or para-hydrogen; by counting the accessible rotational states, one can calculate p (J: even $\leftarrow I = \frac{1}{2}$) $= \Sigma_{J\text{even}} (2J + 1)/\Sigma_{\text{all}J} (2J + 1)$, the probability that the $I = \frac{1}{2}$ complex decays to para-hydrogen. Thus, at low energies below the $J = 2$ threshold ($E < 510$ K) the probability of ortho–para conversion i.e. $j = 1 \rightarrow 0$ is $\frac{1}{12} = (\frac{4}{6} \times 0 + \frac{2}{6} \times \frac{1}{4})$ whereas at high energy we get the asymptotic value of $\frac{1}{6} = (\frac{4}{6} \times 0 + \frac{2}{6} \times \frac{1}{2})$. This approximate conversion rule for I and M_I is seen to manifest itself in the product internal state distributions measured at several angles and energies (Gerlich and Bohli 1981).

Gerlich (1988) has used the MDB statistical theory and has included specifically the anisotropy of the potential to calculate the various ΔJ transitions induced in H_2 by collisions with protons. The value of the $1 \rightarrow 0$ rate coefficient is found to be nearly constant with temperature and approximately equal to 2×10^{-10} cm^3 s^{-1} which compares well with previous assumptions (Dalgarno *et al.* 1973). The $\Delta J = 1$ rates relating higher values of J are, however, quite different from the value deduced from $1 \rightarrow 0$ and the relation (12.27) and can be found in Table 12.4.

Ortho–para conversion of H_2 can also occur at higher temperatures relevant for shock studies ($T \gtrsim 1000$ K) through reactive collisions with H where an activation barrier is present. However, little is known about the rate of this process for which only an estimate can be found (Sternberg and Dalgarno 1989).

Table 12.4. $H^+ + H_2$ (J) → $H^+ + H_2$ (J′) *MDB calculations*

T (K)	Reaction rate coefficients in 10^{-10} cm^3 s^{-1}							
	1→0	2→1	3→2	4→3	5→4	6→5	7→6	8→7
10	2.41	10.7	3.16	8.69	2.56			
30	2.30	10.3	3.0	8.37	2.44	6.52	1.96	5.24
50	2.24	10.1	2.92	8.22	2.38	6.42	1.92	5.16
70	2.16	9.92	2.86	8.12	2.33	6.36	1.89	5.10
90	2.11	9.80	2.81	8.06	2.30	6.32	1.87	5.06
110	2.08	9.73	2.79	8.02	2.29	6.29	1.85	5.03
130	2.05	9.65	2.77	8.00	2.27	6.28	1.84	5.01
160	2.01	9.53	2.75	7.97	2.26	6.27	1.84	4.99
160	1.99	9.47	2.74	7.96	2.25	6.26	1.83	4.98
190	1.98	9.41	2.73	7.95	2.25	6.26	1.83	4.97
210	1.94	9.28	2.71	7.93	2.24	6.26	1.82	4.96
230	1.90	9.13	2.69	7.90	2.23	6.25	1.82	4.95
250	1.89	9.10	2.69	7.89	2.23	6.25	1.82	4.94
270	1.88	9.00	2.69	7.88	2.23	6.26	1.82	4.94

Table taken from Gerlich (1988) (private communication).

12.9. Concluding remarks

We have focused our attention on quantal scattering calculations relevant to systems of astrophysical interest. Such calculations are crucial since few experiments are available for such systems. Whenever experimental data exist, as in the case of NH_3, CO or H_2O, they have always stimulated fruitful exchanges leading frequently to a revision of the relevant potential energy surface. The correct expansion of the accurate *ab initio* energy surface has been shown to be critical for obtaining reliable theoretical results on scattering. Useful scattering approximations have been developed and allow the excitation calculations to be performed for the higher energies which are relevant for interstellar shock studies. Much work has been devoted to the comparison between He and H_2 as perturbers for excitation processes. Various studies show that a potential surface for He is generally a reasonable representation for H_2 in its ground rotational state.

When ortho-hydrogen is considered (essentially $J = 1$ for interstellar conditions), additional long range potential terms involving the quadrupolar moment of H_2 have to be taken into account together with an increase in the complexity of the angular momentum algebra.

Acknowledgements

This review would not have been possible without the help of previous ones. Gianturco's (1980) paper was especially useful for the theoretical formalism whereas Flower (1987) and Takayanagi (1987) have reviewed collision processes for interstellar purposes. The author is also much indebted to D. Gerlich for communicating results before publication.

REFERENCES

Allison A. C., Dalgarno A. (1967). *Proc. Phys. Soc. London* **90**, 609

Allison A. C., Dalgarno A. (1969). *Astrophys. J.* **158** 423

Andresen P., Haüsler D, Lülf H. W. (1984). *J. Chem. Phys.* **81**, 571

Arthurs A. M., Dalgarno A. (1960). *Proc. Roy. Soc. London A* **256**, 540

Bernstein R. B. (1979). Ed. *Atom – molecule collision theory,* Plenum Publ. Co. New York

Billing G. D., Diercksen G. H. F. (1985). *Chem. Phys. Lett.* **121**, 94

Billing G. D., Diercksen G. H. F. (1986). *Chem. Phys.* **105**, 145

Bishop D. M., Shih S. K. (1976). *J. Chem. Phys.* **64**, 162

Black J. H. (1987). In *Interstellar Processes,* eds. Hollenbach D. J. and Thronson M. A., Jr. Reidel D. Publ. Co., Dordrecht 731

Black J. M., Dalgarno A. (1976). *Astrophys. J.* **203**, 132

Black J. M., Dalgarno A. (1977). *Astrophys. J. Suppl. Ser.* **34**, 405

Bochetta C. J., Gerratt J., Guthrie G. (1988). *J. Chem. Phys.* **88**, 975

Brink D. M., Satchler G. R. (1968). *Angular momentum,* Clarendon, Oxford

Broquier M., Picard-Bersellini A. (1985). *Chem. Phys. Lett.* **121**, 437

Broquier M., Picard-Bersellini A., Aroui H., Billing G. D. (1988). *J. Chem. Phys.* **88**, 1551

Buckingham A. D. (1968). *Adv. Chem. Phys.* **12**, 107

Chambaud G., Gerard-Ain M., Kassab E., Levy B., Pernot P. (1984). *Chem. Phys.* **90**, 271

Chapman S., Green S. (1984). *Chem. Phys. Lett.* **112**, 436

Chu Shih-I. (1975). *J. Chem. Phys.* **62**, 4089

Chu Shih-I., Dalgarno A. (1975). *J. Chem. Phys.* **62**, 4009

Corey G. C., Alexander M. H. (1988). *J. Chem. Phys.* **88**, 6931.

Cummins S., Green S., Thaddeus P., Linke R. A. (1983). *Astrophys. J.* **266**, 331

Curtiss C. F., Adler F. T. (1952). *J. Chem. Phys.* **20**, 249

Dalgarno A. (1961). *Proc. Roy. Soc. London A* **262**, 132

Dalgarno A. (1968). *Advances in Chem. Phys.* **12**, 143

Dalgarno A., Rudge M. R. H. (1965). *Proc. Roy. Soc. London A* **286**, 519

Dalgarno A., McCray R. A. (1972). *Ann. Rev. Astr. Ap.* **10**, 375

Dalgarno A., Black J. M., Weisheit J. C. (1973). *Astrophys. Lett.* **14**, 77

Daly P. W., Oka T. (1970). *J. Chem. Phys.* **53**, 3272

Danby G. (1983). *J. Phys. B: Atom Mol. Phys.* **16**, 3393

Danby G., Flower D. R., Kochanski E., Kurdi L., Valiron P., Diercksen G. H. F. (1986). *J. Phys. B: At. Mol. Phys.* **19**, 2891

Danby G., Flower D. R., Valiron P., Kochariskic E., Kurdi L., Diercksen G. H. F. (1987a). *J. Phys. B: At. Mol. Phys.* **20**, 1039

Danby G., Flower D. R., Monteiro T. S. (1987b). *MNRAS* **226**, 739

Danks A. C., Lambert D. L. (1983). *Astron. Astrophys.* **124**, 188

Dewangan D. P., Flower D. R. (1981). *J. Phys. B: Atom. Mol. Phys.* **14**, 2179

Dewangan D. P., Flower D. R. (1983). *J. Phys. B: Atom. Mol. Phys.* **16**, 2157

Dewangan D. P., Flower D. R. (1985). *J. Phys. B: Atom. Mol. Phys.* **18**, L137
Dewangan D. P., Flower D. R., Danby G. (1986). *J. Phys. B: Atom. Mol. Phys.* **19**, L747
Dewangan D. P., Flower D. R., Alexander M. H. (1987). *MNRAS* **226**, 505
van Dishoeck E., de Zeeuw T. (1984). *MNRAS* **206**, 383
Draine B. T., Roberge W. G., Dalgarno A. (1983). *Astrophys. J.* **264**, 485
Flower D. R. (1987). In *Interstellar Processes,* eds. Holenbach D. J. and Thronson M. A. Jr, Reidel D. Publ. Co., Dordrecht 745
Flower D. R., Launay J. M. (1977a). *J. Phys. B: Atom. Mol. Phys.* **10**, 3673
Flower D. R:, Launay J. M. (1977b). *J. Phys. B: Atom. Mol. Phys.* **10**, L229
Flower D. R., Launay J. M., Roueff E. (1977). *Proc. 21st. Inst. Astrophys. Coll.* Université Liège, 137
Flower D. R., Launay J. M. (1985). *MNRAS* **214**, 271
Garrison B. J., Lester W. A. Jr, Miller W. H. (1976). *J. Chem. Phys.* **65**, 2193
Garrison B. J., Lester W. A. Jr (1977). *J. Chem. Phys.* **66**, 531
Gerlich D. (1989). *J. Chem. Phys.,* submitted
Gerlich D., Nowotny U., Schlier C., Teloy E. (1980). *Chem. Phys.* **47**, 245
Gerlich D., Bohli H. J. (1981). *E.C.A.P.* eds. Kowalski J., Zu Putlitz G. and Weber H. G. 930
Gianturco F. A. (1980). Ed., *Atomic and Molecular Collision Theory*, Plenum Publ. Co. New York 315
Giese C. F., Gentry W. R. (1974). *Phys. Rev. A.* **10**, 2156
Glassgold A. E. (1963). *Phys. Rev.* **132**, 2144
Goldflam R., Green S., Kouri D. J. (1977). *J. Chem. Phys.* **67**, 4149
Gray C. G. (1968). *Can. J. Phys.* **46**, 135
Green S. (1974). *Physica* **76**, 609
Green S. (1975a). *J. Chem. Phys.* **62**, 2271
Green S. (1975b). *Astrophys. J.* **205**, 766
Green S. (1976). *J. Chem. Phys.* **64**, 3463
Green S. (1979). *J. Chem. Phys.* **70**, 816
Green S. (1980). *J. Chem. Phys.* **73**, 2740
Green S. (1981). *NASA Technical Memorandum* **83**, 869
Green S. (1985). *J. Phys. Chem.* **89**, 5289
Green S. (1986). *Astrophys. J.* **309**, 331
Green, S. (1989). *Astrophys. J. Supp. Ser.* **70**, 1214
Green S., Thaddeus P. (1976). *Astrophys. J.* **205**, 766
Green S., Chapman S. (1978). *Astrophys. J. Supp. Ser.* **37**, 169
Green S., Garrison B. J., Lester W. A. Jr, Miller W. H. (1978). *Astrophys. J. Supp. Ser.* **37**, 321
Green S., Truhlar D. G. (1979). *Astrophys. J. Lett.* **231**, L101
Green S., DeFrees D. J., McLean A. D. (1987). *Astrophys. J. Supp. Ser.* **65**, 175
Guilloteau S., Wilson T. L., Martin R. N., Batrla W., Pauls T. A. (1983). *Astron. Astrophys.* **124**, 322
Ho P. T. P., Townes C. H. (1983). *Ann. Rev. Astron. Astrophys.* **21**, 239
Itakawa Y., Takayanagi K. (1972). *J. Phys. Soc. Japan* **32**, 1605
Keene J., Blake G. A., Phillips T. G. (1983). *Astrophys. J.* **271**, L27
Klaassen D. B. M., Reijnders J. M. H., ter Meulen J. J., Dynamus A. (1982). *J. Chem. Phys.* **76**, 3019
Klaassen D. B. M., ter Meulen J. J., Dynamus A. (1983). *J. Chem. Phys.* **78**, 767
Kolos W., Wolniewicz L. (1965). *J. Chem. Phys.* **43**, 2429
Launay J. M. (1976). *J. Phys. B: Atom. Molec. Phys.* **9**, 1823

Launay J. M., Roueff E. (1977a). *J. Phys. B: Atom. Molec. Phys.* **10**, 879
Launay J. M., Roueff E. (1977b). *Astron. Astrophys.* **56**, 289
Lavendy H., Robbe J. M., Chambaud G., Levy B., Roueff E. (1989). To be submitted
Leavitt R. P. (1980). *J. Chem. Phys.* **72**, 3472
Lefebvre-Brion M., Field R. W. (1986). *Perturbations in the spectra of diatomic molecules,* Academic Press, Inc. New York 14–24
Mauersberger R., Wilson T. L., Batrla W., Walmsley C. M., Henkel C. (1985). *Astron. Astrophys.* **146**, 168
McGuire P., Kouri D. J. (1974). *J. Chem. Phys.* **60**, 2488
Messiah A. (1961). *Mecanique quantique,* Dunod, Paris
Meyer W., Schaefer J., Liu B. (1980). Unpublished
Mies F. H. (1973). *Phys. Rev. A* **7**, 942
Monchick L., Schaeffer J. (1980). *J. Chem. Phys.* **73**, 6153
Monteiro T. (1984a). *MNRAS* **210**, 1
Monteiro T. (1984b). *MNRAS* **211**, 257
Monteiro T. (1985). *MNRAS* **214**, 419
Monteiro T. S., Stutzki J. (1986). *MNRAS* **221**, 33pp
Monteiro T., Flower D. (1987). *MNRAS* **228**, 101
Monteiro T. S., Flower D. R., Pineau des Forêts G., Roueff E. (1988). *MNRAS* **234**, 863
Oka T. (1980). *Proc. IAU symp. 87 on interstellar molecules,* ed. B. H. Andrew, Dordrecht, Reidel, 221
Pack R. T. (1972). *Chem. Phys. Let.* **14**, 393
Pack R. T. (1974). *J. Chem. Phys.* **60**, 633
Palma A. (1987). *Astrophys. J. Supp. Ser.* **64**, 565
Palma A., Green S. (1987). *Astrophys. J.* **316**, 830
Purcell E., Field G. B. (1956). *Astrophys. J.* **124**, 542
Quack M. (1977). *Mol. Phys.* **34**, 477
Rose M. E. (1957). *Elementary theory of angular momentum,* Wiley, New York
Schinke R., Bowman J. M. (1983). In *Molecular collision dynamics* Ch 4, ed. Bowman J. M., Springer, Berlin
Schinke R., Andresen P. (1984). *J. Chem. Phys.* **81**, 5644
Schinke R, Engel V., Buck U., Meyer H., Diercksen G. H. F. (1985). *Astrophys. J.* **299**, 939
Schlier C. G. (1980). In *Energy storage and redistribution in molecules*, ed. Hinze J., Plenum Press, New York, p. 585
Secrest D. (1975). *J. Chem. Phys.* **62**, 710
Solomon P. M., Jefferts K. B., Penzias A. A., Wilson R. W. (1971). *Astrophys. J.* **168**, L107
Solomon P. M., Penzias A. A., Jefferts K. B., Wilson R. W. (1973). *Astrophys. J.* **185**, L63
Sternberg A., Dalgarno A. (1988). *Astrophys. J.,* in press
Storey J. V., Watson D. M., Townes C. H. (1981). *Astrophys. J.* **244**, L27
Stutzki J., Winnewisser G. (1985). *Astron. Astrophys.* **144**, 1
Takayanagi K. (1987). *IAU Symposium 120 Astrochemistry,* eds. Vardya M. S. and Tarafdar S. P., Reidel Publ. Co., Dordrecht, 31
Townes C. H., Genzel R., Watson D. M., Storey J. W. V. (1983). *Astrophys. J.* **269**, L11
Viscuso P. J., Chernoff D. F. (1988). *Astrophys. J.* **327**, 364
Viscuso P. J., Stacey G. J., Fuller C. E., Kurtz N. T., Harwitt M. (1985a). *Astrophys. J.* **296**, 142

Viscuso P. J., Stacey G. J., Harwitt M., Hass M. R., Erickson E. F., Duffy P. B. (1985b). *Astrophys. J.* **296**, 149
Watson D. M., Genzel R., Townes C. H., Storey J. W. V. (1985). *Astrophys. J.* **248**, 316
Weaver H., Williams D. R. W., Dieter N. H., Lum W. T. (1965). *Nature* **208**, 29
Weinreb S., Barrett A. H., Meeks M. L., Henry J. C. (1963). *Nature* **200**, 829

13

Neutral reactions at low and high temperatures

MARGARET M. GRAFF

School of Physics, Georgia Institute of Technology, Atlanta, Georgia, USA

13.1. Introduction

Astrophysical environments offer chemical modelers the unusual challenge of predicting reaction rate coefficients for temperature ranges well outside those commonly encountered in the laboratory (250–2000 K). The direct extrapolation of thermal laboratory data involves implicit, often incorrect, assumptions about the dynamics of chemical reactions at extreme temperatures. Furthermore, the internal energy distributions of constituent atoms and molecules in astrophysics are frequently nonthermal, and direct application of thermal reaction rate data may not be appropriate. An understanding of the physics of chemical reactions is essential for accurate modeling (Dalgarno 1985). Elucidation of the role of various forms of energy in chemical reactions – translational, vibrational, rotational, and fine structure – now affords improved predictions of rate coefficients for astrophysical situations.

In the low-temperature environment of cold interstellar clouds, the chemical kinetics are dominated by reactions whose potential energy surfaces are without barriers. Such reactions are characterized by large exothermicities and long-range electrostatic interactions that suppress chemical barriers that may occur as collision systems approach short distances. Current models of cold interstellar cloud chemistry illustrate the importance of exothermic ion–molecule reactions (cf. Dalgarno and Black (1976), Black and Dalgarno (1977), van Dishoeck and Black (1986)). Many neutral systems are also likely to react rapidly at very low temperatures and may be important in the chemical kinetics of interstellar clouds (Graff 1989). The following section discusses dynamical characteristics of neutral reactions that are likely to be fast at low temperatures.

At high temperatures, reactions with activation barriers may also be fast. High-temperature chemical kinetics favor the conversion of energetically ‘expensive' radicals such as O, C, OH, and CH to stable species such as CO and H_2O; in regions of high photon flux, photodissociation may slow or reverse the conversion. Reactions with and without barriers show very different dependences on temperature and on internal energy. Reasonable extrapolation of thermal measurements, where available, or estimation of rate coefficients for astrophysics requires an understanding of the basic dynamics controlling the various types of reactions.

13.2. Low-temperature reactions

Reactions may occur at low temperatures in strongly exothermic systems with attractive long-range electrostatic interactions (cf. Clary (1984, 1988)). The importance of ion–molecule reactions, promoted by charge–dipole ($V(R) \sim -R^{-2}$) and charge–induced-dipole ($V(R) \sim -R^{-4}$) interactions, has long been recognized in cold cloud chemistry. Current models devote little attention to atom–radical systems with attractive dipole–quadrupole ($V(R) \sim -R^{-4}$) and quadrupole–quadrupole ($V(R) \sim -R^{-5}$) interactions, which are also potentially fast at low temperatures.

Low-temperature neutral reactions are certainly among those most difficult to study experimentally. Recent technical advances have made possible the determination of ion–molecule reaction rate coefficients below 20 K in traps (cf. Luine and Dunn (1985)) and supersonic jets (cf. Rowe *et al.* (1985)) and below 50 K in drift tubes (cf. Böhringer and Arnold (1982)). For neutral systems, the difficulties of working at low temperatures are compounded by those of detecting neutrals and of studying collisions between two reactive species. The most successful method for direct measurement of rate coefficients has been laser photolysis/resonance fluorescence, reviewed recently by Smith (1988) (see also Lewis and Watson (1980), Messing *et al.* (1980, 1981)). The experiments generally examine rate coefficients above 250 K and are most useful, for the astrophysical context, when used to corroborate theoretical predictions that extend to much lower temperatures.

Clary (1984) has developed the adiabatic capture (AC) method to study radical–radical reactions dominated by long-range electrostatic interactions. The method assumes a capture radius at short internuclear distances and, in its simplest form, seeks the maximum value J_{max} of the total angular momentum for which the collision energy exceeds the barrier maximum in the effective potential $V_{eff}(R, \theta, J)$,

$$V_{eff}(R, \theta, J) = V(R, \theta) + \frac{J(J+1)\hbar^2}{2\mu R^2} \tag{13.1}$$

The angle θ represents the general dependence of the electrostatic potential

$V(R, \theta)$ on atomic and molecular orientations relative to the separation axis. The AC cross section is then given by

$$\sigma(E, \theta) = \frac{\pi}{k_E^2} (J_{\max} + 1)^2 \tag{13.2}$$

where k_E is the wave number for the collision. Reaction is assumed to occur on a single potential surface that corresponds, for simple systems without curve crossings, to the lowest adiabatic state. For low temperatures, the AC method is expected to produce very realistic rate coefficients for atom–molecule reactions, subject to several qualifications (Graff and Wagner 1989): (1) Reaction is likely to occur only if collisions originating from the lowest fine-structure level of the system (i.e. the occupied level at low temperatures) reach the reactive surface at small distances. Transitions among adiabatic states may be an important consideration in systems with curve crossings. (2) The lowest adiabatic surface is attractive from all angles of approach. Viewed classically, the lowest surface corresponds to the orientation of the atomic quadrupole moment that produces, for each angle of approach to the molecular dipole, the most attractive electrostatic potential. (3) Atomic quadrupole moments and thus long-range potential surfaces $V(R, \theta)$ depend on the atomic fine-structure level and M_J sublevel (Gentry and Giese 1977). Long-range adiabatic states may be expressed as expansions of atomic and molecular asymptotic states, with expansion coefficients dependent on the angle of approach and the atom–molecule separation. The adiabatic long-range reaction surface is calculated by diagonalization of the long-range electrostatic and spin–orbit Hamiltonian or, to a good approximation, by optimization of the classical surface using the appropriate quadrupole moment. (4) The chemical potential at short range must include a reaction path sufficiently attractive that collisions are likely to result in reaction rather than reflection back to reactants. For the general reaction $A + BC \rightarrow AB + C$, the most attractive geometry at short range is likely to correspond to that of the stable or metastable reaction intermediate ABC^*.

An important astrochemical reaction that has become a prototype for reactions dominated by long-range interactions is $O + OH \rightarrow O_2 + H$. The work of Clary (1984) and Clary and Werner (1984) has been extended in a study of long-range adiabatic potential curves and interactions (Graff and Wagner 1989). The lowest adiabatic surface is approximately that of a classical electrostatic dipole–quadrupole/quadrupole–quadrupole interaction, with the orientation of the atomic quadrupole moment optimized at each angle of approach. For collision energies below 10 meV (E/k = 115 K), barrier maxima occur at distances sufficiently large that the quadrupole–quadrupole interaction may be negelected, and analytic expressions may be obtained for reaction cross sections and rate

Table 13.1. *Low-temperature rate coefficients for simple neutral reactions*

Reaction	$k(T = 0)$ ($cm^3 s^{-1}$)
$O + OH \rightarrow O_2 + H$	7.9×10^{-11}
$O + CH \rightarrow CO + H$	1.9×10^{-10}
$C + CH \rightarrow C_2 + H$	nonreactive
$C + OH \rightarrow CO + H$	nonreactive

coefficients. The results may be generalized to other exothermic reactions of atomic oxygen with dipolar open-shell molecules. Reactions of atomic oxygen with radicals are likely to be fast at low temperatures primarily because the lowest fine-structure level has a nonzero quadrupole moment: a single surface is most attractive throughout the long-range region, no avoided crossings occur, and the reaction surface can be accessed directly from the lowest spin–orbit state of the collision system. Table 13.1 shows rate coefficients for two reactions (Graff 1989).

Reactions of atomic carbon with radicals, by contrast, are likely to be inhibited at low temperatures. The quadrupole moment of the lowest fine-structure level of atomic carbon $C(^2P_0)$ is zero; the lowest surface for atomic carbon–radical reactions is therefore flat at large distances. An attractive surface correlating to the first excited fine-structure level $C(^2P_1)$ crosses the flat surface at relatively large separation ($\sim 12\ a_0$ for collision partners CH, OH, and others with similar dipole moments) and is likely to lead to reaction at chemical distances. Transitions between the two surfaces are improbable because coupling in the crossing region is very weak. Low-temperature collisions, which originate in the lowest fine-structure level, are therefore likely to be nonreactive.

The importance of fast neutral reactions involving atomic oxygen may be comparable to that of ion–molecule systems in the reaction networks of cold interstellar clouds, and inclusion of these reactions in cloud models may alter predictions of dominant chemical sequences. Neutral reactions with O constitute a major loss mechanism for the important radicals OH and CH; atomic oxygen may similarly attack many other open-shell systems. Clarification of the role of neutral reactions at low temperatures requires further experimental and theoretical study.

13.3. High-temperature reactions

Many interesting astrochemical environments, such as shocked regions or photodissociation regions, have temperatures that exceed the range commonly accessed by laboratory studies. The simple Arrhenius form, $k = A \exp(-B/T)$, is generally used to extrapolate laboratory measurements to high temperatures.

This functional form works well for limited ranges of temperature but cannot be extrapolated with confidence beyond ~2000 K. More accurate temperature dependences are available only for the most important or most easily studied reactions. Wagner and Graff (1987) calculated high-temperature rate coefficients for the major oxygen–hydrogen reactions and found deviations of an order of magnitude or more from the Arrhenius form. As experimental and theoretical data are accumulated, better predictions can also be made about nonthermal effects that may be important in warm regions.

13.3.1. Nonthermal environments

The extreme conditions encountered in astrophysical environments often produce highly nonthermal populations in constituent atoms and molecules. The infrequency of collisions in interstellar clouds results in subthermal internal energy distributions, particularly in high-temperature regions. Internal energy populations of atoms and molecules are determined by spontaneous emission rates, number density, and efficiency of collisional excitation and de-excitation for the individual species. Polar molecules decay rapidly to their lowest rotational/vibrational levels via electric dipole transitions. Atomic and molecular fine-structure distributions are cold except at high densities. Homonuclear diatomic molecules such as H_2, which decay by electric quadrupole transitions, have populations that are more nearly thermal, but large departures from equilibrium may still occur at low densities. Regions of high photon flux can produce superthermal internal energy distributions in H_2 (Sternberg and Dalgarno 1988). Accurate modeling of chemical processes in astrophysics thus requires an understanding of how different types of internal energy affect reactivity. General results and predictions are obtainable for the role of vibrational and fine-structure excitation in some types of reactions; other cases must be analyzed individually.

13.3.2. Effects of nonthermal internal energy in oxygen–hydrogen systems

For many endothermic ion–molecule systems, reactions occur via the potential wells of metastable collision complexes. Energy is shared among the available internal modes, and the role of energy (translational and internal) in the reaction may be treated by statistical methods (cf. the treatment of the reaction $C^+ + H_2 \rightarrow C^+ + H$ by Herbst and Knudson (1981)). Endothermic neutral reactions, by contrast, typically have potential surfaces with barriers that constrain nuclear motion; internal modes tend to evolve in specific ways as reactions progress. Some of the simple oxygen–hydrogen reactions have become benchmark systems for the study of neutral reactions with barriers in the potential surface. Recent theoretical and experimental advances have greatly improved our understanding of these systems (Haug *et al.* 1987, Robie *et al.* 1987, Bowman *et al.* 1984, and references

Table 13.2. *Nonthermal correction factors for rate coefficients of oxygen–hydrogen reactions at* T = *2500 K*, T_{rot} = *1000 K*, T_{vib} = T_{fs} = *0 K*

Reaction	F_{vib}	F_{rot}	F_{fs}
$O + H_2 \rightarrow OH + H$	0.65	0.60	
$H + OH \rightarrow H + H_2$	0.67	0.45	
$OH + H_2 \rightarrow H_2O + H$	0.76		
$H + H_2O \rightarrow OH + H_2$	0.62		
$OH + OH \rightarrow H_2O + O$	0.38		0.93
$O + H_2O \rightarrow OH + OH$	0.34		
$O + OH \rightarrow O_2 + H$			3.3

therein). Cohen and Westberg (1983) have compiled results for many of the major reactions and have recommended rate coefficients.

In recent studies the effects of molecular internal energy on the major oxygen–hydrogen reactions have been surveyed (Wagner and Graff 1987) and the results have been applied in a kinetic model of shocked interstellar clouds (Graff and Dalgarno 1987). Table 13.2 gives examples of nonthermal effects for conditions typical of a *J*-type shock of velocity v_0 = 10 km s^{-1} through a cloud of intermediate density ($n_0 = 10^3$ cm^{-3}): $T = 2500$ K, $T_{vib} = 0$ K, $T_{rot} = 1000$ K, and $T_{fs} = 0$ K. (See Wagner and Graff (1987) for a full discussion of rate coefficients in the general case.) The net correction for nonthermal internal energy distributions is given by the product of individual correction factors for nonthermal vibration, F_{vib}, rotation, F_{rot}, and fine structure, F_{fs}. A qualitative discussion of important effects follows.

Molecular vibration

Vibrationally enhanced reactivity has been observed in flow systems for reactions $O + H_2(v = 1) \rightarrow OH + H$ (Light 1978) and $OH + H_2(v = 1) \rightarrow H_2O + H$ (Spencer, Endo, and Glass 1977, Zellner and Steinert 1981, Glass and Chaturvedi 1981). Those results have been used to evaluate theoretical models describing the enhancement of reaction due to vibrational excitation (Lee *et al.* 1982, Truhlar and Isaacson 1982). The role of vibration in reactions with barriers is now relatively well understood in the context of simple transition-state theory (TST). Individual vibrational levels of separated reactants are assumed to correlate adiabatically to quantum levels at the transition state. The effective barrier and, therefore, the rate coefficient depend on the relative vibrational spacings of reactants and transition state. Rate coefficients for excited vibrational levels can be considerably larger than for the ground level at temperatures around 300 K (cf.

Light (1978) and Lee *et al.* (1982) for $O + H_2$). High-temperature corrections for nonthermal vibration are minor, however, for most astrochemical reactions. Vibrationally excited molecules are minor constituents in a thermal gas at 2500 K (~10% for OH and H_2), so the thermally averaged rate coefficient generally does not differ greatly from the vibrationally cold value. For the oxygen–hydrogen systems listed in Table 13.2, effects of cold vibration are expected to be large only for the reaction $OH + OH \rightarrow O + H_2O$, a heavy–light–heavy collision system with low vibrational frequencies at the transition state.

Vibrational effects may be of greater importance in the chemistry in regions of high ultraviolet photon flux where H_2 vibrational populations are superthermal. Sternberg and Dalgarno (1988) modeled H_2 level populations, temperature and density profiles, and thermal chemistry in photodissociation regions at the boundaries of molecular clouds. The reaction $O + H_2 \rightarrow OH + H$ may be vibrationally enhanced at relatively large cloud depths where temperatures drop to several hundred Kelvin: the reaction is inhibited for the lowest vibrational level but remains relatively fast for $v = 1$. The possible enhancement of reactions by vibration has been discussed by Stecher and Williams (1974) and by Hartquist (1986).

Molecular rotation

Our present knowledge of the effects of rotation on high-temperature reactions is rather fragmentary. No experimental data exist for the role of rotational excitation in the relatively well-studied oxygen–hydrogen systems listed in Table 13.2. Theoretical studies on individual reactions have achieved some qualitative agreement, but as yet no unifying model of rotational effects (analogous to the TST model of vibrational effects) has emerged. A modified TST approach (Wagner 1988, Wagner and Graff 1987) has been used to calculate corrections for nonthermal rotational distributions in the reactions $O + H_2 \leftrightarrow OH + H$.

For the reaction $OH + H_2 \rightarrow H_2O + H$, initial trajectory studies (Schatz 1981, Rashed and Brown 1985) predicted that rotation inhibits reactivity. Recent studies (Harrison and Mayne 1987, 1988) indicate that rotational effects are weaker than previously determined: the reaction probability is relatively constant at translational energy $E/k = 2300$ K for J levels of similar or lower energy; reactivity decreases with increasing OH or H_2 rotational level at translational energy $E/k = 5800$ K. The calculated effects are sensitive to details of the potential surface. These results indicate that nonthermal rotational populations should have little effect on the rate coefficient near 2500 K; no correction factor is listed.

The study of the reaction $O + OH \rightarrow O_2 + H$ by Clary and Werner (1984) indicated that the rate coefficient for the lowest rotational level is ~30% larger than the value for thermally populated rotational levels, a result due largely to the anisotropy of the long-range interaction potential used in that work. A recent study (Graff and Wagner 1989) has shown that the long-range potential surface is relatively isotropic, and rotational effects are expected to decrease accordingly. Table 13.2 offers no rotational correction factor for this reaction.

A general model of rotational effects at high temperatures requires much further work by both experimentalists and theorists. The need for accurate potential surfaces has been demonstrated. Existing theoretical predictions remain to be tested by alternative theoretical methods and, more importantly, by experimental measurement.

Atomic and molecular fine structure

No experimental data exist on effects of fine structure in the oxygen–hydrogen systems. Most transition-state models assume a simple form, presented below, for fine-structure effects in reactions with a barrier. A recent study of the reaction O + OH → O_2 + H (Graff and Wagner 1989) provides insight into the influence of fine-structure excitation in exothermic 'long-range' reactions.

The role of fine structure in reactions may be examined by expressing the total rate coefficient as a weighted average over coefficients $k_i(T)$ of individual fine-structure levels:

$$k(T_{fs}, T) = \frac{\Sigma g_i \exp(-\varepsilon_i/kT_{fs})k_i(T)}{Q_e(T_{fs})} \tag{13.3}$$

The partition function $Q_e(T_{fs}) = \Sigma g_j \exp(-\varepsilon_j/kT_{fs})$ reflects the population of fine-structure levels at temperature T_{fs}; g_i and ε_i are the total electronic degeneracy and energy of fine-structure level i of the collision system ($g_i = g_{1i}g_{2i}$, $\varepsilon_i = \varepsilon_{1i} + \varepsilon_{2i}$). For low- to intermediate-density astrophysical environments, the temperature describing fine-structure populations is effectively zero.

For reactions with a barrier in the potential surface, corrections for nonthermal fine-structure populations are relatively minor. The general TST model assumes that the reaction surface is accessed with statistical probability from all asymptotic fine-structure levels and that the fine-structure splitting at the transition state is small compared to kT. The predicted correction factor is then (Graff and Wagner 1989)

$$\frac{k(T_{fs} = 0, T)}{k(T_{fs} = T, T)} = \frac{Q_e(T)}{\Sigma g_i} \tag{13.4}$$

At high temperatures, the partition function $Q_e(T)$ approaches Σg_i and the correction factor approaches unity. Corrections are significant only for intermediate temperatures for systems with relatively small activation energies (e.g. OH + OH in the oxygen–hydrogen network).

Effects are more pronounced for exothermic reactions without barriers in the potential surface. Wagner and Graff (1987) recommended the correction factor

$$\frac{k(T_{fs} = 0, T)}{k(T_{fs} = T, T)} = \frac{Q_e(T)}{g_0} \tag{13.5}$$

where g_0 refers to the total degeneracy of the lowest fine-structure level of the collision system. Equation (13.5) applies to collision systems, such as those including atomic oxygen, where reaction is likely to occur from the lowest fine-structure level. The nonthermal correction, which simply reflects the relative probability that a collision is initiated from the reactive level, can be large at high temperatures. Systems that may be nonreactive from the lowest fine-structure level, such as those involving atomic carbon, may exhibit very different fine-structure dependences. These remain to be explored.

13.3.3. High-temperature behavior of reactions without barriers

Most astrophysical models assign the functional form $k(T) = A(T/300)^{0.5}$ to rate coefficients of exothermic reactions without barriers in the potential energy surface (Prasad and Huntress 1980). Experimental and theoretical studies of the prototypic reaction O + OH strongly suggest a qualitatively different temperature dependence: the thermal rate coefficient decreases with temperature over the range $200 < T < 2500$ K (Cohen and Westberg 1983). The decrease is due to the combined effects of a decrease in fine-structure excitation (Graff and Wagner 1989) and the reflection of flux to nonreaction (Miller 1986), which is particularly important at high collision energies. A more conservative approach suggested by Leen and Graff (1988) applies the rate coefficient for 300 K at all higher temperatures for reactions without barriers. Further dynamics studies are needed to clarify the temperature dependence of exothermic reactions.

13.4. Summary

This chapter has presented examples illustrating the importance of reaction dynamics to astrochemical studies and has suggested areas deserving further study. For exothermic reactions without potential barriers, the question of which fine-structure level(s) lead to reaction is of major importance to rate coefficients at both low and high temperatures. Qualitative differences may exist between reactions of atomic oxygen vs atomic carbon. The high-temperature behavior of these reactions requires further study. Reactions of atomic oxygen with radicals may be fast at low temperatures and should be included in the chemical networks of cold clouds. Further work is also needed on reactions with barriers in the potential surface. Little is known about rotational effects for these reactions;

experimental work would be particularly valuable. An investigation of nonthermal chemical kinetics in photon-heated regions may reveal vibrationally-enhanced reactivity. An understanding of the dynamics of various classes of reactions can greatly improve estimates of reaction rate coefficients appropriate to the extreme environments commonly encountered in astrophysics.

Much of this work was done in collaboration with Al Wagner, whose contributions are gratefully acknowledged. I sincerely thank Alex Dalgarno for helpful discussions and guidance.

REFERENCES

Black, J. H. & Dalgarno, A. (1977). *Ap. J. Suppl.* **34**, 405.
Bowman, J. M., Wagner, A. F., Walch, S. P. & Dunning, T. H., Jr. (1984). *J. Chem. Phys.* **81**, 1739.
Böhringer, H. & Arnold, F. (1982). *J. Chem. Phys.* **77**, 5534.
Clary, D. C. (1984). *Mol. Phys.* **53**, 3.
Clary, D. C. (1988). In *Reactive Rate Coefficients in Astrophysics,* eds. T. J. Millar & D. A. Williams, in press.
Clary, D. C. & Werner, H.-J. (1984). *Chem. Phys. Letters* **112**, 346.
Cohen, N. & Westberg, K. R. (1983). *J. Phys. Chem. Ref. Data* **12**, 531.
Dalgarno, A. (1985). In *Molecular Astrophysics: State of the Art and Future Directions,* eds. G. H. F. Diercksen, W. F. Huebner & P. W. Langhoff, pp. 3–22. Dordrecht: D. Reidel.
Dalgarno, A. & Black, J. H. (1976). *Rep. Prog. Phys.* **39**, 573.
Gentry, W. R. & Giese, C. F. (1977). *J. Chem. Phys.* **67**, 2355.
Glass, G. P. & Chaturvedi, B. K. (1981). *J. Chem. Phys.* **75**, 2749.
Graff, M. M. (1989). *Ap. J.* **339**, 239.
Graff, M. M. & Dalgarno, A. (1987). *Ap. J.* **317**, 432.
Graff, M. M. & Wagner, A. F. (1988). *J. Chem. Phys.,* submitted.
Graff, M. M. & Wagner, A. F. (1989). *J. Chem. Phys.,* submitted.
Harrison, J. A. & Mayne, H. R. (1987). *J. Chem. Phys.* **87**, 3698.
Harrison, J. A. & Mayne, H. R. (1988). *J. Chem. Phys.* **88**, in press.
Hartquist, T. W. (1986). *Quart. J. Roy. Astron. Soc.* **27**, 71.
Haug, K., Schwenke, D. W., Truhlar, D. G., Zhang, Y., Zhang, J. Z. H. & Kouri, D. J. (1987). *J. Chem. Phys.* **87**, 1892.
Herbst, E. & Knudson, S. (1981). *Ap. J.* **245**, 529.
Lee, K. T., Bowman, J. M., Wagner, A. F. & Schatz, G. C. (1982). *J. Chem. Phys.* **76**, 3583.
Leen, T. M. & Graff, M. M. (1988). *Ap. J.* **325**, 411.
Lewis, R. S. & Watson, R. T. (1980). *J. Phys. Chem.* **84**, 3495.
Light, G. C. (1978). *J. Chem. Phys.* **68**, 2831.
Luine, J. A. & Dunn, G. H. (1985). *Ap. J.* **299**, 67.
Messing, I., Carrington, T., Filseth, S. V. & Sadowski, C. M. (1980). *Chem. Phys. Letters* **74**, 56.
Messing, I., Filseth, S. V., Sadowski, C. M. & Carrington, T. (1981). *J. Chem. Phys.* **74**, 3874.
Miller, J. A. (1986). *J. Chem. Phys.* **84**, 6170.
Prasad, S. S. & Huntress, W. T., Jr. (1980). *Ap. J. Suppl.* **43**, 1.

Rashed, O. & Brown, N. J. (1985). *J. Chem. Phys.* **82**, 5506.
Robie, D. C., Arepalli, S., Presser, N., Kitsopoulos, T. & Gordon, R. J. (1987). *Chem. Phys. Letters* **134**, 579.
Rowe, B. R., Marquette, J. B., Dupeyrat, G. & Ferguson, E. E. (1985). *Chem. Phys. Letters* **113**, 403.
Schatz, J. C. (1981). *J. Chem. Phys.* **74**, 1133.
Smith, I. W. M. (1988). In *Reactive Rate Coefficients in Astrophysics,* eds. T. J. Millar & D. A. Williams, in press.
Spencer, J., Endo, H. & Glass, G. P. (1977). In *Proc. 16th Symp. (Intl.) Combustion,* p. 829.
Stecher, T. P. & Williams, D. A. (1974). *Mon. Not. Roy. Astron. Soc.* **168**, 51.
Sternberg, A. & Dalgarno, A. (1988). *Ap. J.,* submitted.
Truhlar, D. G. & Isaacson, A. D. (1982). *J. Chem. Phys.* **77**, 3516.
van Dishoeck, A. F. & Black, J. H. (1986). *Ap. J. Suppl.* **62**, 109.
Wagner, A. F. (1989). In preparation.
Wagner, A. F. & Graff, M. M. (1987). *Ap. J.* **317**, 423.
Zellner, R. & Steinert, W. (1981). *Chem. Phys. Lett.* **81**, 568.

[illegible] (1985), J. [illegible] Phys. 82, 2500.
Robie, D. C., Arepalli, S., Presser, N., Kitsopoulos, T. & Gordon, R. J. (1987), Chem. Phys. [illegible] 134, [illegible]7.
Rowe, B. R., Marquette, J. B., Dupeyrat, G. & Ferguson, E. E. (1985), Chem. Phys. Letters 1[illegible]3.
Schultz, [illegible] (19[illegible]), [illegible] 14, 1154.
Smith, I. W. M. (198[illegible]), in Rate Coefficients in Astrophysics, eds. T. J. Millar & D. A. Williams, [illegible]
[illegible] (197[illegible]), [illegible] (196[illegible]) [illegible] 23[illegible]
[illegible], T. [illegible] & Wickham, D. A. (19[illegible]) [illegible]
[illegible] (198[illegible]), [illegible] submitted.
[illegible] (198[illegible]), [illegible] 77, 5510.
[illegible] & Black, J. H. (198[illegible]), [illegible] Suppl. 62, 109.
[illegible] in preparation.
[illegible] M. (1979), [illegible] 2[illegible]
[illegible] (19[illegible]) Chem. [illegible] Lett. 81, 28[illegible]

V

Atomic species in dense clouds

14

Observations of atomic species in dense clouds

GARY J. MELNICK

Harvard-Smithsonian Center for Astrophysics, Cambridge, Massachusetts, USA

14.1. Introduction

Over the past ten years, observations of the far-infrared and submillimeter fine structure emission from neutral oxygen and carbon, and singly ionized carbon and silicon, have revealed the presence of a dense ($n_H \geq 10^3$ cm^{-3}), intermediate temperature ($60 \leq T_{gas} \leq 1500$ K) component of the interstellar medium in which a substantial fraction of the gas is atomic (see Table 14.1). In addition, observations of atomic fine structure emission from regions which are mostly molecular have helped to define the nature of the shocks which frequently accompany star formation and to challenge long-held beliefs about the structure and chemistry of molecular clouds.

Emission from atomic and ionic species with ionization potentials less than that of hydrogen arises predominantly in diffuse HI clouds, the photodissociated surfaces of molecular clouds, and the warm gas downstream of passing shock waves, generated deep within molecular clouds by the outflows from newly formed stars. The latter two regions, of primary interest in this contribution, are characterized by gas temperatures of 60–2000 K. Since H and He have no low-lying levels which can effectively cool the gas in this temperature range, the fine structure transitions of OI (63, 146 μm), C^+ (158 μm), and Si^+ (35 μm) generally dominate the cooling and thus serve as important diagnostics of these regions. An indication of the cooling power in just the [OI] 63 μm and [CII] 158 μm lines is given in Table 14.2; between 0.1 and 1% of the *total* luminosity of these sources escapes in these two lines. To a lesser extent atomic gas is also mixed with the mostly molecular gas deeper inside molecular clouds. Within this colder gas, cooling results from fine structure emission of C° (609, 370 μm).

In this brief review of the observations of atomic species in dense clouds, we shall first discuss the physics involved in the interpretation of fine structure lines

Table 14.1. *Average parameters of the interstellar gas*

Component	Density (cm^{-3})	Temperature (K)	Total mass ($M_\odot$)
Ionized			2×10^8
Coronal gas between clouds	10^{-2}–10^{-3}	6×10^5	
Low density HII regions	$3 - 10^2$	8000	
Compact HII regions	$10^3 - 10^6$	8000	
Atomic			1.5×10^9
HI between clouds	0.1–1	6000	
HI in diffuse clouds	20	80	
HI in dense clouds	$10^3 - 10^6$	100–1500	
Molecular			2×10^9
Giant molecular clouds	10^3	15	
Giant molecular cloud cores	5×10^5	35	
Dark clouds	10^3	10	
Dark cloud cores	2×10^4	10	

Table 14.2. *Fine structure luminosities from selected regions*

Source	OI 63 μm ($L_\odot$)	CII 158 μm ($L_\odot$)	L_{bol} ($L_\odot$)	References
M42	600	80	3×10^5	1, 2, 3
M17	3000	2000	6×10^6	1, 4
Sgr A	10^5	10^4	3×10^7	5, 6, 7
M82 (nucleus)	10^8	2.5×10^7	10^{10}	8,6

Table after Tielens and Hollenbach (1985)
References. (1) Melnick, Gull and Harwit (1979); (2) Storey, Watson and Townes (1979); (3) Russell *et al.* (1980); (4) Russell *et al.* (1981); (5) Genzel *et al.* (1984); (6) Crawford *et al.* (1985); (7) Lester *et al.* (1981); (8) Watson *et al.* (1984).

Table 14.3. *Important fine structure transitions arising in neutral regions*

Species/Transition	$\lambda(\mu m)$	A (s^{-1})	n_{cr} (cm^{-3})	Excitation range (eV)
C^0				
$2p^2:{}^3P_1 \rightarrow {}^3P_0$	609.135	7.93×10^{-8}	1.7×10^2	0–11.26
$2p^2:{}^3P_2 \rightarrow {}^3P_1$	370.415	2.68×10^{-7}	7.1×10^2	
C^+				
$2p^1:{}^2P_{3/2} \rightarrow {}^2P_{1/2}$	157.741	2.36×10^{-6}	3.0×10^3	11.26–24.38
O^0				
$2p^4:{}^3P_1 \rightarrow {}^3P_2$	63.184	8.95×10^{-5}	9.8×10^5	0–13.62
${}^3P_0 \rightarrow {}^3P_1$	145.526	1.70×10^{-5}	1.5×10^5	
Si^+				
$3p^1:{}^2P_{3/2} \rightarrow {}^2P_{1/2}$	34.815	2.13×10^{-4}	3.4×10^5	8.15–16.35

Table after Watson (1985).

and then we shall examine their observation from: (1) photodissociation regions, (2) shocked gas regions, and (3) the interiors of molecular clouds.

14.2. Atomic fine structure lines

Because fine structure transitions are often the lowest energy transitions in many atoms and ions and because their energy above the ground state is typically between about 20 and 600 K, these are the strongest lines to arise from atomic gas regions. Fine structure lines are produced by magnetic dipole transitions within multiplets of given spin and orbital angular momentum whose states of different total angular momentum are split by the spin–orbit interaction. Atoms and ions with 1, 2, 4 or 5 p-electrons in their valence shells have ^{2}P or ^{3}P ground states, and hence have astrophysically important fine structure lines. A list of such lines from several of the more abundant atomic and ionic species is given in Table 14.3 and an energy level diagram showing the fine structure transitions for Si^+, O^0, C^0, and C^+ is shown in Figure 14.1.

The interpretation of fine structure line intensities is made simple by the fact that: (1) unlike their optical and near-infrared counterparts, these lines suffer negligible dust extinction, and (2) in most instances these fine structure lines are

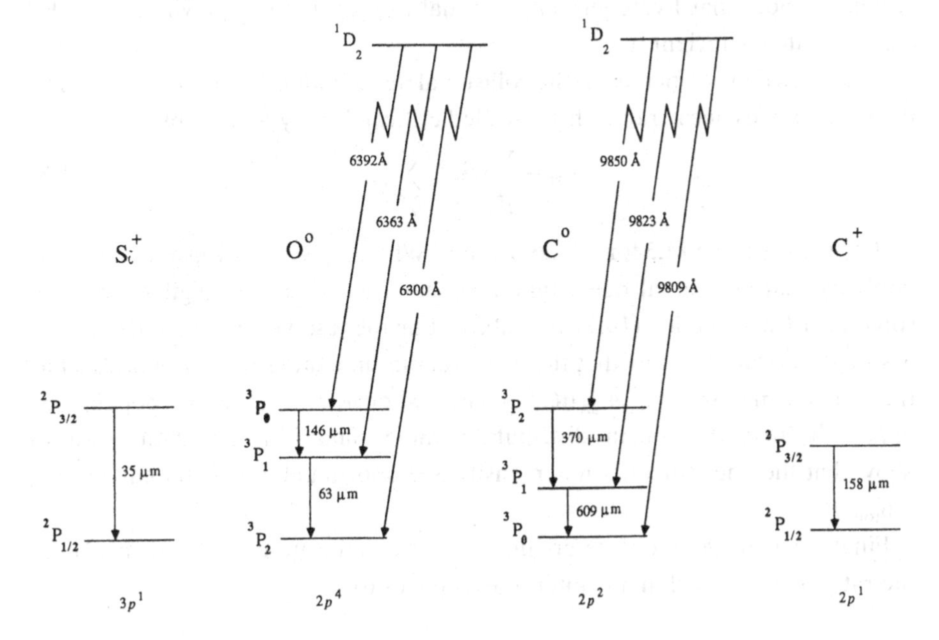

Figure 14.1. Ground-state energy level diagram for the lowest terms of Si^+, O^0, C^0, and C^+. The optical and near-infrared emission lines of O^0 and C^0 are also shown (the Si^+ $3p^1$ and the C^+ $2p^1$ states have no optical or near-infrared transitions).

optically thin. Under such conditions, the emitted power per unit area and solid angle can be written as:

$$I_{i\to l} = (h\nu_{il}/4\pi)A_{il}\int f_i n_{\rm ion}\,d\ell, \tag{14.1}$$

where ν_{il} is the frequency of the transition, A_{il} is the probability for spontaneous emission from the upper state i to the lower state l, f_i is the fraction of all atoms or ions in the upper state, $n_{\rm ion}$ is the number density of the atomic or ionic species, and ℓ is the path length along the line of sight. For species of interest here, the population of the ith fine structure level attains a steady state in roughly

$$t = \left[\sum_{l<i}\right]A_{il}^{-1} \le 2\times 10^7\ {\rm s} \tag{14.2}$$

The lifetime of bright ultraviolet producing stars and even the time scale for the flow through a shock front greatly exceed t, and to an excellent approximation the fractional populations f_i are steady state solutions of the equations of statistical equilibrium,

$$f_i\left[\sum_{l<i}(A_{il}+C_{il})+\sum_{u>i}C_{iu}\right] = \sum_{u>i}f_u(A_{ui}+C_{ui})+\sum_{l<i}f_lC_{li} \tag{14.3}$$

subject to the normalization condition

$$\sum_i f_i = 1 \tag{14.4}$$

The energy levels are labeled such that $E_{\rm u} > E_i > E_l$. C_{il} is the total rate (s^{-1}) of collision-induced transitions from level i to level l, assumed due to collisions with atomic or molecular hydrogen, and is equal to $n_{\rm H}\gamma_{il}$ (or $n_{\rm H_2}\gamma_{il}$) where γ_{il} is the collision rate coefficient ($cm^3\ s^{-1}$).

As the gas density increases, the collisional deexcitation rate becomes equal to the spontaneous decay rate. This so-called critical density is given by

$$n_{\rm cr} = \sum_{l<i}A_{il}\Big/\sum_{l\neq i}\gamma_{il} \tag{14.5}$$

and is listed in Table 14.3 for each transition. When $n_{\rm H} \ll n_{\rm cr}$ for a given transition, implying that collisional deexcitation of that transition is negligible, then by solving for f_i in Equation (14.3) and substituting the result into Equation (14.1), it is straightforward to show that the fine structure line intensity is proportional to the emission measure, $\int n_{\rm H}n_{\rm ion}\,d\ell$. Similarly, when $n_{\rm H} \gg n_{\rm cr}$, the level populations approach their Boltzmann distribution values and it is straightforward to show that the fine structure line intensity is proportional to the column density, $\int {\rm n_{ion}}\,d\ell$.

Finally, when two fine structure lines from the same atom or ion are available, the ratio of optically thin line intensities reduces to

$$\frac{I_{u \to i}}{I_{i \to l}} = \frac{\nu_{ui} A_{ui} f_u}{\nu_{il} A_{il} f_i} = \text{constant} \times (f_u / f_i) \qquad (14.6)$$

which is independent of both source geometry and ionic or atomic abundance. A similar expression also applies to the ratio of two different atoms or ions provided both lines are optically thin, the emitting regions are coextensive, and the abundance ratio is known. As will be discussed below, arguments of this type are used to interpret observations of the [OI] 63 μm and 146 μm and the [CII] 158 μm lines.

14.3. Photodissociation regions

Far-ultraviolet (FUV) ($\lambda > 912$ Å) radiation will escape ionized gas regions and illuminate molecular clouds in the vicinity, photodissociating the molecular material at their exposed surfaces and ionizing those atomic species with ionization potentials less than 13.6 eV. Though only a few galactic sources have been studied in more than one fine structure line, the overall results to date have been remarkably consistent: (1) the [OI] 63 and 146 μm, [CII] 158 μm, and [SiII] 35 μm emissions come from the same gas, (2) the peak emission from these lines arises between the fully ionized and the cold molecular gas, and (3) the atomic gas is dense ($\geq 10^3$ cm^{-3}) and warm ($100 < T_{gas} < 1500$ K).

Three regions which have received particular scrutiny, Orion, M17, and Sgr A West, exemplify points (1) and (2) above and are illustrated in Figure 14.2. In the case of Orion (Figure 14.2(*a*)), the fine structure line emission peaks at BN-KL, the Trapezium, and the ionization bar, 2′ SE of the Trapezium. The line emission near BN-KL is almost certainly due to shock excitation (discussed below), but the emission at the Trapezium and the bar are strongly correlated with regions of high ionization. Since there is an insufficient abundance of O°, C^+, and Si^+ in these highly ionized regions to account for the measured line strengths (most O, C, and Si will be in higher states of ionization – see Russell *et al.* (1980)), it is assumed that the bulk of the [OI], [CII], and [SiII] emission comes from gas immediately adjacent to these HII regions. Direct measurements of the relative spatial distributions of the highly ionized, atomic, and molecular gas in Orion are difficult since both the HII and HI regions lie along our line of sight. However, the fact that at the Trapezium the observed width (FWHM) and V_{LSR} of the [OI] 63 μm and the [CII] 158 μm lines are similar (7 ± 2 and 8 km s^{-1}, respectively, for the [OI] 63 μm line, and 5.0 and 8.6 km s^{-1}, respectively, for the [CII] 158 μm line; Crawford *et al.* (1986), Boreiko, Betz, and Zmuidzinas (1988)), and that they are similar to the width and V_{LSR} of the C109α radio recombination line (7.6 ± 1.5 and 8.4 km s^{-1}, respectively; Jaffe and Pankonin (1978)) argues strongly that these lines are largely coextensive.

This geometry in M17 is more favorable for studying the relative positions of ionized, atomic and molecular gas; in this region the HII/molecular cloud interface is nearly perpendicular to our line of sight. Strip scans made across the ionization front in M17 in the lines of several species are shown in Figure 14.2(*b*). These scans qualitatively show a distribution expected for a photodissociation region seen almost edge on; the C^+ layer lies next to the HII region, while further

Figure 14.2. (*a*) Strip scan across KL, the Trapezium, and the Ionization Bar in Orion (after Genzel and Stacey (1985), and Haas, Hollenbach, and Erickson (1986)). (*b*) Strip scan made across the HII/molecular cloud interface in M17 (from Stutzki *et al.* (1988)). (*c*) Strip scans made across the Sgr A West HII region ($\Delta l, \Delta b \approx 0, -10''$ from IRS 16) (after Genzel *et al.* (1985), and Graf *et al.* (1988)). In all panels, the straight line superimposed on the contour drawings to the left indicates the path of the strip scans, while the small × marks the zero point of each scan.

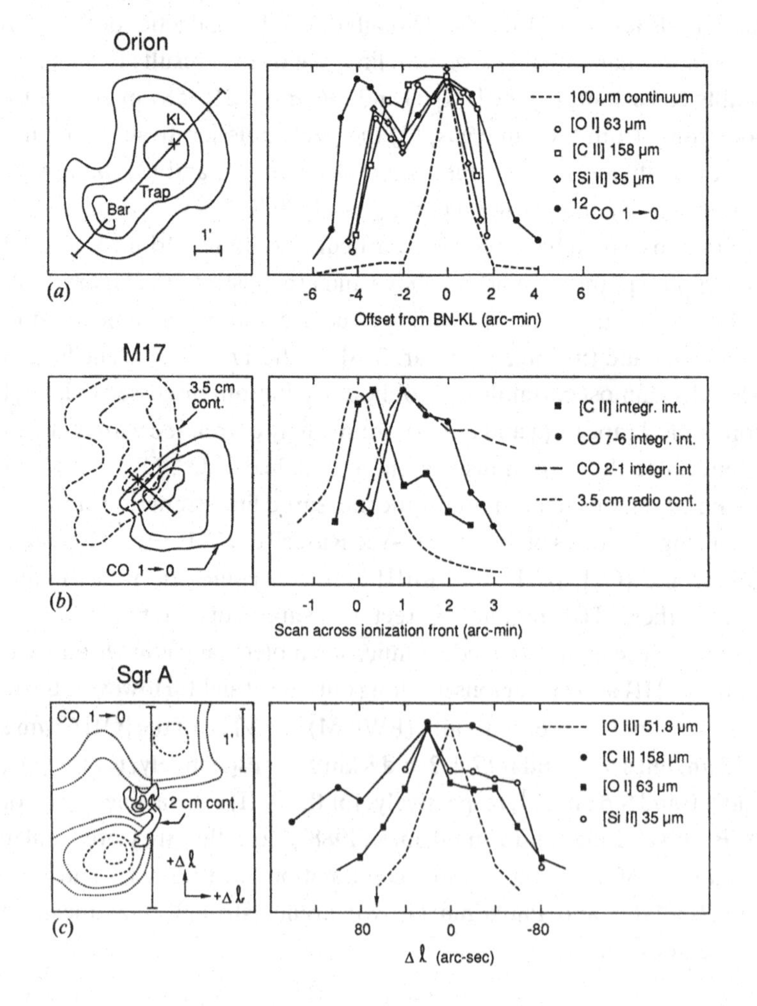

into the molecular clouds is a warm molecular layer, seen in CO $J = 7 \rightarrow 6$, followed by colder molecular material, seen in CO $J = 2 \rightarrow 1$.

The galactic center has been mapped in the fine structures lines of [OI], [CII], and [SiII] (Genzel *et al.* 1985, Graf *et al.* 1988). Strip scans through the Sgr A West HII region are shown in Figure 14.2(*c*). As in M17, the atomic gas in Sgr A West is clearly extended beyond the ionized gas measured by the 2 cm radio continuum free–free emission (Ekers *et al.* 1983) and the [OIII] 51.8 μm emission (see Genzel *et al.* 1985). Two-dimensional maps in the lines of [OI], [CII], and [SiII] show the emission to be double-lobed, centered on Sgr A West, and inclined by $\sim -20°$ to the galactic plane. This distribution is interpreted as evidence for a neutral gas disk or torus with an inner diameter of 1.7 pc and an outer diameter of 10 pc (assuming a distance to the galactic center of 10 kpc). The peak of the [OI] and [SiII] emission, between the Sgr A West HII region and the molecular clouds farther out, suggests that this disk is probably being heated by non-ionizing photons from the sources near IRS 16. The broader distribution of [CII] emission may be indicative of a decrease in density and/or temperature in the outer portions of this neutral disk since the strength of the [CII] line drops less markedly with temperature between 300 and 100 K and density between 10^5 and 10^3 cm^{-3} than either the [OI] 63 μm or [SiII] 35 μm lines.

The physical conditions in these photodissociation regions have been determined from the relative intensities of the [OI] 63 and 146 μm lines and [CII] 158 μm line. If these lines are optically thin, then from the discussion in Section 14.2 it is evident that between the critical density for the [CII] 158 μm transition, $\sim 10^3$ cm^{-3}, and the critical density for the [OI] 63 μm transition, $\sim 10^6$ cm^{-3}, the [OI] 63 μm to [CII] 158 μm intensity ratio is mainly sensitive to density. Conversely, since the critical densities of the [OI] 63 and 146 μm lines are similar, the density dependence of their intensity ratio will be small (about a factor of 2 between 1 and 10^7 cm^{-3} for fixed temperature) and the intensity ratio of these lines will be predominantly sensitive to gas temperature. A plot of the these line ratios as functions of temperature and density is shown in Figure 14.3 on which are superimposed the results for a number of different sources. As is shown, the temperatures and densities of the photodissociation regions associated with these regions fall within a relatively small range of values: $10^3 \leq n_H \leq 10^5$ cm^{-3} and 300 $\leq T_{gas} \leq 1500$ K. If these lines have significant optical depths (i.e. $\tau \geq 1$), as appears likely for the [OI] 63 μm and [CII] 158 μm lines from the Orion Trapezium region (Crawford *et al.* 1986, Boreiko *et al.* 1988), then the gas temperatures indicated in Figure 14.3 will be reduced (for example, the implied gas temperature for the Trapezium photodissociation region in Orion would drop to 130–180 K).

The most successful explanation for the observed line intensities and spatial distributions is provided by a class of models which assumes that a molecular cloud, with a nominal gas-to-grain mass ratio (~ 100), is subject to an intense FUV flux (see Tielens and Hollenbach (1985a) and references therein). In these models the FUV penetrates the cloud surface to a depth of several A_v before being largely attenuated by dust absorption. Within this surface region, this FUV field has three principal effects: (1) it photodissociates molecules and photoionizes those species with ionization potentials less than that of hydrogen, (2) it is absorbed by dust grains, heats the grains (T_{dust} ~ 75–120 K), and is reradiated as infrared continuum, and (3) about 0.1–1% of the FUV flux heats the gas via photoelectric

Figure 14.3. Optically thin ratios of [OI] 63 μm/[C II] 158 μm and [OI] 63 μm/[OI] 146 μm intensities as a function of density and temperature in a neutral gas of cosmic abundance ([O]/[C] ~ 2). Shown are the results for the galactic HII region, Orion, including measurements obtained with a beam size (5′) sufficient to include both the photodissociation and shocked gas regions and a smaller beam size (50″) measurement which includes only the Trapezium region. Also plotted are the results for the planetary nebula, NGC 7027, the galactic center, Sgr A, and the galaxies M82 and IC 342 (from Genzel and Stacey (1985)).

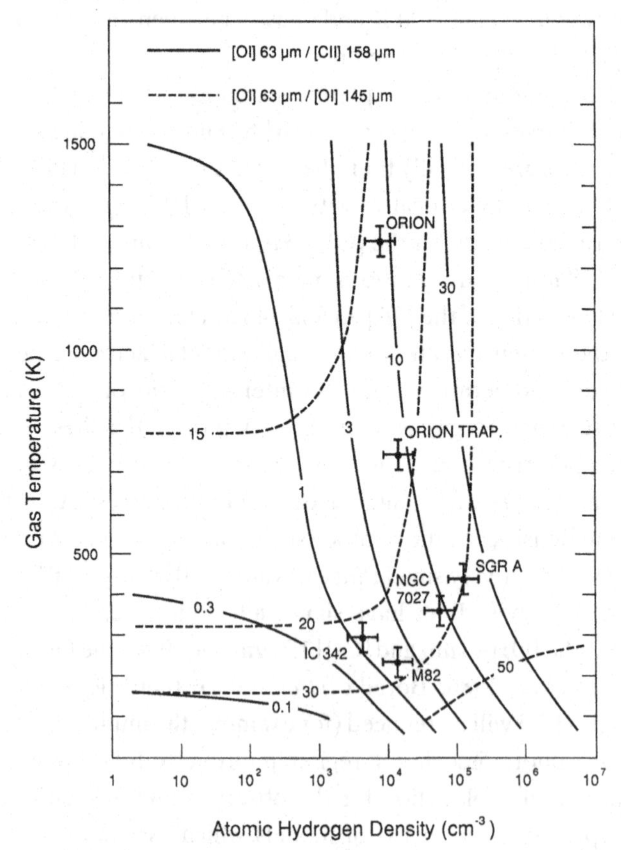

heating, i.e., absorption of FUV photons by neutral dust grains which causes the ejection of energetic (~4 eV) electrons into the gas. Within this zone, cooling is dominated by [OI] 63 μm and [CII] 158 μm line emission. Deeper into the cloud, where the FUV flux is greatly reduced, the gas temperature drops below that of the dust ($T_{\text{dust}} \sim 75$ K) and heating of the gas occurs mainly through line absorption of dust continuum photons and direct gas–grain collisions. Gas cooling within this region occurs chiefly through CO rotational emission. A schematic of a photodissociation region based on the model by Tielens and Hollenbach (1985a) is shown in Figure 14.4.

Figure 14.4. A schematic drawing of a photodissociation region showing the atomic and molecular abundances (relative to the number density of hydrogen nuclei, n) and the major coolants as a function of visual extinction into a molecular cloud with $n = 2.3 \times 10^5$ cm^{-3} and an incident FUV flux equivalent to 10^5 times the ambient interstellar value. The major coolants are shown in the bottom panel (after Tielens and Hollenbach (1985a)).

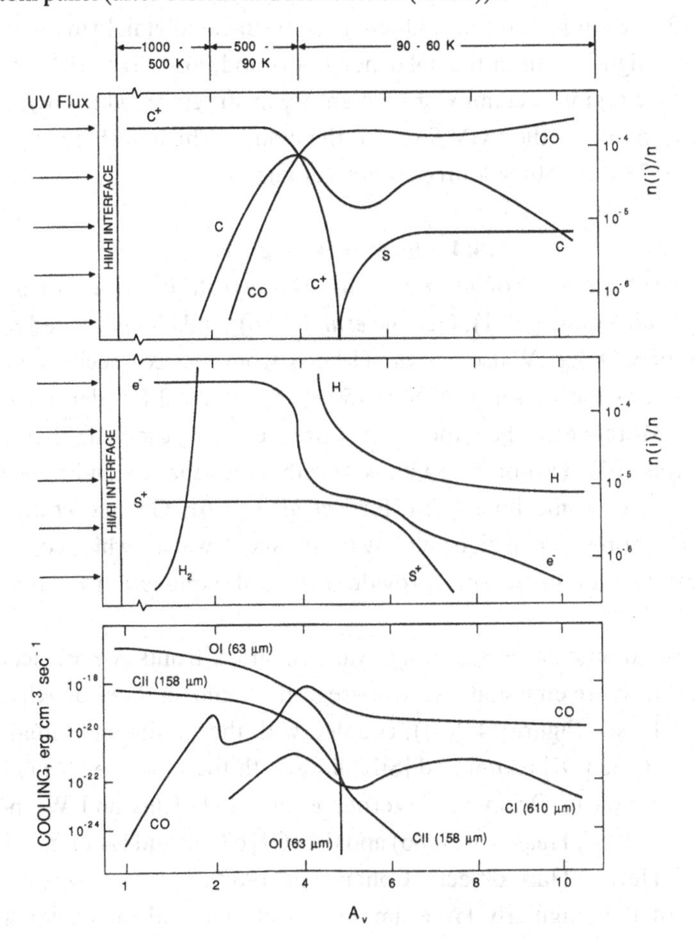

Finally, while the model described above provides a satisfactory accounting for the most intense fine structure emission, there is evidence provided by extended, low intensity [CII] emission that the uniform, one-dimensional geometry assumed in these models may need modification. (As discussed in Section 14.5, extended, strong emission is also observed in the [CI] lines, possibly pointing to the same problem.) Soon after its initial detection in the central regions of NGC2024 and Orion (Russell *et al.* 1980), weaker, extended [CII] emission was found associated with M17, NGC2024, W3, and W51 (Russell *et al.* 1981, Kurtz *et al.* 1983, Melnick *et al.* 1986, Stutzki *et al.* 1988). A detailed analysis of the strength of the [CII] 158 μm emission far removed (>3 pc) from the HII/molecular cloud interface in M17 (Stutzki *et al.* 1988) shows that the [CII] line intensity is about 20 times higher than can be produced by this single photodissociation region. Three possible ways of reconciling the data and the photodissociation model are suggested: (1) there are a number of unobserved B stars embedded in the M17 molecular cloud which could provide the ultraviolet radiation necessary to produce the extended [CII] emission, (2) the strength of the ambient galactic ultraviolet field in the vicinity of M17 may be higher than in the solar neighborhood, and (3) the HII/molecular cloud interface region is clumpy or filamentary in structure, which would permit greater penetration of the FUV flux into the cloud. Which of these explanations applies in M17 or the other sources is not yet known.

14.4. Shocked gas regions

The presence of a number of intense, highly excited rotation-vibration lines of H_2 (cf. Knacke and Young (1981), Geballe *et al.* (1986)) and highly excited rotational transitions of CO (cf. Watson *et al.* (1985)) from sources such as Orion-KL strongly suggests the presence of hot ($750 \leq T_{gas} \leq 2500$ K), dense ($n_{H_2} \geq 10^6$ cm^{-3}) gas. Moreover, when these lines are velocity resolved, they typically display widths (FWHM) of 30–50 km s^{-1} with wings that extend to as much as ± 100 km s^{-1} for some lines (Crawford *et al.* (1986), Geballe *et al.* (1986)), indicating their presence in high velocity flows. Shock waves, which compress and heat the gas as they propagate, provide a natural explanation for both observations.

That dense atomic gas exists in the wake of shock fronts is evidenced by the intense fine structure emission seen from regions of known shock activity, such as Orion BN-KL (see Figure 14.2(*a*)), coupled with the spatial and radial velocity coincidence of the [OI] 63 μm and [SiII] lines with the $v = 1 \rightarrow 0$ S(1) H_2 2 μm emission observed in Orion-KL (Werner *et al.* 1984, Ellis and Werner 1985, Crawford *et al.* 1986, Haas *et al.* 1986) and the [OI] 63 μm and H_2 2 μm emission in a number of Herbig–Haro objects (Cohen *et al.* 1988).

Because of the singularly large amount of atomic and molecular line data obtained from the shocked gas region in Orion-KL, this region has served as the prototype for understanding the detailed nature of postshock gas. In this source, the shock is believed to arise where the high velocity outflow from the young star IRc 2 impacts the quiescent molecular gas which surrounds this source. To date, strong [OI] 63 μm and 145 μm and [SiII] 35 μm emission has been detected from Orion-KL; the strength of the [OI] 63 μm line alone, 50 ± 25 $L_\odot$ (Werner *et al.* 1984), is only 3–4 times smaller than the total luminosity from *all* H_2 infrared lines ($\sim$200 $L_\odot$; cf. Scoville *et al.* (1982), Beckwith *et al.* (1983)), or the luminosity of the far-infrared CO lines (150 $L_\odot$). The strength of the [OI] 63 μm and 146 μm lines and the [OI] 63 μm line profile permit several general conclusions to be drawn about the atomic gas region in Orion-KL: (1) the width of the [OI] 63 μm line, $\sim$ 30 km s^{-1}, implies a negligible opacity in this line ($\tau \sim 10^{-2}$), (2) the [OI] 63 μm to 146 μm intensity ratio implies that emission comes from dense gas ($n_H \approx 10^{5\pm1}$ cm^{-3}) at high temperatures ($T \approx$ 300–1500 K), and (3) based on the derived column density in the [OI] emitting region ($N_H \approx 7 \times 10^{20}$ cm^{-2}), the mass of the gas in this region is $\sim$0.15 $M_\odot$, greater than that derived for the 2000 K gas giving rise to the H_2 emission ($\sim$0.01 $M_\odot$; Beckwith *et al.* (1983)), but less than that derived for the 500–1000 K gas sampled by the high-*J* CO lines ($\sim$0.4–1.3 $M_\odot$; Watson *et al.* (1985)). The [SiII] 35 μm emission at the shocked gas position has a line luminosity of ≥ 2 $L_\odot$, which is consistent with emission from gas with no significant silicon depletion (i.e. approximately solar abundance of silicon).

Together, the atomic and molecular data have been used to infer several properties of the shock wave around IRc2. First, the presence of strong molecular emission, such as from H_2 and CO, indicates that a so-called 'molecular' or C-type shock is present. Such shocks are characterized by: velocities $\leq$ 50 km s^{-1}, they are non-dissociative, they occur only in magnetized, weakly ionized gas, and heating, which is due to collisions between ions tied to the moving magnetic field lines and slower moving neutrals, results in peak postshock temperatures of 3000–5000 K (cf. Draine (1980)). Cooling behind C-shocks occurs primarily via molecular (H_2, CO, OH, and H_2O) vibrational and rotational infrared transitions and atomic fine structure transitions, particularly [OI] 63 μm emission. The C-shock models which best reproduce the [OI] 63 μm and molecular line intensities from Orion-KL assume a preshock density of (2–7) $\times 10^5$ cm^{-3}, a shock velocity of 36–38 km s^{-1}, and magnetic field strength perpendicular to the shock front of 0.45–1.5 mG (Draine and Roberge (1982), Chernoff, Hollenbach and McKee (1982)).

Second, the high [SiII] 35 μm line intensity from Orion-KL, which is not reproduced by the C-shock models, may indicate the presence of an additional, J-type shock in this region (Haas *et al.* 1986). Unlike C-shocks, J-shocks can propagate at velocities $\geq$ 50 km s^{-1}, they are either dissociative or occur in atomic gas with a significant ion fraction, and the peak postshock temperatures are high – approximately 1.4×10^5 (v_{shock}/100 km s^{-1}) K. As such, a J-shock would dissociate and ionize any preexisting molecular silicon and sputter solid phase silicon from grains into the gas. As the initially highly ionized Si cools and recombines, FUV photons from the hot upstream gas would maintain the Si as Si^+, thus producing the strong [SiII] 35 μm emission. To account for the observed [SiII] 35 μm intensity, J-shock models require a preshock density of almost 10^6 cm^{-3} and v_{shock} = 100 km s^{-1}.

In a scenario in which both C and J-shocks are present, the J-shock would be established at the point at which the high velocity outflow collides with the inner edge of the shocked, expanding material. The C-shock occurs where the outer edge of the expanding material strikes the ambient molecular cloud.

14.5. Interiors of molecular clouds

With temperatures above the ground state of 24 and 63 K, the [CI] 609 μm and 370 μm transitions, respectively, are capable of significant excitation within warm molecular clouds. However, prior to its detection, little [CI] emission was anticipated from within molecular clouds since neutral carbon was expected to exist in measurable quantities only in a thin layer between the surface, where most carbon is in C^+, and the cloud interior, where almost all of the carbon was predicted to be locked in CO. In fact, the observed [CI] line strengths greatly exceed these early predictions and, as [CI] 609 μm strip scans through a number of molecular clouds show (see Figure 14.5(*a*) and (*b*)), rather than being confined to a narrow region, the [CI] emission extends throughout these clouds. In M17, the

Figure 14.5. Comparison of CI and CO antenna temperatures in (*a*) M17 and (*b*) S140 integrated over velocity. The ionization front is to the left in both plots (from Keene *et al.* (1985)). (*c*) Ratio of N(CI) to N(CO) as a function of A_v for the dense cloud associated with the star ρ Oph (from Keene (1988)).

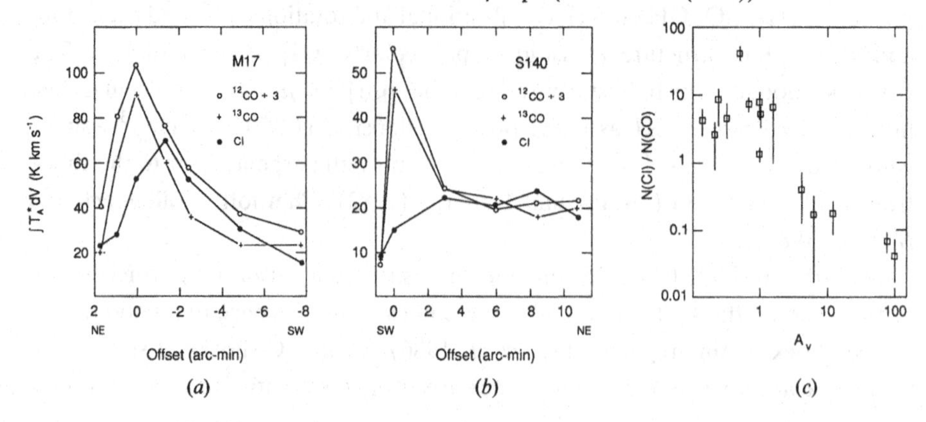

peak [CI] column density occurs at an $A_v \geq 60$ into the cloud, while in S140, another source with an edge-on HII/molecular cloud interface, the [CI] column density peaks at an $A_v \sim 30$ into the molecular gas (Keene *et al.* 1985).

More recent observations of the [CI] 609 μm line (Keene 1988) and the 370 μm line (Zmuidzinas *et al.* 1988) indicate that the [CI] lines are optically thin in most sources and that the [CI]/CO column density ratio in the clouds surveyed is typically about 0.1. Studies of the dense cloud associated with ρ Oph further show that this ratio persists at a value of ~0.1 to an A_v of ~100 into the cloud (Figure 14.5(*c*).

A number of models have been proposed to account for the large observed [CI] column densities. These models, some of which are reviewed by Keene *et al.* ((1985), see references therein) include: (1) time-dependent chemistry, in which it is suggested that the time scale for conversion of C^0 to CO in a cloud core exceeds the age of the clouds observed, (2) dynamical-chemical models in which the conversion of C^0 to CO preferentially occurs at high densities, a state achieved only during the last phases of cloud collapse, (3) turbulence, in which cells of gas from the interior of the molecular cloud are constantly exchanged with gas relatively abundant in atomic carbon from the cloud surface, (4) production of C^0 in the cloud interior through photodissociation of CO by either ultraviolet photons generated by internal shocks and/or cosmic ray excitation of Lyman and Werner bands of H_2 (Gredel, Lepp and Dalgarno 1987), or X-rays from embedded T Tauri and other pre-main-sequence stars, (5) residual atomic carbon resulting from a C/O abundance ratio > 1 in dense clouds, and (6) penetration of CO-photodissociating FUV radiation deep into the cloud due either to lower dust opacity (Tielens and Hollenbach 1985b) or to a clumpy or filamentary cloud structure (Stutzki *et al.* 1988). It is tempting to ascribe the extended [CII] and [CI] emission to the same cause, but until higher spatial resolution studies are able to better correlate the intensity distribution from both species, it is not yet clear how they may be related. It is clear, however, that future models of molecular clouds which do not take account of the extended [CI] 609 μm and 370 μm and [CII] 157 μm observations must be viewed as incomplete.

14.6. Summary

Observations of atomic species in dense clouds have: (1) demonstrated the presence of warm atomic layers situated between hot HII regions and adjacent cold molecular gas, (2) helped establish the presence of magnetized, molecular C-shocks around the archetypal star forming region in Orion-KL and may also indicate the presence of a second, J-shock in this region, and (3) shown that models of molecular cloud structure and/or chemistry need to be revised in order to account for the extended [CII] 158 μm and [CI] 609 μm emission.

REFERENCES

Beckwith, S., Evans, N. J., Gatley, I., Gull, G., and Russell, R. W. 1983, *Ap. J.*, **264**, 152.

Boreiko, R. T., Betz, A. L., and Zmuidzinas, J. 1988, *Ap. J.* (*Letters*), **325**, L47.

Chernoff, D. F., Hollenbach, D. J., and McKee, C. F. 1982, *Ap. J.* (*Letters*), **259**, L97.

Cohen, M., Hollenbach, D. J., Haas, M. R., and Erickson, E. F. 1988, *Ap. J.*, **329**, 863.

Crawford, M. K., Genzel, R., Townes, C. H., and Watson, D. M. 1985, *Ap. J.*, **291**, 755.

Crawford, M. K., Lugten, J. B., Fitelson, W., Genzel, R., and Melnick, G. J. 1986, *Ap. J.*, **303**, L57.

Draine, B. T. 1980, *Ap. J.*, **241**, 1021.

Draine, B. T., and Roberge, W. G. 1982, *Ap. J.* (*Letters*), **259**, L91.

Ekers, R. D., van Gorkom, J. H., Schwarz, U. J., and Goss, W. 1983, *Astr. Ap.*, **122**, 143.

Ellis, H. B., and Werner, M. W. 1985, in preparation.

Geballe, T. R., Persson, S. E., Simon, T., Lonsdale, C. J., and McGregor, P. J. 1986, *Ap. J.*, **302**, 500.

Genzel, R., and Stacey, G. J. 1985, *Mittg. der Astr. Gesellschaft*, **63**, 215.

Genzel, R., Watson, D. M., Crawford, M. K., and Townes, C. H. 1985, *Ap. J.*, **297**, 766.

Genzel, R., Watson, D. M., Townes, C. H., Dinnerstein, H. L., Hollenbach, D., Lester, D. F., Werner, M. W., and Storey, J. W. V. 1984, *Ap. J.*, **276**, 551.

Graf, P., Herter, T., Gull, G. E., and Houck, J. R. 1988, *Ap. J.*, **330**, 803.

Gredel, R., Lepp, S., and Dalgarno, A. 1987. *Ap. J.* (*Letters*), **323**, L137.

Haas, M. R., Hollenbach, D. J., and Erickson, E. F. 1986, *Ap. J.* (*Letters*), **301**, L57.

Jaffe, D. T., and Pankonin, V. 1978, *Ap. J.*, **226**, 869.

Keene, J. 1988, preprint.

Keene, J., Blake, G. A., Phillips, T. G., Huggins, P. J., and Beichman, C. A. 1985, *Ap. J.*, **299**, 967.

Knacke, R. F., and Young, E. T. 1981, *Ap. J.* (*Letters*), **249**, L65.

Kurtz, N. T., Smyers, S. D., Russell, R. W., Harwit, M., and Melnick, G. 1983, *Ap. J.*, **264**, 538.

Lester, D. F., Werner, M. W., Storey, J. W. V., Watson, D. M., and Townes, C. H. 1981, *Ap. J.* (*Letters*), **248**, L109.

Melnick, G. J., Gull, G. E., and Harwit, M. 1979, *Ap. J.* (*Letters*), **227**, L29.

Melnick, G. J., Stacey, G. J., Viscuso, P. J., and Fuller, C. E. 1986, *Ap. J.*, **303**, 638.

Russell, R. W., Melnick, G., Gull, G. E., and Harwit, M. 1980, *Ap. J.* (*Letters*), **240**, L99.

Russell, R. W., Melnick, G., Smyers, S. D., Kurtz, N. T., Gosnell, T. R., Harwit, M., and Werner, M. W. 1981, *Ap. J.* (*Letters*), **250**, L35.

Scoville, N. Z., Hall, D. N. B., Kleinmann, S. G., and Ridgway, S. T. 1982, *Ap. J.*, **253**, 136.

Storey, J. W. V., Watson, D. M., and Townes, C. H. 1979, *Ap. J.*, **233**, 109.

Stutzki, J., Stacey, G. J., Genzel, R., Harris, A. I., Jaffe, D. T., and Lugten, J. B. 1988, preprint.

Tielens, A. G. G. M., and Hollenbach, D. J. 1985a, *Ap. J.*, **291**, 722.

Tielens, A. G. G. M., and Hollenbach, D. J. 1985b, *Ap. J.*, **291**, 747.

Watson, D. M. 1985, *Physica Scripta*, **T11**, 33.

Watson, D. M., Genzel, R., Townes, C. H., and Storey, J. W. V. 1985, *Ap. J.*, **298**, 316.

Watson, D. M., Genzel, R., Townes, C. H., Werner, M. W., and Storey, J. W. V. 1984, *Ap. J.* (*Letters*), **279**, L1.
Werner, M. W., Crawford, M. K., Genzel, R., and Hollenbach, D. J. 1984, *Ap. J.* (*Letters*), **282**, L81.
Zmuidzinas, J., Betz, A. L., Boreiko, R. T., Goldhaber, D. M. 1988, preprint.

15

Ultraviolet radiation in molecular clouds

W. G. ROBERGE

Department of Physics, Rensselaer Polytechnic Institute, USA

15.1. Introduction

Ultraviolet radiation is a crucial ingredient in any theory of interstellar chemistry. In the interplay of molecule formation and destruction processes, ultraviolet photons adopt a multiple role, destroying neutral species on the one hand, while creating chemically reactive ions and depositing thermal energy on the other. It has long been recognized (e.g. Stief *et al.* (1972)), that dust in a cloud's outer layers attenuates ambient Galactic ultraviolet starlight, thereby enhancing the survival of molecules against photodestruction. Unfortunately, however, the degree of attenuation is sensitive to the grain scattering properties, which are not well determined at ultraviolet wavelengths (Sandell and Mattila 1975, Leung 1975, Whitworth 1975, Bernes and Sandqvist 1977, Sandell 1978, Flannery, Roberge, and Rybicki 1980). Since even a small amount of ultraviolet radiation has profound consequences in dark regions, the chemical and ionization balance of such regions has remained uncertain.

The early studies of dust shielding may have been overly pessimistic about uncertainties, however, as noted by Chlewicki and Greenberg (1984a,b). This is due in part to the existence of strong constraints on grain properties that follow from secure observational data, and also to the discovery of the chemical consequences of the ultraviolet emission associated with gas-cosmic ray interactions (Prasad and Tarafdar (1983); see also Chapter 16). The interaction produces an ultraviolet field in clouds which, at great depths, destroys molecules more rapidly than cosmic rays or attenuated starlight. As a result, the role of starlight is restricted to a relatively narrow region near a cloud's surface, where the effects of uncertainties in grain properties are moderate. It therefore becomes reasonable to make quantitative predictions of photodestruction rates in molecular, as well as diffuse, clouds. The following brief review is intended as a practical guide to such calculations.

15.2. The radiative transfer problem

15.2.1. Ultraviolet continuum radiation

Essential features of the radiative transfer problem are illustrated by the plane-parallel cloud model (Figure 15.1) which, though highly idealized, describes the gross structure of many real cloud surfaces. The equation of radiative transfer assumes the form,

$$\mu \frac{\partial I(\tau,\mu)}{\partial \tau} = I(\tau,\mu) - S_e(\tau) - \frac{\omega}{2(1+\varepsilon_g)} \int_{-1}^{+1} F(\mu,\mu') I(\tau,\mu')\, d\mu' \quad (15.1)$$

provided that the specific intensity, I (photons cm^{-2} s^{-1} sr^{-1} Å^{-1}), is azimuthally symmetric about the normal direction. The independent variables in (15.1) are the direction cosine, μ, defined in Figure 15.1, and the extinction optical depth, τ. At ultraviolet wavelengths, continuum opacity is dominated by dust scattering and absorption, with an effective extinction cross section Σ_{ext} per hydrogen nucleus. Thus the optical depth is related to the geometrical depth z by

$$d\tau/dz = n_H \Sigma_{ext} (1 + \varepsilon_g) \quad (15.2)$$

Figure 15.1. Definition of the variables in Equation (15.1). Specific intensity depends on position, indicated by point P, and direction, indicated by the solid arrow. The former is specified by linear distance, z, or optical depth, τ, measured from the upper cloud surface. Direction is specified by the angle, $\cos^{-1}\mu$ as shown. The slab extends to infinity in directions parallel to the slab surface.

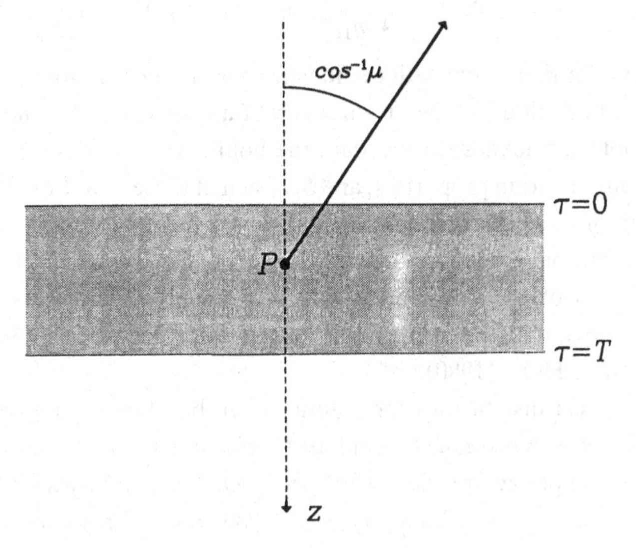

where n_H is the number density of hydrogen nuclei and the quantity ε_g is a correction, frequently small, to account for the opacity of atomic carbon and other gas-phase species. Since scattering in the continuum is effectively coherent, Equation (15.1) can be solved independently for I at each wavelength, λ; the λ-dependence of various quantities has therefore been suppressed.

The integral term in (15.1) is the source function associated with dust scattering. The albedo, ω, is the ratio of the grain scattering cross section to the extinction cross section and $F(\mu, \mu')$ is directly related to the scattering phase function (see Chandrasekhar (1960), Section 11). It is conventional in astrophysics to adopt the expression

$$F(\mu, \mu') = \sum_{l=0}^{\infty} (2l + 1) g^l P_l(\mu) P_l(\mu') \qquad (15.3)$$

(Henyey and Greenstein 1941) in which the P_l are Legendre polynomials and g, the 'asymmetry parameter', is chosen independently at each wavelength. As g varies between zero and unity, Expression (15.3) simulates scattering that varies from being isotropic to purely forward.

The remaining undefined term in (15.1) is $S_e(\tau)$, the source function associated with true emission by the gas. Emission at ultraviolet wavelengths is produced by cosmic ray–H_2 interactions, by collisional excitation of atoms and molecules in hot gas, and by discrete embedded sources such as stars and protostars. (Note, however, that Equation (15.1) describes a one-dimensional geometry, which precludes the treatment of *isolated* stars in or outside the medium.) If the emission occurs isotropically at a rate $j(\tau)$ (photons $cm^{-3}\,s^{-1}\,Å^{-1}$), then

$$S_e(\tau) = \frac{1}{4\pi n_H} j(\tau)[\Sigma_{ext}(1 + \varepsilon_g)]^{-1} \qquad (15.4)$$

It is apparent that, even with the idealizations in the foregoing discussion, the radiation field in clouds depends on several functions and parameters, including the total optical thickness of the slab, the boundary conditions giving the cloud illumination, the grain properties, and S_e. Each of these quantities is a function of wavelength and, as discussed below, is subject to some uncertainty. Once the preceding data have been prescribed, however, Equation (15.1) can be solved using a variety of numerical methods. A comprehensive discussion of various techniques, including extensive tables of numerical results, can be found in the book by van de Hulst (1980).

Some physical insight into the solutions can be obtained by considering the *spherical harmonics method*, a technique for obtaining exact solutions to transport problems in simple geometries (Mark 1947, Davison and Sykes 1957, Case and Zweifel 1967, Dave 1975, Flannery, *et al.* 1980, Karp, Greenstadt, and Fillmore

1980, Benassi *et al.* 1984). One approximates the intensity as a truncated series in Legendre polynomials:

$$I(\tau, \mu) = \sum_{l=0}^{L} (2l + 1) f_l(\tau) P_l(\mu) \tag{15.5}$$

Substitution of (15.3) and (15.5) into the transfer equation reduces the latter to a set of L coupled, linear, ordinary differential equations for the depth-dependent moment functions $f_l(\tau)$. An explicit, analytic solution can be found for arbitrary ω, phase function and $S_e(\tau)$ (Roberge 1983). The rates of photon-induced processes depend on the mean intensity, $f_0 \equiv J$, which, when L is odd, takes the form

$$J(\tau) = \sum_{m=1}^{M} \left\{ C_{-m} \exp(-k_m \tau) + C_{+m} \exp[-k_m(T - \tau)] + R_{m0}^{-1} \frac{k_m}{1-\omega} \int_0^T S_e(\tau') \exp(-k_m |\tau - \tau'|) \, d\tau' \right\} \tag{15.6}$$

where $M \equiv (L + 1)/2$. The quantities C_m, R_{m0}^{-1}, and k_m are all constants that can be determined given ω, ε_g, g, and L. For a given m-value, the exponential terms represent the attenuation of radiation incident upon each of the two cloud boundaries and the integral represents contributions to $J(\tau)$ from true emission. The sum represents (roughly speaking) an average over rays propagating at M different angles to the normal direction. Analogous expressions for the higher moments exist.

The eigenvalues k_m are all positive and may be ordered such that $0 < k_1 < k_2 < \ldots < k_M$. According to (15.6), the mean intensity in a very thick cloud with $S_e = 0$ varies as $J \sim \exp(-k_1 \tau)$ for sufficiently large optical depths. Values of the asymptotic decay rate k_1 are given in Figure 15.2(*a*) versus ω and g. (ε_g was set to zero and L was increased until k_1 had converged to four decimal places.) The approximation $J \sim \exp(-k_1 \tau)$ is good whenever $\tau \gg 1/(k_2 - k_1)$. The quantity $\tau_\infty \equiv 1/(k_2 - k_1)$ is given in Figure 15.2(*b*). For $\omega > 0.2$ and $g < 0.9$, a constraint that interstellar grains probably satisfy throughout the ultraviolet (Chlewicki and Greenberg 1984a), one finds that $\tau_\infty \leq 10$. Consequently, simple exponential formulae for $J(\tau)$ should be good at moderately large depths inside thick clouds without internal sources. An example, which is surprisingly accurate at small depths as well, appears in the paper by Mathis, Mezger, and Panagia (1983). Other useful approximations are discussed in Chlewicki and Greenberg (1984b) and Canfield, McClymont and Puetter (1984). Finally, notice from Figure 15.2(*a*) that the attenuation rate is sensitive primarily to uncertainties in the grain albedo and, to a much lesser degree, in the asymmetry parameter.

Next, consider the mean intensity in a cloud with embedded sources, but no external illumination. In this case the solution reduces to

$$J(\tau) = \int_0^T S_e(\tau')K(|\tau - \tau'|)\, d\tau' \tag{15.7a}$$

where the kernel,

$$K(\tau) = \frac{1}{1-\omega} \sum_{m=1}^{(L+1)/2} R_{m0}^{-1} k_m \exp(-k_m \tau) \tag{15.7b}$$

Figure 15.2. (*a*) Contour plot of the smallest eigenvalue, k_1, in (15.6) versus the grain albedo, ω, and asymmetry parameter, g. The interval between contours is 0.05. Gas opacity was neglected ($\varepsilon_g = 0$) and eigenvalues were computed by increasing the truncation parameter, L, until the results had converged to four decimal places. Note that the contour for $k_1 = 1$ coincides with $\omega = 0$. (*b*) Contour plot of the quantity $1/(k_2\text{–}k_1)$, versus ω and g. The interval between contours is unity.

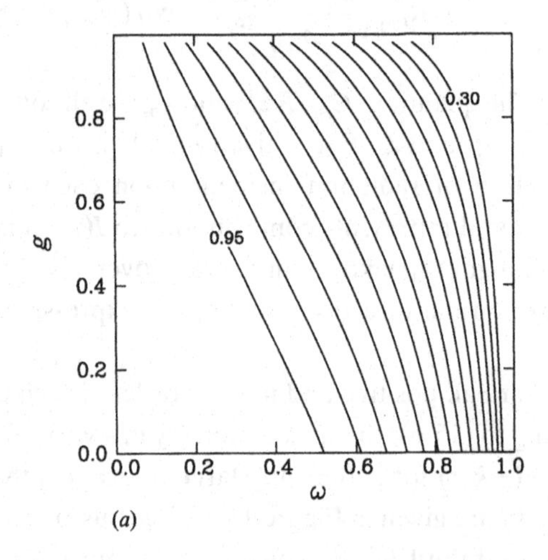

(*a*)

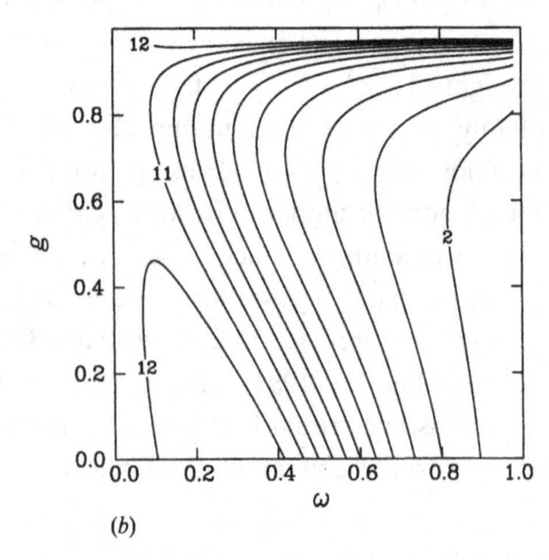

(*b*)

depends parametrically on the grain scattering properties and L. Equations (15.7) resemble the Schwarzschild–Milne solution to the equation of transfer, (Mihalas 1978, Section 2.2) but there are two important differences: First, K, is not the Schwarzschild–Milne kernel, one-half times the first exponential integral. This is related to the second difference: the integrand in (15.7a) is the source function due to true emission alone – the effects of scattering have been subsumed in the τ-dependence of K. In Figure 15.3, $K(\tau)$ has been plotted for a few choices of ω, g, and L. By comparing the various curves one sees explicitly the sensitivity of solutions to grain properties noted above. In particular, treating the extinction as

Figure 15.3. The kernel function defined in Equation (15.7), as computed in the $L = 9$ approximation. The solid curves were computed assuming isotropic scattering ($g = 0$) and the dashed curves assuming $g = 0.8$. The numbers refer to different values of ω, with $\omega = 0$ (curve 1), $\omega = 0.4$ (curves 2 and 4), or $\omega = 0.8$ (curves 3 and 5). Dotted curve: $\log |1-2K(\tau)/E_1(\tau)|$, where $K(\tau)$ was computed in the $L = 9$ approximation with $\omega = 0$ and $g = 0$. This quantity, which measures the error in the approximate kernel $K(\tau)$, goes to infinity at $\tau = 0$ because $K(\tau)$ is finite while $E_1(\tau)$ diverges. Note that pure absorption corresponds to the case where the spherical harmonics method is least suitable.

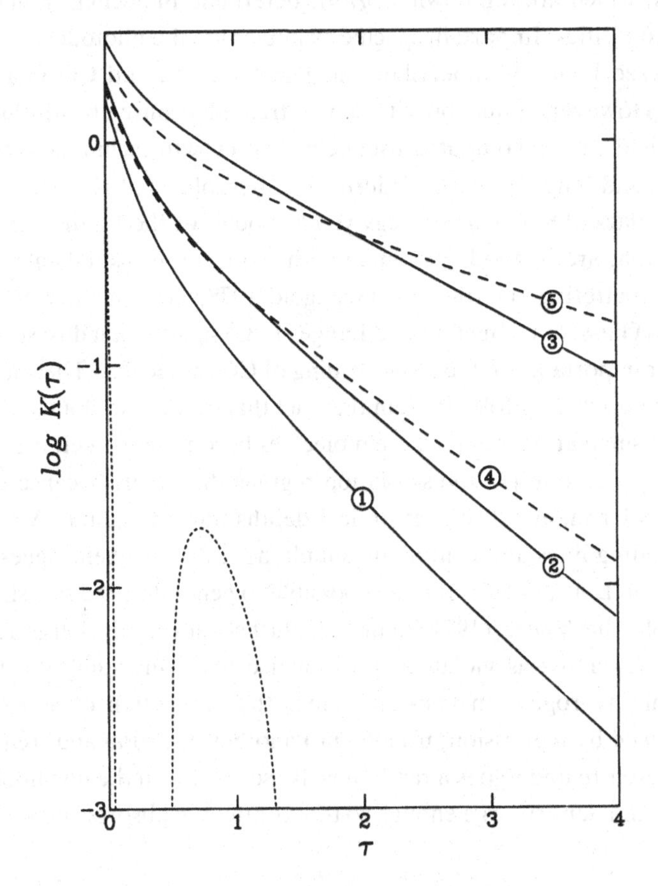

pure absorption leads to large errors in the mean intensity away from a source of radiation.

15.2.2. Ultraviolet lines

For some purposes it suffices to neglect ultraviolet lines inside interstellar clouds, e.g. by smoothing the spectrum over a sufficiently large wavelength interval. However, there are many photon-induced processes in clouds where the lines must be treated explicitly. The methods outlined above carry over to such problems if (1) the line opacity and emissivity are prescribed functions of z and (2) scattering in the lines is negligible. One simply incorporates the opacity in ε_g, the emissivity in j, and solves (15.1) on a fine grid of wavelengths. Important examples to which this strategy applies are molecular line emission induced by cosmic ray–H_2 interactions and the irradiation of clouds by forbidden lines emitted in HII regions and shocks. The following problems are two important counterexamples.

Assumption (1) above fails to describe the photodestruction of molecules by discrete transitions to dissociating electronic states (van Dishoeck 1987, 1988); the line intensities, which are unknown *a priori*, determine molecular abundances and, thus, line opacities. Important specific examples are the photodestruction of H_2 (Solomon, see Field, Sommerville, and Dresler 1966) and CO (Bally and Langer 1982). However, if dust opacity can be treated as pure absorption, then the radiation field can be computed using curve of growth techniques (Hollenbach, Werner, and Salpeter 1971, Federman, Glassgold, and Kwan 1979, van Dishoeck and Black 1986). This strategy yields models of the H_2 distribution in diffuse clouds that are in good agreement with more sophisticated calculations including dust scattering (Gerola and Glassgold 1978, Federman *et al.* 1979). Recent studies (Glassgold, Huggins, and Langer 1985) suggest that dust scattering is also of minor importance in the self-shielding of CO molecules. The success of this approach undoubtedly follows from that fact that, in diffuse clouds, the dust optical depth is small at the point where molecules become self-shielding. This is unlikely to be the case in photodissociation regions, however, because the gas becomes molecular many ultraviolet optical depths into the cloud. An *ad hoc* method for treating lines and continuum simultaneously has been suggested by van Dishoeck and Black (1986). It is also possible to generalize the self-shielding function of Hollenbach *et al.* (1971) to include dust scattering (Roberge 1989).

Assumption (2) above is violated by resonance lines, including the Lyman series of atomic hydrogen. In resonance lines the absorption of a photon is promptly followed by reemission, usually into another direction and frequency. The photon therefore undergoes a random walk, scattering in the line until it has traversed a column density large enough so that continuum absorption occurs. To see the relative importance of scattering versus absorption, consider a semi-infinite cloud whose surface is illuminated by resonance line emission. Photons emitted in shocks or HII regions escape at large Doppler shifts, of order 100 km s^{-1}, relative to absorbers in the cloud (Draine and Salpeter 1978, Shull and Draine 1987). Photons scatter in the Lorentz line wings, with mean free path

$$L_{sca} = 2.2 \times 10^{17} w_7^2 [(g_l/g_u) f_{lu}^2 A_X n_H]^{-1} \text{ cm} \tag{15.8}$$

Here w_7 is the Doppler shift, in units of 100 km s^{-1}, relative to the line center frequency of an atom in the cloud, g_u and g_l are statistical weights of the upper and lower levels, respectively, f_{lu} is the absorption oscillator strength, and A_X is the abundance of the emitting atom relative to hydrogen. The mean free path for absorption by dust is

$$L_{abs} = 10^{21} [n_H (1 - \omega + \varepsilon_g) \Sigma_{-21}]^{-1} \text{ cm} \tag{15.9}$$

where Σ_{-21} equals Σ_{ext} in units of 10^{-21} cm^2, evaluated at the line wavelength. It is evident that, except for lines of very rare species, scattering overwhelms absorption.

Solving the transfer equation accurately for strong scattering lines is a formidable numerical problem (Auer 1968). However a simple estimate of the column density of gas affected by the photons can be obtained if line scattering is assumed to be coherent. The predominance of scattering ensures that the radiation field is nearly isotropic, and the transfer problem can be solved in the two-stream approximation (cf. Rybicki and Lightman (1979), Section 1.8). The mean intensity at depth z from the illuminated surface is

$$J(z) \approx \frac{\sqrt{3} F_{inc}}{2\pi} \exp\left(\frac{-z}{L_{dif}}\right) \tag{15.10}$$

where F_{inc} is the flux at the cloud surface and the diffusion depth, $L_{dif} = [L_{abs} L_{sca}/3]^{1/2}$, equals

$$L_{dif} = 8.6 \times 10^{18} (w_7/n_H) [(g_l/g_u f_{lu}^2 A_X (1 - \omega + \varepsilon_g) \Sigma_{-21}]^{-1/2} \text{ cm} \tag{15.11}$$

One should note that, while coherent scattering is a reasonable approximation in the extreme Lorentz wings of resonance lines (Hummer 1963), (15.11) is probably a very crude approximation for $z \gg L_{dif}$, since a photon scatters roughly $(L_{dif}/L_{sca})^2$ times in penetrating each additional diffusion depth.

15.3. Sources of ultraviolet radiation

15.3.1. The interstellar radiation field

A lower limit on the ultraviolet intensity at cloud surfaces is imposed by the interstellar radiation field (ISRF), a background continuum due mainly to Galactic starlight. Because the average interstellar extinction is large in the Galactic disk (about 8 mag kpc^{-1} at 1000 Å, Spitzer (1978)), the far ultraviolet flux

Figure 15.4. Solid curve: approximate values of the average interstellar ultraviolet flux, obtained by using spline interpolation in Table A.3 of Mathis *et al.* (1983). For accurate values one should consult the original reference. Dotted curve: the flux that would be measured at a distance of 100 pc from an O5 star assuming (unrealistically) no attenuation by intervening dust. Stellar fluxes are from Bradley and Morton (1969) assuming a surface gravity, log $g = 4$, and stellar radius, log $(R/R_\odot) = 1.25$.

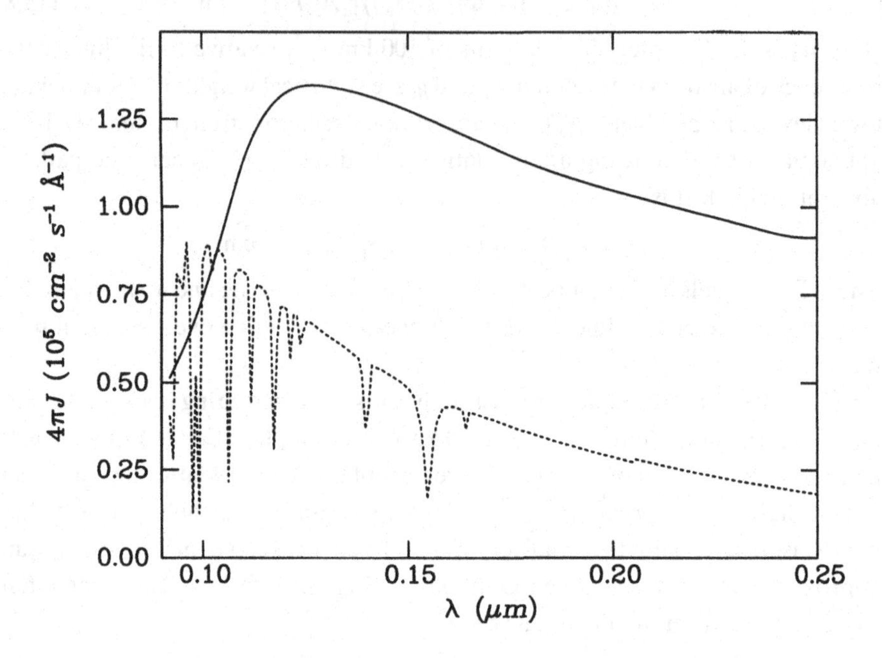

at a typical location depends only on sources within a few hundred parsecs. In the solar neighborhood, for example, about 50% of the ISRF at 1000 Å comes from two dozen nearby O and B stars (Opal and Weller 1984). Thus the local flux, $4\pi J_\odot$, can be determined accurately by adding together the fluxes of stars in existing catalogs, with appropriate corrections for extinction (Witt and Johnson 1973, Jura 1974, Henry 1977). The results, which agree well with numerous photometric observations of $J_\odot$ in selected fields, have been summarized by several authors† (see the review by Paresce and Jakobsen (1980)). Figure 15.4 shows fluxes for a typical point in the disk at galactocentric distance $D = 10$ kpc, obtained by interpolating in Table A.3 of Mathis *et al.* (1983). The flux is presumed to vanish for $\lambda < 912$ Å, due to the large opacity of interstellar HI. Also shown, for

† However the analytic formula in Roberge, Dalgarno, and Flannery (1980) is inaccurate for $\lambda <$ 1000 Å, owing to an insufficient number of significant digits in certain numerical coefficients. The photodestruction rates in that paper, which were calculated using an accurate radiation field, are unaffected.

comparison, is a plot of the ultraviolet flux 100 pc away from an O5 star assuming (unrealistically) no extinction by intervening material. For accurate values of the ISRF shown one should consult the original reference, which also discusses the dependence of J on D. See also Mezger, Mathis, and Panagia (1982).

15.3.2. Interstellar ultraviolet line emission

Less information exists on emission lines in the interstellar radiation field. Ultraviolet lines are produced by collisional excitation in coronal gas (e.g. Stern, Wang, and Bowyer (1978), Gaetz and Salpeter (1983)), shock waves (e.g. Raymond (1979), Shull and McKee (1979)), gaseous nebulae, and nebulae ionized by X-rays (Kallman and McCray 1982, Kallman 1984). The coronal gas in standard Galactic models makes a negligible contribution to the total ISRF integrated over ultraviolet wavelengths (Jakobsen and Parsce 1981, Paresce, Monsignori Fossi and Landini 1983). The combined effects of other sources have not been assessed, but are unlikely to be large. Spectroscopic observations have revealed only a few far-ultraviolet lines with fluxes greater than 10^5 cm^{-2} s^{-1} (Feldman, Brune, and Henry 1981). A possible exception is the Lyman α line of atomic hydrogen, which is emitted copiously by gaseous nebulae. Draine and Salpeter (1978) estimated an upper limit of about 2×10^7 cm^{-2} s^{-1} for the Lyα flux, at a typical point in the interstellar medium, by considering Lyα production in diffuse HII regions and shock waves. According to (15.11), however, Lyα photons reaching a cloud will penetrate to depths where the visual extinction is only $\sim$0.01 mag if $w_7 = 0.5$ and the dust scattering properties are as in Draine and Lee (1984). (This estimate neglects scattering of Lyα by molecular hydrogen. For a discussion of this process, see Chapter 21.)

15.3.3. Ultraviolet sources embedded in clouds

A variety of internal photon sources may contribute to the ultraviolet field inside clouds. In photodissociation regions, the close proximity of a cloud and OB association enhances the continuum field enormously. At the edge of an ionization-bounded HII region, the continuum may be enhanced relative to the average ISRF by factors approaching 10^5 (Tielens and Hollenbach 1985a,b) and the Lyα flux is

$$F \approx 6 \times 10^6 \dot{N}_{48} n_e^{4/3} f_{esc}\ \mathrm{s}^{-1} \tag{15.12}$$

(Draine and Salpeter 1978). Here n_e is the electron density of the ionized gas, 10^{48} $\dot{N}_{48}$ s^{-1} is the luminosity of the ionizing source in the Lyman continuum, and f_{esc} is the probability that a Lyα photon emitted in the nebula eventually escapes. Escape probabilities depend on n_e and $\dot{N}_{48}$ (see Bonilha *et al.* (1979), but typically $f_{esc} \sim 0.1$).

Another potentially important ultraviolet source is emission from shock waves driven by supernovae inside clouds (see the review by Shull and Draine (1987)). At the shock front, the flux of photons emerging in line j is $F = n_0 v_s \Phi j$, where n_0 is the preshock number density, v_s is the shock speed, and Φ_j, the number of photons emitted in line j per particle streaming into the shock, depends on the line and on details of the shock structure (e.g. Shull and McKee (1979), Hollenbach and McKee (1979)). For shocks with $v_s \approx 100\,\mathrm{km\,s^{-1}}$, Φ is approximately unity for Lyα and of order 0.01 for other strong ultraviolet lines. Of larger potential significance is radiation emitted in the recombination and two-photon continua of hydrogen and helium, which is attenuated only by dust (Shull and McKee 1979, Brugel, Shull, and Seab 1982).

As a final possibility, consider the effects of a hypothetical point source embedded deep inside a molecular cloud. Suppose that the object emits ultraviolet photons at a rate $10^{43}\,\dot{N}_{43}\,\mathrm{s^{-1}}$, and that a fraction ε of such photons escapes destruction by material in the object's immediate environment. A conservative estimate of the photon flux a distance r from the source is obtained if one treats dust opacity as pure absorption. The result is

$$F \sim 10^6 \frac{\varepsilon \dot{N}_{43} \exp\left(-1.8 n_3 r_{18}\right)}{r_{18}^2}\ \mathrm{cm^{-2}\,s^{-1}} \qquad (15.13)$$

where $r_{18} \equiv r/(10^{18}\,\mathrm{cm})$, $n_3 \equiv n_{\mathrm{H}}/(10^3\,\mathrm{cm^{-3}})$ is the ambient density in the cloud, and the argument of the exponential was evaluated assuming grain properties from Draine and Lee (1984) at $\lambda = 1000$ Å. Expression (15.13) is to be compared to the lower limit on the ultraviolet flux imposed by cosmic ray–H_2 interactions which is of order $10^3\ \mathrm{cm^{-2}\ s^{-1}}$. One sees that point sources of the postulated luminosity could dominate the ionization balance of molecular clouds provided a significant fraction of the ultraviolet escapes from each source, the cloud is not too dense, and the number density of sources is at least a few per cubic parsec.

15.4. Discussion

Some of the foregoing results are summarized in Figure 15.5, which gives the continuum ultraviolet field at various depths in a semi-infinite plane-parallel, molecular cloud. This plot illustrates the long-standing point of view, that all of the ultraviolet inside clouds is due to penetration of the ISRF. The cloud in Figure 15.5 is illuminated isotropically by the average interstellar radiation field shown in Figure 15.4, enhanced by a wavelength-independent factor $G_0 = 10^5$, the latter corresponding to a strong photodissociation region like the Orion nebula (Tielens and Hollenbach 1985b). Since Equation (15.1) is linear in the intensity, the field for other G_0 values can be obtained by scaling. The opacity due to atomic carbon, H_2, and other gas-phase species has been neglected; including carbon with the abundance suggested by recent far-infrared observations (Keene *et al.* 1985) would increase the opacity by about 10% for $\lambda < 1100$ Å.

In computing these results it has been necessary to choose values for the grain properties. Given the absence of a definitive choice, the plots shown here *for illustrative purposes* were generated using properties chosen, somewhat arbitrarily, from Draine and Lee (1984). With minor exceptions, the Σ_{ext} values reproduce the average interstellar extinction curve of Savage and Mathis (1979).

Figure 15.5. Solid and dashed curves: The continuum ultraviolet flux at various depths inside a plane-parallel, semi-infinite cloud. The cloud is illuminated isotropically by the average interstellar radiation field, as in Figure 15.4, enhanced by a factor 10^5 at every wavelength. Curves are labeled by depth measured inward from the cloud surface, in visual magnitudes. Solid curves were computed assuming Σ_{ext}, ω, and g values from Draine and Lee (1984). Dashed curves were computed using Σ_{ext} from Draine and Lee, but ω and g values from Chlewicki and Greenberg (1984b, Model 1). Gas opacity has been neglected in all calculations ($\varepsilon_g = 0$). Dotted curve: the unattenuated interstellar ultraviolet flux, as in Figure 15.4. The shaded band is defined in the text.

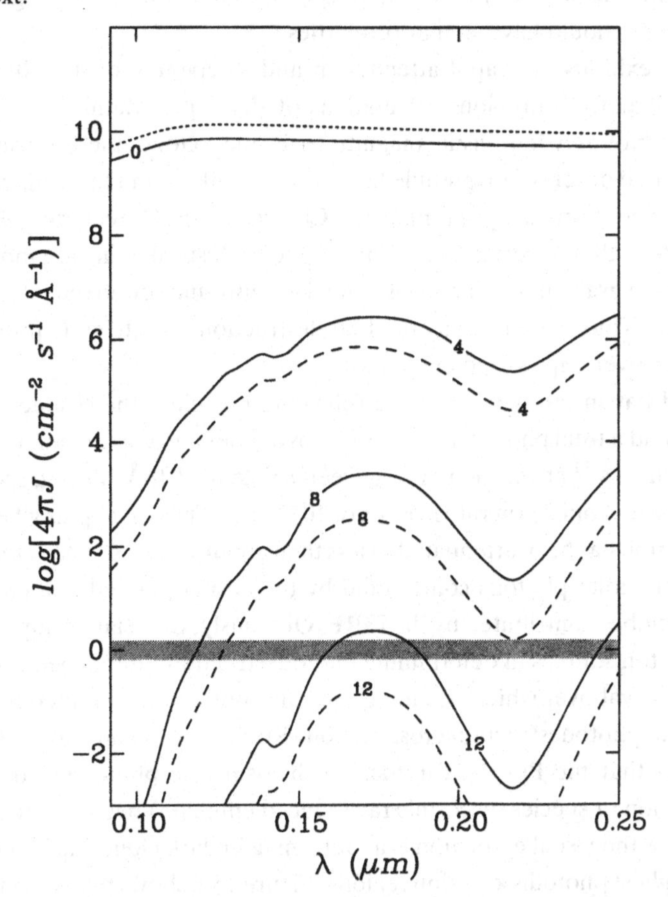

The grain albedo ranges over $\omega \approx 0.4$–0.6 and the g-factor is $g \approx 0.5$–0.6 in the wavelength interval shown. One should note that grain models of the type used by Draine and Lee have been challenged by Greenberg and Chlewicki (1983), who question their ability to reproduce observed fluctuations in the interstellar extinction. The mixture of grain sizes and compositions studied by Draine and Lee also lacks very small particles (with sizes of order 10–100 Å) in sufficient numbers to explain the near-infrared spectral features observed in protostars, reflection nebulae and other objects (Tielens and Allamandola 1987). The main effect of including such small particles would presumably be to lower the grain albedo, since the total extinction of a revised mixture could not increase without conflicting with observations, and the small particles are in the Rayleigh scattering limit at ultraviolet wavelengths. The figure also includes calculations performed with ω and g values from grain model 1 of Chlewicki and Greenberg (1984b), which is consistent with observed interstellar extinction, and has $\omega \approx 0.2, g \approx 0.8$ in the far-ultraviolet. Finally, one should emphasize that the grain models adopted here describe the diffuse interstellar medium. It has been argued, however, (Mathis *et al.* 1983) that at depths less than about $A_v = 25$, grains in molecular clouds should have similar properties.

Figure 15.5 exhibits the rapid attenuation and steepening of the ultraviolet spectrum familiar from previous calculations of this type (Mathis *et al.* 1983). Thus, at a depth of only four visual magnitudes inside a cloud, the spectrum falls off by about three orders of magnitude between $\lambda = 1500$ Å and the Lyman limit. The figure also confirms the point made by Chlewicki and Greenberg (1984a,b) and noted above, that uncertainties in the degree of dust shielding are moderate provided that the grain models are constrained to reproduce observed interstellar extinction data. Thus the sensitivity of photodestruction rates to grain properties may have been overemphasized in the past.

The shaded bar in the figure has the following meaning: the cosmic ray–H_2 interactions yield a total photon flux, summed over lines and continuum, of about $1 \times 10^3\ \zeta_{-17}\ \mathrm{cm^{-2}\ s^{-1}}$ at wavelengths between 850 and 1750 Å. Here ζ_{-17} is the primary cosmic ray ionization rate in units of $10^{-17}\ \mathrm{s^{-1}}$. Thus, at depths where the ISRF illumination has been attenuated to less than about $1\ \mathrm{cm^{-2} s^{-1} Å^{-1}}$, the total number of ultraviolet photons contributed by the cosmic ray induced emission exceeds the number contributed by the ISRF. Obviously, the relative importance of the two photon sources in determining photodestruction rates depends on the wavelength interval over which one integrates the continuum and, therefore, on the appropriate photodestruction cross section. Roughly speaking, however, the figure suggests that the Prasad–Tarafdar mechanism (the photoionization and photodissociation of species by cosmic ray induced emission) begins to dominate at depths where the visual extinction is about 5 mag in dark clouds, and about 10 mag in the brighest photodissociation regions. Figure 15.6 shows this effect for the photoionization of atomic carbon. Since carbon is ionized by a relatively broad range of wavelengths, an abrupt transition occurs to the limiting rate imposed by the Prasad–Tarafdar mechanism. For other species the transition will occur more gradually, and somewhat deeper into the cloud. In any case, since the effects occur at moderate depths, they should be considered in photodestruction rate calculations.

The simplicity of the model in Figure 15.5 and its counterparts elsewhere naturally invites skepticism. In fact, models based on homogeneous, isotropically

Figure 15.6. The depth-dependent photoionization rate of atomic carbon in a semi-infinite molecular cloud. Solid curve: an estimate of the rate including effects of the Prasad–Tarafdar mechanism. Dotted curve: the rate neglecting the Prasad–Tarafdar process, from Table 3 of van Dishoeck (1988). Both curves assume grain properties from Model 2 of Roberge, Dalgarno and Flannery (1981). The solid curve was computed by assuming a primary cosmic ray ionization rate $\zeta = 10^{-17}$ s^{-1}, and estimating that 10% of the ultraviolet photons generated by the cosmic ray–H_2 interactions are emitted at wavelengths $\lambda < 1100$ Å (see Sternberg, Dalgarno, and Lepp (1987), Figures 1 and 2).

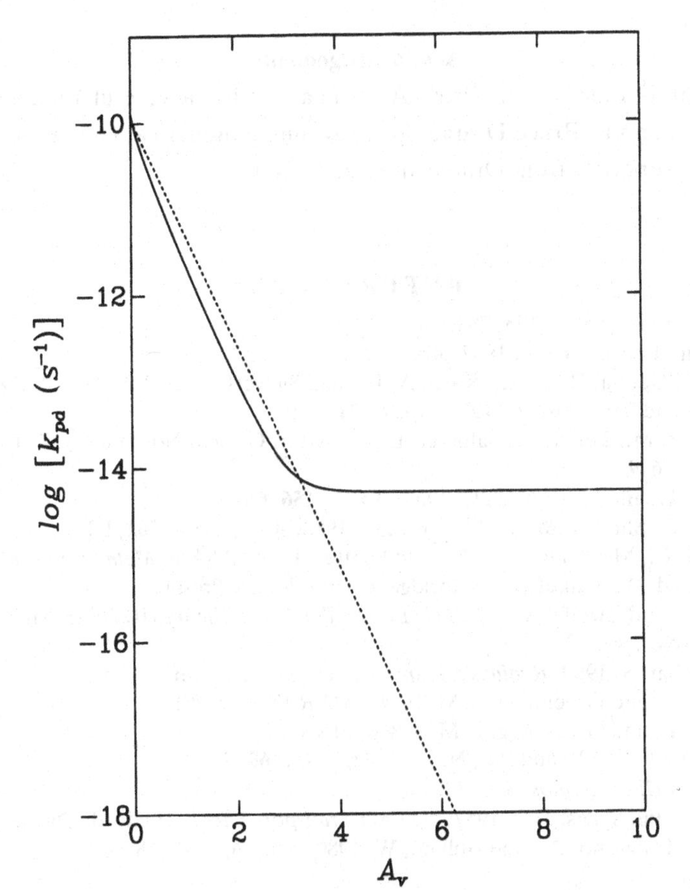

illuminated clouds have been successful in accounting for many observations of molecular cloud photochemistry (e.g. Langer (1976), Clavel, Viala, and Bel (1978), de Jong, Dalgarno, and Boland (1980), Tielens and Hollenbach (1985a,b)). An important exception may be the pervasiveness of far-infrared emission from warm neutral gas in star-forming regions. Stutzki *et al.* (1988) have suggested that spatially extended [CII] emission in M17 SW requires a clumpy cloud structure and an enhanced ultraviolet field at great depths inside the cloud. Clumpiness can enhance the penetrability of photons generated by external and internal sources and, as Stutzki *et al.* note, one expects massive stars to be present inside such regions. Alternatively, Expression (15.13) shows that less luminous sources, if present in sufficient numbers, will substantially enhance the ultraviolet field inside a dark cloud. In this connection it is worth noting that T Tauri stars, which emit a few percent of their luminosity in the ultraviolet (Giampapa and Imhoff 1985) may exist in star-forming regions with densities of several pc^{-3} (e.g. Montmerle *et al.* (1983)). The exploration of these possibilities, by solving the radiation transport problem for a random, clumpy medium with embedded point sources, is a challenging prospect.

Acknowledgements

The author is indebted to Prof. A. Dalgarno, for a careful reading of the manuscript, and to Bruce Draine, for providing a numerical tabulation of grain scattering properties from Draine and Lee (1984).

REFERENCES

Auer, L. H. 1968, *Ap. J.*, **153**, 783.
Bally, J., and Langer, W. D. 1982, *Ap. J.*, **255**, 143.
Benassi, M., Garcia, R. D. M., Karp, A. H., and Siewert, C. E. 1984, *Ap. J.*, **280**, 853.
Bernes, C., and Sandqvist, A. 1977, *Ap. J.*, **217**, 71.
Bonilha, J. R. M., Ferch, R., Salpeter, E. E., Slater, G., and Noerdlinger, P. D. 1979, *Ap. J.* **233**, 649.
Bradley, P. T., and Morton, D. C. 1969, *Ap. J.*, **156**, 687.
Brugel, E. W., Shull, J. M., and Seab, C. G. 1982, *Ap. J. Lett.*, **262**, L35.
Canfield, R. C., McClymont, A. N., and Puetter, R. C. 1984, in *Methods in Radiative Transfer*, ed. W. Kalkofen (Cambridge: Cambridge U. Press).
Case, K. M., and Zweifel, P. F., 1967, *Linear Transport Theory* (Reading, Mass: Addison-Wesley).
Chandrasekhar, S. 1960, *Radiative Transfer* (New York: Dover).
Chlewicki, G., and Greenberg, J. M. 1984a, *MNRAS*, **210**, 791.
Chlewicki, G., and Greenberg, J. M. 1984b, *MNRAS*, **211**, 719.
Clavel, J., Viala, Y. P., and Bel, N. 1978, *Astr. Ap.*, **65**, 435.
Dave, J. V. 1975, *J. Atmos. Sci.*, **32**, 790.
Davison, B., and Sykes, J. B. 1957, *Neutron Transport Theory* (Oxford: Clarendon).
de Jong, T., Dalgarno, A., and Boland, W. 1980, *Astr. Ap.*, **91**, 68.

Draine, B. T., and Salpeter, E. E. 1978, *Nature,* **271**, 730.
Draine, B. T., and Lee, H. M. 1984, *Ap. J.,* **285**, 89.
Federman, S. R., Glassgold, A. E., and Kwan, J. 1979, *Ap. J.,* **227**, 466.
Feldman, P. D., Brune, W. H., and Henry, R. C. 1981, *Ap. J. Lett.,* **249**, L51.
Field, G. B., Sommerville, W. B., and Dressler, K. 1966, *Ann. Rev. Astr. Ap.,* **4**, 226.
Flannery, B. P., Roberge, W., and Rybicki, G. B. 1980, *Ap. J.,* **236**, 598.
Gaetz, T. J., and Salpeter, E. E. 1983, *Ap. J. Suppl.,* **52**, 155.
Gerola, H., and Glassgold, A. E. 1978, *Ap. J. Suppl.,* **37**, 1.
Giampapa, M. S., and Imhoff, C. L. 1985 in *Protostars and Planets* II, eds. D. C. Black and M. S. Matthews (Tucson: U. Arizona Press), p. 386.
Glassgold, A. E., Huggins, P. J., and Langer, W. D. 1985, *Ap. J.,* **290**, 615.
Greenberg, J. M., and Chlewicki, G. 1983, *Ap. J.,* **272**, 563.
Henry, R. C. 1977, *Ap. J. Suppl.,* **33**, 451.
Henyey, L. G., and Greenstein, J. L. 1941, *Ap. J.,* **93**, 70.
Hollenbach, D. J., and McKee, C. F. 1979, *Ap. J. Suppl.,* **41**, 555.
Hollenbach, D. J., Werner, M. W., and Salpeter, E. E. 1971, *Ap. J.,* **163**, 165.
Hummer, D. G. 1963, *MNRAS,* **125**, 21.
Jakobsen, P., and Paresce, F. 1981, *Astr. Ap.,* **96**, 23.
Jura, M. 1974, *Ap. J.,* **191**, 375.
Kallman, T. R. 1984, *Ap. J.,* **280**, 269.
Kallman, T. R., and McCray, R. A. 1982, *Ap. J. Suppl.,* **50**, 263.
Karp, A. H., Greenstadt, J., and Fillmore, J. A. 1980, *J. Quant. Spectrosc. Rad. Transf.,* **24**, 391.
Keene, J., Blake, G. A., Phillips, T. G., Huggins, P. J., and Beichman, C. A. 1985, *Ap. J.,* **299**, 967.
Langer, W. 1976, *Ap. J.,* **206**, 699.
Leung, C. M. 1975, *Ap. J.,* **199**, 340.
Mark, J. C. 1947, *Phys. Rev.,* **72**, 558.
Mathis, J. S., Mezger, P. G., and Panagia, N. 1983, *Astr. Ap.,* **128**, 212.
Mathis, J. S., Rumpl, W., and Nordsieck, K. H. 1977, *Ap. J.,* **217**, 425.
Mezger, P. G., Mathis, J. S., and Panagia, N. 1982, *Astr. Ap.,* **105**, 372.
Mihalas, D. 1978, *Stellar Atmospheres* (San Francisco: W. H. Freeman).
Montmerle, T., Koch-Miramond, L., Falgarone, E., and Grindlay, J. E. 1983, *Ap. J.,* **269**, 182.
Opal, C. B., and Weller, C. S. 1984, *Ap. J.,* **282**, 445.
Paresce, F., and Jakobsen, P. 1980, *Nature,* **288**, 119.
Paresce, F., Monsignori Fossi, B. C., and Landini, M. 1983, *Ap. J. Lett.,* **266**, L107.
Prasad, S. S., and Tarafdar, S. P. 1983, *Ap. J.,* **267**, 603.
Raymond, J. C. 1979, *Ap. J. Suppl.,* **39**, 1.
Roberge, W. G. 1983, *Ap. J.,* **275**, 292.
Roberge, W. G. 1989, in preparation.
Roberge, W. G., Dalgarno, A., and Flannery, B. P. 1981, *Ap. J.,* **243**, 817.
Rybicki, G. B., and Lightman, A. P. 1979, *Radiative Processes in Astrophysics* (New York: J. Wiley).
Sandell, G. 1978, *Astr. Ap.,* **69**, 85.
Sandell, G., and Mattila, K. 1975, *Astr. Ap.,* **42**, 357.
Savage, B. D., and Mathis, J. S. 1979, *Ann. Rev. Astr. Ap.,* **17**, 73.
Shull, J. M., and Draine, B. T. 1987, in *Interstellar Processes,* eds. D. J. Hollenbach and H. A. Thronson, Jr (Dordrecht: Reidel).
Shull, J. M., and McKee, C. F. 1979, *Ap. J.,* **277**, 131.

Stern, R., Wang, E., and Bowyer, S. 1978, *Ap. J. Suppl.*, **37**, 195.
Spitzer, L., Jr 1978, *Physical Processes in the Interstellar Medium* (New York: Wiley-Interscience).
Sternberg, A., Dalgarno, A., and Lepp, S. 1987, *Ap. J.*, **320**, 676.
Stief, L. J., Donn, B., Glicker, S., Gentieu, E. P., and Mentall, J. E. 1972, *Ap. J.*, **171**, 21.
Stutzki, J., Stacey, G. J., Genzel, R., Harris, A. I., Jaffe, D. T., and Lugten, J. B. 1988, *Ap. J.*, **332**, 379.
Tielens, A. G. G. M., and Allamandola, L. J. 1987, in *Interstellar Processes*, eds. D. J. Hollenbach and H. A. Thronson, Jr (Dordrecht: Reidel).
Tielens, A. G. G. M., and Hollenbach, D. J. 1985a, *Ap. J.*, **291**, 722.
Tielens, A. G. G. M., and Hollenbach, D. J. 1985b, *Ap. J.*, **291**, 747.
van de Hulst, H. C. 1980, *Multiple Light Scattering*, Vols. 1 and 2 (New York: Academic Press).
van Dishoeck, E. F. 1987, in *IAU Symp. 120, Astrochemistry*, eds. M. S. Vardya and S. P. Tarafdar (Dordrecht: Reidel).
van Dishoeck, E. F. 1988, in *Rate Coefficients in Astrochemistry*, eds. T. J. Millar and D. A. Williams, in press.
van Dishoeck, E. F., and Black, J. H. 1986, *Ap. J. Suppl.*, **62**, 109.
Whitworth, A. P. 1975, *Ap. Space Sci.*, **34**, 155.
Witt, A. N., and Johnson, M. W. 1973, *Ap. J.*, **181**, 363.

16

Cosmic ray induced photodissociation and photoionization of interstellar molecules

ROLAND GREDEL

European Southern Observatory, Santiago, Chile

16.1. Introduction

The interior of a dense molecular cloud is efficiently shielded from the ultraviolet photons of the interstellar radiation field by the grains. Accordingly, the effects of the ultraviolet photons on the physical and chemical states of dense clouds are in general, neglected. However, several internal sources of ultraviolet photons may be present, including the emission from embedded stars and internal shocks driven by mass loss from young stars. Here the discussion is focused on a diffuse source of internal ultraviolet photons arising from energetic cosmic rays capable of penetrating dense clouds. Cosmic ray particles with energies between 10 and 100 MeV ionize molecular hydrogen in the interior of the clouds and generate secondary electrons with a mean energy of around 30 eV (Cravens and Dalgarno 1978). Because the fractional ionization is generally low, the secondary electrons degrade mainly through excitations of various electronic states of H_2. Typical excitation energies are 10–15 eV. Ionization of H_2 by the secondary electrons may also occur. The subsequent decays of the electronically excited states of H_2 produce ultraviolet photons within the clouds.

The idea of molecular hydrogen emission inside dense clouds was invoked by Prasad and Tarafdar (1983) to explain the large abundance of atomic carbon which exists in several molecular clouds. Observations of emission in the $^3P_1$–$^3P_0$ and $^3P_2$–$^3P_1$ fine-structure lines of neutral atomic carbon in dense interstellar clouds (Phillips and Huggins 1981, Keene *et al.* 1985, Zmuidzinas, Betz and Goldhaber 1986) have demonstrated the existence of C with abundances relative to CO exceeding 0.1. The abundance ratios calculated in steady state models of cloud composition with solar abundances are orders of magnitude smaller. In the scheme proposed by Prasad and Tarafdar (1983) CO is photodissociated

$$CO + h\nu \rightarrow C + O \tag{16.1}$$

by the ultraviolet photons. The ultraviolet photons also ionize C,

$$C + h\nu \rightarrow C^+ + e \tag{16.2}$$

The internal photons generated inside the molecular clouds have enough energy to cause the photodissociation and photoionization of other molecules as well. In the following sections, a quantitative evaluation of the mechanism suggested by Prasad and Tarafdar (1983) is carried out. The cosmic ray induced photodestruction of interstellar molecules is calculated and included in chemical reaction networks. The changes in the steady state abundances are discussed.

16.2. The H_2 spectrum

The collisional excitation of rotation-vibration levels in the various electronic states of H_2 by the secondary electrons followed by the emission of a photon with energy $h\nu$ is illustrated by

$$X, v_0 J_0 \overset{e}{\rightarrow} i, v'J' \overset{h\nu}{\rightarrow} f, v''J'' \tag{16.3}$$

X denotes the $X^1\Sigma_g^+$ electronic ground state of H_2 and i any excited electronic state. The final electronic state f is either the $X^1\Sigma_g^+$ ground state or the triplet $b^3\Sigma_u^+$ state depending on whether a singlet or a triplet state is excited. Direct collisional excitations into the triplet $b^3\Sigma_u^+$ state are followed by dissociation of H_2 without the emission of photons. J and v are the rotation and vibration quantum numbers, respectively. It is assumed that initially the total H_2 population density resides in the X, $v_0 = 0$, $J_0 = 0$ level.

The energy degradation process of the 30 eV secondary electrons is calculated following the method introduced by Cravens, Victor and Dalgarno (1975). Collisional excitations into the $B^1\Sigma_u^+$, $B'^1\Sigma_u^+$, $B''^1\Sigma_u^+$, $C^1\Pi_u$, $D^1\Pi_u$ and $D'^1\Pi_u$ Rydberg states, into the valence E, $F^1\Sigma_g^+$ and $a^3\Sigma_g^+$ states, into the repulsive $b^3\Sigma_u^+$ state and into vibrational levels of the ground state are considered. The total energy loss of the 30 eV secondary electrons may be expressed in terms of excitations $e(i, v')$ into vibrational levels v' in the various electronic states i. Excitation cross sections were provided by Shemansky (1987). It is assumed that the collisional excitation into a particular rotational level J' can be described in terms of Honl–London factors $S(J', J_0)$. The number of photons $N(i, \nu)$ emitted in a particular transition $i, v'J' \rightarrow f, v''J''$ with frequency ν (cf. (16.3)) is then given by

$$N(i, \nu) = \frac{e(i, v')S(J', J_0)}{\Sigma_{vJ} e(i, v)S(J, J_0)} \times \frac{A(i, v'J'; f, v''J'')}{\Sigma_{vJ} A(i, v'J'; f, vJ)} \tag{16.4}$$

where the $A(i, v'J'; f, v''J'')$s are the spontaneous transition probabilities for the various rotation-vibration transitions in the i–f systems. Note that the E, $F^1\Sigma_g^+$

state decays into $B^1\Sigma_u^+$. The cascading leads to an enhancement of the effective number of excitations $\tilde{e}(B, v')$ into $B^1\Sigma_u^+$ which may be described by

$$\tilde{e}(B, v') = e(B, v') + \sum_v e(\mathrm{EF}, v)q(v, v') \tag{16.5}$$

where $q(v, v')$ are the Franck–Condon factors for the E, $F^1\Sigma_g^+$–$B^1\Sigma_u^+$ transition (Lin 1974).

Vibrational levels in a number of Rydberg states have predissociation efficiencies $\eta(v')$ greater than zero. The $v' > 0$ levels in $B''^1\Sigma_u^+$ and the $v' > 3$ levels in $D'^1\Pi_u$ have $\eta(v')$ near 1, and the $v' > 3$ levels in $D^1\Pi_u$ have $\eta(v') \approx 0.5$ (Ajello *et al.* 1984). In the case where predissociation is important the additional factor $(1 - n(v'))$ is included in (16.4).

If $e(\mathrm{i}) = \Sigma_v\, e(i, v)$ is the total efficiency for excitations into i, the number of photons $N(\nu)$ with frequency ν produced per ionization is given by

$$N(\nu) = \sum_i N(i, \nu)e(i)/IZ \tag{16.6}$$

where IZ is the number of ionizations per secondary electron. Figure 16.1 reproduces the final H_2 spectrum $N(\lambda)$ versus wavelength λ; $N(\lambda)$ is the probability

Figure 16.1. The emission spectrum from the $B^1\Sigma_u^+$, $B'^1\Sigma_u^+$, $B''^1\Sigma_u^+$, $C^1\Pi_u$, $D^1\Pi_u$ and $D'^1\Pi_u$ Rydberg states to the $X^1\Sigma_g^+$ ground state and from the $a^3\Sigma_g^+$ state to the $b^3\Sigma_u^+$ state in H_2 produced by a cosmic ray secondary electron with an initial energy of 30 eV.

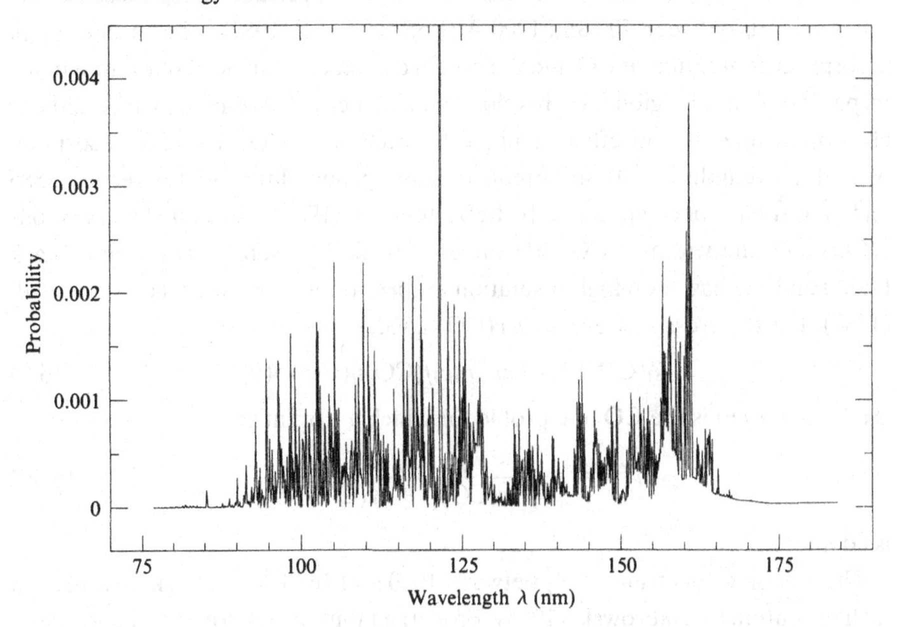

for photon emission between λ and $\lambda + d\lambda$ with $d\lambda = 1$ Å. The spectrum consists of numerous discrete lines emitted in the various band systems, mainly in the Lyman and Werner bands. Superimposed is a weak continuum emission from the $a^3\Sigma_g^+–b^3\Sigma_u^+$ system and from transitions into the vibrational continuum of the ground state. Significant emission occurs between 850 and 1750 Å.

16.3. Cosmic ray induced photodestruction of interstellar molecules

The photodissociation or photoionization rate $R(m)$ cm^{-3} s^{-1} of a species m with density $n(m)$ cm^{-3} corresponding to a total cosmic ray ionization rate of ζ s^{-1} is given by

$$R(m) = \zeta n(m) \int \frac{\sigma(m, \nu) N(\nu)}{\sigma(\text{tot}, \nu)} d\nu \tag{16.7}$$

$\sigma(m, \nu)$ is the photodissociation or photoionization cross section of species m at frequency ν and $\sigma(\text{tot}, \nu)$ is the total photon absorption cross section. It may be written in the form

$$\sigma(\text{tot}, \nu) = \sigma(g, \nu)(1 - \omega) + \sum_m \varkappa(m)\sigma(m, \nu) \tag{16.8}$$

$\sigma(g, \nu)$ is the grain extinction cross section per hydrogen nucleus, ω is the grain albedo and $\varkappa(m) = n(m)/n(\text{H})$. In the following, the dimensionless rates $R'(m) = R(m)/\zeta n(m)$ are used throughout.

The photodissociation of CO is induced through discrete line absorption followed by predissociation. The line absorption spectrum into predissociating states extends between 700 and 1200 Å (Letzelter *et al.* 1987). The absence of an absorption continuum of CO and the absence of an H_2 emission continuum in the respective spectral region requires close coincidences of H_2 emission lines and CO absorption lines for an efficient photodestruction of CO. The CO absorption oscillator strengths $f(\text{CO})$ and predissociation probabilities $\eta(\text{CO})$ between 885 and 1150 Å have been measured by Letzelter *et al.* (1987). Accurate line positions $\nu(\text{CO})$ and line widths $\gamma(\text{CO})$ have been provided by van Dishoeck and Black (1987) and are based on high resolution absorption measurements of Stark *et al.* (1987). For the cross sections near $\nu(\text{CO})$ a value of

$$\sigma(\text{CO}, \nu) = (\pi e^2/mc) f(\text{CO}) \eta(\text{CO}) \phi(\nu) \tag{16.9}$$

is adopted. $\phi(\nu)$ is the CO line profile for which a Lorentzian

$$\phi(\nu) = \frac{\gamma(\text{CO})/4\pi^2}{[\nu - \nu(\text{CO})]^2 + [\gamma(\text{CO})/4\pi]^2} \tag{16.10}$$

is adopted.

The emission spectrum of H_2 between 1000 and 1650 Å has been measured at high resolution by Dabrowski (1984). From the tabulated energies of the rotation-

Table 16.1. *Dimensionless cosmic ray induced photodestruction rates* R′ *of interstellar molecules* $\omega = 0$

Reaction			R' (m)
C	$\rightarrow C^+$	+ e	510
C_2	$\rightarrow$ C	+ C	200
CH	$\rightarrow$ C	+ H	770
CH^+	$\rightarrow$ C	+ H^+	180
CO	$\rightarrow$ C	+ O	5–23
CN	$\rightarrow$ C	+ N	9160
OH	$\rightarrow$ O	+ H	490
O_2	$\rightarrow$ O	+ O	750
	$\rightarrow O_2^+$	+ e	100
NO	$\rightarrow$ N	+ O	450
H_2O	$\rightarrow$ OH	+ H	1200
CO_2	$\rightarrow$ CO	+ O	1490
	$\rightarrow CO_2^+$	+ e	10
HCN	$\rightarrow$ CN	+ H	3120
HCO	$\rightarrow HCO^+$	+ e	1110
C_3	$\rightarrow C_2$	+ C	1060
NH_2	$\rightarrow$ NH	+ H	80
	$\rightarrow NH_2^+$	+ e	570
NH_3	$\rightarrow NH_2$	+ H	1160
	$\rightarrow$ NH	+ H_2	450
	$\rightarrow NH_3^+$	+ e	510
CH_4	$\rightarrow CH_2$	+ H_2	2670
C_2H_2	$\rightarrow C_2H$	+ H	6680
	$\rightarrow C_2H_2^+$	+ e	1140
C_2H_4	$\rightarrow C_2H_2$	+ H_2	3670
H_2CO	$\rightarrow$ CO	+ H_2	2630

vibration levels of the $X^1\Sigma_g^+$, $B^1\Sigma_u^+$ and $C^1\Pi_u$ electronic states the H_2 emission frequencies may be computed to a probable accuracy of 0.1 cm^{-1}.

A total of 69 H_2–CO line pairs with wavelength separations less than 5 cm^{-1} is found. The large number of close line pairs reduces the sensitivity of the dimensionless photodissociation rate R'(CO) to the molecular parameters. It increases from 5 at 5 K to 20 at 40 K. The temperature dependence arises because $\sigma(CO, \nu)$ depends on the fractional population density in the absorbing CO rotational level J. If a Voigt profile is employed in (16.9), R'(CO) does not change unless the velocity dispersion is of the order of 10 km s^{-1}. This is because the intrinsic CO line widths due to predissociation are of the order of a few kilometers per second. At 10 km s^{-1}, the dimensionless photodissociation rate increases to 15 and 23 at 5 K and 40 K, respectively.

Table 16.1 lists the photodissociation and photoionization rates of C, C_2, CN, CH, CH^+, CO, OH, NO, O_2, HCN, HCO, NH_2, NH_3, CO_2, C_3, H_2O, H_2CO,

CH_4, C_2H_2 and C_2H_4. In contrast to CO most of these molecules have broad absorption continua in the 850–1750 Å region. Accordingly, they have much higher photodestruction rates which are insensitive to the temperature and the velocity dispersion.

16.4. Steady state abundances of interstellar molecules

The chemical model employed was described by Sternberg, Dalgarno and Lepp (1987). It was supplemented by the nitrogen chemistry using the reaction rates tabulated by Prasad and Huntress (1980). Steady state abundances are calculated for a cloud with $T = 30$ K, $n(\mathrm{H}) = 10^4\,\mathrm{cm}^{-3}$, $\zeta = 10^{-17}\,\mathrm{s}^{-1}$ and a metal abundance ratio of 1.5×10^{-8}. The initial chemical abundances of Graedel, Langer and Frerking (1982) are adopted.

The molecular equilibrium abundances are presented in Table 16.2. In Model I the effects of the ultraviolet photons are neglected. In Models II and III, the cosmic ray induced photodestructions are included for grain albedos of $\omega = 0.5$ and 0.8, respectively.

If chemical equilibrium is attained, atomic carbon is mainly formed through the cosmic ray induced photodissociation of CO (Gredel, Dalgarno and Lepp 1987). The C/CO ratio of 2×10^{-5} in Model I increases to 4×10^{-3} in Model II, which is large but still less than the values suggested by the observations (Keene *et al.* 1985). A much higher C/CO ratio can be obtained if CO is severely depleted, because only a small fraction of CO is photodissociated. The enhancement is, however, limited by the destruction of C by reactions with H_3^+. At a grain albedo of 0.5, a C/CO ratio of 0.015 is obtained if CO is depleted by 0.1 and a C/CO ratio of 0.06 is obtained if CO is depleted by 0.01.

The equilibrium abundance of C^+ is low and increases by a factor of 2 in Model II. The large enhancement of neutral carbon does not severely influence the steady state value of C^+ because C^+ is efficiently removed by reactions with other species. The equilibrium abundance of CO is not affected by the ultraviolet photons.

The large increase in atomic carbon counteracts the photodestruction of many carbon bearing molecules. The abundances of CH^+, CN, HCN, CH, H_2CO increase by factors of 140, 40, 6, 5 and 4, respectively, in Model II. Their abundances decrease in Model III because of the decrease in atomic carbon and their more efficient photodestruction. The increase of CH_4 in Model II is not very pronounced and reflects its large photodissociation rate. The photodestruction of C_2H_2 is efficient enough to lead to a decrease of its abundance by an order of magnitude. Its dissociation product C_2H increases by a factor of 2. At $\omega = 0.8$, C_2H_2 decreases by a factor of 35. The abundance of the interstellar ring molecule C_3H_2 decreases in Models II and III because it forms through reactions with C_2H_2.

Table 16.2. *Steady state abundances of interstellar molecules*[a]

Molecule	Model I no photons	Model II $\omega = 0.5$	Model III $\omega = 0.8$
C	2.9 (−9)[b]	5.5 (−7)	3.9 (−7)
C^+	8.7 (−11)	1.9 (−10)	3.0 (−10)
CO	1.5 (−4)	1.4 (−4)	1.4 (−4)
CH	1.7 (−12)	7.9 (−12)	5.9 (−12)
CH^+	5.5 (−18)	7.7 (−16)	6.7 (−16)
CN	1.1 (−11)	4.6 (−10)	4.0 (−10)
HCN	7.5 (−11)	4.6 (−10)	3.4 (−10)
HCO^+	3.4 (−10)	8.5 (−10)	1.3 (−9)
O	8.5 (−5)	1.3 (−4)	1.6 (−4)
O_2	4.1 (−5)	2.6 (−5)	1.9 (−5)
OH	1.6 (−9)	5.1 (−9)	7.2 (−9)
H_2O	3.9 (−5)	2.0 (−5)	1.4 (−5)
H_2CO	2.4 (−10)	1.1 (−9)	9.0 (−10)
CO_2	9.9 (−8)	3.6 (−8)	2.1 (−8)
CH_4	6.8 (−8)	7.2 (−8)	2.7 (−8)
C_2H	1.1 (−9)	2.2 (−9)	9.4 (−10)
C_2H_2	4.5 (−9)	5.9 (−10)	1.3 (−10)
C_3H_2	2.6 (−11)	1.9 (−11)	9.3 (−12)
NO	2.3 (−8)	2.4 (−8)	3.5 (−8)
NH_2	6.9 (−9)	4.9 (−9)	4.6 (−9)
NH_3	3.2 (−8)	2.9 (−8)	2.6 (−8)

[a] See text for adopted physical parameters.
[b] Abundances relative to H_2; $n(C) = 2.9 \times 10^{-9} n(H_2)$.

The cosmic ray induced photodissociation of H_2O is the main source of OH in interstellar clouds if OH does not form through dissociative recombination of H_3O^+ (Bates 1986). OH increases by factors of 3 and 5 in Models II and III, respectively, and H_2O decreases by factors of 2 and 3. The photodissociation of O_2 results in a decrease of its abundance by a factor of about 2 and an increase in O by a similar factor. NO forms through reactions with OH and increases by some 50% in Model III. NH_2 and NH_3 are rapidly removed through reactions with other species, and their equilibrium abundances are not affected by the ultraviolet photons.

REFERENCES

Ajello, J. M., Shemansky, D., Kwok, T. L., and Yung, Y. L. 1984, *Phys. Rev.*, **29**, 636.
Bates, D. R. 1986, *Ap. J.* (*Letters*), **306**, L45.
Cravens, T., and Dalgarno, A. 1978, *Ap. J.*, **219**, 750.
Cravens, T. E., Victor, G. A., and Dalgarno, A. 1975, *Planet Spa. Sci.*, **23**, 1059.
Dabrowski, I. 1984, *Canadian J. Phys.*, **62**, 1639.

Graedel, T. E., Langer, W. D., and Frerking, M. A. 1982, *Ap. J. Suppl.*, **48**, 321.
Gredel, R., Dalgarno, A., and Lepp, S. 1987, *Ap. J.* (*Letters*), **323**, L137.
Keene, J., Blake, G. A., Phillips, A. G., Huggins, P. J., and Beichman, C. A. 1985, *Ap. J.*, **299**, 967.
Letzelter, C., Eidelsberg, M., Rostas, F., Breton, J., and Thieblemont, B. 1987, *Chem. Phys.*, **114**, 273.
Lin, C. S. 1974, *J. Chem. Phys.*, **60**, 4660.
Phillips, T. G., and Huggins, P. J. 1981, *Ap. J.*, **251**, 533.
Prasad, S. S., and Huntress, W. T. 1980, *Ap. J. Suppl.*, **43**, 1.
Prasad, S. S., and Tarafdar, S. P. 1983, *Ap. J.*, **267**, 603.
Shemansky, D. E. 1987, private communication.
Stark, G., Smith, P. L., Yoshino, K., and Parkinson, W. H. 1987, private communication.
Sternberg, A., Dalgarno, A., and Lepp, S. 1987, *Ap. J.*, **320**, 676.
van Dishoeck, E. F., and Black, J. H. 1987, private communication and in preparation.
Zmuidzinas, J., Betz, A., and Goldhaber, D. M. 1986, *Ap. J.* (*Letters*), **307**, L75.

17

Chemistry in the molecular cloud Barnard 5

S. B. CHARNLEY

Max Planck Institute for Physics and Astrophysics, Institute for Extraterrestrial Physics, Garching, FRG

D. A. WILLIAMS

Mathematics Department, UMIST, Manchester, UK

17.1. Introduction

Interstellar chemistry began to be studied in a fairly serious way when, in the late 1960s and early 1970s, it was demonstrated that a wide variety of molecules existed in dense molecular clouds. At first the main effort was in identifying the main chemical routes by which molecules were formed and destroyed. It was realized that, even in dark molecular clouds where starlight is excluded, cosmic rays may penetrate and cause ionizations which drive a chemistry which would otherwise 'run down'. This chemistry would, therefore, be largely one of positive ions and molecules. This early recognition met with great success and – although the level of ionization in molecular clouds remains uncertain – the detection of interstellar ions such as HCO^+ and N_2H^+ is strong support for positive ion–neutral molecule chemistry. Models of interstellar chemistry involving hundreds or even thousands of reactions are now routinely studied: some of these reactions may be important.

These early studies, understandably, concentrated on the chemistry. They deliberately made the dynamics as simple as possible. Thus, uniform density and temperature were usually invoked, in geometrically convenient shapes such as semi-infinite slabs, or spheres. Steady-state calculations were often performed, without a full consideration of the applicability of steady-state. Later studies showed that it might take around 30 million years to achieve steady-state in molecular clouds, and it was realised that such extended periods might not be available in interstellar clouds.

At the same time, detailed observations were indicating that molecular clouds were far from simple objects. Molecular line profiles were often seen to be complex and wide, indicating the presence of superthermal motions. Molecular abundances and total gas density were found to vary substantially from point to point in the cloud, where telescopes had sufficient angular resolution to make such measurements. In particular, molecular clouds were found to be sites of star formation, and it became apparent that stars would have profound effects on the gas in their vicinity. Thus, the gas should not be viewed separately from the stars, nor should its motion and density be regarded as irrelevant to its chemistry. Describing the evolution of the physical conditions of the situation correctly had become as significant as identifying the chemical pathways.

This chapter is concerned with the physical and chemical evolution of one particular molecular cloud: Barnard 5 (B5), as revealed by a series of recent observations, and with the chemistry in the varying conditions implied by those observations. In describing B5, one may recognise that the work has application to a whole class of molecular clouds in which low mass stars are found.

17.2. Observations of B5, and inferences

B5 appears to be a fairly typical nearby dark cloud which until recently was thought to be fairly quiescient. Its proximity has enabled some very detailed studies to be made, and it is now recognized as an active region, possibly the prototype of many molecular clouds.

A survey by the Infrared Astronomical Satellite detected four compact sources towards (and probably within) this object. From the infrared luminosities and temperatures, these objects were inferred to be low mass stars still enshrouded in dust shells. The precise location of these sources with respect to the gas has been made clear by a survey of the 1–0 rotation emission of the minor isotope $^{12}C^{18}O$ of carbon monoxide (see Figure 17.1). A similar map in $^{13}C^{16}O$ shows essentially the same features. The cloud is clumpy: the stars are not necessarily located in the clumps. Some data concerning B5 and its clumps are given in Table 17.1.

Observations in the $J = 1$–0 line of $^{12}C^{16}O$, the major isotope, show a variety of unusual line profiles, indicating the existence of high velocity molecular flows in the vicinity of the stars. An evolutionary sequence is proposed: IRS1 is still embedded in a dense clump; it has the highest speed of flow and yet the flow is smallest in extent. At the other extreme IRS4 appears to have created a cavity; the gas surrounding is still slowly moving with respect to the ambient gas, and this is the most extended molecular flow. Goldsmith, Langer and Wilson (1986), who carried out the CO observations of B5, also note that the mechanical input rate into the flows may, over the lifetime of the flows, be capable of dissipating the dense clump of material in which the young stars have formed (see Table 17.2). Thus, the formation of a single low mass star might halt further star formation from that clump.

In this interpretation, the gas in B5 is in continual motion, stirred by the winds generated by the low mass stars. The sequence of events is as follows: a dense

Table 17.1. *Observed properties of the fragments in B5*

Fragment designation	Size (pc)	Peak $N(H_2)$ (10^{21} cm^{-2})	Mass ($M_\odot$)	$\langle n(H_2) \rangle$ (10^3 cm^{-3})	U_{grav} (10^{43} erg)
C	0.50	11	54	7.1	55
S	0.25	5	7	6.4	2
SW	0.50	5	14	3.2	4
E	0.35	4	22	3.5	13
NE	0.50	5	12	3.2	3

Table taken from Goldsmith *et al.* (1986).

Table 17.2. *Observed properties of the molecular outflows in B5*

Outflow	Velocity (km s^{-1})	Mass ($M_\odot$)	Momentum ($M_\odot$ km s^{-1})	Energy (10^{43} erg)	Size (pc)	Timescale (10^4 yr)
IRS1	31	0.02	0.3	5	0.3	1
IRS2	10	0.4	2	15	1.4	14
IRS3	10	0.1	0.4	2	0.9	9
IRS4	6	5.0	16	39	1.4	24

Table taken from Goldsmith *et al.* (1986).

Figure 17.1. Map of B5 in the $J = 1\text{–}0$ transition of $^{12}C^{18}O$. The properties of the designated molecular fragments are given in Table 17.1. (From Goldsmith, Langer and Wilson (1986).)

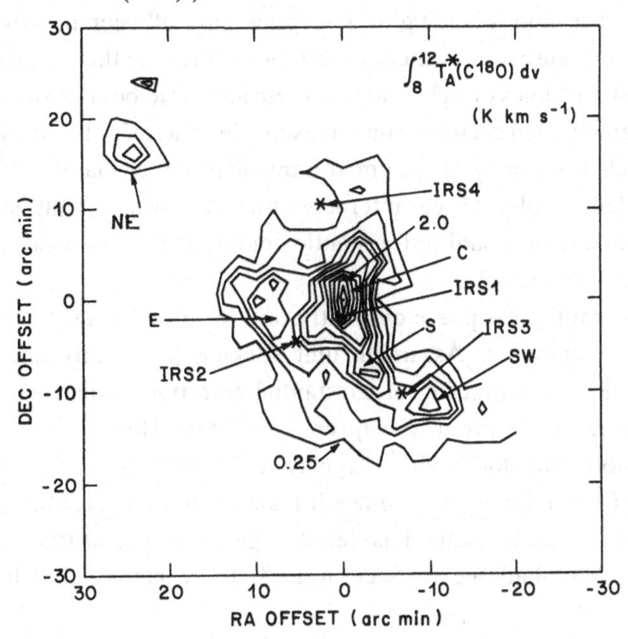

clump of gas gives rise to a low mass star whose winds erode the clump, and carry it away in a low density form. This mass loaded wind will ultimately be brought to local rest and will accumulate, eventually forming a clump from which a new star will form. This cycling of gas between low and high density state is sufficiently rapid that chemistry in the gas never achieves steady-state. In addition, the changes of state that occur may be so abrupt that shocks occur. A model to explore the chemistry in such a situation must, therefore, take account of the variety and duration of the physical conditions experienced by a parcel of gas throughout the cycle. Such a study must follow the evolution of the chemistry through a succession of these events.

17.3. Physical and chemical processes in the model

17.3.1. Accretion and the formation of molecular mantles

Dense interstellar clouds are generally cool, and the dust grains mixed with the gas are also normally at a low temperature, typically about 10 K. Laboratory experiments and theoretical studies both suggest that heavy molecules, on collision with the dust, will freeze out to form a molecular mantle. While this phenomenon was largely ignored in early interstellar work, the detection of molecular mantles of amorphous H_2O ice (probably containing NH_3) and of solid CO by means of solid state band absorptions, had lead to a re-evaluation of these ideas and an acceptance that accretion of molecules onto dust grains must be an efficient process. Of course, there may also be processes which continually return solid material to the gas phase: photodesorption, cosmic ray heating of mantle material, and reactions leading to desorption have all been proposed as mantle inhibiting mechanisms. Any theory must, however, note that substantial mantles are seen on many lines of sight and that these appear to be consistent with growth by simple unimpeded accretion on a reasonable timescale. In the cycle model of molecular clouds it may be assumed immediately that mantle removal occurs whenever dense molecular gas interacts with stellar winds. Thus, star formation provides a natural mechanism for mantle limitation, and the local star formation rate is the crucial factor.

The effect on the gas phase chemistry on loss of molecules by accretion onto dust is quite dramatic. Assuming unit sticking probability and a canonical gas : dust ratio, the timescale for substantial accretion of heavy molecules onto dust is about $3 \times 10^9/n_H$ yr, where $n_H(\mathrm{cm}^{-3})$ is the total hydrogen nucleon density. In typical molecular clouds where n_H may be $\sim 3 \times 10^3$ cm^{-3}, this timescale is on the order of one million years. Since it takes up to 30 times longer than this to achieve chemical steady-state, it is clear that gas phase chemistry is constrained to be 'young'. We shall see in Section 17.4, for example, that high C^0 atom abundances occur because there is insufficient time for the full processing of C^0 into CO.

There remains the intriguing possibility that some clumps do not proceed to form a star which would disrupt them. If supported against collapse, perhaps by magnetic fields, such clumps might exist as a gas of H_2, He and mantled grains. Such evolved clumps would be unobservable in the usual molecular gas phase lines, though they might be detectable in the absorption bands of solid state species.

17.3.2. Mass loading of winds and erosion of dense clumps

A fast, neutral stellar wind impinging on a dense clump of gas will suffer a shock. If the shock is sufficiently strong, the wind will be partially ionized and this higher level of ionization may be mixed with the flow. A wind of low density, hot gas flowing over cool dense molecular gas is unstable at the interface. The turbulent mixing that may occur transports molecular material (not necessarily through a shock) into the flow. The efficiency of turbulent mixing across a magnetic boundary layer is highly uncertain, ambipolar diffusion or even reconnection driven by ambipolar diffusion may be efficient. At the same time, hot material mixed with the cool molecular gas will cause chemical changes in that gas. Thus, the interface is a complex region where cool material is accelerated and exposed to hot gas: in such a situation molecular mantles may be rapidly removed. As mass loading continues, the wind velocity declines. Thus, ionization of the wind occurs nearest to the star, and helps to destroy molecules released from the mantles of dust grains that have been eroded from the clumps.

17.4. Calculations of mass loading and clump chemistry

To illustrate the effects of the processes described in Section 17.3 we describe briefly the chemistry arising in a mass loaded wind brought to local rest by a reverse shock, in a stellar wind-blown bubble, and accumulating in a dense shell. This flow structure is shown schematically in Figure 17.2. We assume that the wind encounters and erodes many clumps of dense gas and dust before it is decelerated. The earliest encounters of wind with clumps will generate shocks powerful enough to ionize the wind. The erosion of clumps loads the wind with atomic and molecular material so that, in general, the gas that ultimately accumulates after passing through the weak reverse shock, which arrests the flow, is composed of partially ionized hydrogen, helium, heavy atoms and molecules, and dust grains.

The weak reverse shock causes a modest temperature rise to 1000–2000 K or greater depending on the shock speed. Conventional shock chemistry involving

Figure 17.2. A schematic representation for the flow structure of a wind-driven shell in a clumpy medium.

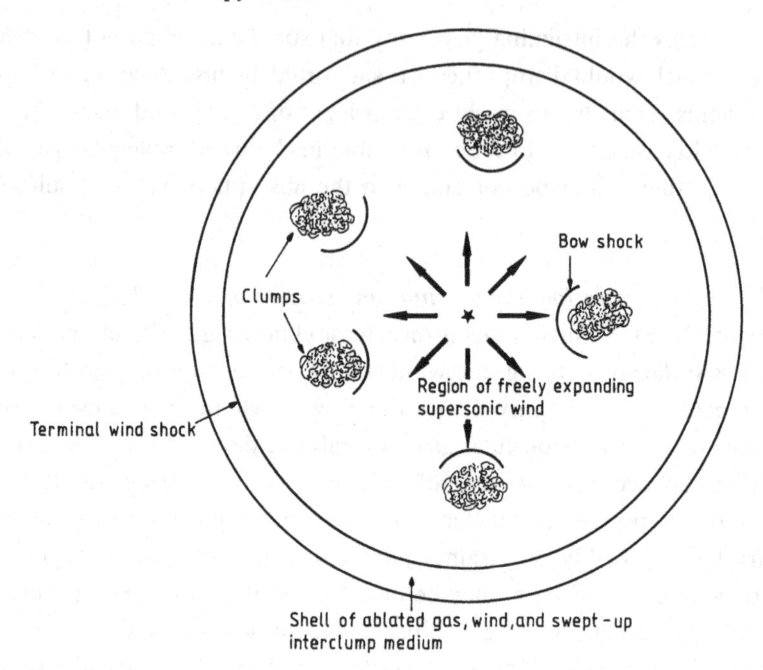

Figure 17.3. Chemical evolution of a shocked mass loaded wind. The model parameters are given in Table 17.3. This calculation has v_s = 6 km s^{-1}, $n(H^+)$ = 7.5 cm^{-3}, $n(He^+)$ = 0.75 cm^{-3} and n_c = 5 × 10^3 cm^{-3}. (From Charnley *et al.* (1988a).)

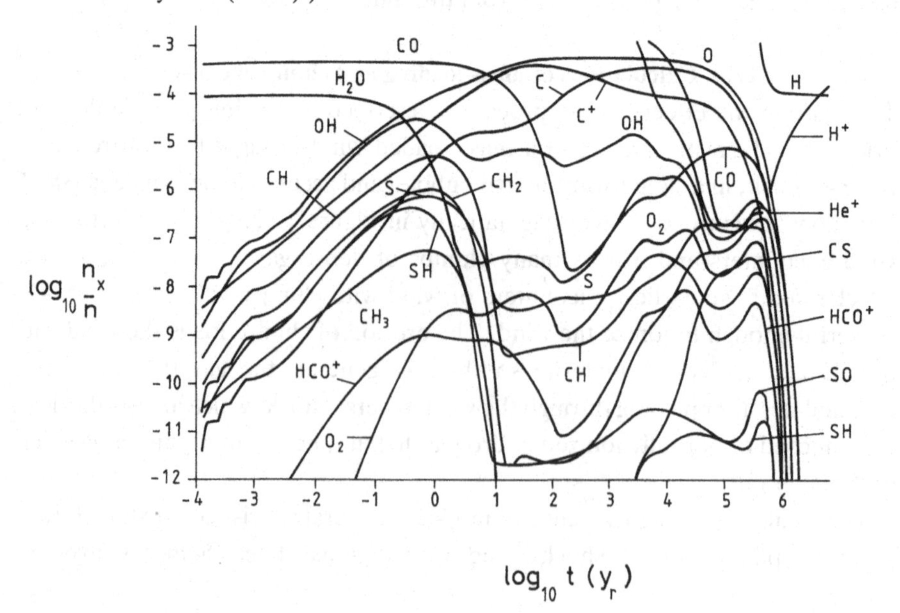

Table 17.3. *Representative preshock parameters for modelling a shocked mass loaded flow*

$n(H_2)$ (cm^{-3})	T (K)	v_s ($km\ s^{-1}$)	$n(H^+)$ (cm^{-3})	$n(He^+)$ (cm^{-3})	n_c ($10^3\ cm^{-3}$)	τ_c (yr)
150	10	6 or 11	7.5–0.15	0.75–0.015	5–50	$\sim$ few $\times 10^2$

The shock speed is v_s, τ_c is the cooling time until the prescribed density compression limit, n_c, is attained. The chemical initial conditions are typically: all hydrogen molecular with solar abundances of the heavy elements, present either in atomic form or in saturated molecules and CO.

mildly endothermic reactions occurs in this phase. The shocked gas cools rapidly, accompanied by substantial density enhancements. Table 17.3 indicates typical physical values appropriate for such a model. Cool phase chemistry proceeds in the normal way in this dense gas, to be interrupted when accretion of heavy atoms and molecules depresses their gas phase abundances to insignificant levels.

The results of an illustrative calculation are shown in Figure 17.3. These are given for a limiting case in which, preshock, all the heavy atoms are contained in hydrogenated species and carbon monoxide. This is representative of mass loading a wind from clumps which have experienced a complete freeze-out of the heavy elements, and in which no mantle removal occurs either at the wind–clump interface or by sputtering in the bow shocks. Here, the hot postshock phase is evidenced in the steady temporal rise of simple molecules such as OH and CH, for times up to about one year postshock. Then, the high abundances of ions created in the wind attack these molecules and even the CO (which in this calculation survives even in the flow regions) is substantially destroyed. However, from about 100 yr postshock, conventional cool dense gas phase chemistry increases molecular abundances from this initial state. This continues until the effect of accretion becomes significant, at times $\sim 10^6$ yr postshock. Thereafter, heavy molecules and atoms are efficiently removed from the gas.

The chemistry here is particularly sensitive to the value of the density compression limit, since this controls the accretion timescale. For higher densities the accretion timescale is shorter so that the time available for conventional cold cloud chemistry is even more severely truncated. Thus, if postshock evolution attains higher densities, molecular abundances will be *reduced* from the values illustrated, the reverse of expectation.

Other parameters, necessary for the implementation of the model, turn out not to be so crucial. Variations in the level of ionization attained in the preshock era appear not to affect the chemistry to a significant degree, as long as the enhanced ion abundances are above some minimum value. The inclusion of a radiation field from the low mass stars also affects the chemistry only slightly. The temperature attained in the reverse shock does modify the chemistry but these effects are relatively minor and the postshock evolution has the general form of Figure 17.3.

The main conclusion of this study is that chemistry occurring in denser clumps formed during the cooling of the mass loaded wind never achieves steady-state. The accretion of molecules onto grain surfaces effectively terminates chemical activity at a 'young' stage. The products of this chemistry are themselves subsequently affected by the clump erosion process. One effect particularly noticeable in the exploratory calculation illustrated is that C^+, C^0 and CO are all comparable in abundance at times $\sim 10^6$ yr. Such a result is apparently required by the observations; conventional chemistry in static clouds cannot account for it.

17.5. Dynamical cycling scenarios for B5

The cloud dynamics (shocks, collapse, stellar wind interactions) and the cloud chemistry will be intimately linked. One may therefore ask what is the effect of continual recycling, due to low mass star formation, upon the chemical evolution of a dense molecular cloud. Since a chemical steady-state will not be achieved (Section 17.3.1) it is not clear that the chemical concentrations in a parcel of gas will repeat at similar times between any two cycles. Mathematically, the question we wish to answer is this: what is the effect of short timescale (relative to the chemical equilibrium time) perturbations on the long-time behaviour of the system of differential equations governing the chemical evolution? Are chemical abundances repeated between cycles, exhibiting a 'limit cycle' behaviour, or will it be impossible to predict abundances, implying a transition to 'chemical chaos'?

From observations, and from theoretical considerations of the likely effects of low mass star formation, it is possible to construct simple dynamical–chemical models of the cycling. In the following subsections we describe two such models for a dense cloud chemistry which never attains steady-state. In both these models one follows the chemistry in a parcel of gas as it collapses, expands, accretes onto grains, and is shocked. In each of these models species are allowed to stick, and in some cases react, on the dust, the grains being 'cleaned' by periodic shock processing. The differential equations describing the chemical evolution are identical for both models, however the imposed analytic density and temperature profiles, and hence the cycling period, are quite distinct.

17.5.1. Rapid cycling model

From $C^{18}O$ observations of B5, Goldsmith *et al.* (1986) estimate the time for recycling of material between dense clumps and the more diffuse interclump medium as $\sim 6 \times 10^5$ yr. Thus the density varies rapidly during the chemical evolution and several such cycles may occur in a cloud lifetime. A schematic diagram representing the density structure throughout one such rapid cycle is shown in Figure 17.4. This scenario is somewhat different from that discussed in Section 17.4 in that the clump material is swept out by a star in the centre of the clump. Gas starting in the interclump phase undergoes an isothermal collapse, in about a free-fall time, to form a clump having $n(H_2) \sim 7 \times 10^3$ cm^{-3}, as inferred from observation. This cool clump phase persists for $\sim 5 \times 10^5$ yr until it is terminated by a shock driven by a stellar wind-blown bubble. In this shock the neutral gas is heated and compressed to temperatures and densities consistent with those of a C-type MHD shock. In this shock all grain mantles are assumed to be removed instantaneously and any accreted atoms and the radicals OH, SH, CH, CH_2, CH_3 are removed in water, hydrogen sulphide and methane, corresponding to hydrogenation on grain surfaces. After a time consistent with the radiative cooling of such a shock, the shocked clump material moves with the expanding wind-driven shell. The density falls continously and the material eventually becomes incorporated in the interclump medium which forms the starting point of the subsequent cycle. Ten such cycles are considered, each of 6×10^5 yr duration, giving a total processing time of 6×10^6 yr.

Figure 17.4. A schematic representation of the fast cycle model. The gas remains at 10 K as n_H increases from 2×10^3 cm^{-3} at point A to 1.4×10^4 cm^{-3} over an interval of 4×10^5 yr and as n_H stays at 1.4×10^4 cm^{-3} for another 1×10^5 yr until point B is reached. The gas is then shocked, and instantaneously T and n_H rise to 800 K and 4×10^4 cm^{-3} and all heavy elements which accreted onto the grains are injected into the gas phase. T and n_H are then constant for 1×10^3 yr. Over the subsequent interval of 1×10^3 yr, the gas cools and becomes more dense until $n_H = 10^5$ cm^{-3} and $T = 10$ K at point C. T remains at 10 K as n_H drops for 10^5 yr to 2×10^3 cm^{-3}. (From Charnley *et al.* (1988b).)

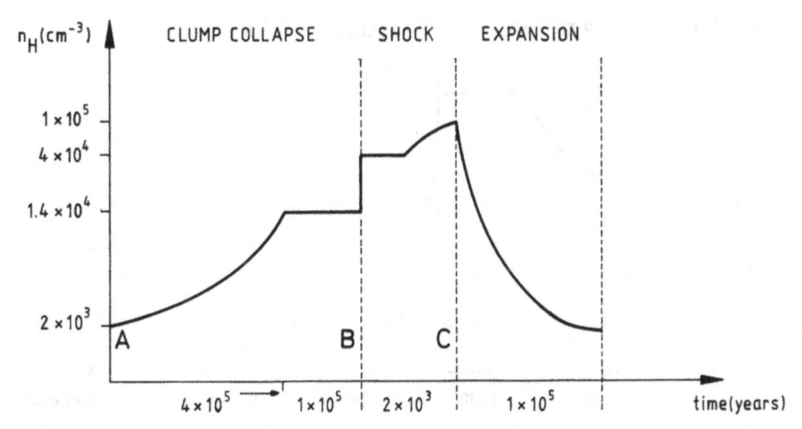

17.5.2. *Slow cycling model*

The rapid cycling of the previous model requires that the winds from young T Tauri stars carry sufficient momentum to alter the dynamics on short timescales. Simple estimates suggest that the observed spatial structure in B5 may not be generated over such timescales by wind-blown bubbles. They may not carry sufficient momentum and the timescales for the bubbles to disrupt the cloud, and the corresponding cycle period, could be significantly longer. The density structure for such a slow cycling model is shown schematically in Figure 17.5. The period for the cycle is $\sim 6 \times 10^6$ yr, an order of magnitude longer than that of Section 17.5.1. As before the cycle begins with a collapse, although in this case much longer than free-fall due to the assumed effects of magnetic fields. Once a clump has formed it is eventually eroded by the wind from a nearby star producing a mass loaded flow as discussed in Section 17.3.2. During this ablation phase the density falls and the ions are injected at constant volume rates into the clump gas. For simplicity the interface is assumed isothermal and clump material is not assumed to be heated as it is ablated and picked up by the wind. The accreted

Figure 17.5. A schematic representation of the slow cycle model. The gas remains at 10 K as n_H increases from 2800 cm^{-3} at point A to 1.4×10^4 cm^{-3} over an interval of 4×10^6 yr and for a further 2×10^6 yr, during which n_H stays at 1.4×10^4 cm^{-3}, to point B. The gas is then ablated from the clumps and traverses the interior of the wind-blown bubble for a total of 2.5×10^4 yr. During the first 5×10^3 yr of that interval, H^+ and He^+ mix with the gas and n_H drops to 400 cm^{-3} at point C. During the final 2×10^4 yr of the interval, n_H is constant to point D. The gas then passes through a shock at the interior of the shell, and instantaneously n_H and T increase to 1200 cm^{-3} and 4000 K and the heavy elements which have accreted onto the grains are injected into the gas phase. The gas cools to 10 K and n_H increases to 2800 cm^{-3} over an interval, τ_c, which is calculated with a one-fluid nonmagnetic shock model in which the density is limited artificially. (From Charnley *et al.* (1988b).)

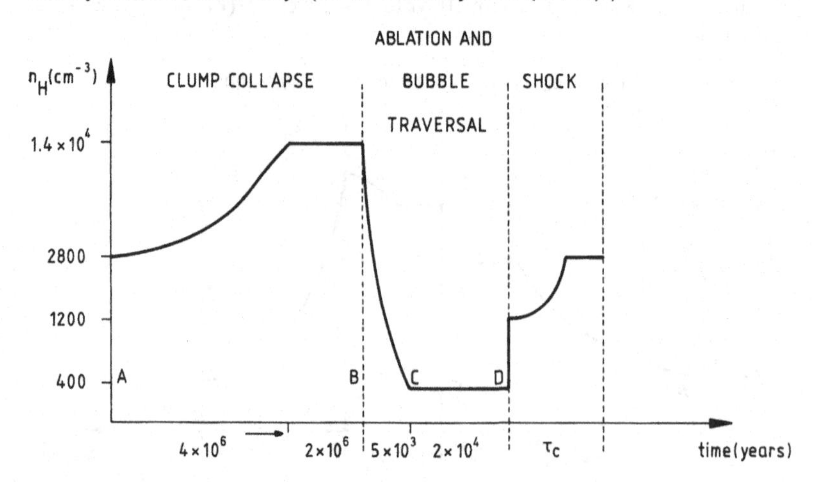

mantles are thus not removed at this point in the cycle. In fact, more detailed calculations of the wind–clump interface show that hot ions from the wind can heat the clump material to temperatures of ~ few $\times 10^3$ K, in a few thousand years, although we ignore any effects of this heating here.

Assuming that the process is efficient, mixing of appreciable abundances of He^+ rapidly destroys molecular clump material as the wind traverses the bubble interior. After ion injection has terminated H^+ and He^+ are removed by electron recombination and, as long as the bubble traversal time is sufficiently short (~ few $\times 10^4$ yr), enhanced abundances of these ions will be present in the postshock gas. The model here thus produces self-consistently the preshock fractional ionization and gas abundances for the calculation described in Section 17.4. Immediately postshock the grain mantles formed during the clump phase are dumped into the gas and the subsequent shock cooling performed using a one-dimensional steady shock model. This has to be done self-consistently since the cooling rate, and also the rates of several important endothermic reactions (e.g. $C + H_2 \rightarrow CH + H$) will be dependent on the preshock abundances of H^+ and He^+ and on their rates of removal in the postshock gas.

The most novel feature of the slow cycling model is then the mixing of stellar wind material with cool clump material. After the shocked gas has cooled, the gas collapses again to form the next generation of clumps. The cycling is followed through three such clump generations yielding a total processing time of 1.8×10^7 yr.

17.6. Cycles in B5

The two cycling scenarios described in Section 17.5 have been examined using the same dense cloud chemical network of 51 species and 512 reactions. The first cycle of both models begins with a collapse from the same diffuse atomic gas. The results for ten rapid cycles and three slow cycles show that, for both models, following the initial cycle there is not much variation in the chemical abundances at the same point between cycles. We thus conclude that approximate 'chemical limit cycles' develop in models of cyclic processing, each chemical species tends towards a limiting abundance value.

Figures 17.6 and 17.7 show the chemical evolution of selected species during the final cycle of the rapid and slow models respectively. In Figure 17.6, near the end of the clump phase, molecules are lost from the gas phase by accretion. In the postshock gas the accreted (and grain-catalysed) mantles are removed, leading to enhancements of the preshock (clump) abundances of saturated molecules in subsequent cycles. The peak shock temperature is ≈ 800 K and is sufficiently low to reduce the contribution of several endothermic rections in this phase. Figure 12.7, for the slow model, shows that more accretion occurs, removing substantial amounts of gas species onto the dust (e.g. CO). During the ablation phase the gas component, remaining after the clump evolution, is attacked by hydrogen and helium ions mixed from stellar winds leading to significant molecule destruction. As the mass loaded wind traverses the bubble interior, the excess ions recombine with electrons, also mixed with the clump material, and form again heavy molecular species. On entering the hydrodynamic reverse shock, the gas consists mainly of H_2, He, H^+, He^+, C^0, C^+, O^0, S^+, S^0, and dust grains with thick mantles. In the hot postshock gas the mantles are removed adding high abundances of species such as H_2O, H_2S, CH_4 and CO to the above mixture. The enhanced helium ion abundance can depress molecule formation in the hot gas ($T_{peak} \approx 4000$ K). As the gas cools and the ions are removed one obtains the initial abundances of the next cycle.

Unfortunately no detailed chemical abundance determinations have been made for B5. The proximity of this dark cloud offers the promise of detailed mapping of the spatial variations of various molecules and ice features. Predicted abundances from the models should correspond to well-defined regions in B5 and may allow testing of the proposed dynamics. Table 17.4 shows a comparison of the predicted abundances of species, some of which should be observable in B5, at similar points for both the rapid and slow cycling models.

Whether or not the precise details and assumptions of these models are correct (e.g. rate coefficients, density and thermal structures), it is clear that they

Figure 17.6. Chemical evolution during the tenth cycle of the fast cycle model. The time axis is broken where the gas is shocked. Different points in the cycle are marked A, B and C as in Figure 17.4. (From Charnley *et al.* (1988b).)

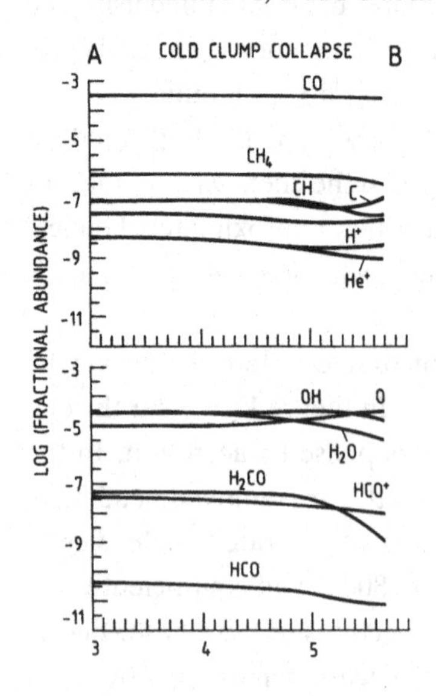

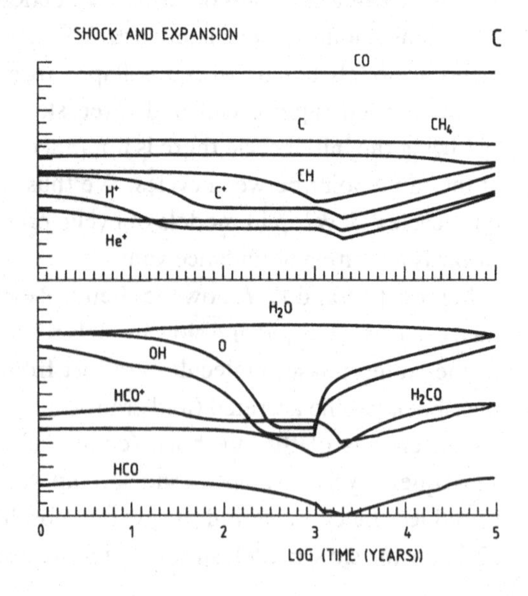

Figure 17.7. Chemical evolution during the third cycle of the slow cycle model. The time axis is broken where ablation occurs and where the gas is shocked. Different points in the cycle are marked A, B, C and D as in Figure 17.5. (From Charnley *et al.* (1988b).)

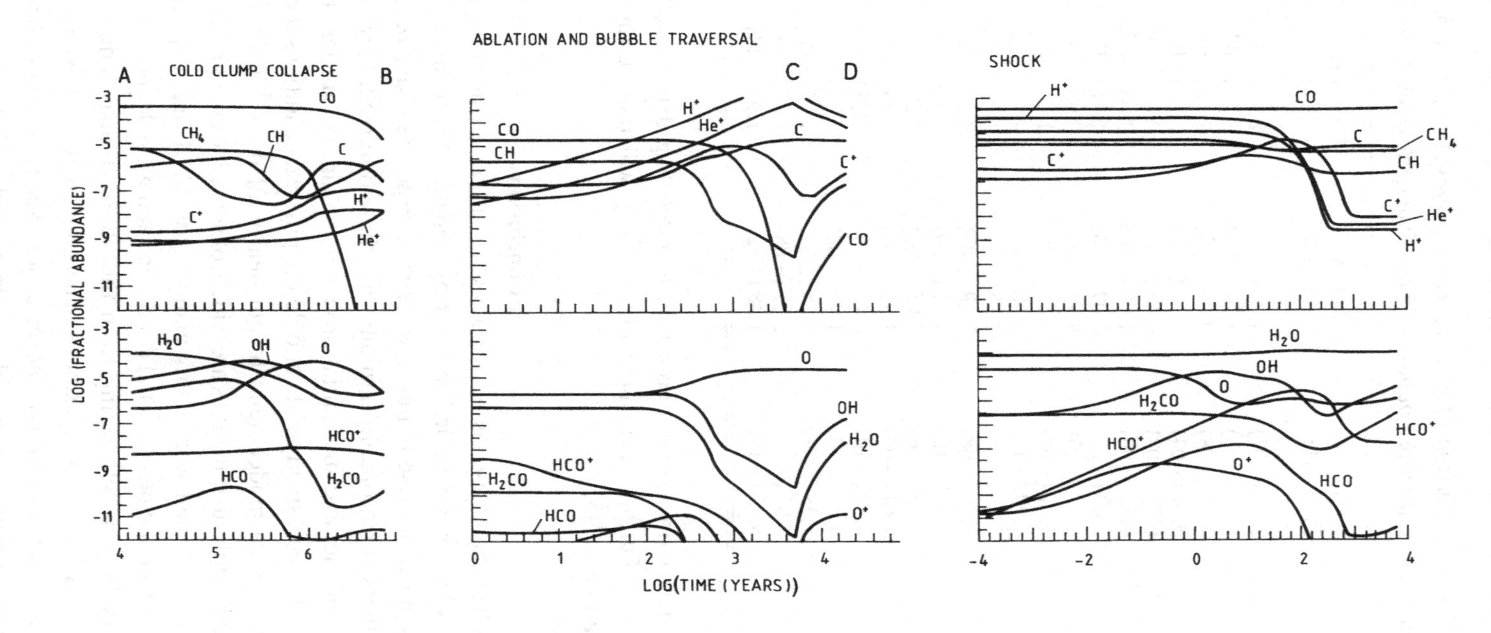

Table 17.4. *Comparison of abundances at points A and B in each of the models. The abundances are those obtained in the final cycle of each model*

Species	Point A Fast cycle	Point A Slow cycle	Point B Fast cycle	Point B Slow cycle
e^-	3.46 (−07)	4.24 (−07)	2.39 (−07)	4.45 (−07)
C	8.48 (−08)	8.24 (−06)	1.26 (−07)	2.54 (−07)
C^+	3.32 (−08)	9.08 (−09)	2.30 (−08)	7.17 (−08)
CO	3.51 (−04)	3.44 (−04)	2.82 (−04)	1.61 (−05)
CH	6.35 (−08)	6.43 (−07)	3.03 (−08)	2.28 (−06)
O	9.30 (−06)	1.05 (−06)	3.05 (−05)	1.94 (−06)
OH	2.62 (−05)	3.09 (−06)	1.31 (−05)	2.27 (−06)
H_2O	2.58 (−05)	9.11 (−05)	2.98 (−06)	5.66 (−07)
HCO^+	3.81 (−08)	1.53 (−08)	1.00 (−08)	4.25 (−09)
HCO	3.90 (−11)	3.50 (−12)	7.46 (−12)	2.66 (−12)
H_2CO	3.78 (−08)	2.83 (−07)	1.08 (−09)	1.36 (−10)
S	8.49 (−08)	3.18 (−07)	7.24 (−08)	3.03 (−08)
H_2S	5.66 (−12)	1.41 (−09)	3.34 (−12)	1.99 (−12)
CS	1.05 (−06)	1.05 (−06)	9.29 (−07)	7.78 (−08)
OCS	2.64 (−07)	1.81 (−08)	1.55 (−07)	1.29 (−09)
HCS^+	4.87 (−10)	2.13 (−10)	2.37 (−10)	7.58 (−11)
H_2CS	4.68 (−10)	1.49 (−08)	7.22 (−11)	2.59 (−13)
SO	1.38 (−08)	3.66 (−09)	1.30 (−08)	1.33 (−09)
SO_2	1.03 (−07)	3.71 (−08)	8.18 (−08)	2.79 (−10)
CH^+	3.12 (−14)	1.48 (−12)	1.03 (−14)	4.89 (−13)

represent a more realistic approach to modelling dense cloud evolution than heretofore. Further refinement of this work will require input from detailed observations of B5.

17.7. Conclusions

Interstellar chemistry has been successful in identifying some of the routes by which molecules are formed and destroyed in the interstellar medium. The simple models in which these routes have been incorporated have been able to predict abundances in reasonable harmony with those observed. However, such models are not unique, and substantial problems remain. Is sufficient time available to achieve chemical steady-state and do the physical conditions remain uniform for this time? Such models tend to put nearly all the C^0 in CO; why is this not observed? Molecular mantles are widely observed: can this be incorporated in the global models, and what limits mantle growth? If time-dependent models are used, what should be the initial conditions and what defines the age of a cloud?

The cycling models arising from the observational studies of B5 provide a natural answer to these questions. The cyclic nature of events ensures that the question of initial conditions never arises. The accretion of heavy species onto dust grains curtails in a natural way the period available to gas phase chemistry and restricts its products to those characteristic of a 'young' chemistry. The accretion timescale is, for typical clouds, close to the collapse timescale and star formation may be expected to follow the accretion process. Winds generated by the star provide a natural mechanism of mantle limitation and reduction of chemical complexity. They also drive the superthermal motions observed.

B5 has therefore proved to be a most instructive case study. Given that infrared compact sources have been detected in many molecular clouds it is likely that B5 is a paradigm of a whole class of molecular clouds. It demonstrates clearly the need for dynamics and chemistry in molecular clouds to be intimately linked, indeed regarded as complementary aspects of the cloud state. The calculations in Sections 17.5 and 17.6 show, in particular, that chemistry after several cycles is repeatable, even if the cycle period is much shorter than the steady-state time. Chemical chaos does not occur, and chemical limit cycles are attained. This result is of the greatest importance for both observational and theoretical studies: it shows that the chemistry is a proper tracer of conditions in molecular clouds and that the goal of interstellar chemistry is attainable.

REFERENCES

Charnley, S. B., Dyson, J. E., Hartquist, T. W. and Williams, D. A. (1988a). *MNRAS*, **231**, 269.

Charnley, S. B., Dyson, J. E., Hartquist, T. W. and Williams, D. A. (1988b). *MNRAS*, **225**, 1257.

Goldsmith, P. F., Langer, W. D. and Wilson, R. W. (1986). *Astrophys. J.*, **303**, L11.

Hartquist, T. W. and Dyson, J. E. (1988). *Astrophys. Space Sci.*, **144**, 615.

Keene, J., Blake, G. A., Phillips, T. G., Huggins, P. J. and Beichman, C. A. (1985). *Astrophys. J.*, **299**, 967.

Norman, C. and Silk, J. (1980). *Astrophys. J.*, **238**, 138.

Prasad, S. S. and Huntress, W. T., Jr (1982). *Astrophys. J.*, **260**, 590.

Williams, D. A. and Hartquist, T. W. (1984). *MNRAS*, **210**, 141.

18

Molecular cloud structure, motions, and evolution

P. C. MYERS

Harvard-Smithsonian Center for Astrophysics, Cambridge, Massachusetts, USA

18.1. Introduction

This chapter describes some aspects of the structure, motions, energies, and evolution of molecular clouds, from the viewpoint of observations and simple models. The emphasis is on observations of nearby clouds, and on physical, rather than chemical properties. The approach is partly pedagogical and partly a summary of recent results. Recent reviews on related subjects are those of Blitz (1987; diffuse molecular clouds), Scalo (1987; turbulence), Mouschovias (1987; magnetic effects), Shu, Adams and Lizano (1987; cloud physics and star formation), and Larson (1988; large-scale aspects of cloud and star formation). A review with a similar viewpoint to this one, but with older information, appears in Myers (1987).

We consider a molecular cloud to be a collection of interstellar gas and dust, whose gas has a substantial molecular component, and whose mean density inside an observationally definable boundary exceeds that outside the boundary. In this chapter we do not discuss galaxies, or molecular clouds in galaxies other than the Milky Way. Further information on these subjects is in Chapter 2.

The main constituents of molecular clouds, other than molecules, are stars, the 'cores' or condensations that form stars, dust grains, ions, and atoms. In this chapter we focus on the role of stars and cores.

The material in this chapter is organized into discussions of cloud structure and kinematics (Section 18.2), cloud energetics (Section 18.3), and time scales and evolution (Section 18.4).

18.2. Cloud structure and kinematics

18.2.1. Observational techniques and limitations

Our knowledge of molecular cloud structure and motions is derived mainly from maps and spectra of molecular emission lines, made with radio telescopes operating at centimeter and millimeter wavelengths. For filled aperture ('single-dish' telescopes, maps are usually made one pixel at a time, by sequentially stepping the telescope beam a fixed number (usually one) of FWHM beam widths, in a two-dimensional grid. The smallest visible angular scale is that of the diffraction-limited beam, or 0.2–2 arcmin for most telescopes, depending on line frequency and diameter of the primary reflector. For a cloud at 500 pc, the smallest visible linear scale is then 0.03–0.3 pc. The largest visible angular scale is set by the step size and the number of times the telescope beam is stepped: this depends on many factors, including the signal-to-noise ratio of the spectral line, the available observing time, and the strategy adopted by the observer. A 'minimal' map strategy is to map the emission to define the half-maximum intensity contour, e.g., one beam beyond the half-maximum contour. This method can underestimate the extent of a cloud, especially if its emission consists of multiple peaks separated by deep valleys. Such a pattern of emission is often called a 'complex' or 'cloud complex'. Most molecular cloud line maps are 3–30 beams in extent, or 0.1–10 pc for clouds at 500 pc.

For maps made with interferometers (or 'aperture synthesis' telescopes), the smallest visible angular scale is set by the fringe spacing in the multiple aperture diffraction pattern. The 'synthesized beam' is then usually a few arcseconds for most molecular cloud maps made with the VLA or with millimeter-wavelength interferometers. This corresponds to 0.008 pc at 500 pc. Most interferometer maps are made without stepping of each antenna's primary beam with respect to the cloud, in contrast to the normal single-dish map procedure. Then the largest visible angular scale is about half the width of the primary beam. Emission on larger scales is 'resolved out' and is not detected. For the 25-m-diameter VLA antennae at the 1.3-cm wavelength lines of NH_3, or the 10-m-diameter Owens Valley Interferometer antennae at the 0.26-cm line at CO, this largest visible scale is about 25″, or 0.06 pc at 500 pc.

At present, most of the available information on molecular clouds comes from single-antenna telescopes. We rely here on data from observations of the 2.6-mm $J = 1 \rightarrow 0$ rotational line of ^{12}CO to define molecular clouds, and on data from observations of the 1.3-cm $(J, K) = (1, 1)$ rotation–inversion line of NH_3 to define 'dense cores', or condensations within clouds. Use of these lines is convenient because they are extensively observed. However, other lines, such as the rotational lines of ^{13}CO, $C^{18}O$, and CS, also reveal molecular cloud properties. We emphasize that in a given cloud the density, column density, line width, size, temperature, and mass indicated by a line varies from one line to the next, primarily because each line traces a different mean density of collision partners (mainly H_2 molecules and He atoms) needed for excitation. In turn, this density defines a size scale, over which the quantities derived from the line observations are averages. The CO and NH_3 lines are useful because they trace size scales separated by a large factor – 10^1–10^3 – and because these size scales reveal several aspects of cloud physics and star formation. For a discussion of 'clumps' with an intermediate size of ~1 pc traced by the ^{13}CO line, see Elmegreen (1985). For a discussion of cloud structure in terms of hierarchical organization, see Scalo (1988).

18.2.2. Types of clouds and cores

Table 18.1 summarizes properties of three 'types' of molecular clouds – diffuse, dark, and giant clouds, as revealed by ^{12}CO line observations. The characteristic size R is the FWHM size of the emission line map. The mass M within R is obtained by integrating the column density in each beam over the map area within R. The mean number density within R is obtained from M, R, and a model of a uniform sphere. The FWHM line width Δv is due mainly to Doppler shifts in the molecular motions and is generally greater than the thermal width by a significant factor. The gas kinetic temperature T is derived from the brightness temperature of the line: the ^{12}CO $J = 1 \rightarrow 0$ line is optically thick, so the line brightness is close to the transition excitation temperature, which in turn is a measure of the kinetic temperature. For diffuse clouds, T is much less certain than for dark and giant clouds.

The diffuse, dark, and giant clouds progress in size roughly from 1 to 10 to 100 pc, and in mass roughly from 10^2 to 10^4 to 10^6 $M_\odot$. The diffuse clouds have mean visual extinction $\lesssim$1 mag and can be seen on Palomar Sky Survey photographs by their faint reflection of starlight. They are most easily recognized at high galactic latitudes, where they reflect the diffuse starlight from the plane. Diffuse clouds can vary appreciably in their molecular content; they have been classified as 'CO-rich' and 'CO-poor' by Lada and Blitz (1988). Those discussed here are known by their CO properties. Some diffuse clouds are much larger than the representative value ~1 pc given in Table 18.1. A few diffuse clouds are known to have low-mass cores from their NH_3 emission (L1489 and Bernes 48, Myers and Benson (1983), and MBM21, Stacy, Myers and de Vries (1988)). A few are also known to have associated low-mass stars. Also, a large number of diffuse clouds have small, opaque knots on optical photographs, indicating the probable presence of low-mass cores.

Dark clouds have mean extinctions of several magnitudes and are therefore recognized more by their absorption of background starlight than by reflection. Like the diffuse clouds, dark clouds have a great range of size and visual appearance. The properties given in Table 18.1 refer to the well-studied complexes in Taurus and Ophiuchus. Many smaller and more isolated dark clouds ('globules') are also known (Leung 1985). Dark clouds also vary widely in the number and distribution of their cores and stars: the Taurus–Auriga complex has

Table 18.1. *Molecular cloud properties based on the* J = *1* → *0 Line of* ^{12}CO

Type	R (pc)	n (cm^{-3})	M ($M_{\odot}$)	Δv (km s^{-1})	T (K)	Cores and stars
Diffuse	0.3–3	30–500	0.5–100	0.7–1.5	10?	Low-mass
Dark	3–10	10^{2-3}	10^{3-4}	1–3	10	Low-mass
Giant	20–100	10–300	10^{5-6}	5–15	10–20	Massive (and low-mass)

more than 100 visible T Tauri stars (Cohen and Kuhi 1979, Herbig and Bell 1988) and a comparable number of pre-main-sequence stars detectable via their x-ray emission (Walter *et al.* 1988). In contrast, the dark cloud complex in Aquila, whose most prominent member is Lynds 673, has only a few T Tauri stars (Herbig 1962) and NH_3 cores (Benson and Myers 1988). The proportion of massive (OB) stars in nearby dark clouds is low: a few examples are AB Aur in Taurus-Auriga; BD + 30° 549 in Perseus; HD 147889 in Ophiuchus; TY CrA in Corona Australis; and HD 200775 in Lynds 1172. It is unclear whether the general tendency for low-mass clouds to make low-mass stars reflects a difference in relevant physical conditions such as magnetic fields, temperature, density, structure, and ionization between low-mass and massive clouds, or rather merely a difference in the mass available to make stars of all types.

The 'giant' molecular clouds are known from CO observations of optical emission nebulae such as M42 (Orion) and its associated dark cloud Lynds 1641; from CO observations of more distant radio HII regions such as W49; and from CO surveys of the galactic plane. For giant clouds, the mean density n is significantly less than the ~300 cm^{-3} needed for excitation because the volume filling factor is typically much smaller than unity: the cloud is 'clumpy'. The complex, clumpy and filamentary structure of the CO emission in the Orion giant cloud illustrates this property; it is evident in the maps of Schloerb *et al.* (1986) and Bally *et al.* (1987).

Properties of giant clouds are summarized by Scoville *et al.* (1987) and Solomon *et al.* (1987). Stellar content is discussed by Myers *et al.* (1986) and Waller *et al.* (1987). The association of giant clouds and massive stars is well known, since the high luminosity of OB stars makes their identification relatively easy. The presence of lower-mass and lower-luminosity stars is expected but is much less certain. Many T Tauri and other low-luminosity stars are known in Orion (Herbig 1982, Strom, Strom and Newton 1988), but in the more distant clouds where other OB stars are known, current sensitivity limitations prevent detection of T Tauri stars. Similarly, the incidence of low-mass cores in giant clouds is uncertain: few are known (e.g., Harris *et al.* 1983), and poor sensitivity makes detection difficult.

Table 18.2. *Dense core properties based on the* (J, K) = *(1, 1) line of* NH_3

Type	R (pc)	n (cm^{-3})	M ($M_\odot$)	Δv ($km\ s^{-1}$)	T (K)	Stars
Low-mass	0.05–0.2	10^{4-5}	0.3–10	0.2–0.4	10	T Tauri
Massive						
(large)	0.3–0.6	10^{4-5}	30–10^4	1–2	10–30	OB
(small)	0.01–0.03	10^{6-7}	0.3–300	1–3	30–100	OB

A survey for NH_3 cores near Orion IRAS sources indicates that detected sources have significantly broader lines than in Taurus (Wouterlout, Walmsley and Henkel 1988). Thus, these Orion cores are probably more massive than their Taurus counterparts.

Table 18.2 summarizes the same properties as in Table 18.1 (size R, density n, mass M, line width Δv, and temperature T) but for dense cores within molecular clouds, based on NH_3 line observations. Here R and Δv are determined from maps and spectra as before; while n and T are generally deduced from excitation analysis of the $(J, K) = (1, 1)$, $(2, 2)$, and higher-excitation lines; and M is obtained from n, R, and the model of a uniform sphere.

The 'low-mass' cores are known in dark and diffuse clouds within a few hundred parsecs from the Sun. They are remarkable in their narrow, nearly thermal line widths, and in their relationship to low-mass stars. In 12 low-mass cores without stars analyzed by Benson (1986), the mean kinetic temperature is $T = 10$ K, and the mean FWHM line width is $\Delta v = 0.27$ km s^{-1}. Hence, the ratio of nonthermal to thermal velocity dispersion for a molecule of mean molecular weight (2.3 m_H, allowing for 10% helium) is only 0.5. In contrast, the same ratio is 14 for a massive core with $\Delta v = 2$ km s^{-1} and $T = 50$ K. Low-mass cores in dark clouds have associated low-luminosity ($\lesssim 10\ L_\odot$) IRAS sources in about half of the $\sim$100 cases examined by Beichman *et al.* (1986). This propensity of low-mass cores to form low-mass stars, and the similarity of the mass within the typical NH_3 core line map to 1 $M_\odot$, suggest that the typical molecule in a starless core has a relatively high likelihood of adding its mass to that of a forming star. Thus molecules in low-mass cores are 'destined for stardom', and low-mass core properties are often considered as initial conditions for models of low-mass star formation (e.g., Shu *et al.* (1987)).

Massive cores are listed in Table 18.2 as 'large' or 'small' according to whether they have been observed with filled aperture antennae (typical resolution 40–80″), or with the VLA (5–10″). As discussed earlier, interferometric observations tend to detect structures with high brightness contrast on the scale of the synthesized beam and to resolve out more extended emission. Thus, the small massive cores can be considered as those subregions of large massive cores having the highest brightness contrast, which in turn arises in regions with the highest absolute surface brightness. This high brightness probably arises from the locally increased gas temperature and/or density near an associated OBA star. (A similar distinction probably exists for 'large' and 'small' low-mass cores, but too few VLA detections are presently available.) Massive cores generally have greater line width, temperature, and mass than do low-mass cores. Massive cores also tend to have associated OBA stars – more massive and much more luminous than the GKM T Tauri stars seen in or near low-mass cores.

18.2.3. Internal motions

The internal motions in molecular clouds are complex, and our knowledge of them is still relatively crude. The most prevalent internal motion is evident in the supersonic line widths in all molecular clouds. This apparently chaotic activity cannot be explained easily by systematic radial or rotational motions on the scale of the cloud. The line broadening is usually called 'turbulent' and is modeled as a superposition of random motions. The supersonic motions in molecular clouds have several interesting properties, first noted by Larson (1981) and extended by Leung, Kutner and Mead (1982), Myers (1983), Dame *et al.* (1986), Falgarone and Perault (1987), Solomon *et al.* (1987), Scoville *et al.* (1987), and Myers and Goodman (1988b). More than 100 clouds and cloud cores, ranging in size from $R = 0.1$ pc to 100 pc, typically satisfy virial equilibrium,

$$\frac{GM}{5R} = \frac{\Delta v^2}{8 \ln 2} \tag{18.1}$$

(this formulation is for an isolated uniform sphere, supported by random motions against self-gravity); and each of three additional trends,

$$\Delta v = aR^{1/2} \tag{18.2a}$$

$$n = bR^{-1} \tag{18.2b}$$

$$\Delta v = cn^{-1/2} \tag{18.2c}$$

where $a \approx 1.4$ km s^{-1} pc^{-1}, $b \approx 1800$ cm^{-3} pc, and $c \approx 59$ km s^{-1} cm$^{-3/2}$. Among Equations (18.2a), (18.2b), and (18.2c), only two are independent: $c = ab^{1/2}$. Further, Equation (18.1) (virial equilibrium) and only one of Equations (18.2a)–(18.2c) are independent: together these two generate the remaining two equations. A possible explanation of these trends in terms of magnetic support is discussed in Section 18.3.

Among systematic motions in molecular clouds, the most prevalent known are the molecular outflows associated with winds from young stars. First detected in the 2.6-mm line of ^{12}CO (Snell, Loren and Plambeck 1980), outflows are now also evident in lines of ^{13}CO, HCO^+, CS, H_2O, OH, SO, SO_2, H_2, and in the optical, radio, and far infrared continuum. Outflows have been detected over spatial scales ~0.01–1 pc, velocities ~1–100 km s^{-1}, and with kinetic energies 10^{42}–10^{47} erg. They are generally believed to be a common feature of very early stellar evolution and to play an important role in the way a star becomes optically visible. They are usually recognized by the presence of line 'wings' or 'shoulders' at velocities exceeding the gravitational escape speed, and often by a distinct spatial separation of their maps of high and low-velocity emission. In this latter case, they are called 'bipolar flows'. Detailed reviews have been given by Lada (1985), Rodríguez (1987), and Snell (1987). The wide variety of outflow phenomena is illustrated in the meeting proceedings edited by Pudritz and Fich (1988).

Cloud rotation can be identified by a continuous increase or decrease of the spectral line central velocity across the map or by a corresponding progression in the asymmetry of the line profile. The signature of rotation is thus similar to that of bipolar flows, but the gravitationally bound rotational velocities are significantly smaller than the gravitationally unbound outflow velocities. In most molecular clouds, the spectral line velocities vary by less than one line width and do not fit a well-defined rotational pattern. In clouds and cloud cores that show a distinct rotational pattern, the shift in velocity across the map usually is about one line width, or less. The corresponding 'velocity gradient' or angular velocity is 1–3 km s^{-1} pc^{-1}, or $0.3–1 \times 10^{-13}$ rad s^{-1} (Arquilla and Goldsmith 1986). This degree of rotation is too small to account, by itself, for the typical supersonic line width. Two possible trends in the data on rotating clouds need more investigation: (1) some cloud cores observed with interferometers, and thus with finer angular resolution, corresponding to size $\lesssim 0.1$ pc, have velocity gradients of order 10 km s^{-1} pc^{-1}, distinctly higher than at larger scales (e.g., Harris *et al.* (1983), Wadiak *et al.* (1985), Zheng *et al.* (1985)); and (2) cores with stars appear to have more evidence for rotation than cores without stars; some of these cases are suggestive of rotating circumstellar disks (Bieging 1984, Torrelles *et al.* 1986, Sargent and Beckwith 1987, Jackson, Ho, and Haschick 1988). Rotation is not a unique interpretation of a velocity gradient in a map: shear motions and other types of flow may be involved, especially if the rotational period implied by the data significantly exceeds the age of the cloud, or cloud core.

Radial infall is the least well known of all cloud internal motions, but the expectation that infall arises from gravitational collapse has generated great interest in detection of infall as a means of studying star formation. Reports of infall have been based on measurement of foreground gas at velocities redshifted with respect to a velocity taken as the stellar or systemic velocity. Reid *et al.* (1980) reported OH masers in front of the optically thick compact HII region W3(OH) at velocities greater by several kilometers per second than that of the H109α recombination line from the HII region. Scoville *et al.* (1983) presented CO absorption against the Becklin–Neugebauer star in Orion at ~10 km s^{-1} greater than the estimated stellar velocity. Walker *et al.* (1986) showed CS self-absorption in the $J = 5 \rightarrow 4$ line toward IRAS 16293–2422, with the absorption at ~1 km s^{-1} greater than the peak velocity of the CS $2 \rightarrow 1$ line. Keto, Ho and Haschick (1987) reported NH_3 absorption toward the compact HII region G10.6–0.4, at ~3 km s^{-1} greater than the centroid of the NH_3 emission. Welch *et al.* (1987) reported HCN absorption toward the HII region W49 at ~10 km s^{-1} greater than the HCN and CO emission peaks. These measurements are all suggestive of gravitational infall, but this interpretation is not unique and has been questioned in some cases. As more signatures of infall are found in emission lines, it should become possible to observe a two-dimensional pattern of gravitational infall motions. Such a pattern might be more convincing than evidence based on absorption along a single line of sight.

18.3. Cloud energetics

Table 18.3 presents a comparison of energy densities and equilibrium magnetic field strengths for the clouds and cloud cores described in Tables 18.1 and 18.2. These comparisons reveal that diffuse clouds have internal pressure comparable to that of the intercloud medium; that all objects considered, except diffuse clouds, are gravitationally bound; that the kinetic energy density of all objects considered, except low-mass cores, is primarily nonthermal; and that many clouds and cloud cores probably have nearly equal magnetic and nonthermal kinetic energy density. For each type of cloud and cloud core in Tables 18.1 and 18.2, representative values of the listed cloud parameters were used to calculate the 'nonthermal kinetic energy' density

$$K_{NT} = 3mn\sigma_{NT}^2 \tag{18.3}$$

where m is the mean molecular mass, $2.3\,m_H$; the 'thermal kinetic energy' density,

$$K_T = 3mn\sigma_T^2 \tag{18.4}$$

and the gravitational potential energy density

$$U_G = \frac{3GMmn}{5R} \tag{18.5}$$

Here the 'kinetic energy' densities are each twice the true kinetic energy densities, as in the virial theorem, and all notation follows that of Myers and Goodman (1988a). Each energy density in Table 18.3 is expressed in terms of 10^4 times Boltzmann's constant, a measure of the energy density of the intercloud medium. This value lies within a factor ~3 of most estimates (Kulkarni and Heiles 1987).

Table 18.3. *Observed energy densities and equilibrium magnetic field strengths of typical clouds and cloud cores*

Object	Energy density: Nonthermal kinetic	Thermal kinetic	Gravitational	Equilibrium magnetic field strength
	(1.38×10^{-12} erg cm^{-3})			(μG)
Diffuse cloud	2	0.3	0.08	6
Dark cloud	40	3	50	30
Giant cloud	30	0.06	20	20
Low-mass core	20	60	70	20
Massive core				
(large)	400	60	400	70
(small)	2(5)	4(4)	2(5)	2000

Each energy density is calculated from typical values of cloud and core properties in Tables 18.1 and 18.2, according to equations (2)–(4) of Myers and Goodman (1988b). Each 'kinetic energy' density is twice the three-dimensional kinetic energy density, as in the virial theorem. Each energy density is given in units of 1.38×10^{-12} erg cm^{-3}, or 10^4 times Boltzmann's constant, taken as a measure of the intercloud energy density. The equilibrium magnetic field strength is ($4\pi \times$ nonthermal kinetic energy density)$^{1/2}$.

The equilibrium magnetic field strength

$$B = (4\pi K_{\mathrm{NT}})^{1/2} \tag{18.6}$$

is calculated by equating the magnetic and nonthermal kinetic energy terms of the virial theorem.

The energy densities in Table 18.3 increase monotonically from those of diffuse clouds to those of massive cores. Among these, only the diffuse clouds have a kinetic energy density comparable to the intercloud value. The diffuse clouds are also distinctive in that they are the only objects that are far from gravitational binding: their kinetic energy density exceeds their gravitational energy density by an order of magnitude. Thus, the typical diffuse cloud is not bound by self-gravity (Magnani, Blitz and Mundy 1985) but may be close to pressure balance with the intercloud gas (Keto and Myers 1986, Elmegreen 1988).

Apart from the diffuse clouds, all of the objects in Table 18.3 appear to be gravitationally bound, in that $K_{\mathrm{NT}} + K_{\mathrm{T}}$ lies within a factor 2 of U_{G}. In each of these objects, K_{NT} exceeds K_{T} by at least a factor of 5, except for the low-mass cores, where $K_{\mathrm{NT}}/K_{\mathrm{T}} = 0.3$. Thus, the low-mass cores are supported against gravity primarily by thermal motions, while the other objects are dominated by nonthermal motions.

The equilibrium field strengths in Table 18.3 are similar, within factors of 2–3, to available magnetic field strengths, suggesting an important role for magnetic energy in cloud dynamics and evolution. For diffuse clouds the equilibrium field strength is 6 μG, within a factor 2 of most estimates of the 'background' field strength throughout the Galaxy (Heiles 1987). This similarity implies that diffuse clouds that are in 'pressure balance' with the intercloud gas may be largely in magnetic pressure balance; and that diffuse clouds that are overpressured and expanding may not expand freely but instead may be retarded or constrained by various magnetic interactions. The equilibrium field strengths in dark clouds, giant clouds, and low-mass cores, 20–30 μG, are remarkably similar. A possible explanation for this similarity involves 'photoionization-regulated star formation' (McKee and Lin 1988). Some five Zeeman effect measurements of field strength in these objects have been reported, and these lie within a factor 2 of the equilibrium values (Crutcher, Kazès, and Troland 1987, Myers and Goodman 1988a, Goodman *et al.* 1988). Similarly, some five field strengths, typically ~100 μG, are known for large massive cores, and more than ten, typically ~3000 μG, are known from OH masers for small, massive cores (Myers and Goodman 1988a).

If the available measured field strengths in dark clouds, giant clouds, and low-mass cores described above are representative, the empirical trends summarized in Equations (18.2a)–(18.2c) have a simple explanation. The nonthermal kinetic energy density, the gravitational energy density, and the magnetic energy density are then essentially equal, so Equations (18.3), (18.5), and (18.6) yield

$$\sigma_{NT} = (G/45)^{1/4} B^{1/2} R^{1/2} \tag{18.7}$$

and

$$n = \left(\frac{5}{G}\right)^{1/2} \frac{B}{4\pi m R} \tag{18.8}$$

These equations can be expected to match the observed correlations, provided that the relative range of B is small compared to the relative range of R. Since the correlations are known for three decades in R, from 0.1 to 100 pc, even a factor of 30 scatter in B allows a discernible correlation (Myers and Goodman 1988b).

18.4. Time scales and evolution

Description of molecular clouds in evolutionary terms is a subject where a great variety of interesting theoretical ideas are limited by only a relatively small number of observational facts. A good sampling of the issues in the evolution of clouds and cloud cores is evident in the papers by Shu *et al.* (1987), Silk (1987), Elmegreen (1987, 1988), Turner (1988), and McKee and Lin (1988). Here we present a few time scales that may be relevant to cloud formation and destruction, and brief discussions of dynamical cycles, and of the relative roles of kinetic, magnetic, and gravitational energy.

Table 18.4. *Typical time scales for cloud and core age and life expectancy*

Object	Log age (yr)	Log life expectancy (yr)
Diffuse cloud	—	6–7
Dark cloud	7	7–8
Giant cloud	7	7–8
Low-mass core	6	5–6
Massive core		
(large)	6	5–6
(small)	5	4–5

Table 18.4 lists some crude estimates of time scales for the typical clouds and cloud cores listed in Tables 18.1–18.3. The cloud 'age' and 'life expectancy' represent the elapsed and future time intervals during which the cloud possessed, or will possess, most of the defining observational characteristics it has today. The age is estimated for objects in virial equilibrium as at least the time needed to reach equilibrium, or a few signal crossing times, $3R/\sigma$, where σ is the quadrature sum of σ_T and σ_{NT}. In most cases, this estimate agrees, within order of magnitude, with the ages of the oldest associated stars. This estimate also corresponds to the free-fall time of a uniform sphere of constant mass, whose initial size is greater than at present by a factor 3.3.

The life expectancy of each object depends more explicitly on the assumed destruction mechanism than the age depends on the formation mechanism. For diffuse clouds, unbound by gravity, it is assumed that each cloud expands in a time between that of free expansion, R/σ, and that of 'ambipolar diffusion', or motion limited by the friction of ion–neutral collisions as neutrals try to move across stationary magnetic field lines. The time scale for ambipolar diffusion can be written

$$\tau_{AD} = 4\pi\gamma CR^2(mn)^{3/2}B^{-2} \tag{18.9}$$

where γ is the drag coefficient associated with momentum exchange in ion–neutral collisions, and C relates the ionic and neutral gas densities n_i and n, and particle masses m_i and m, according to $n_i = Cn^{1/2}m^{1/2}m_i^{-1}$ (Shu *et al.* 1987). Dark and giant clouds are probably dissipated by a combination of winds and radiation from embedded stars and/or external supernovae, tidal stresses from cloud–cloud collisions, and gravitational interactions between clouds and larger features of galactic structure, such as spiral arms or 'superclouds'. The range of life expectancy given in Table 18.4 corresponds to the ages of the oldest stars known in typical clouds (e.g., $\sim 10^7$ yr in Taurus-Auriga) and to the ages of the oldest stars associated with molecular gas (e.g., $\sim 10^8$ yr in some open clusters: Leisawitz (1988)). For cloud cores, the expectancy ranges from the dynamical time of associated molecular outflows, for cores with stars, to that of a few signal crossing times, taken as a measure of the time to form a star. These estimates are obviously crude and cannot be considered reliable in every case.

Three notable cyclical schemes of star–cloud dynamical interaction have been proposed. Elmegreen and Lada (1977) proposed 'sequential star formation' whereby an OB star forms an HII region, which expands and compresses neighboring neutral gas, which becomes gravitationally unstable, and forms a new OB star. Norman and Silk (1980) suggested that winds from low-mass stars create spherical shells, which intersect, compress ambient gas, and thereby form new low-mass stars. Goldsmith, Langer, and Wilson (1986) proposed that in regions of low-mass star formation such as Barnard 5, material cycles between clumps and a tenuous interclump medium; the clumps form by gravitational collapse and are destroyed by winds from the stars that form in the clumps. Charnley *et al.* (1988), elaborating on this picture, predict distinct changes in chemical abundance over the cycle. These and other schemes of dynamical activity require numerous observational studies to assess their degree of validity. It should be noted that such schemes that depend on significant stellar density cannot be expected to provide a general mechanism for supersonic line widths and cloud support: a significant fraction of dark clouds have no associated stars according to IRAS (Emerson 1987).

How do the kinetic, magnetic, and gravitational energies discussed in Section 18.3 develop and maintain their apparent approximate equality? One may speculate based on the following facts: (1) In the intercloud medium as a whole, estimates of mean kinetic energy density and magnetic energy density indicate approximate balance, even along lines of sight where no self-gravitating clouds are known. (2) Some diffuse molecular clouds, whose self-gravity is too weak to bind their observed motions on the scale of ~ 1 pc, have embedded low-mass cores, which are gravitationally bound on the scale of 0.1 pc (see Section 18.3 and Table 18.3). (3) Many molecular clouds with greater column density than in diffuse clouds appear to have equal gravitational, magnetic and kinetic energy on scales of 1–100 pc. (4) Such a cloud cannot bind all the gas that surrounds it within an arbitrarily large radius: beyond some radius, the gas 'belongs' instead to the intercloud medium. This gas is not bound to any cloud, yet it probably has comparable kinetic and magnetic energy density. In short, clouds or cloud regions that have gravitational, magnetic, and kinetic energy balance may be surrounded by regions with magnetic and kinetic energy balance, but weaker self-gravity.

The foregoing picture suggests that in the history of a cloud, regions with magnetic and kinetic energy balance may arise earlier, and may extend further, than regions with magnetic, kinetic, and gravitational energy balance: self-gravitating regions may develop later and from the 'inside out'. In this picture the non-self-gravitating gas must not dissipate before the time needed for the self-gravitating part to condense. This requirement may be inconsistent with free expansion in some cases and may call for slower expansion, or simultaneous accretion and mass loss. Detailed observational studies of clouds are needed to test this concept, in molecular lines that trace widely different densities.

REFERENCES

Arquilla, R., and Goldsmith, P. F. (1986). *Astrophysical Journal,* **303**, 356.

Bally, J., Langer, W. D., Stark, A. A., and Wilson, R. W. (1987). *Astrophysical Journal (Letters)*, **312**, L45.

Beichman, C. A., Myers, P. C., Emerson, J. P., Harris, S., Mathieu, R., Benson, P. J., and Jennings, R. E. (1986). *Astrophysical Journal,* **307**, 337.

Benson, P. J. (1986). In *Masers, Molecules, and Mass Outflows in Star Forming Regions,* ed. A. D. Haschick, p. 55. Westford, MA: Haystack Observatory.

Benson, P. J., and Myers, P. C. (1988). *Astrophysical Journal,* submitted.

Bieging, J. H. (1984). *Astrophysical Journal,* **286**, 591.

Blitz, L. (1987). In *Physical Processes in Interstellar Clouds,* eds. G. E. Morfill and M. Scholer, p. 35. Dordrecht: Reidel.

Charnley, S. B., Dyson, J. E., Hartquist, T. W., and Williams, D. A. (1988). *Monthly Notices of the Royal Astronomical Society,* **235**, 1257.

Cohen, M., and Kuhi, L. V. (1979). *Astrophysical Journal Supplements (Series)*, **41**, 743.

Crutcher, R. M., Kazès, I., and Troland, T. H. (1987). *Astronomy and Astrophysics,* **181**, 119.

Dame, T., Elmegreen, B., Cohen, R., and Thaddeus, P. (1986). *Astrophysical Journal,* **305**, 892.

Elmegreen, B. G. (1985). In *Protostars and Planets* II, eds. D. C. Black and M. S. Mathews, p. 33. Tucson: University of Arizona.

Elmegreen, B. G. (1987). In *Galactic and Extragalactic Star Formation,* eds. R. Pudritz and M. Fich, p. 215. Dordrecht: Kluwer Academic Publishers.

Elmegreen, B. G. (1988). In *Interstellar Matters,* eds. J. M. Moran and P. T. P. Ho, p. 55. New York: Gordon & Breach.

Elmegreen, B. G., and Lada, C. J. (1977). *Astrophysical Journal,* **214**, 725.

Emerson, J. (1987). In *Star Forming Regions,* eds. M. Peimbert and J. Jugaku, p. 19. Dordrecht: Reidel.

Falgarone, E., and Perault, M. (1987). In *Physical Processes in Interstellar Clouds,* eds. G. E. Morfill and M. Scholer, p. 59. Dordrecht: Reidel.

Goldsmith, P. F., Langer, W. D., and Wilson, R. W. (1986). *Astrophysical Journal (Letters)*, **303**, L11.

Goodman, A. A., Crutcher, R. M., Heiles, C., Myers, P. C., and Troland, T. H. (1988). *Astrophysical Journal (Letters)*, submitted.

Harris, A. W., Townes, C. H., Matsakis, D., and Palmer, P. (1983). *Astrophysical Journal (Letters)*, **265**, L63.

Heiles, C. (1987). In *Interstellar Processes,* eds. D. J. Hollenbach and H. A. Thronson Jr, p. 171. Dordrecht: Reidel.

Herbig, G. H. (1962). *Advances in Astronomy & Astrophysics,* **1**, 47.

Herbig, G. H. (1982). *Annals of the New York Academy of Sciences,* **395**, 64.

Herbig, G. H., and Bell, K. R. (1988). *Lick Observatory Bulletin* No. 1111.
Jackson, J. M., Ho, P. T. P., and Haschick, A. D. (1988). *Astrophysical Journal* (*Letters*), in press.
Keto, E. R., and Myers, P. C. (1986). *Astrophysical Journal,* **304**, 466.
Keto, E. R., Ho, P. T. P., and Haschick, A. D. (1987). *Astrophysical Journal,* **318**, 712.
Kulkarni, S. R., and Heiles, C. (1987). In *Interstellar Processes,* eds. D. J. Hollenbach and H. A. Thronson, p. 87. Dordrecht: Reidel.
Lada, C. J. (1985). *Annual Review of Astronomy and Astrophysics,* **23**, 267.
Lada, E. A., and Blitz, L. (1988). *Astrophysical Journal* (*Letters*), in press.
Larson, R. B. (1981). *Monthly Notices of the Royal Astronomical Society,* **194**, 809.
Larson, R. B. (1988). In *Galactic and Extragalactic Star Formation,* eds. R. Pudritz and M. Fich, p. 459. Dordrecht: Kluwer Academic Publishers.
Leisawitz, D. (1988). NASA Reference Publication 1202.
Leung, C. M. (1985). In *Protostars and Planets* II, eds. D. C. Black and M. S. Mathews, p. 104. Tucson: University of Arizona.
Leung, C., Kutner, M., and Mead, K. (1982). *Astrophysical Journal,* **262**, 583.
Magnani, L., Blitz, L., and Mundy, L. (1985). *Astrophysical Journal,* **295**, 402.
McKee, C. F., and Lin, J.-Y. (1988). In *Origin, Structure, and Evolution of Galaxies,* ed. L. Z. Fang. Singapore: World Scientific, in press.
Mouschovias, T. Ch. (1987). In *Physical Processes in Interstellar Clouds,* eds. G. E. Morfill and M. Scholer, p: 453. Dordrecht: Reidel.
Myers, P. C. (1983). *Astrophysical Journal,* **270**, 105.
Myers, P. C. (1987). In *Interstellar Processes,* eds. D. J. Hollenbach and H. A. Thronson, p. 71. Dordrecht: Reidel.
Myers, P. C., and Benson, P. J. (1983). *Astrophysical Journal,* **266**, 309.
Myers, P. C., Dame, T. M., Thaddeus, P., Cohen, R. S., Silverberg, R. F., Dwek, E., and Hauser, M. G. (1986). *Astrophysical Journal,* **301**, 398.
Myers, P. C., and Goodman, A. A. (1988a). *Astrophysical Journal* (*Letters*), **326**, L27.
Myers, P. C., and Goodman, A. A. (1988b). *Astrophysical Journal,* **329**, 392.
Norman, C. J., and Silk, J. (1980). *Astrophysical Journal,* **238**, 158.
Pudritz, R., and Fich, M., eds. (1988). *Galactic and Extragalactic Star Formation.* Dordrecht: Kluwer Academic Publishers.
Reid, M. J., Haschick, A. D., Burke, B. F., Moran, J. M., Johnston, K. J., and Swenson, G. (1980). *Astrophysical Journal,* **239**, 89.
Rodríguez, L. F. (1987). In *Star Forming Regions,* ed. M. Peimbert and J. Jugaku, p. 239. Dordrecht: Reidel.
Sargent, A. E., and Beckwith, S. (1987). *Astrophysical Journal,* **323**, 294.
Scalo, J. (1987). In *Interstellar Processes,* eds. D. J. Hollenbach and H. A. Thronson, p. 349. Dordrecht: Reidel.
Scalo, J. (1988). In *Molecular Clouds in the Milky Way and External Galaxies,* eds. R. Dickman, R. Snell and J. Young. Dordrecht: Reidel, in press.
Schloerb, F. P., Snell, R. L., Goldsmith, P. F., and Morgan, J. A. (1986). In *Interstellar Processes* (*Abstracts of Contributed Papers*), eds. D. J. Hollenbach and H. A. Thronson, Jr, p. 25. Laramie: University of Wyoming.
Scoville, N. Z., Kleinmann, S. G., Hall, D. N. B., and Ridgway, S. T. (1983). *Astrophysical Journal,* **275**, 201.
Scoville, N. Z., Yun, M. S., Clemens, D. P., Sanders, D. B., and Waller, W. H. (1987). *Astrophysical Journal Supplements* (*Series*), **63**, 821.
Shu, F. H., Adams, F. C., and Lizano, S. (1987). *Annual Review of Astronomy and Astrophysics,* **25**, 23.

Silk, J. (1987). In *Star Forming Regions,* eds. M. Peimbert and J. Jugaku, p. 663. Dordrecht: Reidel.
Snell, R. L. (1987). In *Star Forming Regions,* eds. M. Peimbert and J. Jugaku, p. 213. Dordrecht: Reidel.
Snell, R. L., Loren, R. B., and Plambeck, R. L. (1980). *Astrophysical Journal* (*Letters*), **239**, L17.
Solomon, P. M., Rivolo, A. R., Barrett, J., and Yahil, A. (1987). *Astrophysical Journal,* **319**, 730.
Stacy, J. G., Myers, P. C., and de Vries, H. W. (1988). In *Molecular Clouds in the Milky Way and External Galaxies,* eds. R. Dickman, R. Snell, and J. Young. Dordrecht: Reidel, in press.
Strom, K. M., Strom, S. E., and Newton, G. (1988). *Bulletin of the American Astronomical Society,* **20**, 693.
Torrelles, J. M., Ho, P. T. P., Rodríguez, L. F., and Cantó, J. (1986). *Astrophysical Journal,* **305**, 721.
Turner, B. G. (1988). In *Galactic and Extragalactic Radio Astronomy,* eds. G. L. Verschuur and K. I. Kellerman, p. 154. Berlin: Springer-Verlag.
Wadiak, E. J., Wilson, T. L., Rood, R. T., and Johnston, K. J. (1985). *Astrophysical Journal* (*Letters*), **295**, L43.
Walker, C. K., Lada, C. J., Young, E. T., Maloney, P. R., and Wilking, B. A. (1986). *Astrophysical Journal* (*Letters*), **309**, L47.
Waller, W. H., Clemens, D. P., Sanders, D. B., and Scoville, N. Z. (1987). *Astrophysical Journal,* **314**, 397.
Walter, F., Brown, A., Mathieu, R., Myers, P. C., and Vrba, F. (1988). *Astronomical Journal,* **96**, 297.
Welch, W. J., Dreher, J. W., Jackson, J. M., Terebey, S., and Vogel, S. N. (1987). *Science,* **238**, 1550.
Wouterlout, J. G., Walmsley, C. M., and Henkel, C. (1988). *Astronomy and Astrophysics,* in press.
Zheng, X. W., Ho, P. T. P., Reid, M. J., and Schneps, M. H. (1985). *Astrophysical Journal,* **293**, 522.

VI

H_2 in regions of massive star formation

19

Infrared observations of line emission from molecular hydrogen

T. R. GEBALLE

Joint Astronomy Centre, Hilo, Hawaii, USA

19.1. Introduction

Observations of infrared line emission from molecular hydrogen in astronomical sources have gone from the novel to the commonplace in a time that is short relative to most timescales for the advancement of astrophysical knowledge. In the face of the current onslaught of observations of H_2, it may seem surprising that only a dozen years ago, when the 2 μm lines were first detected, it is reported to have taken their discoverers several months to identify the emitting species. Excited H_2 continues to be detected in new and surprising places. In the Galaxy, H_2 line emission is found at interfaces between young stellar winds and the interstellar medium, where it first was discovered by Gautier *et al.* (1976), in reflection nebulae (Gatley *et al.* 1987), in supernova remnants (e.g., Burton *et al.* (1988)), even including the Crab Nebula (Graham, Wright, and Longmore 1989), in planetary and proto-planetary nebulae (Treffers *et al.* 1976), and in the nucleus (Gatley *et al.* 1984). Beyond the Galaxy H_2 line emission is found in Seyfert, as well as starburst galaxies (Thompson, Lebofsky, and Rieke 1978, Joseph *et al.* 1986, Fischer *et al.* 1987), in individual HII regions of normal spiral galaxies (Israel *et al.* 1989), and in interacting and merging galaxies (e.g., Joseph *et al.* 1986). Closer to home, H_2 line emission has been detected very recently in the aurora of Jupiter (Trafton *et al.* 1988). In all of these examples, because of the unique physical conditions which must be satisfied in order that its infrared line emission be detectable, H_2 lines provide important information about environments that are difficult to study by other means.

This chapter is an attempt to summarize infrared astronomical observations of H_2 line emission, hopefully more from physical and spectroscopic viewpoints than from astronomical and morphological ones. Such an observational summary cannot be made without occasional reference to theoretical work. Many of the observations that I will mention have been stimulated by theoretical work and predictions of Alexander Dalgarno, his students, and co-workers. Our current understanding of the phenomena connected with H_2 line emission, both on a molecular scale and on astronomical scales, is due in large part to their efforts. Likewise, I am certain that the observational data, many of them surprising (to both astronomers and physicists), push our understanding of many astrophysical environments, and even of molecular physics.

19.2. The infrared spectrum of H_2 – an observer's guide

The quantum mechanical selection rules for vibration-rotation transitions in H_2 differ from those of most other astrophysically important molecules. Dipole transitions are not allowed; thus the strongest infrared transitions are electric quadrupole, corresponding to $\Delta J = 2$. Triplet (ortho) states correspond to odd J levels, singlets (para) to even J. As these states are radiatively isolated from one another, the ratio of triplets to singlets is determined by processes which occur during the formation of H_2 and by subsequent atom-exchange interactions, which, at sufficiently high temperatures, produce an equilibrium value of 3. In such cases the emission lines linking states of odd J will be three times stronger than those between even J levels.

One of the interesting aspects of the study of H_2 line emission is that the combined vibrational and rotational spectum of this lightest of molecules resembles that of no other molecule. The spring constant binding the two hydrogen nuclei is not the primary cause of this, although it is greater than those of most, if not all, other molecules. The uniqueness rather is due to the moment of inertia of H_2, which is an order of magnitude smaller than that of other astrophysically observed molecules. As a result the vibration-rotation bands of H_2 extend over unusually large spectral intervals and the pure rotational spectrum, which occurs in the millimeter and sub-millimeter wavebands for most molecules of astrophysical interest, extends from the mid-infrared (28 μm) into the short wavelength infrared (almost to 3.3 μm), passing through a part of the fundamental vibration-rotation band (see Figure 19.1). The large spacings of vibrational and rotational levels mean that H_2 will not be collisionally excited except at very high temperatures. Until the middle of the last decade, such temperatures were not expected to exist in interstellar molecular clouds.

The quadrupole H_2 transitions are associated with long upper state lifetimes, typically 10^6–10^7 s for the upper levels of lines in the 2–5 μm band (Turner, Kirby-Docken and Dalgarno 1977). In contrast, the typical lifetimes of excited vibration-rotation levels are less than one second for non-homonuclear diatomic molecules, such as CO. On the other hand, collisional relaxation times for vibrational and rotational states of H_2 are not widely dissimilar from those of

Figure 19.1. Spectrum of Peak 1 in Orion, from 3.0 to 3.6 μm, obtained at a resolving power of ~400. All lines are from H_2. (Geballe 1986.)

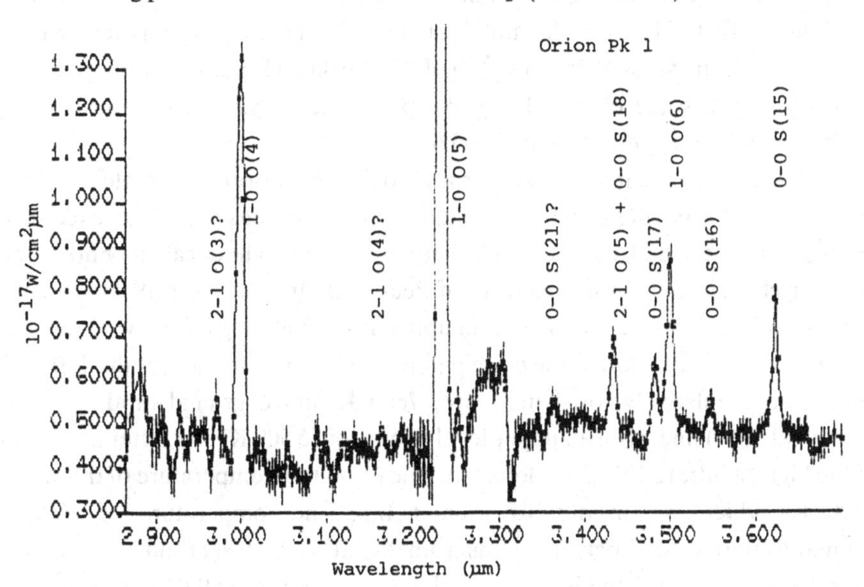

other molecules. As a result the vibrational levels of H_2 are in collisional equilibrium at much lower densities (~10^6 cm^{-3}) than are the levels of other astrophysically important molecules. Therefore, H_2 can have a rich vibrational spectrum in hot and dense regions of molecular clouds.

19.3. Excitation of H_2

Observational and theoretical studies have demonstrated clearly that the H_2 infrared line emission observed in the interstellar medium occurs in some cases following the collisional excitation of the molecule and in others after absorption by the molecule of ultraviolet photons.

19.3.1. Collisional excitation

The first of the above mechanisms to be observed was collisional excitation, which was identified because the relative intensities of the detected H_2 lines were approximately those which would be observed from an optically thin gas in LTE at ~2000 K. The first detection, as well as numerous subsequent ones, were deep within molecular clouds, near known or suspected locations of star formation. As an isolated molecular cloud or portion thereof at 2000 K cools rapidly, it was quickly realized that the H_2 line emission must be a transient event. Hence the role of shock waves was established and winds from young stars were implicated as the driving agents. Theoretical work demonstrated that a shock-heated slab of molecular gas passes through this temperature regime as it cools, so long as the shock velocity is large enough (>6 km s^{-1}; e.g., Kwan (1977), London, McCray, and Chu (1977), Hollenbach and Shull (1977)). For shocks having velocity discontinuities in excess of 25 km s^{-1} all of the shocked H_2 is dissociated. In higher velocity shocks, some of the H_2 will be preserved if the acceleration is made gradual, e.g. by magnetic precursors (Draine 1980).

Observations of the line emission from shocked gas normally sample a column of H_2 molecules having a range of temperatures and densities. Thus excitation temperatures derived from the relative intensities of two lines can depend on the line pair chosen. The 3 μm spectrum of Peak 1 in Orion (Figure 19.1) is a good example of this. In it we see lines from upper states covering a very wide range in energy. Of particular note is the close pair near 3.5 μm, consisting of the 1–0 O(6) line, from a relatively low-lying state, 7600 K above ground, and the pure rotational S(17) line, with an upper level energy of 25 500 K. The latter line is one of the highest infrared H_2 lines detected. The excitation temperature of this pair is 2700 K, which is considerably greater than the temperature of 2000 K commonly derived in shocked regions from the ratio of 1–0 S(1) and 2–1 S(1) lines near 2 μm. At 2000 K the intensity of the S(17) line is only 0.04 that of the 1–0 O(6) line, eight times less than the observed intensity ratio. If, as is suspected, the line emission occurs in cooling gas, some of that gas must be at temperatures well in excess of 2700 K.

19.3.2. Radiative excitation

Although the existence of line emission from shocked H_2 was initially a surprise, the existence of fluorescent H_2 line emission was predicted two decades before it was found. The reasons for the subsequent lack of detections are related to limited infrared sensitivities and to selection effects. Fluorescent line emission occurs at the surfaces of molecular clouds or within diffuse clouds. Detections at infrared wavelengths at both of these locations are difficult, because of the large extents of the clouds and because of the low surface brightness of the emission. The first detection, reported by Gatley *et al.* (1987), in the reflection nebula NGC 2023, was made only recently. Since then detections have been reported near a number of molecular clouds that are illuminated by ultraviolet radiation (e.g., by Hayashi *et al.* (1985), Sellgren (1986)), and subsequently in a proto-planetary nebula (Dinerstein *et al.* 1988) and in a number of extragalactic objects (Israel *et al.* 1989, Puxley, Hawarden, and Mountain 1988).

An example of line emission from fluorescent H_2 is shown in Figure 19.2. The source is a giant HII region in the nearby spiral galaxy M33. This HII region is known to border on a molecular cloud and the fluorescence is believed to be occurring at the interface of the two. The observational determinants of fluor-

Figure 19.2. The 2.0–2.3 μm spectrum of the giant HII region, NGC 604, in the galaxy M33. (Israel *et al.* 1988.)

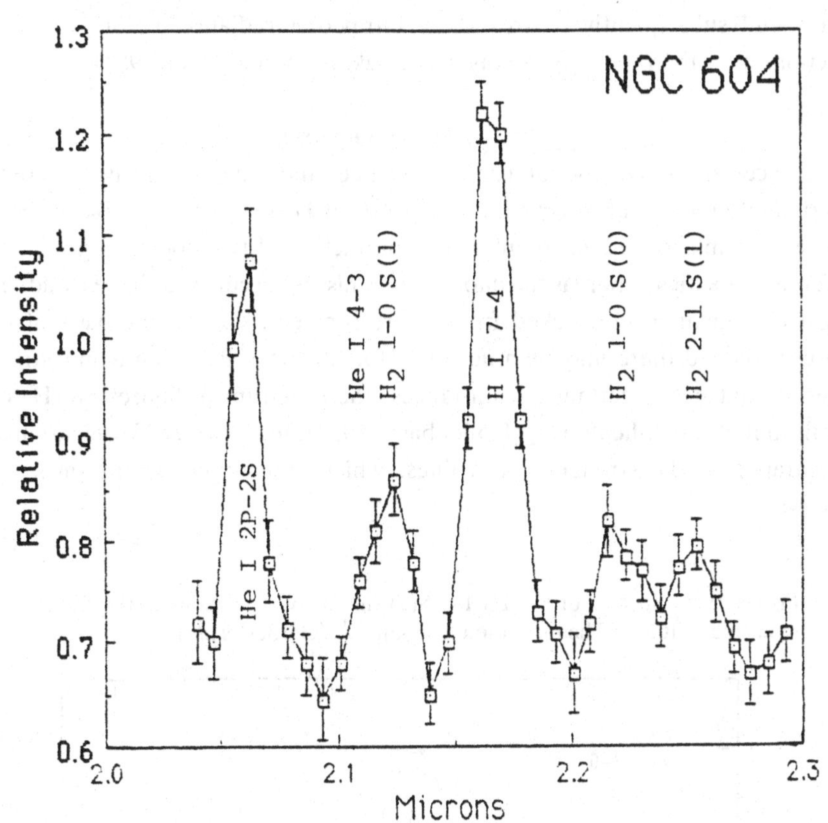

escence in this case (and in most others reported to date) are the enhanced intensities, relative to the 1–0 S(1) line at 2.122 μm, of the 1–0 S(0) and (especially) 2–1 S(1) lines at 2.224 μm and 2.248 μm, resectively. The enhancement factors are ~2 and ~5, respectively, over the values observed in shocked regions. As the observed fluorescence is part of a cascade down the vibrational levels, following absorption of ultraviolet photons by cold H_2 molecules, one measures high vibrational excitation temperatures and low rotational excitation temperatures. The observational differences between collisional and ultraviolet excitation of H_2 also are pronounced in the 1.1–1.7 μm spectrum, but because of the weakness of the line emission, only one spectrum of a portion of this region has been reported (Gatley *et al.* 1987).

An interesting result from the study of fluorescent line emission in the reflection nebula NGC 2023 (Hasegawa *et al.* 1987) is that the ortho/para ratio of H_2 populations there is 2 or less. This is significantly lower than the 'expected' value of 3, which has been found in regions of shocked gas. The low value in NGC 2023 may be due to the H_2 being formed with this ratio on low temperature (60–70 K) grains and subsequently destroyed by ultraviolet radiation before exchange reactions alter the ratio to 3 (Takayanagi, Sakimoto, and Onda 1987).

19.3.3. Discriminators

Differences in relative line intensities have been the principal means to date for discriminating between shocked and fluorescent H_2 line emission. Because line emission from shocked gas requires the interaction of two bodies of gas having different velocities, whereas the fluorescent emission occurs in a single cloud, one might also distinguish shocked from fluorescent emission on the basis of line profiles. Indeed there may be a necessity for this, as it has been demonstrated recently that under certain circumstances, the spectrum of fluorescent H_2 can mimic that of shock-heated H_2 (Hollenbach (1988); see Chapter 22) at least in the appearance of the stronger 2 μm lines, which commonly are the only ones observed.

Figure 19.3. Spectra of the H_2 1–0 S(1) line at two positions in the DR21 molecular outflow. The resolution is 35 km s^{-1}. (Garden *et al.* 1989).

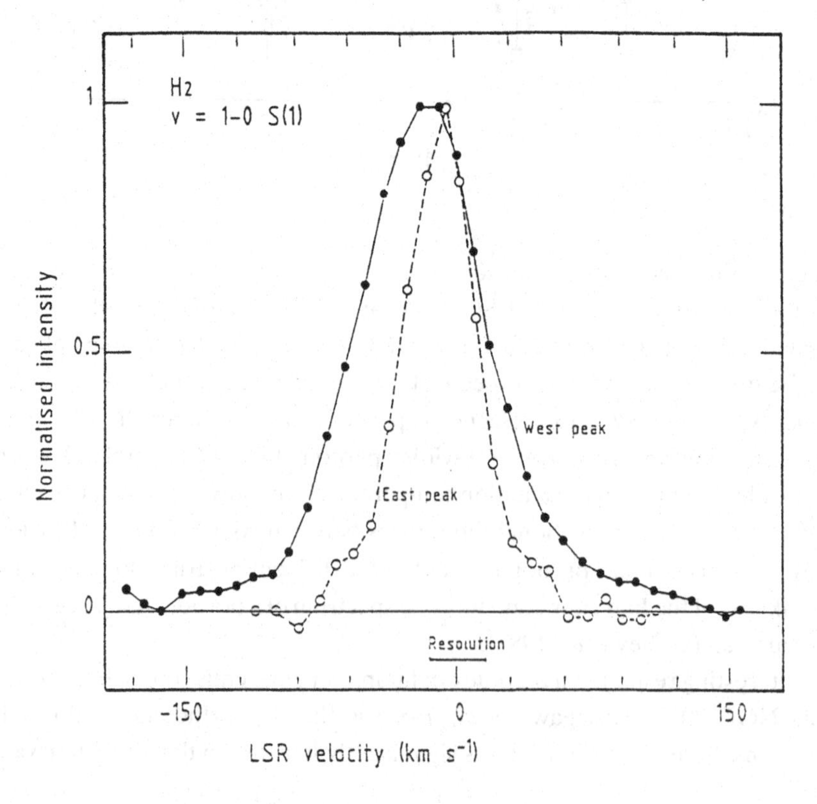

In practice lines of shocked H_2 can be embarrassingly broad. The case of the Orion outflow is well known, but very wide lines also are found in several other sources. Figure 19.3 shows spectra of the 1–0 S(1) line at two locations in the DR21 molecular outflow. At one of these positions the full width at zero intensity (FWZI) of the line is ~150 km s^{-1}; the line is fully as wide as at Peak 1 in Orion. In the proto-planetary nebula CRL 618, Burton and Geballe (1986) have observed an S(1) line with a FWZI of 250 km s^{-1}. In contrast, Burton *et al.* (1989) set upper limits of ~10 km s^{-1} to the H_2 linewidths in several regions where the emission is known to be fluorescent.

For the more distant, extragalactic environments, even small apertures normally include regions which may be undergoing a variety of large scale motions. For this reason it is likely that the distinguishing criterion between collisional and ultraviolet excitation will have to be line intensity ratios, rather than velocity widths.

19.4. Morphology

Mapping and imaging of H_2 line emission do not provide detailed information concerning the molecular physics involved in the emission. Nevertheless, in revealing the spatial distribution of the line emission, they often point to specific astrophysical causes for the emission, and in many cases can discriminate by themselves between shocked and fluorescent line emission. Figures 19.4 and 19.5 are only a small example of the power and usefulness of the new infrared cameras in studying shocks. These images, obtained by IRCAM, the infrared camera recently put into use at the United Kingdom Infrared Telescope, reveal details at the 1 arcsec level of the shocked line emission in the Herbig–Haro objects HH7–11 and similar details as well as the enormous extent of the molecular outflow in DR21. In Figure 19.4 note the bow shock in HH7. In Figure 19.5, the jet-like appearance of the H_2 line emission in DR21 clearly points to shock waves driven by a well-collimated wind as the cause of the emission (Garden *et al.* 1986, 1989).

Spectroscopy of H_2 lines may also help determine the morphology of the line-emitting gas. Figure 19.6 demonstrates this for the planetary nebula NGC 7027. The H_2 1–0 S(1) line profile is double-peaked, whereas at the same resolution the HI 7–4 (Br γ) line is single-peaked. Each profile was obtained in a small beam passing through the center of the nebula. The difference in the profiles is not due solely to the greater thermal Doppler broadening of the atomic hydrogen line, but is due in significant part to different spatial and velocity distributions of the emitting species. The H_2 line profile suggests that the shocked gas is located in an expanding shell. Previous mapping of these lines in NGC 7027 showed that the H_2 is external to the ionized gas (Beckwith *et al.* 1980). The separation of the peaks in the S(1) profile implies that the shell is expanding at 15 km s^{-1}. As the expansion

Figure 19.4. The Herbig–Haro objects HH7–11, observed with IRCAM through a narrow-band filter transmitting the 1–0 S(1) line of H_2. The star SSV13, thought to be the source of the stellar wind, is at top right, surrounded by S(1) line emission and adjacent to a foreground star. The rest of the emission is from the S(1) line. Note the bow shock in HH7, at lower left. Scales are in arc-seconds. (Garden, Russell, and Burton 1989.)

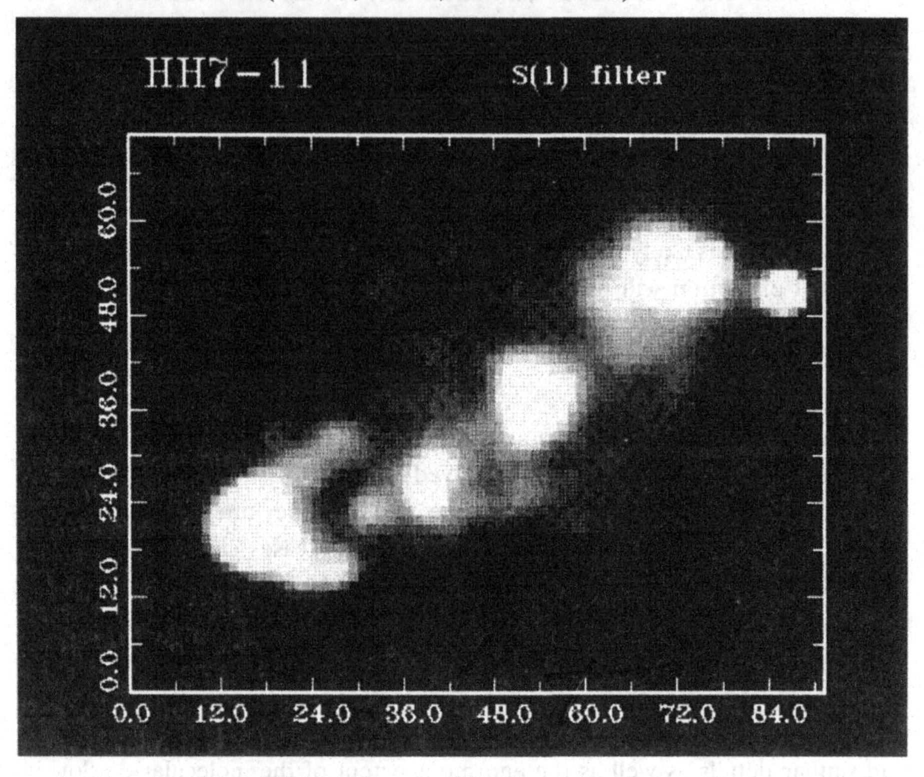

velocity of the ionized gas in NGC 7027 is believed to be ~25 km s^{-1}, these results are at least in superficial agreement with models of a planetary nebula in which a high velocity wind overtakes and shocks molecular material expelled during an earlier (red giant) phase of mass loss (e.g., Kwok (1982)).

Deconvolution of the ultrabroad S(1) line in the proto-planetary nebula CRL 618 (Burton and Geballe 1986) implies that its line emission does not occur in a single shell of gas, but rather in several clumps of gas embedded in and shocked by a fast stellar wind. Higher spectral resolution observations (Burton and Geballe 1989) at least partially bear this out. At Peak 1 in Orion, observations by Brand *et al.* (1989) at high spectral and spatial resolution reveal a line profile which is as broad (FWZI = 140 km s^{-1}) and as smooth as those observed with larger apertures and lower resolution (e.g., Nadeau and Geballe 1979). It is difficult to explain the velocity extent in terms of a single shock and it is also difficult to explain the smoothness in terms of discrete shocked clumps, unless a very large

Figure 19.5. IRCAM mosaic of the DR21 outflow, imaged in the 1–0 S(1) line. The total linear extent of the line emission is 5 pc. The source of the outflow is undetected, but is believed to be located in the gap between the two jets, near the middle of the image. (Garden, Russell, and Burton 1989.)

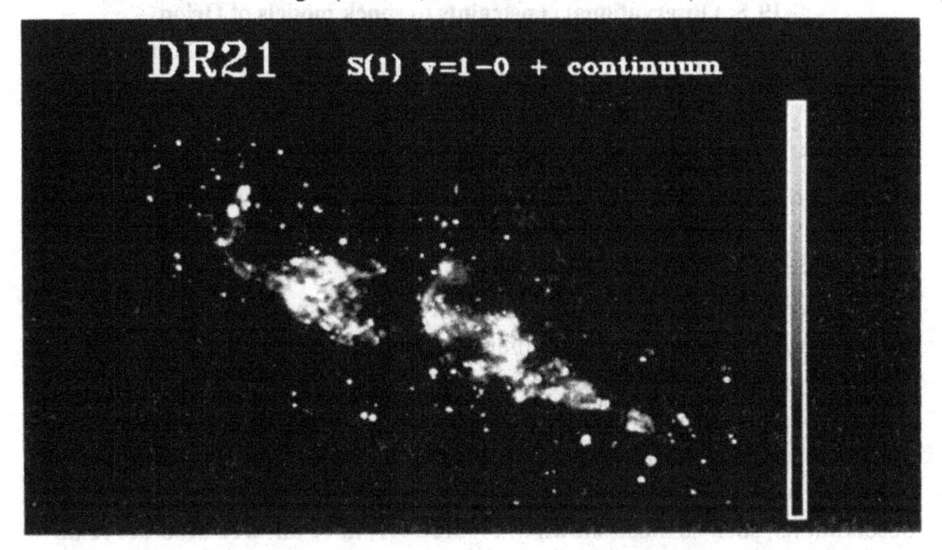

Figure 19.6. Spectra of the planetary nebula NGC 7027 in the H_2 1–0 S(1) and HI 7–4 (Br γ) lines both obtained at 12 km s^{-1} resolution. (Smith and Geballe 1989.)

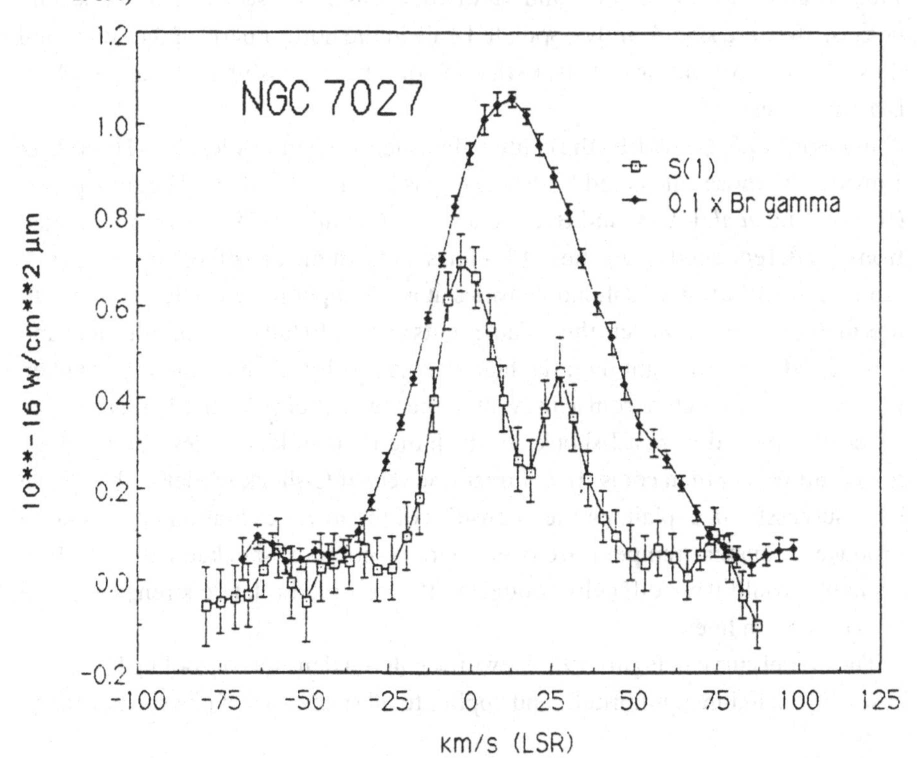

number of them (>40), with velocities covering the observed range, are within the beam.

19.5. Observational constraints to shock models of Orion

Because of its high H_2 line brightnesses, Orion has long been the test bed for models of shock waves in molecular clouds. The early observations of H_2 lines in Orion were explainable on the basis of the simple jump-shock (J-shock) models referred to earlier. Subsequent measurements of an extremely supersonic range of emission velocities in Orion (Nadeau and Geballe 1979), which subsequently was found in several additional objects and which, at least in the case of Orion, appears to be due to actual gas motions (Geballe *et al.* 1986), helped to prompt the development of continuous shock (C-shock) models, in which the shock speeds are increased above the 25 km s^{-1} limit for J-shocks (Chernoff, Hollenbach, and McKee 1982, Draine, Roberge, and Dalgarno 1983). However, even the current C-shock models cannot explain the full velocity width of the line emission.

Until recently, shock models could be tested by measurements of only a few H_2 lines, which were almost entirely from the $v = 1$ and $v = 2$ levels. However, recent observations, such as those shown in Figure 19.1, have allowed accurate relative column densities to be derived for a large number of states, including several $v = 0$ states with high J and a number of $v = 3$ and 4 states of low J (Beck and Beckwith 1983, Brand *et al.* 1988, Oliva and Moorwood 1988). These states are important because they are significantly populated only in the hottest parts of the shock and thus contain information specific to that region, the successful modelling of which is a critical test.

In Figure 19.7 are plotted the column densities of 17 energy levels of H_2 at Peak 1 divided by those calculated for a slab of gas in LTE at 2000 K. The data points (from Brand *et al.* (1988) and Beck, Lacy, and Geballe (1979)) were determined from the dereddened intensities of H_2 lines and assume an ortho/para ratio of 3. An LTE distribution of column densities at any temperature would be a straight line in this figure. However, these data points appear to fall on a smooth curve and cannot be fit by a straight line; the high energy levels are more highly populated than would be expected from a linear fit to the lower energy levels. This is the type of result expected from a J-shock, in which the H_2 is suddenly elevated to a high temperature and then cools. In contrast, the recent C-shock models, which have been successful in explaining the intensities of the lower excitation lines, tend to produce a constant temperature over a large column of shocked gas and, thus, probably would have difficulty producing the observed relative strengths of the high excitation lines.

The model curve in Figure 19.7 shows the column densities of cooling H_2 for the case of Boltzmann equilibrium, and applies to all such cases of postshock cooling

Figure 19.7. The ratios of observed, dereddened column densities of H_2 at Orion Peak 1 to those for a Boltzmann distribution at 2000 K (normalized so that the 1–0 S(1) ratio is unity), versus upper energy level. The data point represented by a box is from Beck *et al.* (1979); the remaining points and the J-shock model predictions (solid curve) are from Brand *et al.* (1988).

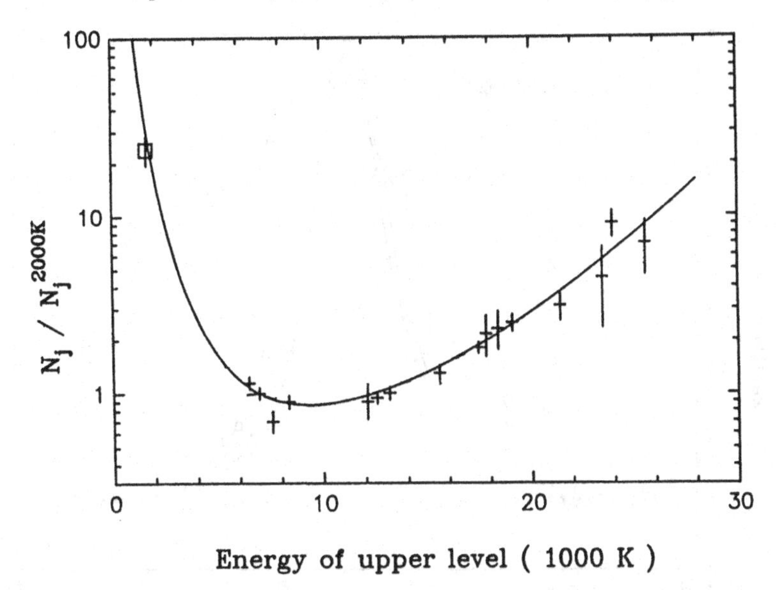

in which the H_2 is heated above ~3500 K (and not completely dissociated). The good fit to the data suggests that if the J-shock model is valid, then all of the H_2 levels are in LTE. Although collisional excitation rates for H_2 are not accurately known, the J-shock explanation would probably require very high densities ($>10^7$ cm^{-3}) of the postshock gas in Orion.

Superficially, the simple J-shock, LTE model of the H_2 line emission can explain the observed column densities of both high and low energy levels, as is demonstrated in Figure 19.7. One must remember, however, that the column densities plotted in Figure 19.7 are based on integrated line intensities. For most of these lines, velocity information is not yet available. The model used in Figure 19.7 predicts that all line profiles should be identical. Recent spectra of H_2 lines of widely different excitation (Figure 19.8) bear on this issue. The profiles of the 1–0 and 2–1 lines correspond closely, with a slight indication that the 2–1 line is broader. The spectrum of the 3–2 line is contaminated by a contribution from at least one other H_2 line; its broader profile is in large part due to this. Thus, the differences in the profiles appear to be quite small, again implying overall shock behavior more similar to J-shocks than to C-shocks.

The above discussion is greatly lacking in detail, and parts of it may be

Figure 19.8. Velocity profiles of three lines of H_2, at Orion Peak 1. All three are close in wavelength (near 2.2 μm), and were measured in a 12 arc sec aperture at 35 km s^{-1} resolution. The 3–2 S(3) profile is contaminated by a contribution from the 4–3 S(5) line, denoted by an arrow, and shifted ~ −60 km s^{-1} with respect to the former line. (Moorhouse *et al.* 1989.)

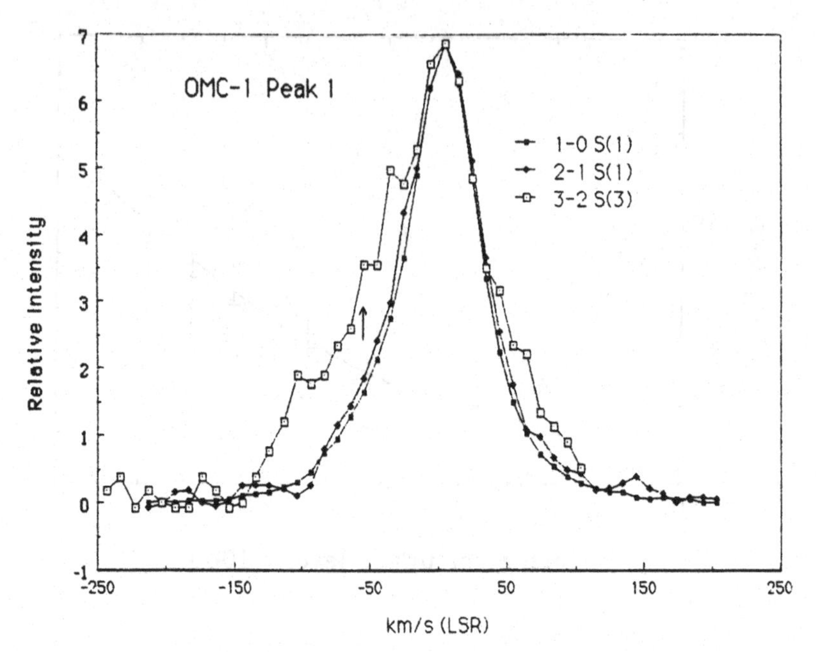

somewhat controversial. However, the data in Figures 19.1, 19.7, and 19.8 demonstrate that new and important diagnostics of shock types now are being obtained. These will lead to a more detailed understanding of shock waves in molecular clouds.

19.6. Extragalactic H_2 line emission

Observations of line emission from H_2 have become a fundamental part of several branches of extragalactic astronomy. Luminous H_2 line emission has been detected in three types of external galaxies: Seyfert galaxies, galaxies undergoing bursts of star formation, and interacting and merging galaxies. In Seyferts the excitation mechanism appears to be collisional in most, although perhaps not in all cases (e.g., Fischer *et al.* (1987)), and may be due to the interaction of a nuclear wind with surrounding molecular clouds. In starburst galaxies recent observations suggest that the phenomenon is frequently one of fluorescence (Puxley *et al.* 1988). The situation in these galaxies may be similar to that discussed above for NGC 604 (Figure 19.2), in which the ultraviolet photons from large numbers of young and massive stars are radiatively exciting the H_2 in nearby molecular clouds.

Figure 19.9. The 2.0–2.5 μm spectrum of the merging galaxy system NGC 6240, obtained at a resolving power of 120. (Joseph *et al.* 1986.)

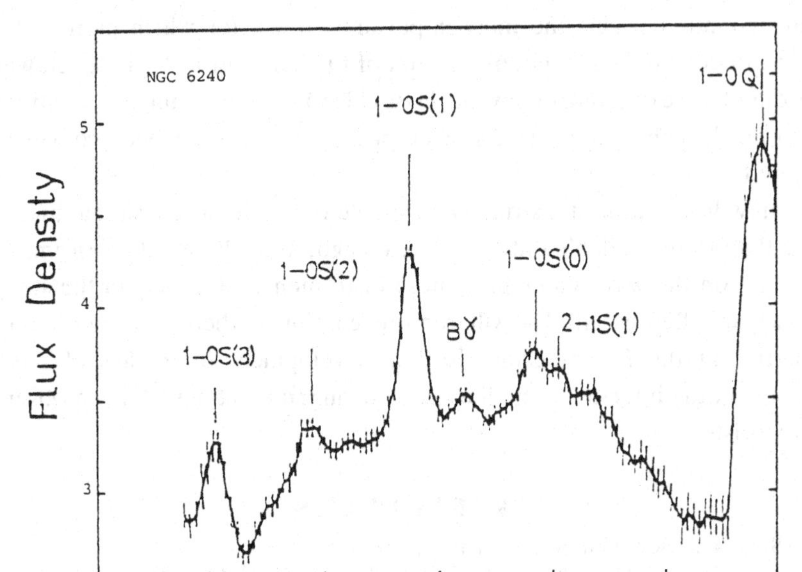

Perhaps the most intriguing and spectacular extragalactic H_2 line emission originates in interacting and merging spiral galaxies. Very little was known about these galaxies, until they were singled out by IRAS as ultraluminous at far-infrared wavelengths. Subsequent near-infrared spectroscopy has provided dramatic evidence of their activity. The 2 μm spectrum of the most extreme case, NGC 6240, is shown in Figure 19.9; the spectrum is dominated by line emission from shock-excited H_2. The luminosity from all of the H_2 lines in this galaxy is probably 3×10^9 $L_\odot$, which is roughly $\frac{1}{2}$% of the total luminosity of the galaxy and is equivalent to $\sim 10^7$ Orions. The most plausible mechanisms for the emission are cloud–cloud impacts resulting from the galaxy collisions or an intense burst of star formation, set off by the collision and resulting in wind–cloud interactions as in Orion. The latter explanation seems the less likely in NGC 6240, because the pre-main-sequence winds are thought to be short-lived, and it is thus unlikely that 10^7 Orion-like events could be so well synchronized.

19.7. Conclusion

Observations of H_2 line emission have taught astrophysicists a great deal about the physical processes in the interstellar medium, as well as about the evolution of molecular clouds, star formation, some of the late stages of stellar evolution, and large scale phenomena in galaxies. As usual, however, there remains a great deal to understand. Perhaps the most important subject for which more detailed physical understanding is required is that of the behavior of H_2 in shock waves. For example, the question of how the H_2 is able to survive to emit at velocities that are so much higher than the dissociation limit has not yet been convincingly answered.

We now have infrared instruments capable of much more accurate measurements than were available only a few years ago, and additional major improvements are on the way. To employ these instruments effectively in the study of shocked and fluorescent H_2 will require continued theoretical work on the properties of the H_2 molecule, further development of models of physical processes in the interstellar medium, and strong interactions between observers and theorists.

REFERENCES

Beck, S. C., and Beckwith, S. 1983, *Ap. J.*, **271**, 175.
Beck, S. C., Lacy, J. H., and Geballe, T. R. 1979, *Ap. J.* (*Letters*), **234**, L213.
Beckwith, S., Persson, S. E., Neugebauer, G., and Becklin, E. E. 1980, *Astron. J.*, **85**, 886.
Brand, P. W. J. L., Toner, M. P., Geballe, T. R., and Webster, A. S. 1989, *MNRAS*, **237**, 1009.
Brand, P. W. J. L., Moorhouse, A., Burton, M. G., Geballe, T. R., Bird, M., and Wade R. 1988, *Ap. J.* (*Letters*), **334**, L103.
Burton, M. G., and Geballe, T. R. 1986, *MNRAS*, **223**, 13P.
Burton, M. G., and Geballe, T. R. 1989, in preparation.
Burton, M. G., Geballe, T. R., Brand, P. W. J. L., and Webster, A. S. 1988, *MNRAS*, **231**, 617.
Burton, M. G., Geballe, T. R., Moorhouse, A., and Brand, P. W. J. L. 1989, *Ap. J.*, in press.
Chernoff, D. F., Hollenbach, D. J., and McKee, C. F. 1982, *Ap. J.* (*Letters*), **259**, L97.
Dinerstein, H. L., Lester, D. F., Carr, J. S., and Harvey, P. M. 1988, *Ap. J.* (*Letters*), **327**, L27.
Draine, B. T. 1980, *Ap. J.*, **241**, 1021.
Draine, B. T., Roberge, W. G., and Dalgarno, A. 1983, *Ap. J.*, **264**, 485.
Fischer, J., Geballe, T. R., Smith, H. A., Simon, M., and Storey, J. W. V. 1987, *Ap. J.*, **320**, 667.
Garden, R., Geballe, T. R., Gatley, I., and Nadeau, D. 1986, *MNRAS*, **220**, 203.
Garden, R., Geballe, T. R., Gatley, I., Hayashi, M., Hasegawa, T., and Nadeau, D. 1989, *Ap. J. Suppl.*, in press.
Garden, R., Russell, A. P. G., and Burton, M. G. 1989, in preparation.
Gatley, I., Jones, T. J., Hyland, A. R., Beattie, D. H., and Lee, T. J. 1984, *MNRAS*, **210**, 565.
Gatley, I., *et al.* 1987, *Ap. J.* (*Letters*), **318**, L73.
Gautier, T. N., III, Fink, U., Treffers, R. R., and Larson, H. P. 1976, *Ap. J.* (*Letters*), **207**, L29.

Geballe, T. R. 1986, in *Summer School on Interstellar Processes,* NASA TM 88342, p. 129.
Geballe, T. R., Persson, S. E., Simon, T., Lonsdale, C. J., and McGregor, P J. 1986, *Ap. J.,* **200**, 693.
Graham, J. G., Wright, G. S., and Longmore, A. J. 1989, in preparation.
Hasegawa, T., Gatley, I., Garden, R. P., Brand, P. W. J. L. B., Ohishi, M., Hayashi, M., and Kaifu, N. 1987, *Ap. J.* (*Letters*), **318**, L77.
Hayashi, M., Hasegawa, T., Gatley, I., Garden, R., and Kaifu, N. 1985, *MNRAS,* **215**, 31p.
Hollenbach, D. J. 1988, *Astron. Lett. & Comm.*, **26**, 191.
Hollenbach, D. J., and Shull, J. M. 1977, *Ap. J.,* **216**, 419.
Israel, F. P., Hawarden, T. G., Wade, R., Geballe, T. R., and van Dishoeck, E. F. 1989, *MNRAS,* **236**, 89.
Joseph, R. D., Wright, G. S., Wade, R., Graham, J. G., and Gatley, I. 1986, in *Star Formation in Galaxies*, ed. C. J. Lonsdale Persson (NASA CP2466), p. 421.
Kwan, J. 1977, *Ap. J.,* **216**, 713.
Kwok, S. 1982, *Ap. J.,* **258**, 280.
London, R., McCray, R., and Chu, S. 1977, *Ap. J.,* **217**, 442.
Moorhouse, A., Brand, P. W. J. L., Geballe, T. R., and Burton, M. G. 1989, in preparation.
Nadeau, D., and Geballe, T. R. 1979, *Ap. J.* (*Letters*), **230**, L169.
Oliva, E., and Moorwood, A. F. M. 1988, *Astron. Ap.*, **197**, 261.
Puxley, P. J., Hawarden, T. G., and Mountain, C. M. 1988, *MNRAS,* **234**, 29P.
Sellgren, K. 1986, *Ap. J.,* **305**, 399.
Smith, M. G., and Geballe, T. R. 1989, in preparation.
Takayanagi, K., Sakimoto, K., and Onda, K. 1987, *Ap. J.* (*Letters*), **318**, L81.
Thompson, R. I., Lebofsky, M. J., and Rieke, G. H. 1978, *Ap. J.* (*Letters*), **222**, L49.
Trafton, L., Carr, J., Lester, D., and Harvey, P. 1988, in *Proceedings of the Conference on Time Variable Phenomena in the Jovian System*, in press.
Treffers, R. R., Fink, U., Larson, H. P., and Gautier, T. N., III 1976, *Ap. J.,* **209**, 793.
Turner, J., Kirby-Docken, K., and Dalgarno, A. 1977, *Ap. J. Suppl.*, **35**, 281.

20

Shocks in dense molecular clouds

DAVID F. CHERNOFF

Center for Radiophysics and Space Research, Cornell University, USA
(Presidential Young Investigator)

CHRISTOPHER F. McKEE

University of California at Berkeley, USA

20.1. Introduction

Shock waves are ubiquitous in the interstellar medium (ISM) because efficient radiative cooling allows interstellar gas to cool to temperatures low enough that the sound speed is small compared to the velocities of disturbances in the ISM, such as cloud–cloud collisions, bipolar outflows, expanding HII regions, and supernova explosions. Shock waves in dense molecular gas are almost always radiative: The relative kinetic energy of the shocked and unshocked gas is converted into radiation, and since the radiating gas is dense, it is very bright. Because much of the mass in molecular clouds is obscured by dust, the emission from shocks provides a powerful probe of energetic activity occurring in these clouds. In particular, stars inject large amounts of energy into their surroundings in the process of formation, giving rise to bipolar outflows with velocities in excess of 100 km s^{-1}, characteristic of stellar escape velocities (Lada 1985). Intense maser emission in the 1.35 cm line of water is also observed to be associated with newly formed stars, particularly massive stars, with velocities of tens to hundreds of kilometers per second (Genzel 1986). Understanding the structure and spectrum of the shocks associated with these high velocity flows in dense molecular gas is thus a prerequisite for unraveling the complex physical processes attending the birth of stars.

Early studies of shocks in molecular clouds assumed that the neutrals and ions were tied together into a single fluid, and that the shock front was an abrupt transition on the scale of the molecular mean free path (e.g., Field *et al.* (1968), Hollenbach and McKee (1979)). It is now known that such a treatment is valid for the dissociative shocks discussed by Neufeld in Chapter 21. However, for shocks which are too slow to dissociate the pre-existing H_2 (i.e., for $v_s \lesssim$ 40–50 km s^{-1}), the dissipation in the shock is effected by the drift of the ions and magnetic field through the neutral gas. Mullan (1971) first studied the possibility of such two-fluid shocks in the context of atomic gas. Draine (1980) pointed out that the distinction between the two cases, which he termed J-shocks (for 'jump') and C-shocks (for 'continuous'), would have important astrophysical consequences if a substantial amount of radiation were emitted from the transition region where the relative drift occurred. In particular, J-shocks emit most of their radiation in the ultraviolet region of the spectrum, whereas C-shocks emit primarily in the infrared; for this reason, the intense H_2 emission centered on the luminous embedded protostar IRc2 in Orion (Nadeau and Geballe 1979) has been identified as arising in a C-shock (Chernoff, Hollenbach, and McKee 1982, Draine and Roberge 1982). The chemistry that occurs in such shocks is discussed by Hartquist *et al.* in Chapter 5. A recent introductory review of astrophysical shocks has been given by McKee (1987).

Here, we shall briefly address two aspects of shocks in dense molecular clouds: First, we shall discuss the conditions under which shocks are J-type, C-type, or have some intermediate structure. Second, we shall discuss the possibility that the intense water masers observed in star forming regions are produced by shocks. It is a pleasure to dedicate this chapter to Alex Dalgarno in view of his many contributions to our understanding of the interstellar medium, including the interstellar shocks discussed here (Roberge and Dalgarno 1982, Draine, Roberge, and Dalgarno 1983).

20.2. Shock structures

A shock always results when a medium is disturbed at a speed greater than that of all relevant signal speeds. There are several signal speeds in dense molecular material: (1) The *neutral speed of sound* is

$$c_s = (5kT/3\mu)^{1/2} \approx 0.77 \times 10^5 T_2^{1/2} \text{ cm s}^{-1} \tag{20.1}$$

where T is the gas temperature, $T_2 = T/(10^2 \text{ K})$, and $\mu = 3.9 \times 10^{-24}$ g is the average mass per particle in molecular gas with a helium abundance $n(\text{He})/[n(\text{H}) + 2n(\text{H}_2)] = 0.1$. Sound waves are compressional disturbances, with gas pressure serving as the restoring force. (2) The *Alfven speed* is

$$c_A = (B^2/4\pi\rho)^{1/2} \approx 1.8 \times 10^5 b \text{ cm s}^{-1} \tag{20.2}$$

where B is the magnetic field and ρ is the total mass density of the molecular material. The variation of B with density in the interstellar medium may be approximately parameterized as $B = 10^{-6} n_0^{1/2} b$ G, where n_0 is the density of hydrogen nuclei. In molecular clouds, the Alfven velocity is of order a few kilometers per second (Myers and Goodman 1988), so b is of order unity. Pure Alfven waves are transverse disturbances in which magnetic tension serves as a restoring force. Compressional waves in which both the magnetic pressure and gas pressure provide the restoring force are termed magnetosonic waves and propagate at a velocity $c_{ms} = (c_s^2 + c_A^2)^{1/2}$. (3) The *ion-Alfven speed* is

$$c_{iA} = (B^2/4\pi\rho_i)^{1/2} = c_A/y_i^{1/2} \tag{20.3}$$

where ρ_i is the ion mass density and $y_i = \rho_i/\rho$ is the ion mass fraction. Ion-Alfven waves propagate in the ion component, not the total ion plus neutral gas. The speed is much larger than the Alfven speed because the inertia is that associated with only the ion component.

A linear analysis of wave propagation in molecular cloud material shows that the gas plus field supports modes which involve all of these characteristic velocities (Chernoff 1987). Usually, $c_s \lesssim c_A \ll c_{iA}$ because the ion fraction is so small in dense molecular material, $y_i \approx 10^{-6}$–10^{-8} (Langer 1985). Assuming the velocity of the shock v_s satisfies the ordering $c_s \lesssim c_A < v_s < c_{iA}$, then neither normal sound waves nor Alfven waves in the composite medium are capable of transmitting information of the disturbance ahead of its arrival, but ion-Alfven waves can. However, the ion-Alfven waves are subject to damping through collisions with the neutrals. Thus, when a disturbance propagates through the medium at a velocity v_s, there is a race between two competing effects: First, a signal is transmitted upstream of the disturbance by an ion-Alfven wave which outraces the disturbance because $v_s < c_{iA}$ (the *precursor*). Second, that signal is damped by the ion's collisional coupling to the neutrals, which are very slow to respond to any changes. If the signal propagates sufficiently far upstream, then the neutral flow can be modified (heated and slowed), so that the neutrals effectively respond to the disturbance upstream; otherwise a shock in the neutrals results. This basic competition is responsible for many of the effects we now discuss.

20.2.1. The triple point

Consider a fluid flow of velocity v_s perpendicular to a field of strength B. We work in the shock frame, that Galilean frame in which the disturbance appears at rest. We assume that there is a single ion species with a low total ionization, that the ratio of the gyroperiod to damping time for the charged particles is very small, and that the gas upstream of the shock is cold. The flow is governed primarily by two parameters: the Alfven Mach number of the flow, $M_A = v_s/c_A$; and the ratio of ion–neutral drift heating to radiative cooling in the shock. The drift heating may be written $n^2\Gamma = n_n n_i \langle \sigma v \rangle \mu_{in} \Delta v^2$, where n_n and n_i are the number density of neutrals and ions respectively, μ_{in} is the reduced ion–neutral mass, and Δv is the velocity difference between neutrals and ions. The collision rate coefficient $\langle \sigma v \rangle$ is the rate of momentum transfer between the drifting ion and neutral populations; at low velocities the main interaction is charge exchange and at high velocities elastic collisions predominate. The radiative cooling may be written $n^2\Lambda =$

$n_n^2\langle\sigma v\rangle_R kT$, where $\langle\sigma v\rangle_R kT$ is the cooling rate coefficient in units of erg cm^3 s^{-1}. Define Ψ to be proportional to the ratio of the cooling to the heating:

$$\frac{\Lambda}{\Gamma} = \frac{n_n\langle\sigma v\rangle_R kT}{n_i\langle\sigma v\rangle\mu_{in}\Delta v^2} \equiv \left(\frac{kT}{m\Delta v^2}\right)\Psi \tag{20.4}$$

where m is the neutral mass; then Ψ is inversely proportional to the ionization. For schematic illustration we assume a simple functional dependence,

$$\Psi = \Psi_0 \left(\frac{v_i}{v_s}\right)^\alpha \left(\frac{v_n}{v_s}\right)^\beta \left(\frac{T_n}{T_s}\right)^\gamma \tag{20.5}$$

where T_s is the shock temperature of a flow with the same velocity, v_i and v_n are the ion and neutral velocities measured in the shock frame, and α, β and γ are constants designed to mock up realistic cooling rate coefficients. A variety of values has been studied.

Figure 20.1 illustrates how the qualitative nature of the flow solutions varies as a function of M_A and Ψ_0 for the specific set of parameters $\alpha = 0$, $\beta = -0.5$ and $\gamma = 1.5$. These choices correspond to assuming that the rate of cooling scales like n_n^2 (i.e. the low density limit, as opposed to the LTE limit) and that the ionization n_i/n_n is determined by a local balance between ionization (an ionization rate per neutral) and electron recombination. When there is no cooling at all ($\Psi = 0$), we

Figure 20.1. Qualitative division of solution space as a function of M_A, the Alfven Mach number, and ψ_0, a schematic measure of the rate of cooling to heating within the flow.

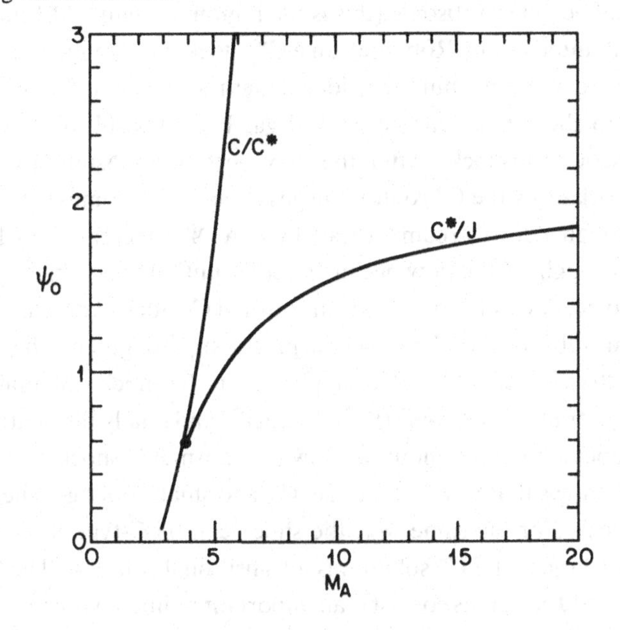

have an energy conserving flow. If the magnetic field is sufficiently strong in such a flow ($M_A < 2.76$), then the flow is continuous; this has been termed a C-shock by Draine (1980). Neutral gas particles run into the perturbed ions and field far upstream of the disturbance; they are slowed and heated to the postshock temperature. The final state of the gas is subsonic relative to the magnetosonic velocity, but supersonic relative to c_s. Neutral viscosity is not needed to effect the transition. For larger values of M_A, ion–neutral dissipation alone is insufficient fully to heat and slow the neutrals; a viscous subshock appears at a unique location and the gas flow is henceforth subsonic relative to c_s (J-shock). In the energy conserving limit, the division between C and J-shocks is particularly clear and can be derived analytically (Mullan 1971, Draine 1980).

However, in the presence of strong radiative cooling ($\Psi \gtrsim 1$), C-shocks are possible even at large M_A. Consider a J-shock at high M_A and low Ψ, in which most of the dissipation in the shock occurs in a strong viscous subshock where the neutrals make an abrupt transition to subsonic flow ($v_n < c_s$). As Ψ increases, the compression across the viscous subshock decreases because: (1) cooling upstream increases the number density of neutrals and hence the rate of dissipation, so more entropy is created in the precursor and less is necessary at the shock; and (2) the compression of the cooling postshock gas increases the strength of the magnetic field directly behind the viscous subshock, and the field is then able to penetrate further upstream. Both effects mean that the precursor to the viscous subshock contributes more heating and dissipation as cooling processes become more efficient. Eventually, the viscous shock is completely eliminated, though the neutral gas still becomes subsonic (this is the region of Figure 20.1 marked C*, a solution first pointed out by Roberge). In a C*-shock, ion–neutral heating, acting like an effective viscosity but over long length scales, raises the neutral gas temperature to the point that the neutral gas becomes subsonic relative to c_s without a viscous subshock. After the flow becomes subsonic (whether by a viscous subshock or by the C* route), the gas cools and becomes supersonic, so it must pass through a downstream critical point. As Ψ is increased further, the gas everywhere throughout the flow becomes cooler until it is no longer possible for the neutrals to become subsonic; this is the normal C-shock. Thus, as Ψ increases, first the viscous subshock weakens and the precursor strengthens (the shock is still J-type); then the viscous subshock disappears, but the shock still retains a region in which the neutrals are subsonic (the C*-shock); and finally the neutrals become completely supersonic throughout the flow (the normal C-shock).

Figure 20.1 shows that flows with large M_A and small cooling-to-heating ratios Ψ_0 give J-shocks. For the same M_A, the shock can be C-type if the cooling-to-heating ratio is large. The C* solution is a transitional solution. The triple point where C, C*, and J solutions coexist is an important point in parameter space: For

Ψ_0 less than the critical value, the flow structure is primarily determined by the strength of the magnetic field, whereas above the critical point the role of cooling is crucial.

20.2.2. *Results for astrophysical shocks*

Three problems must be faced before realistic calculations of astrophysical shocks can be made: First, the chemistry must be followed in order to determine both the ionization, which governs the heating rate, and the molecular abundances, which govern the cooling rate. Second, it is numerically difficult to find the J and C* solutions because of the presence of critical points in the flow; it now appears that the simplest procedure may be to integrate the time-dependent equations directly. Finally, charged grains are present and these interact within the shock like heavy ions. The treatment of these grains has hitherto been based on the approximation that there is an equilibrium between drag and magnetic forces (Draine 1980). However, the inertial forces on the grains are also important and, if the flow is superalfvenic relative to the charged grains, then the grains can shock. The situation is further complicated by the presence of a spectrum of grain sizes, so that the charge-to-mass ratio of the grains varies, and by the fact that not all grains are well tied to field lines.

Detailed calculations of the shock structure have been performed for C-shocks (Chernoff, Hollenbach and McKee 1988). Our work is based on a code which employs a four-fluid description (ions, electrons, neutrals and grains) of the shocked gas. There are three temperatures (ions, electrons and neutrals) and two separate velocities (charged particles and neutrals). The grain velocity is not integrated, but instead is determined by the condition of force balance as described by Draine (1980); hence, there is no velocity dispersion associated with the grains. This approximation is good whenever the flow is subsonic relative to the grain Alfven speed ($c_{gA} = c_A / y_g^{1/2}$, where $y_g = \rho_g / \rho \approx 0.01$ is the dust-to-gas ratio) so that $M_A \lesssim 10$, or if the grains are collisionally dominated. (In the next section we discuss when this assumption fails.) The chemical reaction network includes 22 separate species, focusing on the H, C, and O chemistry, plus a variety of nonreacting metals and ions. An attempt has been made to include all relevant cooling mechanisms; all relevant heating mechanisms, including that due to the fact that ions produced by charge-exchange have a two-dimensional velocity distribution which is unstable and heats the electrons; and all relevant ionization mechanisms, including exotic processes such as fast neutrals which ricochet off drifting grains and collide with oncoming neutrals.

Figure 20.2 illustrates the region over which C-shocks have been found as a function of shock velocity v_s and preshock hydrogen nuclei density n_0, for $b = 1$. The line is the 'breakdown' line (Chernoff *et al.* 1982). C-shocks are found to the

Figure 20.2. Extent of C solutions for full scale simulations of shocks in dense molecular cloud material. The magnetic field is given as $B = 10^{-6} n_0^{1/2}$ G (i.e., $b = 1$), where n_0 is the pre-shock hydrogen nuclei density and v_s is the shock speed.

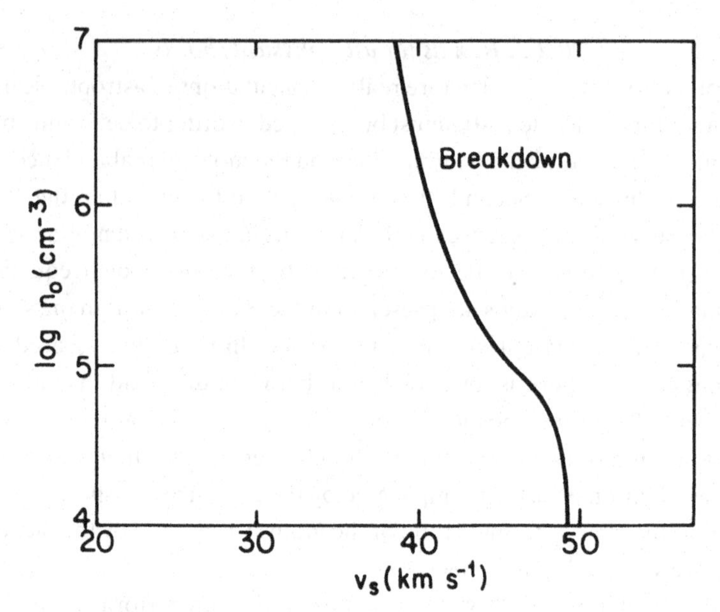

left of the line, C* or J to the right. At a fixed preshock density, as one increases v_s, the maximum temperature in the flow increases. Near 'breakdown' the maximum temperature is about 5000 K and the maximum ionization fraction is increasing dramatically. Dissociation of molecular coolants also occurs. It appears that many processes contribute to the increase in y_i. If the mocked-up models in the last section are an accurate indication, then the flows to the right of the 'breakdown' line are similar to previously calculated J-shocks, in which the main effect of the magnetic field is to limit the compression; C*-shocks appear to occupy a negligible region of parameter space in practice. The variation in the characteristic 'breakdown' velocity with density comes from the change in the cooling capacity of the gas: for $n_0 \gtrsim 10^5$ cm^{-3}, the efficiency of H_2 cooling per molecule decreases, so breakdown occurs at lower velocities. Water dominates the cooling at higher densities and causes the breakdown line to level off at about 40 km s^{-1}.

Figure 20.3 summarizes the various signal velocities with which we have been concerned and illustrates some of the inadequacies of our current understanding. The speed of sound, c_s is the vertical line far to the left; sound waves, not shocks, exist to the left of this line. To the right shocks may be possible. The top (dashed) line, labeled $2.76c_A$ corresponds to the division between C and J-shocks for $\Psi = 0$.

Figure 20.3. Division of parameter space based on characteristic speeds of sound: c_s (solid line) is the normal speed of sound; 2.67 c_A (dashed line) is the division between C and J solutions in the absence of cooling; c_{iA} (dashed-dot line) is the ion-Alfven velocity for a density of $n_0 = 10^4$; and c_{gA} is the grain-Alfven velocity for a gas with a 1% mass fraction of charged grains. An ionization rate of 10^{-15} per nuclei has been assumed; $b(v_s)$ scales as the quarter power of the rate, so the position of the line is relatively insensitive to the choice. The vertical lines with arrows are the breakdown lines as determined by numerical simulations.

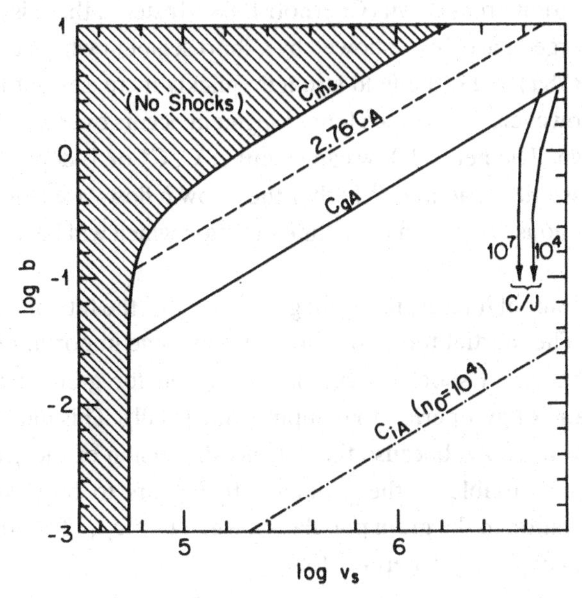

Even in the absence of any cooling, flows above and to the left of this line are continuous. Depending upon the molecular coolants and the ionization fraction, C, C* and J-shocks are all possible below this line. The dashed-dot line is $v_s = c_{iA}$ for an ionization $y_i = 2 \times 10^{-6}$, typical of molecular cloud material at a density of 10^4 cm^{-3} (Langer 1985); according to Draine (1980), under these conditions C-shocks are possible only above this line. The pair of solid, nearly vertical lines to the right, are the limit of C solutions which have been found with our detailed numerical treatment described above. It remains to be proven that the flow to the right is essentially a strong J-shock; in principle, it might be C* and observationally distinct from previously calculated J-shocks, although this is unlikely because of the rapidly increasing ionization. The graph illustrates the steep velocity dependence of the ionization breakdown, which agrees in spirit with Alfven's original notion of a critical velocity for ion-neutral drift set by the energy of ionization (and nothing else).

If the onset of the ionization breakdown at low b occurs at a velocity v_* which is independent of the magnitude of b (as suggested by the breakdown lines plotted in Figure 20.3), there are two points of fundamental interest. First, consider the region $v_* > v_s > c_{iA}$, which is possible only for $b \ll 0.1$, corresponding to $c_A \ll 0.2$ km s^{-1} (Figure 20.3); the flow in this region violates our original ordering that $v < c_{iA}$. Nonetheless, it appears that continuous solutions for the neutral flow are possible here also if the ionization mechanisms are too inefficient to lead to a self-generated ionization breakdown (Chernoff 1988). Instead, there is a collisionless shock in the charged particles, followed by a short segment where the ions are hot ($\beta = (nkT)_i/(B^2/8\pi) > 1$); the field strength quickly increases until it dominates the pressure, however, and a much longer segment occurs over which the neutrals are slowed down. The neutral flow can be either C, C* or J, depending upon the cooling and ionization fraction. Whether these low b flows are closer to continuous or discontinuous flows is an important question which will have observational ramifications.

The second point of interest in the diagram concerns the fate of grains for flows with $v_s > c_{gA}$. The inertial forces on the grains become important in this regime and we find that grain shocks occur in this region for grains which are well magnetized (ratio of gyroperiod to damping time small). This indirectly changes the character of the flow, because the velocity dispersion of the grains becomes another dynamic variable in the problem. It appears to be possible to have betatron acceleration in the grain population when $v_s > c_{gA}$. The solutions in this region are being explored (Chernoff 1988).

20.3. H_2O masers in shocks

20.3.1. Observations of H_2O masers

The water molecule has a rich rotational spectrum which fortuitously contains a near degeneracy between two levels (6_{16} and 5_{23}) 650 K above the ground state. Radio observations at the 1.35 cm wavelength separating these levels have shown that they can have inverted populations and produce the brightest molecular maser emission of any known galactic astronomical source. Genzel (1986) has summarized what is known about the H_2O masers, and we shall extract some of the key points here. The masers are associated with star forming regions of all types, with the brightest masers occurring in regions of OB star formation. Observations of other molecules near the masers show bipolar outflows and dense [$n(H_2) \sim 10^7$ cm^{-3}], hot ($T \sim$ 100–200 K) molecular cores. A given source has of order 10–$10^{2.5}$ velocity components spread out over a range of velocities of 10–$10^{2.5}$ km s^{-1}. Each component typically has a line width of 0.5–1.5 km s^{-1}. It is possible to infer the space velocities of the components from proper motion studies, and it is found that they are flowing outward from one or more sources.

For example, the low and high radial velocity masers in W49, W51, and SgrB2 each seem to have similar space velocities, ranging from about 30 to in excess of 100 km s^{-1}. Features with low radial velocities are almost always much brighter than those with high radial velocities. VLBI observations indicate that the individual maser spots are each $10^{12.5}$–$10^{13.5}$ cm in size, and are spread out over a distance of order 10^{17} cm. Often, the spots are grouped together in subclusters 10^{15}–10^{16} cm in size. The small size and large luminosity of the maser spots (up to 0.08 $L_\odot$) imply brightness temperatures that are often of order 10^{13} K and can reach 10^{15} K. H_2O masers observed in the nuclei of some other galaxies can have luminosities several orders of magnitude greater than those observed for galactic masers, but it is not known whether the brightness temperatures are also higher.

Genzel draws several conclusions from these data. First, the separation from the central source of the outflow is large enough ($\sim 10^{17}$ cm) that infrared pumping of the maser from that source is ruled out. Second, since the separation of the spots in subclusters is only a few times greater than their size, it is likely that 'the masers are filamentary structures which map out favorable low velocity gradient gain paths in a turbulent medium'. This is supported by the fact that low radial velocity features, which would correspond to filaments seen end on, are the brightest. Finally, the fact that subclusters of maser features flare up, expand and fade suggests that they are due to collisions between clumps in the wind from the central star and other clumps, a picture similar to the 'interstellar bullets' proposed by Norman and Silk (1979).

20.3.2. Shock models

The high velocities of the observed masers and the requirement of a local energy source to pump the maser both suggest shocks as the likely site of H_2O masers. Tarter and Welch (1986) have proposed that strong J-shocks in very dense cloudlets ($n_0 \sim 10^9$ cm^{-3}) can radiatively pump the preshock gas and cause it to mase. In order to account for extremely luminous masers, Strelnitskij (1984) has suggested that masers could operate at yet higher densities ($n \sim 10^{11}$ cm^{-3}) and give higher brightness temperatures, provided the electrons and H_2 molecules are maintained at different temperatures; he proposed shocked gas as the natural site of the required high densities. Kylafis and Norman (1986) have suggested that the unequal temperatures could arise in C-shocks, but it remains to be shown that this can occur at such high densities. In these models, the shock velocity must generally be much less than the observed space velocity of the masers since fast shocks give off too much continuum radiation at such high densities: the shocked gas radiates as a blackbody at a temperature $T = 320[n_{09}v_{s7}^3]^{1/4}$ K, where $n_{09} = n_0/(10^9\ \mathrm{cm}^{-3})$ and $v_{s7} = v_s/(100\ \mathrm{km\ s}^{-1})$ (McKee 1987).

An alternative shock model has been proposed by Hollenbach, McKee and

Chernoff (1987) and developed by Elitzur, Hollenbach and McKee (1989; hereafter EHM). In this model, the preshock density is only about 10^6–10^7 cm^{-3}, closer to observed values in interstellar, as opposed to circumstellar, space. The observed size of the maser spots is identified as the shock thickness, whereas the amplification path lies in the shock plane. The greatest amplification will occur along paths which maintain velocity coherence over the greatest distance, as envisioned by Genzel (1986), and so will naturally tend to be roughly cylindrical. As shown by Goldreich and Keeley (1972), cylindrical masers of diameter d and length $l \equiv ad$ beam their radiation into a solid angle $\Delta\Omega \propto 1/a^2$ and therefore have a brightness temperature T_m that scales as $l/\Delta\Omega \propto a^3 d$; numerically, EHM show

$$T_m = 5.0 \times 10^{11} \left(\frac{\eta}{10^{-2}}\right) \left(\frac{q}{10^{-11}\ \mathrm{cm}^3\ \mathrm{s}^{-1}}\right) \left(\frac{a}{10}\right)^3 \frac{n_9^2 d_{13}}{\Delta v_5}\ \mathrm{K} \qquad (20.6)$$

where q is the pump rate, η the fraction of the pumping events which produce a maser photon, $d_{13} = d/(10^{13}$ cm), and Δv_5 is the line width in kilometers per second; a water abundance $n(H_2O)/n = 3 \times 10^{-4}$ has been assumed. For collisionally excited maser emission, a column density $nd \sim 10^{22}$ cm^{-2} (corresponding to $n_9 d_{13} \sim 1$) of hot ($T \sim 500$ K) molecular gas is required to achieve the brightness temperatures observed in H_2O masers.

In principle, such columns can occur in either C-shocks or J-shocks, but detailed calculations have been carried out only for the latter (Hollenbach *et al.* 1987, Hollenbach and McKee 1989). H_2 molecules are dissociated in J-shocks, and then form again on grain surfaces at a rate $\gamma n n(\mathrm{H})$, where $n = n(\mathrm{H}) + 2n(H_2)$. The density is very nearly constant in the H_2 formation region because the pressure is magnetically dominated there; numerically, we have $n/n_0 = 76.7 v_{s7}/b$ (Hollenbach and McKee 1979). Half the hydrogen is converted to molecular form at a column $N_{1/2} = 4.5 \times 10^{21}(b/\gamma_{-17})$ cm^{-2}, where $\gamma_{-17} = \gamma/(10^{-17}$ cm^3 s^{-1}). At low densities, the heat of formation (about 4 eV per H_2) is radiated away, but at high densities ($n \gtrsim 10^5$ cm^{-3} for collisions with H, $n \gtrsim 10^8$ cm^{-3} for collisions with H_2) this energy is transferred to the gas by collisions. The result is a temperature plateau with $T \approx 400(n_{07} v_{s7}/\Delta v_5)^{2/9}$ K (EHM). The code used to carry out the shock calculations is similar to that described in Section 20.2, except that there is only one temperature and one fluid velocity. The results show that the temperature plateau extends to a column of about $2N_{1/2}$. The thickness of the plateau is $d \approx 2N_{1/2}/n \approx 1 \times 10^{13}[b/(\gamma_{-17} n_9)]$ cm, which is comparable to the observed size of maser features if the factor in brackets is of order unity. Fast J-shocks in dense gas thus provide an ideal environment for the observed H_2O masers.

In order to determine whether the level populations in the temperature plateau are sufficiently inverted to produce strong maser emission, EHM calculated the level populations of the 40 lowest levels of ortho-H_2O using recently calculated

H_2O collision cross sections (Palma *et al.* 1988). Inversion of the maser line was found for a wide range of densities ($n \sim 10^6$–10^9 cm^{-3}) and temperatures ($T \gtrsim 300$ K). The inversion is established because levels off the 'backbone', which is the set of lowest levels for each J, are populated primarily by decays from higher backbone levels; the upper level of the maser transition, 6_{16}, is on the backbone whereas the lower level, 5_{23}, is not. The peak brightness temperature of the maser emission is $1.6 \times 10^{12}(a/10)^3$ K, and $T_{\rm m}$ is within a factor 2 of this for $n_9^2 d_{13}/\Delta v_5 \sim 10^{0.5\pm1}$. The most luminous galactic masers can be accounted for by aspect ratios $a \lesssim 50$. The magnetic field in the masing gas is given by the shock jump conditions as $B = 0.28(n_9 b v_{s7})^{1/2}$ G, which is large enough to account for the linear polarization observed in many water masers (Deguchi and Watson 1986). This model may even account for the extremely luminous extragalactic masers: the parameters $a = 70$, $d = 5 \times 10^{14}$ cm, and $\Delta v = 10$ km s^{-1} give an isotropic luminosity $L = (2\pi^2/\lambda^3)kT_{\rm m}d^2\Delta v \approx 40\,{\rm L}_\odot$ for a single maser spot. Such a large shock thickness can be produced only if the preshock magnetic field is also large, $b \gtrsim 10$. Thus, in this model the existence of extremely luminous masers in galactic nuclei depends on the presence of an unusually large field there.

20.4. Conclusions

Interstellar shocks exhibit a rich range of behavior. In an unmagnetized gas, shocks are J-type: there is an abrupt jump in the density and temperature of the gas at the shock front due to dissipation of the relative kinetic energy of the shocked and unshocked gases by the viscosity of the neutrals (if the gas entering the shock front is predominantly neutral) or by plasma instabilities (if it is predominantly ionized). If the gas is magnetized, the structure is similar if the velocity is sufficiently high (Figure 20.2), except that a magnetic precursor precedes the shock front if the ion-Alfven velocity $c_{\rm iA}$ exceeds $v_{\rm s}$. In most cases, J-shocks are dissociative, and the molecules reform at a column density of 10^{21}–10^{22} cm^{-2}. On the other hand, for $v_{\rm s} \lesssim 40$–50 km s^{-1}, the structure is completely different: the viscous subshock disappears and the dissipation is effected by ion–neutral friction. In such a C-shock the neutral velocity remains supersonic everywhere; for strong shocks, this is possible only if the gas is molecular because efficient cooling is required to keep the neutrals supersonic. It is possible to have a C-type shock in which the neutrals make a smooth transition to subsonic flow (the C* solution), but this appears to be relatively unimportant in practice. Problems now under investigation include C* shocks, inclusion of grain dynamics for $M_{\rm A} \gtrsim 10$, and shock structures at low field strengths.

Because shocks in dense gas are generally obscured by dust, infrared emission lines are the primary diagnostic of such shocks. C-shocks emit most of their energy in this band; J-shocks, though weaker in the infrared, are also readily detectable. At high densities, shocks are likely sites of collisionally excited H_2O masers. Detailed calculations of the brightness temperature of the H_2O 6_{16}–5_{23} transition under conditions found in the H_2 formation temperature plateau behind strong J-shocks are in good agreement with observations of interstellar water masers. The rapid improvement in infrared detectors now underway and the advent of the VLBA should open up a new era in our understanding of interstellar shocks in dense molecular gas, and therefore of how stars form.

Acknowledgements

We wish to thank D. Hollenbach and M. Elitzur for their help on the work reported here. D.F.C. acknowledges support by NSF grants AST84-15162 and AST86-57467, and C.F.M. by NSF grant AST86-15177.

REFERENCES

Chernoff, D. F. 1987, *Ap. J.* **312**, 143.
Chernoff, D. F. 1988, in preparation.
Chernoff, D. F., Hollenbach, D. J. and McKee, C. F. 1982, *Ap. J. Lett.*, **259**, L97.
Chernoff, D. F., Hollenbach, D. J. and McKee, C. F. 1988, in preparation.
Deguchi, S. and Watson, W. D. 1986, *Ap. J.*, **302**, 750.
Draine, B. T. 1980, *Ap. J.*, **241**, 1021.
Draine, B. T. and Roberge, W. G. 1982, *Ap. J. Lett.*, **259**, L91.
Draine, B. T., Roberge, W. G. and Dalgarno, A. 1983, *Ap. J.*, **264**, 485.
Elitzur, M., Hollenbach, D. J. and McKee, C. F. 1989, *Ap. J.*, in press.
Field, G. B., Rather, J. D. G., Aanestad, P. A. and Orszag, S. A. 1968, *Ap. J.*, **151**, 953.
Genzel, R. 1986, in *Masers, Molecules and Mass Outflows in Star Forming Regions*, ed. A. D. Haschick (Haystack Observatory), p. 233.
Goldreich, P. and Keeley, D. 1972, *Ap. J.*, **174**, 517.
Hollenbach, D. J. and McKee, C. F. 1979, *Ap. J. Suppl.*, **41**, 555.
Hollenbach, D. J. and McKee, C. F. 1988, *Ap. J.*, to be submitted.
Hollenbach, D. J., McKee, C. F. and Chernoff, D. F. 1987, in *Star Forming Regions, IAU Symp. 115*, eds. M. Peimbert and J. Jugaku (Dordrecht: Reidel), p. 334.
Kylafis, N. D. and Norman, C. 1986, *Ap. J.* (*Letters*), **300**, L73.
Lada, C. J. 1985, *Ann. Rev. Astr. Ap.*, **23**, 267.
Langer, W. D. 1985, in *Protostars and Planets* II, eds. D. Black and M. Matthews (Tucson: University of Arizona Press), p. 650.
McKee, C. F. 1987, in *Spectroscopy of Astrophysical Plasmas*, eds. A. Dalgarno and D. Layzer (Cambridge: Cambridge University Press), p. 226.
McKee, C. F., Chernoff, D. F. and Hollenbach, D. J. 1984, *Galactic and Extragalactic Infrared Spectroscopy*, eds. M. F. Kessler and J. P. Phillips (Dordrecht: D. Reidel).
Mullan, D. J. 1971, *MNRAS*, **153**, 145.
Myers, P. C. and Goodman, A. A. 1988, *Ap. J.* (*Letters*), **326**, L27.
Nadeau, D. and Geballe, T. R. 1979, *Ap. J.* (*Letters*), **230**, L169.
Norman, C. and Silk, J. 1979, *Ap. J.*, **228**, 197.

Palma, A., Green, S., DeFrees, D. and McLean, A. D. 1988, *Ap. J. Suppl.*, **68**, 287.
Roberge, W. and Dalgarno, A. 1982, *Ap. J.*, **255**, 176.
Strelnitskij, V. S. 1984, *MNRAS*, **207**, 339.
Tarter, J. C. and Welch, W. J. 1986, *Ap. J.*, **305**, 467.

21

Dissociative shocks

DAVID A. NEUFELD

University of California at Berkeley, USA

21.1. Introduction

Fast shocks destroy molecules. Whereas the compressive and heating effects of slow shocks profoundly alter the chemical composition of a molecular gas but leave its molecular nature intact, interstellar shocks travelling faster than about 50 km s^{-1} result in the complete and very rapid dissociation of any pre-existing molecules by collisional processes (see Figure 3 in McKee, Chernoff and Hollenbach 1984), and if the shock velocity exceeds 70 km s^{-1}, the atomic dissociation products are then largely ionised. Shocks faster than 80 km s^{-1} generate enough ultraviolet radiation to destroy molecules by photoionisation before they even reach the shock front (Hollenbach and McKee 1989).

Dissociative shocks may be present in the interstellar medium wherever gas is moving supersonically at velocities greater than 50 km s^{-1}. Such velocities of bulk motion may result from the outflow of material from young stellar objects, from stellar winds, or from supernova explosions. Clearly, dissociative shocks are intimately associated with stars: they may be generated during the birth, throughout the life and by the explosive death of stars. Emission from fast shocks may serve as a tracer of active star formation in the Galaxy.

Despite the extremely destructive effects of fast shocks, molecules can reform behind a passing dissociative shock before the shocked gas has cooled to the temperature of the ambient, unshocked medium. Free electrons produced behind the shock front initiate the process by catalysing H_2 formation. The rapid return of a dissociated shocked plasma to the molecular phase is a striking demonstration of the tendency toward molecule formation in the interstellar medium. In clouds of primordial composition where no grain surfaces are available to facilitate molecule formation, the effect of fast shocks may be to create rather than to destroy H_2 (MacLow and Shull 1986, Shapiro and Kang 1987). Such molecule creation is an important factor in some models of primordial galaxy and star formation (Ostriker and Cowie 1981, Pudritz and Silk 1989), because the dynamical evolution of a primordial gas cloud is crucially affected by the molecular fraction, H_2 being the only species capable of cooling the gas below 8000 K.

The reformation and chemical evolution of molecules behind a dissociative shock take place in an environment where the gas is warm, partially ionised, partially molecular, and subject to a destructive ultraviolet radiation field that is generated within the shock itself. These unusual conditions lead to a unique chemical signature. A full treatment (Hollenbach and McKee 1979, McKee *et al.* 1984, Neufeld 1987) of the structure of – and emission from – dissociative shocks draws together many of the elements of molecular astrophysics that have been discussed in this book. Equally evident is the fact that such a treatment rests heavily upon the enormous contribution that Alexander Dalgarno has made to the elucidation of those fundamental physical processes that determine the nature and appearance of astrophysical objects.

21.2. Physical and chemical processes

21.2.1. Hydrodynamics

Shocks convert the energy of a bulk fluid motion into random, thermal energy, and finally, when the shocked gas cools, into radiation that we may hope to observe. Because the cooling timescale in the gas behind a fast, dissociative shock exceeds the ion–neutral collision time, the charged and neutral components of the gas have a common bulk velocity and temperature and may be regarded as a single fluid to which magnetic field lines are frozen. The hydrodynamic equations which govern the flow (Field *et al.* 1968, Hollenbach and McKee 1979) say simply that the mass flux is constant, that the momentum flux is constant, and that the enthalpy flux in the shocked gas decreases according to the net cooling rate. The shock front itself is a thin adiabatic region across which the mass, momentum and energy fluxes remain unchanged but the temperature, density and entropy jump suddenly as bulk energy is converted into thermal energy and the supersonic flow becomes subsonic. The dissociative shock models that have been constructed to date make the simplifying assumption that the flow variables are in steady-state. The stability of the steady-state solution has yet to be fully investigated.

21.2.2. Heating and cooling processes

If the net cooling rate is known at each point, the structure of the shocked gas region may be determined by integrating the hydrodynamic equations. The shocked gas cools primarily by the excitation of electronic energy levels of atoms and atomic ions and of rovibrational levels of molecules. Additional cooling results from the collisional transfer of gas kinetic energy to grains, and from the dissociation of H_2 and the ionisation of H. These cooling processes are partially offset by the heating that results from the absorption of radiation emitted by warmer upstream gas, and from the release of chemical potential energy associated with the formation of molecules. Figure 21.1 shows how the magnitude of the major heating and cooling mechanisms varies with the column density behind the shock front, and indicates the resulting temperature profile. The example presented here applies to a shock of velocity 80 km s^{-1} propagating in gas of pre-shock density 10^5 cm^{-3} and shows the structure of the region cooler than 10^4 K.

The gas temperature immediately behind a shock of velocity 80 km s^{-1} is about 10^5 K. At that temperature molecules are rapidly dissociated and the shocked gas cools by the emission of optical and ultraviolet lines of atoms and atomic ions. The emitted spectrum is dependent upon the shock velocity, though under a large

Figure 21.1. (*a*) Temperature profile and (*b*) cooling and heating rates behind a shock of velocity 80 km s^{-1} propagating in gas of pre-shock density 10^5 cm^{-3}. In (*b*) solid lines: cooling by transitions of atoms, OH, and CO; dashed lines: heating associated with photoionisation and photodissociation by ultraviolet radiation, pumping of H_2 by Ly-α, and formation of H_2 on grains.

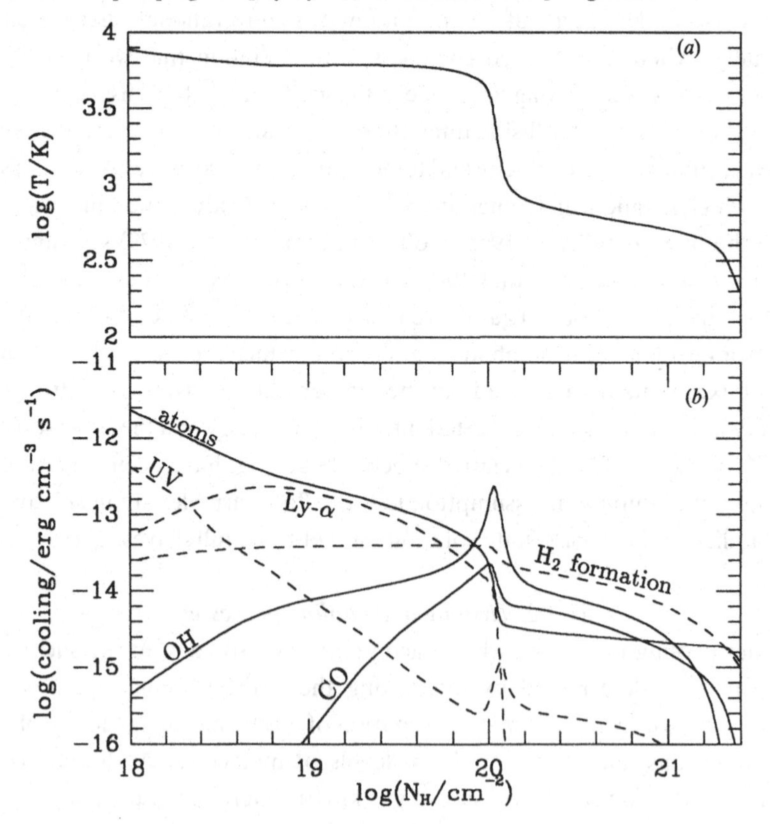

range of conditions much of the shock energy is radiated in the Ly-α line of atomic hydrogen.

Once the gas cools to about 10^4 K, its temperature is temporarily maintained by the heating effects of radiation produced upstream in the hottest part of the shocked gas region. Eventually, the cooling gas is carried far enough behind the shock front that a sufficient column density of material is built up to shield it from that radiation, and the cooling resumes. At this point molecule formation begins, and when the temperature drops below 3000 K molecules become the major coolants. When the temperature falls to a few hundred degrees, the release of chemical potential energy associated with molecule formation becomes a significant heating source. The temperature profile once again shows a plateau, this time until molecule formation is complete.

21.2.3. Chemistry

Because the chemical composition of the shocked gas determines the cooling rate within the molecular region, the hydrodynamic equations must be integrated simultaneously with the coupled set of first order differential equations that describe the chemical evolution. The major molecular coolants are H_2, OH, H_2O, and CO, and their formation and destruction are controlled by about 30 chemical reactions and radiative processes. Figure 21.2 shows how the abundances of these species vary with the column density behind the shock front. Other molecules are of interest not as coolants but as possible chemical diagnostics of dissociative shock activity. Their formation and destruction are governed by a more elaborate network of chemical reactions (e.g. Neufeld (1987), and references therein).

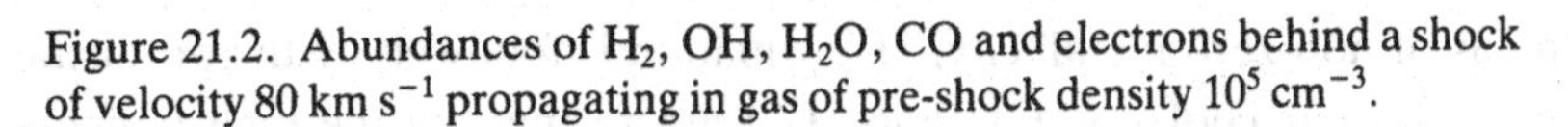

Figure 21.2. Abundances of H_2, OH, H_2O, CO and electrons behind a shock of velocity 80 km s^{-1} propagating in gas of pre-shock density 10^5 cm^{-3}.

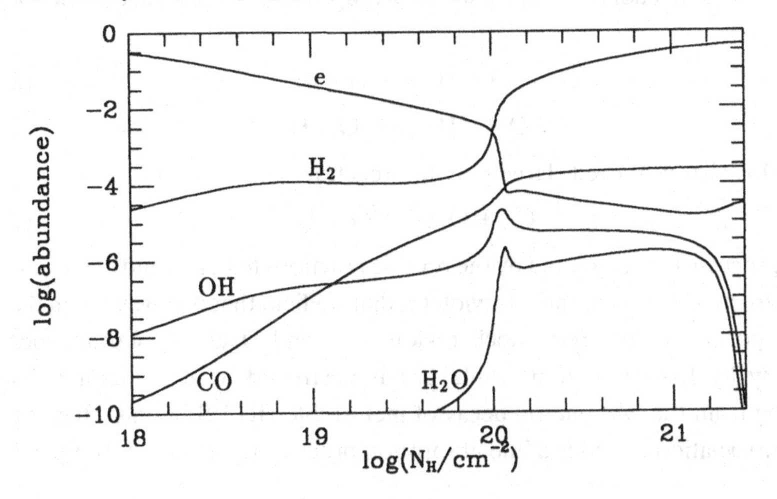

The first molecular system to form is molecular hydrogen. Because the timescale for radiative recombination is long, the shocked gas remains partially ionised at temperatures below 10^4 K, allowing H_2 formation to occur by the sequence

$$H + e \rightarrow H^- + h\nu \tag{21.1}$$

$$H^- + H \rightarrow H_2 + e \tag{21.2}$$

The H^- intermediate may undergo mutual neutralisation with H^+, a reaction which limits the efficiency of H_2 formation. Once the fractional ionisation falls below 2%, the effectiveness of this gas-phase scheme diminishes and grain–surface reactions come into play. Whilst icy grain mantles are photodesorbed in fast shocks, the refractory cores of small grains are heated and eroded but nevertheless survive, even if the shock velocity is as large as 150 km s^{-1} (McKee *et al.* 1987). These grains can catalyse the formation of H_2, although, unfortunately, the efficiency of H_2 formation on the surfaces of warm (~100 K) shock-heated grains can only be guessed at present. The effects of the elevated (~1000 K) gas temperature on the grain–surface reaction rate are also uncertain.

Even if grains are absent, a molecular fraction of 10^{-3} is achieved as a result of gas-phase processes alone. Thus when ionising shocks propagate in low density gas of primordial composition, they generate enough H_2 to cool the post-shock gas below 100 K (MacLow and Shull 1986, Shapiro and Kang 1987). Because primordial gas clouds cannot cool below 8000 K by Ly-α emission alone, the effect of fast shocks is to *facilitate cooling*. However, the effect of non-ionising shocks of intermediate velocities, 20 km $s^{-1} < v <$ 50 km s^{-1}, remains to be considered, and the dependence upon any trace metallicity and upon any external ionising radiation has not yet been fully investigated.

In gas of non-zero metallicity, H_2 formation is followed by the production of other molecular coolants. OH and H_2O are formed by the high temperature reactions

$$O + H_2 \rightarrow OH + H \tag{21.3}$$

$$OH + H_2 \rightarrow H_2O + H \tag{21.4}$$

and CO is also produced, largely by the reaction

$$C^+ + OH \rightarrow CO + H^+ \tag{21.5}$$

These processes are mediated by the reverse reactions to (21.3) and (21.4), and by the destructive effects of the ultraviolet radiation field that is generated in the hot atomic portion of the post-shock region. OH and H_2O are photodissociated primarily by Ly-α radiation, while CO is destroyed by continuum radiation resulting from the two-photon decay of metastable He(2^1S) atoms. Because the photodissociation of CO is a line absorption process, the effects of self-shielding diminish its effectiveness. The photodissociation of CO has been discussed further in Chapters 3 and 16.

When the density of the unshocked gas lies between 10^4 and 10^7 cm^{-3} the conditions in the post-shock region are especially favourable to the formation of OH, and this species is the dominant molecular coolant. OH also has a crucial role as a reactive intermediary in the chemistry of the less abundant elements silicon, sulpur and nitrogen, leading to the formation of oxides via the reactions:

$$X + OH \rightarrow XO + H \tag{21.6}$$

$$X^+ + OH \rightarrow XO^+ + H \tag{21.7}$$

where X is Si, S, or N. If the pre-shock density exceeds 10^7 cm^{-3}, H_2O becomes increasingly significant as a coolant. The formation of water masers behind dissociative shocks has been considered in Chapter 20.

21.2.4. The transfer of Ly-α radiation behind fast shocks

The ultraviolet radiation generated behind dissociative shocks has a profound influence on the chemical evolution in the downstream region of molecule formation, through its photochemical effects. Because much of the shock energy emerges in the Ly-α line, the details of its radiative transfer assume a considerable importance in any treatment of dissociative shocks.

After their original emission in recombining gas at temperatures above 6000 K, Ly-α photons are resonantly scattered a large number of times by ground state hydrogen atoms before arriving in downstream molecular gas where their photo-destructive effects are felt. These resonance line photons are particularly subject to the effects of dust attenuation, because the mean path length that they travel is increased by a factor of 10 or more as a result of the scattering process (Adams 1975, Hummer and Kunasz 1980). Ly-α photons may also be converted into radiation of the two-photon continuum, as a result of collisionally-induced transitions from H(2p) to H(2s).

Furthermore, Ly-α radiation may pump the nearly coincident B-X 1–2 P(5) and 1–2 R(6) lines of molecular hydrogen (Shull 1978, Black and van Dishoeck 1987), resulting in the conversion of Ly-α radiation into H_2 Lyman band emission. Although the pumped $J = 5$ and $J = 6$ levels of the $v = 2$ state lie over 13000 K above the ground state, enough H_2 is produced while the post-shock gas is still warm to allow much of the Ly-α radiation produced in fast molecular shocks to be converted into H_2 Lyman band emission before it is absorbed by dust (Neufeld and McKee (1988): unpublished results). This process heats the gas and modifies the ultraviolet radiation field in the downstream molecular region.

When fast shocks propagate in clouds of modest extinction, there is some prospect of detecting the pumped Lyman band emission. In HH43 and HH47, H_2

emission resulting from pumping in the 1–2 P(5) transition *has* been observed (Schwartz 1983), although it remains unclear whether a single, fast shock is responsible, or whether Ly-α radiation produced in a fast shock pumps H_2 heated in a separate slower, non-dissociative shock. Because the pumped 1–2 P(5) and 1–2 R(6) lines are displaced by different amounts from the Ly-α line centre, the relative strengths of the resulting emission lines provide, in principle, a probe of the Ly-α line profile in the region where H_2 is pumped. Such a probe might allow these two possibilities to be distinguished, though detailed models relevant to H_2 pumping in Herbig–Haro objects have yet to be constructed. Current instrumental sensitivities to H_2 Lyman band emission will be improved upon by the Space Telescope.

21.3. Emission from dissociative shocks

Most of the energy dissipated by dissociative shocks is radiated in optical and ultraviolet lines of atoms and atomic ions, radiation which cannot escape a dense molecular cloud because of dust absorption. Thus in contrast to the emission from non-dissociative molecular shocks, only a few per cent of the energy of dissociative shocks is radiated in infrared rovibrational transitions of molecules and fine-structure transitions of atoms. Nevertheless, such radiation may serve as a diagnostic of fast shock activity.

21.3.1. Chemical diagnostics

Model calculations show that the unique astrophysical environment behind a fast shock gives rise to a characteristic signature of molecular abundances (Neufeld and Dalgarno 1989). Enhanced abundances of SiO and HeH^+ are particularly noteworthy. Silicon monoxide is formed by the sequence,

$$Si^+ + OH \rightarrow SiO^+ + H \tag{21.8}$$

$$SiO^+ + H_2 \rightarrow SiOH^+ + H \tag{21.9}$$

$$SiOH^+ + e \rightarrow SiO + H \tag{21.10}$$

and is destroyed by photoionisation and by reaction with C^+. Conditions in the post-shock region are particularly conducive to SiO formation because of the presence of Ly-α radiation. Ly-α photons (energy 10.2 eV) can photoionise Si (ionisation potential 8.2 eV), producing Si^+, but can neither photoionise SiO (IP 11.6 eV) nor produce the destructive C^+ ion by ionising carbon (IP 11.3 eV). For suitable shock parameters, silicon is largely incorporated into SiO while the gas is still warm, suggesting that high-J SiO emission may be characteristic of fast shocks. Table 21.1 shows how the predicted intensity of the $J = 19$–18 line depends upon the shock velocity. Faster shocks produce less emission because they generate more ultraviolet radiation capable of photoionising SiO. The

Table 21.1. *Predicted line strengths (units of 10^{-5} erg cm^{-2} s^{-1} sr^{-1} averaged over a spherical shell) for shocks with pre-shock density 10^5 cm^{-3}*

	Shock velocity (km s^{-1})				
	60	80	90	100	EDL[a]
HeH^+ $v = 1$–0 P(2) (3.61 μm)	0.5	1.0	1.7	2.7	2
SiO $J = 19$–18 (364 μm)	16	17	4.7	3.1	5

[a] Estimate of the current detection limit.

$J = 19$–18 line, which has yet to be searched for in interstellar gas, lies within an atmospheric transmission window at 364 μm.

To date, HeH^+ has never been detected in interstellar gas. In shocks of velocity 90 km s^{-1} or greater, HeH^+ is formed directly from its constituent atoms by radiative association

$$He^+ + H \rightarrow HeH^+ + h\nu \tag{21.11}$$

and by Penning ionisation of the metastable He(2^3S) atom:

$$He(2^3S) + H \rightarrow HeH^+ + e \tag{21.12}$$

He(2^3S) atoms are produced following radiative recombination of He^+ to triplet states. HeH^+ is destroyed by proton transfer to H and does not undergo rapid dissociative recombination (Roberge and Dalgarno 1982). The conditions in dissociative shocks favour the production of HeH^+ because departures from ionisation equilibrium in the shocked gas permit a substantial (and unusual) coexistence of H and He^+. Furthermore, HeH^+ is formed in shocked gas that is still warm and partially ionised, an environment favourable to the electron-impact excitation of vibrational emission. Table 21.1 shows how the predicted intensity of the $v = 1$–0 P(2) line at 3.607 μm depends upon the shock velocity. The intensity increases with shock velocity because faster shocks produce more He^+.

21.3.2. Shocks in Orion-KL

The best-studied shocked gas region lies in the vicinity of the Kleinmann–Low nebula in the Orion molecular cloud. The postulate of a non-dissociative shock of velocity about 40 km s^{-1} (Chernoff, Hollenbach and McKee 1982, Draine and Roberge 1982, Draine, Roberge and Dalgarno 1983) has been remarkably successful in accounting for the vibrationally excited H_2 (Beckwith *et al.* 1978, Scoville *et al.* 1982) and high-J CO line emission (Watson *et al.* (1985) and references therein) observed in Orion-KL. The broad (FWZI > 200 km s^{-1}) H_2 line wings observed, however, led Chernoff *et al.* (1982) to propose that a second, faster, dissociative shock is also present. This proposal is supported, although not conclusively confirmed, by other observations:

Firstly, the observed line intensity (Werner *et al.* 1984) of the [OI] 63 μm fine-structure line exceeds the predictions of the non-dissociative shock models and might be explained by the additional contribution made by a dissociative shock. However, the discrepancy between the observations and the predictions of the Draine and Roberge (1982) model, in particular, amounts only to a factor of 2–3, and might be accounted for by uncertainties in the rate coefficients of the chemical reactions which produce atomic oxygen.

Secondly, observations of the [SiII] 35 μm fine-structure line (Haas, Hollenbach and Erickson 1986) have revealed enhanced emission at the position of the shocked H_2 and CO emission. Since Si^+ cannot be produced in a non-dissociative shock, there being no internal source of ionising radiation, the additional presence of a dissociative and ionising shock is suggested. However, because [SiII] emission is also produced in an extended photodissociation region (PDR) which overlies the shocked gas region in Orion-KL, the observed intensity profile might simply reflect an enhancement in the PDR emission occurring, by coincidence, at the projected position of the shocked H_2 emission. Similar remarks apply (McKee and Hollenbach 1987) to an observed intensity enhancement at the shocked H_2 position of another tracer of ionised gas, the 51-α recombination line of hydrogen (Hasegawa and Akabane 1984). The [SiII] and H51-α observations were carried out using beams of size 47″ and 30″ respectively, and point to a dissociative shock velocity of between 60 and 80 km s^{-1} and a pre-shock density of between 10^4 and 10^5 cm^{-3}.

Hydrogen recombination lines of the Brackett series permit a search for ionised gas to be carried out at higher spatial resolution. Using a 5″ beam centered at the position of peak H_2 emission, Geballe and Garden (1987) failed to detect Br-α emission. Their upper limit on the line intensity is uncomfortably close to the value expected if a dissociative shock is present; a more sensitive search for the Brackett line emission at the same and nearby positions has now become an urgent priority. The recent advent of infrared CCD array cameras and the future development of infrared echelle spectrometers will greatly facilitate the search for evidence of shock ionised gas in Orion-KL and other molecular regions. If dissociative shocks are present, the detection of HeH^+ may be possible.

Acknowledgements

I am grateful to Alex Dalgarno for his collaboration on much of the research described here. This chapter has been written with the support of a special NASA astrophysics theory program which funds the Center for Star Formation Studies.

REFERENCES

Adams, F. (1975). *Astrophys. J.*, **201**, 350.

Beckwith, W., Persson, S. E., Neugebauer, G., and Becklin, E. E. (1978). *Astrophys. J.*, **223**, 464.

Black, J., and van Dishoeck, E. F. (1987). *Astrophys. J.*, **322**, 412.

Chernoff, D. F., Hollenbach, D. J., and McKee, C. F. (1982). *Astrophys. J. Lett.*, **259**, L97.

Draine, B. T., and Roberge, W. G. (1982). *Astrophys. J. Lett.*, **259**, L91.

Draine, B. T., Roberge, W. G., and Dalgarno, A. (1983). *Astrophys. J.*, **264**, 485.

Field, G. B., Rather, J. D. G., Aannestad, P. A., and Orszag, S. A. (1968). *Astrophys. J.*, **151**, 953.

Geballe, T. R., and Garden, R. (1987). *Astrophys. J.*, **317**, 101.

Haas, M., Hollenbach, D. J., and Erickson, E. F. (1986). *Astrophys. J. Lett.*, **301**, L57.

Hasagawa, T., and Akabane, K. (1964). *Astrophys. J. Lett.*, **287**, L91.

Hollenbach, D. J., and McKee, C. F. (1979). *Astrophys. J. Suppl.*, **41**, 555.

Hollenbach, D. J., and McKee, C. F. (1989). *Astrophys. J.*, **342**, 306.

Hummer, D. G., and Kunasz, P. B. (1980). *Astrophys. J.*, **236**, 609.

MacLow, M. M., and Shull, J. M. (1986). *Astrophys. J.*, **302**, 585.

McKee, C. F., Chernoff, D., and Hollenbach, D. (1984). In *XVI ESLAB Symposium, Galactic & Extragalactic Infrared Spectroscopy*, eds. M. Kessler & J. Phillips, p. 103. Reidel: Dordrecht.

McKee, C. F., and Hollenbach, D. J. (1987). *Astrophys. J.*, **322**, 275.

McKee, C. F., Hollenbach, D. J., Seab, C. G., and Tielens, A. G. G. M. (1987). *Astrophys. J.*, **318**, 674.

Neufeld, D. A. (1987). PhD thesis, Harvard University.

Neufeld, D. A., and Dalgarno, A. (1989). *Astrophys. J.*, **340**, 869.

Ostriker, J. P., and Cowie, L. L. (1981). *Astrophys. J. Lett.*, **247**, L127.

Pudritz, R. E., and Silk, J. (1989). *Astrophys. J.*, **342**, 650.

Roberge, W., and Dalgarno, A. (1982). *Astrophys. J.*, **255**, 489.

Schwartz, R. D. (1983). *Astrophys. J. Lett.*, **268**, L37.

Scoville, N. Z., Hall, D. N., Kleinmann, S. G., and Ridgeway, S. T. (1982). *Astrophys. J.*, **253**, 136.

Shapiro, P. R., and Kang, H. (1987). *Astrophys. J.*, **318**, 32.

Shull, J. M. (1978). *Astrophys. J.*, **224**, 841.

Watson, D. M., Genzel, R., Townes, C. H., and Storey, J. W. V. (1985). *Astrophys. J.*, **298**, 316.

Werner, M. W., Crawford, M. K., Genzel, R., Hollenbach, D. J., Townes, C. H., and Watson, D. M. (1984). *Astrophys. J. Lett.*, **282**, L81.

22

Infrared molecular hydrogen emission from interstellar photodissociation regions

AMIEL STERNBERG

School of Physics and Astronomy, Tel Aviv University, Tel Aviv, Israel

22.1. Introduction

Infrared line emission from vibrationally excited molecular hydrogen (H_2) has been observed in many objects in the Galaxy including planetary and reflection nebulae and various molecular cloud complexes such as those in Orion and the vicinity of the Galactic center. Similar emission has also been detected in diverse extragalactic sources such as Seyfert and interacting galaxies and objects in the Small Magellanic Cloud. Most interstellar molecular hydrogen does not emit substantial vibrational emission since it exists in cold clouds at temperatures less than 100 K while the vibrational energy levels lie many thousands of degrees above the molecular ground state. Considerable infrared vibrational emission is produced, however, in regions which are heated to sufficiently high temperatures where the hydrogen molecules are thermally excited by collisional processes, or in regions where an efficient nonthermal molecular excitation mechanism is operating.

Many of the observed H_2 emitting regions are physically associated with sources of intense ultraviolet radiation. Photons with wavelengths longward of the Lyman limit can escape the ionized clouds of hydrogen gas that usually surround the radiation sources and penetrate into neutral gas clouds called photodissociation regions. The thermal and chemical structures of these regions are critically influenced by the ultraviolet radiation. The molecular hydrogen that is present in these clouds is vibrationally excited by the discrete absorption of ultraviolet photons, and it may also be collisionally excited in warm gas heated by the ultraviolet radiation. The radiative decay of the excited molecular hydrogen produces an infrared spectrum that depends on a variety of physical parameters which may vary considerably from one interstellar photodissociation region to another. An analysis of the H_2 emission provides a wealth of information about the gas densities and temperatures, ultraviolet field intensities, and dust grain properties in such clouds.

22.2. H_2 excitation in photodissociation regions

The gas in photodissociation regions consists primarily of a mixture of atomic and molecular hydrogen gas with densities ranging from $\sim 10^2$ cm^{-3} in the diffuse interstellar clouds to $\sim 10^7$ cm^{-3} at the outer edges of dense star-forming regions. Trace amounts of heavier elements are also present in atomic and molecular form as well as in solid dust grains. The ultraviolet radiation rapidly photodissociates and ionizes molecules and atoms with dissociation and ionization energies less than the Lyman limit. The intensity of the ultraviolet radiation to which such regions are exposed ranges from the average interstellar field intensity of $\sim 10^{-8}$ photons cm^{-2} s^{-1} Hz^{-1} at 100 nm to values many orders of magnitude greater in the vicinity of hot early type stars. More intense radiation fields are usually parameterized by a multiplicative ultraviolet intensity scaling factor, χ, relative to the average interstellar field intensity.

Molecular hydrogen is photodissociated by ultraviolet photons in the 91.2–110.8 nm band. The photons are absorbed in discrete transitions to the excited $B^1\Sigma_u^+$ and $C^1\Pi_u$ electronic states following which the molecule spontaneously decays to the vibrational continuum of the ground $X^1\Sigma_g^+$ electronic state resulting in the dissociation of the molecule (see Figure 3.1). The excellent agreement between theoretical calculations and laboratory measurements of the ultraviolet continuum emitted during the molecular dissociation (Dalgarno, Herzberg and Stephens 1970) is remarkable, and demonstrates how well the interaction of molecular hydrogen with ultraviolet radiation is understood.

Interstellar molecular hydrogen is usually formed by association on the surfaces of dust grains at a rate per unit volume Rnn_1, where $n = n_1 + 2n_2$ is the total density of hydrogen nuclei, and where n_1 and n_2 are the atomic and molecular densities respectively. The rate coefficient R depends on the grain cross section per hydrogen nucleus and the atom–grain collision frequency. For typical gas to dust ratios it is of order $3 \times 10^{-18} T^{1/2} y_F$ cm^3 s^{-1} where T is the gas temperature. The parameter y_F is the efficiency with which molecules are formed per atom–grain collision.

The atomic to molecular hydrogen density ratio varies through the cloud due to the changes in the rates of molecular formation and destruction. At the outer edges of photodissociation regions the gas is almost entirely atomic since the molecular destruction rate is much faster than the molecular formation rate, even in high density clouds exposed to relatively weak ultraviolet fields. As the incident ultraviolet radiation penetrates the cloud it is attenuated by the H_2 line absorptions and continuum absorption by the dust grains and the H_2 photodissociation rate is diminished. At sufficiently large cloud depths the ultraviolet radiation is completely absorbed and the H_2 is destroyed very slowly by collisions with cosmic ray protons and the hydrogen gas is almost entirely molecular.

Molecular hydrogen exposed to ultraviolet radiation is also vibrationally excited (by the same photons which initiate the molecular dissociation) if the downward electronic transitions terminate in bound vibrational levels rather than in the continuum of the ground electronic state (see Figure 3.1). The ground electronic state supports 15 bound vibrational energy levels which are identified by the quantum number v. The energy separation between the lowest pair of these levels is 0.5 eV (5800 K). A series of rotational levels, each of which is identified by the quantum number j are associated with each vibrational level of the molecule. Only slow quadrupole radiative transitions may occur between the rovibrational levels of the ground electronic state because H_2 is a homonuclear molecule. During such transitions the rotational quantum number may change by 2, 0 or −2 (excluding 0–0 transitions) with no restrictions on the change in the vibrational quantum number. These transitions are identified by the vibrational quantum numbers of the upper and lower levels and by the letters S(j), Q(j), and O(j) depending on whether the rotational quantum number j of the lower level of the transition is smaller than, equal to, or greater than that of the upper level of the transition. The commonly observed H_2 line at 2.12 μm is, for example, due to the 1–0 S(1) transition. The ultraviolet pumping of the excited vibrational levels initiates a cascade of quadrupole rovibrational transitions which results in an infrared fluorescent emission line spectrum which was first calculated in detail by Black and Dalgarno (1976). Many fluorescent emission lines have been observed in various objects such as the reflection nebular Parsamyan 18 (Sellgren 1986) and the planetary nebular Hubble 12 (Dinerstein *et al.* 1988).

The gas in photodissociation regions is heated by the incident ultraviolet radiation in various ways. Two important heating mechanisms are the photoejection of energetic electrons from the surfaces of dust grains, and the collisional deexcitation of ultraviolet pumped molecular hydrogen. The gas is cooled primarily by the emission of far-infrared fine-structure emission lines of trace atoms and ions. Of particular importance are the OI $^3P_1$–$^3P_2$ 63 μm and the CII $^2P_{3/2}$–$^2P_{1/2}$ 158 μm cooling lines both of which have been observed in numerous sources such as the neutral gas disk at the Galactic center (Genzel *et al.* 1985). The rates of heating and cooling vary through the cloud, and therefore, so does the equilibrium gas temperature. The thermal structure of photodissociation regions depends on the intensity of the incident ultraviolet radiation, the gas density, the fractional abundances of the trace elements, as well as the abundances and physical properties of the dust grains. Explicit calculations (Sternberg and

Dalgarno 1988) indicate that the gas in the outer boundaries of photodissociation regions can attain temperatures which range from $\sim 10^2$ K to $\sim 10^3$ K.

The molecular hydrogen in photodissociation regions is vibrationally excited by the ultraviolet pumping mechanism as well as by inelastic collisions with other hydrogen particles if the gas is sufficiently warm. The infrared H_2 emission lines that are produced in photodissociation regions are usually optically thin and their intensities are given by

$$I(v\xi \rightarrow v'j') = \frac{1}{4\pi} N_{vj} A(vj \rightarrow v'j') h\nu \text{ erg s}^{-1} \text{ cm}^{-2} \text{ s}^{-1} \tag{22.1}$$

where vj and $v'j'$ are the upper and lower rotational and vibrational quantum numbers of the transition, A is the quadrupole radiative transition probability, ν is the transition frequency, and N_{vj} is the total column density of molecules in level vj along the line of sight. Steady-state plane-parallel models have been constructed by Sternberg and Dalgarno (1988) to compute the excited state column densities for a wide range of cloud conditions. In these models it is assumed that all of the incident ultraviolet radiation is absorbed in the cloud, and the H_2 level populations are calculated as functions of cloud depth and are integrated to yield the excited level column densities N_{vj} and resulting infrared line intensities.

22.2.1. Low density regions

In low density gas ($n \leqslant 10^4$ cm^{-3}) the temperature remains low ($\sim 10^2$ K) and is approximately constant throughout the entire photodissociation region and ultraviolet pumping dominates the H_2 excitation. At these densities and temperatures the quadrupole cascade is unaffected by collisions and a pure, nonthermal, radiative fluorescent emission spectrum is produced (see Figure 1). The *relative* intensities of the fluorescent emission lines depend primarily on the branching ratios of the radiative cascade, which are internal molecular constants, and are insensitive to external cloud conditions. In particular, the intensity ratio of the 2.12 μm 1–0 S(1) and 2.25 μm 2–1 S(1) lines is ~ 2 for such spectra. The excitation temperature of ~ 6000 K that corresponds to this line ratio is much higher than the kinetic temperature of the gas and is a reflection of the nonthermal nature of the ultraviolet pumping excitation mechanism.

The *absolute* intensities of the fluorescent emission lines do depend, however, on the efficiency with which the incident ultraviolet radiation is absorbed by the molecular hydrogen gas and converted into infrared radiation. This efficiency depends simply on the gas density n, the ultraviolet field intensity scaling factor χ, the molecular formation rate coefficient R, and the effective grain ultraviolet continuum absorption cross section σ. In order to understand the dependence of the ultraviolet to infrared conversion efficiency it is useful to consider a simplified model of the H_2 molecule in which there are only two bound vibrational levels in the ground electronic state and a single level in an excited electronic state. The lower vibrational level represents all of the low lying rotational levels from which the H_2 is excited, and the upper vibrational level represents all of the ultraviolet pumped rovibrational levels.

In photodissociation regions which are in steady-state the rates of molecular formation and destruction are equal so that,

$$Rnn_1 = \chi D\Theta n_2 \tag{22.2}$$

where $D = 5.4 \times 10^{-11}\ \mathrm{s}^{-1}$ is the unattenuated H_2 photodissociation rate in the average interstellar field, and Θ is the ultraviolet attenuation function for the single electronic transition of the simplified H_2 molecule. The ultraviolet attenuation function at a given depth into a plane-parallel cloud may be approximated by the product

$$\Theta = f(N_2)\exp\left[-\sigma(N_1 + 2N_2)\right] \tag{22.3}$$

where N_1 and N_2 are the atomic and molecular hydrogen column densities measured from the outer edge of the cloud to the specific cloud depth. The exponential factor accounts for the continuum attenuation by dust scattering and absorption. The value of the effective ultraviolet continuum cross section, σ, depends on the gas to dust ratio and the grain ultraviolet scattering properties (Roberge, Dalgarno and Flannery 1981). It is typically of order $10^{-21}\ \mathrm{cm}^2$ in the 91.2–110.8 nm H_2 photodissociation band. The attenuation of the ultraviolet radiation due to the H_2 line absorption is described by the line self-shielding function, $f(N_2)$, which depends on the column density of H_2 molecules, and it is proportional to the derivative $\mathrm{d}W/\mathrm{d}N_2$ where W is the equivalent width of the ultraviolet absorption line (Federman, Glassgold and Kwan 1979).

Because $n_1/n_2 = \mathrm{d}N_1/\mathrm{d}N_2$ Equation (22.2) is a differential equation relating N_1 and N_2 and it follows that

$$N_1 = \left(\frac{1}{\sigma}\right)\ln\left[\alpha G(N_2) + 1\right]\ \mathrm{cm}^{-2} \tag{22.4}$$

where the dimensionless parameter $\alpha \equiv D\chi/Rn$ is the ratio of the unattenuated photodissociation and molecular formation rates, and where $G(N_2)$ is the dimensionless integral

$$G(N_2) \equiv \int_0^{N_2} \sigma f(N_2')\exp(-2\sigma N_2')\,\mathrm{d}N_2' \tag{22.5}$$

The total column density, N_1^{tot}, of atomic hydrogen in the photodissociation region is obtained from Equation (22.4) by letting $N_2 \rightarrow \infty$. The integral in Equation (22.5) then converges to a constant G, and

$$N_1^{tot} = (1/\sigma) \ln (\alpha G + 1) \text{ cm}^{-2} \quad (22.6)$$

At typical gas to dust ratios the length scales associated with the line and continuum opacities differ by many orders of magnitude. The onset of line self-shielding occurs at an H_2 column density of $\sim 10^{14}$ cm^{-2} which is very small compared with $1/\sigma \sim 10^{21}$, and therefore, the constant $G \ll 1$ and is typically of order 5×10^{-5}. In fact, throughout most of a photodissociation region the H_2 absorption lines are very optically thick and the absorption occurs in the damping wings of the lines where the self-shielding function varies as $N_2^{-1/2}$. The value of G therefore scales with the ultraviolet continuum absorption cross section as $G(\sigma)/G(\sigma') = (\sigma/\sigma')^{1/2}$.

The total intensity of fluorescent emission is proportional to N^*, the total column density of vibrationally excited molecules in the cloud. N^* must be proportional to N_1^{tot} since the vibrational excitation and molecular dissociation are both initiated by the absorption of the same ultraviolet photons. In fact,

$$N^* = \frac{Rn}{A}\left(\frac{P}{D}\right)N_1^{tot} \quad (22.7)$$

where P is the unattenuated ultraviolet rate, so that P/D is the ratio of vibrational excitations to dissociations per ultraviolet photon absorption. The additional proportionality factor Rn/A is the ratio of the rate at which molecules are formed to the rate at which the vibrationally excited molecules are destroyed by quadrupole radiative decay.

The quantity $Y \equiv N^*/\chi$ measures the efficiency with which incident ultraviolet radiation is converted into H_2 infrared line emission, and from the above it follows that

$$Y(\alpha) = \frac{1}{\sigma}\frac{P}{A}\frac{1}{\alpha} \ln (\alpha G + 1) \text{ cm}^{-2} \quad (22.8)$$

The rate of molecular formation on grain surfaces is very slow compared with the rate of unattenuated molecular photodissociation so that $\alpha \gg 1$ for typical values of χ/n. However, since G is very small the product αG can be either much greater than or smaller than unity for realistic values of n, χ and R, and $\alpha G \sim 1$ when $\chi/n \sim 0.01$ cm^3 (for $R = 3 \times 10^{-17}$ cm^3 s^{-1}). When $\alpha G \ll 1$ the conversion efficiency function is independent of α, that is of n, χ, and R, and when $\alpha G > 1$ it decreases as $\ln(\alpha)/\alpha$. Since the infrared intensity is proportional to χY it follows from Equation (22.8) that when $\alpha G \ll 1$ the fluorescent emission intensity is inversely proportional to $\sigma^{1/2}$, is independent of the molecular formation rate Rn, and is proportional to χ. When $\alpha G > 1$ the emission intensity is inversely proportional to σ, is proportional to Rn, and increases only logarithmically with χ. Thus, for typical values of R and σ when $\chi/n \ll 0.01$ cm^3 the fluorescent emission intensity is proportional to the intensity of the incident ultraviolet radiation.

When $\chi/n > 0.01$ cm^3 the emission intensity effectively saturates and is sensitive only to very large changes in the ultraviolet intensity and increases only logarithmically with χ.

In the full multilevel problem, where the ultraviolet pumping occurs through many absorption lines, the ultraviolet to infrared conversion efficiency function is simply a sum over terms like Equation (22.6) (Sternberg 1988). The total fluorescent emission intensity is given by $\chi Y^{\rm tot}$ where for a wide range of the parameters χ, n, R, and σ the function $Y^{\rm tot}$ is well approximated by the expression

$$Y^{\rm tot} = 3.2 \times 10^{-8}\left(\frac{\sigma_0}{\sigma}\right)\left(\frac{R}{R_0}\right)\frac{n}{\chi}\ln\left[90(\sigma/\sigma_0)^{1/2}(R_0/R)\frac{\chi}{n} + 1\right] \mathrm{erg\,s^{-1}\,cm^{-2}\,sr^{-1}} \tag{22.9}$$

where $\sigma_0 = 1.9 \times 10^{-21}$ cm^{-2} and $R_0 = 3 \times 10^{-17}$ cm^3 s^{-1}. The total emission is distributed over the various quadrupole lines with fractional intensities which are insensitive to the cloud conditions. The fractional intensities of the 20 strongest lines of the spectrum displayed in Figure 22.1 are listed in Table 22.1.

Equation (22.9) can be used to estimate the combinations of cloud parameters that are consistent with the intensities of the H_2 fluorescent emission lines that are observed in specific objects. For example, a number of fluorescent emission lines

Figure 22.1. Theoretical H_2 emission spectrum produced in a photodissociation region with a hydrogen density $n = 10^3$ cm^{-3} exposed to an ultraviolet field with $\chi = 10^2$ (Sternberg and Dalgarno 1988).

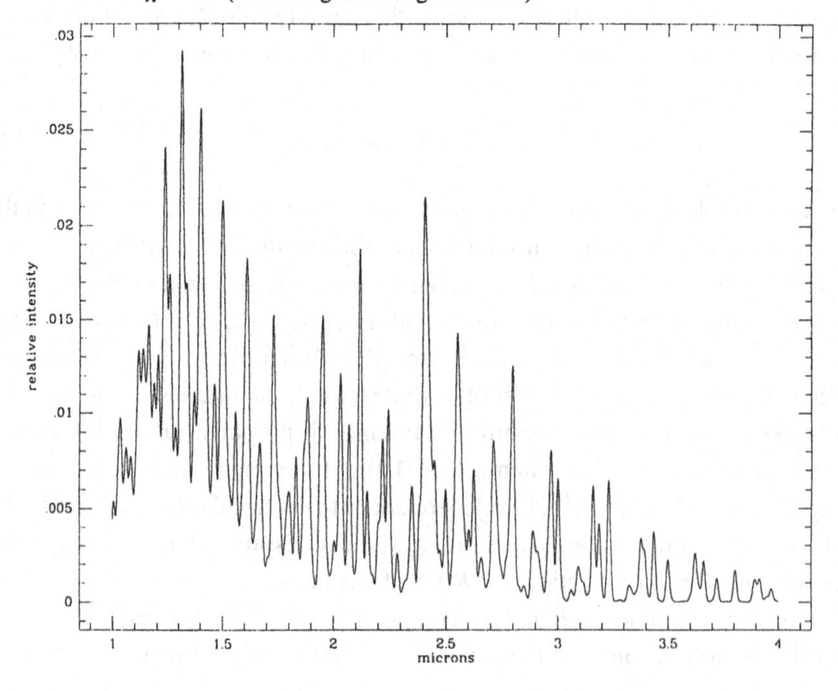

Table 22.1. *H_2 fluorescent emission line intensities*[a]

μm	$100I/I_{tot}$	Line	μm	$100I/I_{tot}$	Line
1.16	0.67	2,0 S(1)	2.12	1.59	1,0 S(1)
1.23	0.85	3,1 S(1)	2.22	0.76	1,0 S(0)
1.31	0.79	4,2 S(1)	2.25	0.89	2,1 S(1)
1.31	0.79	3,1 Q(1)	2.41	1.36	1,0 Q(1)
1.40	0.80	4,2 Q(1)	2.41	0.84	1,0 Q(2)
1.49	0.71	5,3 Q(1)	2.42	1.12	1,0 Q(3)
1.51	0.68	4,2 O(3)	2.55	0.86	2,1 Q(1)
1.83	0.68	1,0 S(5)	2.57	0.65	2,1 Q(3)
1.96	1.26	1,0 S(3)	2.80	1.15	1,0 O(3)
2.03	0.97	1,0 S(2)	2.97	0.74	2,1 O(3)

[a] The fractional intensities of the 20 strongest lines of the H_2 emission spectrum in the 1–4 μm band produced in a photodissociation region with $n_T = 10^3$ cm^{-3} and $\chi = 10^2$, relative to the total intensity $I_{tot} = 4.90 \times 10^{-5}$ erg s^{-1} cm^{-2} s^{-1}. This spectrum is displayed graphically in Figure 22.1.

have been observed in the reflection nebula Parsamyan 18 with intensity ratios which are in good agreement with the theoretically predicted values (Sellgren 1986). The inferred value of the total fluorescent emission intensity in Parsamyan 18 is ~3×10^{-3} erg s^{-1} cm^{-2} s^{-1}. It follows from Equation (22.9) that the minimum value of χ that is consistent with the observed emission is ~$2 \times 10^3(\sigma/\sigma_0)^{1/2}$. An ultraviolet field with this intensity can produce the observed emission only if $Rn/R_0 \gg 10^4(\sigma/\sigma_0)^{1/2}$. The observed line ratios imply, however, that the quadrupole cascade is unaffected by collisional processes so that the hydrogen density, n, cannot be much greater than 10^4 cm^{-3}. Thus, the weakest possible ultraviolet field can produce the Parsamyan 18 emission only if $R \gg R_0$ in this nebula. If, on the other hand, it is assumed that in this object $R \sim R_0$ then $\chi(\sigma/\sigma_0)^{1/2}$ must be much larger than 10^3. However, the star that is illuminating this reflection nebula is probably not producing ultraviolet radiation with an intensity corresponding to $\chi > 10^4$ at the boundary of the fluorescing region, so that if $R \sim R_0$ then σ must be considerably smaller than σ_0. The continuum absorption cross section could be this small if the dust in Parsamyan 18 consists of very forward scattering grains.

22.2.2. High density regions

In high density ($n \gtrsim 10^4$ cm^{-3}) photodissociation regions the H_2 quadrupole cascade and associated emission spectrum are modified by inelastic collisional processes such as

$$H + H_2(vj) \rightarrow H + H_2(v'j')$$

$$H_2 + H_2(vj) \rightarrow H_2 + H_2(v'j')$$

which influence the distribution of the excited rovibrational levels. In dense regions the gas heating efficiency increases due to the collisional deexcitation of the ultraviolet pumped molecular hydrogen. This, combined with the decrease in the cooling efficiency due to the collisional quenching of the fine-structure cooling transitions, results in high equilibrium temperatures exceeding 10^3 K in the outer portions of the clouds. In the warm gas the thermal collisional excitation of the low lying vibrational levels becomes very rapid relative to the ultraviolet pumping rates of these levels. If the size of the warm region is sufficiently large the total H_2 emission is dominated by the thermal emission which is produced in the warm gas, and in particular the intensities of the lines emitted from the $v = 1$ levels are much more intense than all of the other lines. The intensity ratio of the 2–1 S(1) and 1–0 S(1) lines emitted from such clouds typically ranges from 0.1 to 0.01, and the excitation temperature derived from this ratio is close to the average kinetic temperature of the gas in the warm region.

The absolute and relative intensities of the thermal emission lines depend on the temperature and size of the warm region which depend in a complicated way on a variety of parameters including the gas density, the intensity of the incident ultraviolet field, and the molecular formation coefficient whose value is uncertain at high gas temperatures. In high density regions the total column density of vibrationally excited H_2 which is due to ultraviolet pumping is approximately

$$N^* = \frac{1}{\sigma}\frac{R}{q}\frac{P}{D}\ln(\alpha G + 1) \tag{22.10}$$

where in Equation (22.7) the quadrupole radiative rate is replaced by the vibrational collisional deexcitation rate qn, where q is the collisional deexcitation rate coefficient. It follows that collisional processes dominate the production of the total column density of vibrationally excited H_2 when

$$N_2^{\rm w} > 3 \times 10^{15} \exp(6000\ {\rm K}/T)\left(\frac{\sigma_0}{\sigma}\right)\left(\frac{R}{R_0}\right)\left(\frac{q_0}{q}\right)\ln(\alpha G + 1)\ {\rm cm}^{-2} \tag{22.11}$$

where $N_2^{\rm w}$ is the column density of molecules in the warm region, T is the temperature of the warm gas, and where $q_0 = 10^{-10}\,{\rm cm}^3\,{\rm s}^{-1}$. It is evident from this equation that even if the size of the warm region is a very small fraction of the photodissociation region, whose size is of order $1/\sigma$, the thermally excited molecules in the warm region can dominate the total column density of vibrationally excited H_2. Sternberg and Dalgarno (1989) have computed the H_2 emission spectra produced in dense regions for a range of conditions. The H_2 emission spectrum calculated for a photodissociation region with a hydrogen density of 10^6 cm^{-3} which is exposed to an ultraviolet field with $\chi = 10^2$ is displayed in Figure 22.2, and the fractional intensities of the 20 strongest lines are listed in Table 22.2.

Table 22.2. *H_2 thermal emission line intensities*[a]

μm	$100I/I_{tot}$	Line	μm	$100I/I_{tot}$	Line
1.75	0.16	1,0 S(7)	2.44	0.34	1,0 Q(4)
1.83	0.40	1,0 S(5)	2.45	0.62	1,0 Q(5)
1.89	0.26	1,0 S(4)	2.47	0.12	1,0 Q(6)
1.96	1.27	1,0 S(3)	2.50	0.18	1,0 Q(7)
2.03	0.61	1,0 S(2)	2.63	0.59	1,0 O(2)
2.12	2.17	1,0 S(1)	2.80	2.07	1,0 O(3)
2.22	0.59	1,0 S(0)	3.00	0.50	1,0 O(4)
2.41	2.44	1,0 Q(1)	3.23	0.86	1,0 O(5)
2.41	0.65	1,0 Q(2)	3.50	0.14	1,0 O(6)
2.42	1.53	1,0 Q(3)	3.81	0.17	1,0 O(7)

[a] The fractional intensities of the 20 strongest lines of the H_2 emission spectrum in the 1–4 μm band produced in a photodissociation region with $n_T = 10^6$ cm^{-3} and $\chi = 10^2$, relative to the total intensity $I_{tot} = 5.38 \times 10^{-5}$ erg s^{-1} cm^{-2} s^{-1}. This spectrum is displayed graphically in Figure 22.2.

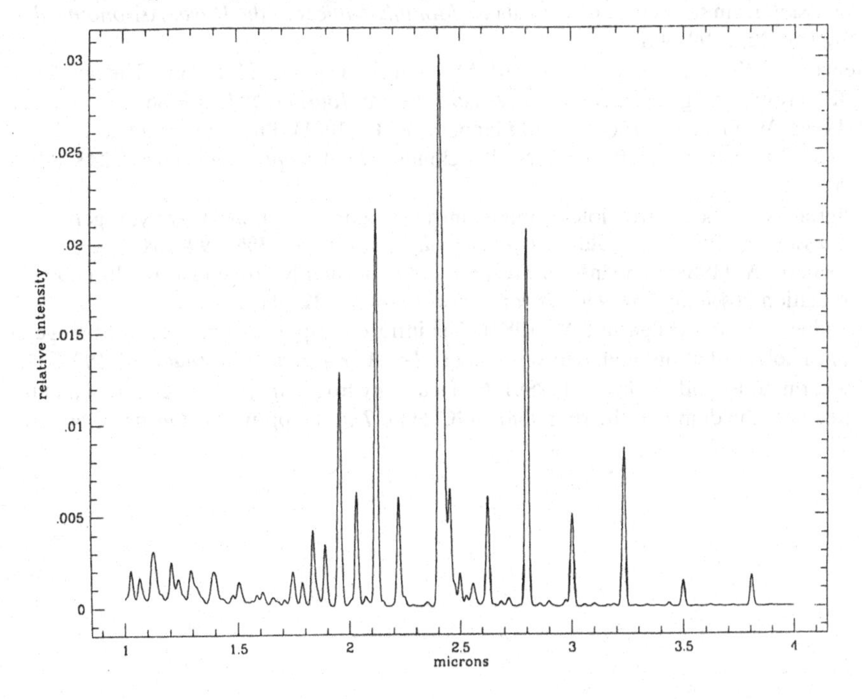

Figure 22.2. Theoretical H_2 emission spectrum produced in a photodissociation region with a hydrogen density $n = 10^6$ cm^{-3} exposed to an ultraviolet field with $\chi = 10^2$ (Sternberg and Dalgarno 1989).

Thermal H_2 emission is also produced in shock heated regions, and it may, in practice, be difficult to determine whether the particular source is being heated dynamically or by ultraviolet radiation. The presence of faint fluorescent emission lines from very high lying vibrational levels, which cannot be rapidly collisionally excited even in warm gas, may be used to distinguish between radiatively heated and shock heated H_2 gas. The thermal H_2 emission which has been observed near sources of very intense ultraviolet radiation such as the neutral clouds at the Galactic center (Gatley *et al.* 1986) and various planetary nebulae (Zuckerman and Gatley 1988) may be produced in radiatively heated gas.

REFERENCES

Black, J. H. and Dalgarno, A. (1976). Interstellar H2: The population of excited rotational states and the infrared response to ultraviolet radiation. *The Astrophysical Journal,* **203**, 132–42.

Dalgarno, A., Herzberg, G. and Stephens, T. L. (1970). A new continuous emission spectrum of the hydrogen molecule. *The Astrophysical Journal Letters*, **162**, L49–53.

Dinerstein, H. L., Lester, D. F., Carr, J. S. and Harvey, P. M. (1988). Detection of fluorescent molecular hydrogen emission in the planetary nebula Hubble 12. *The Astrophysical Journal Letters*, **327**, L27–30.

Federman, S. R., Glassgold, A. E. and Kwan, J. (1979). Atomic to molecular hydrogen transition in interstellar clouds. *The Astrophysical Journal*, **227**, 466–73.

Gatley, I., Jones, T. J., Hyland, A. R., Wade, R., Geballe, T. R. and Krisciunas, K. (1986). The spatial distribution and velocity field of the molecular hydrogen line emission from the centre of the galaxy. *Monthly Notices of the Royal Astronomical Society*, **222**, 299–306.

Genzel, R., Watson, D. M., Crawford, M. K. and Townes C. H. (1985). The neutral gas disk around the galactic center. *The Astrophysical Journal*, **297**, 766–86.

Roberge, W. G., Dalgarno, A. and Flannery, B. P. (1981). Photoionization and photodissociation in diffuse interstellar clouds. *The Astrophysical Journal*, **243**, 817–26.

Sellgren, K. (1986). Ultraviolet-pumped infrared fluorescent molecular hydrogen emission in reflection nebulae. *The Astrophysical Journal*, **305**, 399–404.

Sternberg, A. (1988). The infrared response of molecular hydrogen gas to ultraviolet radiation: a scaling law. *The Astrophysical Journal,* **332**, 400–9.

Sternberg, A. and Dalgarno, A. (1989). The infrared response of molecular hydrogen to ultraviolet radiation: high density regions. *The Astrophysical Journal,* **338**, 197–233.

Zuckerman, B. and Gatley, I. (1988). Molecular hydrogen maps of extended planetary nebulae: the dumbell, the ring, and NGC 2346. *The Astrophysical Journal*, **324**, 501–15.

VII

Molecules near stars and in stellar ejecta

23

Masers in the envelopes of young and old stars

J. M. MORAN

Harvard–Smithsonian Center for Astrophysics, Cambridge, Massachusetts, USA

23.1. Introduction

The first observation of an extraterrestrial radio line from a molecule was made 25 years ago in 1963 with the discovery of the 1665 and 1667 MHz lines of OH in absorption against the powerful background source Cassiopeia A. Shortly thereafter, studies of OH led to the surprise discovery of very strong emission in the directions of HII regions. Maser action was postulated to explain the unusual characteristics of this emission. Figure 23.1 shows a set of OH spectra from eight different transitions towards the first identified maser, W3(OH). The energy level diagram and transitions are shown in Figure 23.2. These spectra have many interesting characteristics: (1) the spectrum of each transition consists of many narrow features with linewidths typical of thermal gas at 4–100 K; (2) the features are spread over $\sim$10 km s^{-1}, indicating considerable dynamic activity; (3) the spectra among the various transitions are quite different, suggesting complex excitation conditions; (4) the features are highly circularly polarized due to the Zeeman effect, revealing milliGauss magnetic fields. VLBI observations of the OH lines show that each spectral feature corresponds to a 'spot' of emission of angular size $0''.01$ ($\sim 10^{14}$ cm). The spots are distributed over $2''$ (10^{16} cm) in front of the compact HII region, as shown in Figure 23.3. The brightness temperature of the emission is $\sim 10^{12}$ K. The only viable mechanism that has been proposed to explain these results, especially the very high brightness temperatures and narrow linewidths, is maser action.

Over the past 23 years, more than a thousand masers have been found, and strong maser action has been established as a common phenomenon in the dusty envelopes of both newly formed massive stars and cool evolved stars. Strong maser action requires an energy source for power, non-LTE conditions to allow efficient pumping, and a large optical depth. Such maser action has been found in

Figure 23.1. Spectra of OH maser lines toward the compact HII region W3(OH). The transitions are labeled by frequency and identified in Figure 23.2. The spectra in solid lines were measured in right circular polarization and those in dashed lines in left circular polarization. The velocity axis is referred to the local standard of rest. The systematic shift between the different circularly polarized spectra in each transition is probably due to the Zeeman effect in a field of ~5 mG. Adapted from Guilloteau (1982).

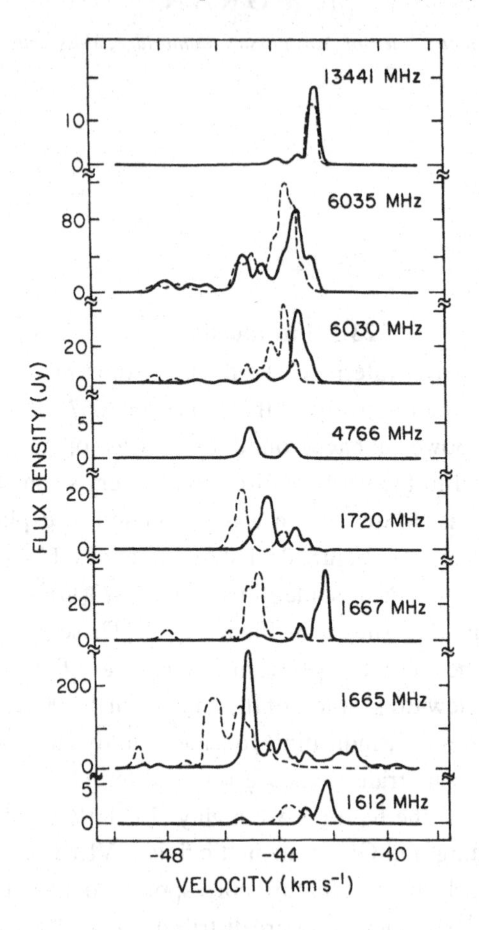

Figure 23.2. Part of the rotational energy level diagram of OH. The rotational ladder has two branches due to spin splitting. Each rotational level is split by lambda doubling and hyperfine interaction. The levels are marked by the rotational quantum number J and by the total angular momentum quantum number F. The known maser transitions are marked. They are: $^2\Pi_{\frac{3}{2}}$, $J = \frac{3}{2}$, $F = 1 \rightarrow 2$ (1612.231 MHz); $J = \frac{3}{2}$, $F = 1 \rightarrow 1$ (1665.402 MHz); $J = \frac{3}{2}$, $F = 2 \rightarrow 2$ (1667.359 MHz); $J = \frac{3}{2}$, $F = 2 \rightarrow 1$ (1720.530 MHz); $J = \frac{5}{2}$, $F = 2 \rightarrow 2$ (6030.747 MHz); $J = \frac{5}{2}$, $F = 3 \rightarrow 3$ (6035.092 MHz); $J = \frac{7}{2}$, $F = 4 \rightarrow 4$ (13441.417 MHz); $^2\Pi_{\frac{1}{2}}$, $J = \frac{1}{2}$, $F = 0 \rightarrow 1$ (4660.420 MHz), and $J = \frac{1}{2}$, $F = 1 \rightarrow 0$ (4765.562 MHz). From Moran (1982).

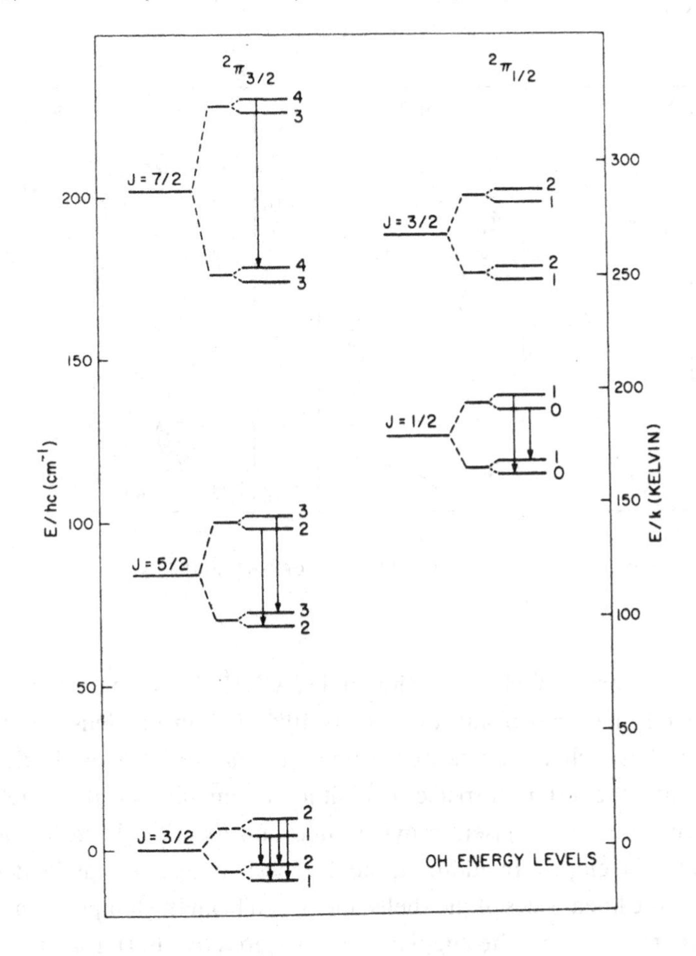

Figure 23.3. (*a*) The OH and CH_3OH masers towards the compact HII region W3(OH). The contours show the thermal radio continuum emission from the HII region. The filled circles mark the positions of the OH masers at 1665 MHz and the filled squares the OH masers at 6035 MHz. Zeeman pairs are identified by the lines linking 6035 MHz maser spots. The letters A, B, and C mark the positions of the OH masers at 4765 MHz. The open squares and crosses mark the positions of the CH_3OH masers at 23.1 GHz and 12.1 GHz, respectively. The dashed box shows the area blown up in (*b*). (*b*) The detailed images of the CH_3OH maser emission with velocities indicated. The 23.1 GHz emission is shown on the left and the 12.1 GHz emission on the right. From Menten *et al.* (1988).

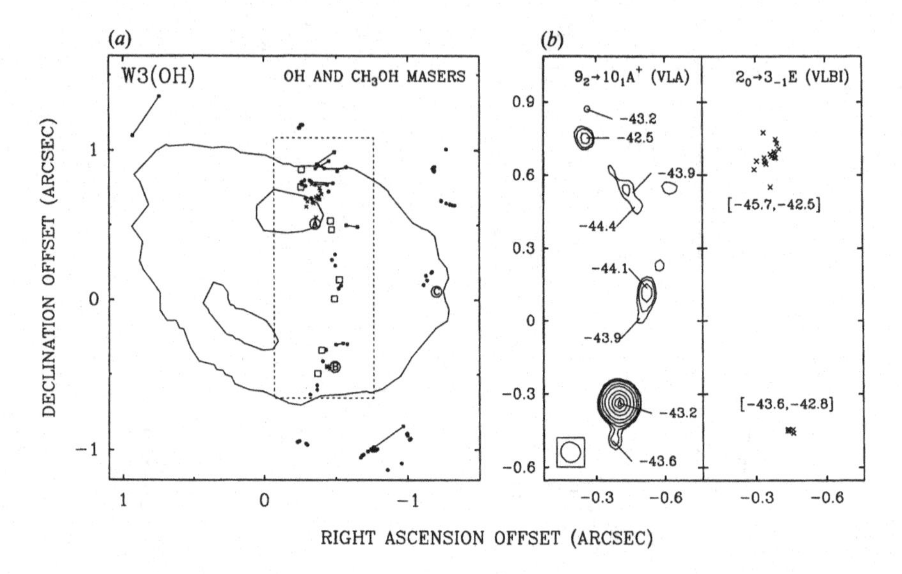

various transitions of OH, H_2O, SiO, and CH_3OH. These masers are probes of the extended envelopes of stars on scales of 10^{13}–10^{17} cm and densities of 10^6–10^{11} cm^{-3}. The basic characteristics of cosmic masers are listed in Table 23.1. In evolved stars, the different species exhibit maser emission in circumstellar shells at different radii. The masers provide important dynamical information about these stellar envelopes. In addition, the distances to the stars can be deduced by comparing the linear sizes of the shells, measured from the temporal variations of the maser emission, and the angular sizes measured by interferometry. In young stellar objects, masers are frequently associated with bipolar outflows, infrared sources, and ultracompact HII regions. The relative motions of H_2O maser features have been tracked by VLBI, thus providing velocity information in three dimensions. The distances to some of these masers have been determined by comparing transverse angular velocities and linear, line-of-sight velocities from the Doppler shifts of various spectral components.

Table 23.1. *Typical parameters of strong masers*

Type	A (s^{-1})	Γ (s^{-1})	T_s^a (K)	F (Jy)	T_B (K)	$\theta_s/\theta_d{}^b$ (mas/″)	$\Delta v/\delta v^c$ ((km s^{-1})/(km s^{-1}))
OH – stellar (1612 MHz)	1.3×10^{-11}	0.03	1×10^{11}	2×10^2	10^8	$10^3/1$	1/30
H_2O – stellar (22.2 GHz)	1.2×10^{-9}	1	3×10^{11}	4×10^3	10^{11}	$10/10^{-1}$	1/20
SiO – stellar (43.1 GHz)	3×10^{-6}	5	2×10^9	2×10^3	10^{10}	$10/10^{-1}$	1/10
OH – interstellar (1665 MHz)	7.2×10^{-11}	0.03	2×10^{10}	2×10^2	10^{12}	10/1	0.1/5
H_2O – interstellar (22.2 GHz)	1.9×10^{-9}	1	3×10^{11}	4×10^4	10^{14}	1/1	1/50
CH_3OH – interstellar (12.1 GHz)	7.9×10^{-9}	1	5×10^{10}	5×10^2	$>10^{12}$	$<2/2$	1/5

[a] Saturation brightness temperature, with $\Omega = 10^{-2}$.
[b] Angular size of maser spots/size of maser region. 1 mas at 1 kpc = 1.5×10^{13} cm.
[c] Velocity width of maser features/velocity range of emission.

There are many excellent reviews of maser theory and observations, as well as of the historical development of the field, in the literature. Some of these are listed in the bibliography. The purpose of this paper is to introduce the reader to the elementary theory of cosmic masers and to present some of the recent results on strong masers associated with young and evolved stars. For historical reasons, these masers are usually referred to as interstellar and circumstellar masers, respectively.

Maser action has also been detected in comets (e.g., dePater, Palmer, and Snyder (1986)) and in planetary atmospheres (Mumma *et al.* 1981). These masers will not be discussed here.

23.2. Cosmic and laboratory masers

For several reasons cosmic masers have little in common with their laboratory counterparts beyond the fact that both have population inversions, which allow stimulated emission to dominate and produce very intense emission. First, cosmic masers are single pass amplifiers, that is, they have no feedback or reflection mechanism, and high gain is achieved by enormous path lengths through the amplifying medium ($\sim 10^{14}$ cm). Second, cosmic masers seem to have little spatial coherence. To understand this, consider two waves propagating in a maser cloud at the same frequency but in slightly different directions. Over a distance L, the phase difference between these waves is

$$\frac{2\pi L}{\lambda}(1 - \cos\theta) \approx \frac{\pi L}{\lambda}\theta^2 \tag{23.1}$$

where θ is the angle between the waves, and λ is the wavelength. This phase difference will be greater than 2π for $L = 10^{14}$ cm and $\lambda = 1.3$ cm, unless $\theta < 10^{-7}$ rad. Cosmic masers have no mechanism for achieving such a precise alignment, and hence they cannot have a high degree of spatial coherence or exhibit phenomena such as coherent pulsing. As a consequence of this situation, the radiation fields of cosmic masers are expected to have Gaussian statistics. Since the power received from many OH and H_2O masers is much greater than the noise power generated in the receivers of the radio telescopes, the statistics of the radiation can be readily studied. The radiation of several masers has been carefully analyzed and found to have Gaussian statistics (Evans *et al.* 1972, Moran 1981, deNoyer and Dodd 1989). Third, cosmic masers show little temporal coherence; the linewidths are generally comparable to those appropriate for thermal broadening in a gas at $\sim$100 K. The line broadening mechanism is inhomogeneous. Laboratory masers achieve high gain through multiple pass amplification in a cavity or system with mirrors. In such systems, a high degree of temporal and spatial coherence can be achieved. An important requirement in the following discussions is that the linewidth of the maser, $\Delta\nu$, must be greater than the stimulated emission rate (also called the microwave rate), W; otherwise the maser could exhibit many of the characteristics of laboratory masers. This condition, $W < \Delta\nu$, seems to be met by all cosmic masers. An important semantic difference in the characterization of laboratory masers and cosmic masers is that laboratory masers are said to be saturated when the pump rate is large enough to dominate the population inversion, thus maintaining exponential amplification, while cosmic masers are said to be saturated when the radiation field is strong enough to affect the population inversion.

23.3. Theoretical background

Many of the features of the radiation fields from cosmic masers can be understood by the analysis of the propagation of radiation in one direction through a one-dimensional medium (see Goldreich and Kwan (1974a), for details of approximations involved in the following analysis). The radiative transfer equation in such a situation is

$$dI_\nu/dz = \varkappa_\nu I_\nu + \eta_\nu \tag{23.2}$$

where I_ν is the specific intensity at frequency ν of a ray propagating in the z direction, and $\varkappa_\nu$ and η_ν are the gain and spontaneous emission coefficients, respectively. These coefficients are given by

$$\varkappa_\nu = (n_2 - n_1)B(h\nu/4\pi)\phi_\nu \tag{23.3}$$

and

$$\eta_\nu = n_2 A(h\nu/4\pi)\phi_\nu \tag{23.4}$$

where n_2 and n_1 are the populations of the upper and lower levels of the maser, h is Planck's constant, ϕ_ν is the line profile function, usually a Gaussian function of width $\Delta\nu_D$, and A and B are the Einstein coefficients. In this discussion, we assume that the statistical weights of the maser levels are the same (hence, $B = Ac^2/2h\nu^3$). The crucial requirement for maser action is that the population difference, $n_2 - n_1$, be greater than zero. This population inversion can be calculated from the appropriate rate equations. For our purposes, consider the transition processes shown in Figure 23.4. In the steady state, the net time rate of change in upper and lower maser level populations will be zero when

$$P_2 - (n_2 - n_1)BJ - n_2\Gamma - n_2C_{21} + n_1C_{12} - n_2A = 0 \tag{23.5a}$$

and

$$P_1 - (n_1 - n_2)BJ - n_1\Gamma + n_2C_{21} - n_1C_{12} + n_2A = 0 \tag{23.5b}$$

where P_1 and P_2 are pump rates per unit volume into maser levels 1 and 2, C_{12} and C_{21} are the collision rates per molecule across the maser levels, Γ is the decay rate

Figure 23.4. Schematic diagram of the energy levels of a maser and the relevant transition rates. The radiative transition rates linking the maser levels are A (spontaneous emission), $BI\Omega/4\pi$ (absorption and stimulated emission); the collision rates are C_{21} and C_{12}; the net pump rates from other levels into the upper and lower maser level are P_2 and P_1, respectively. The decay rates from the maser levels, Γ, are taken to be equal.

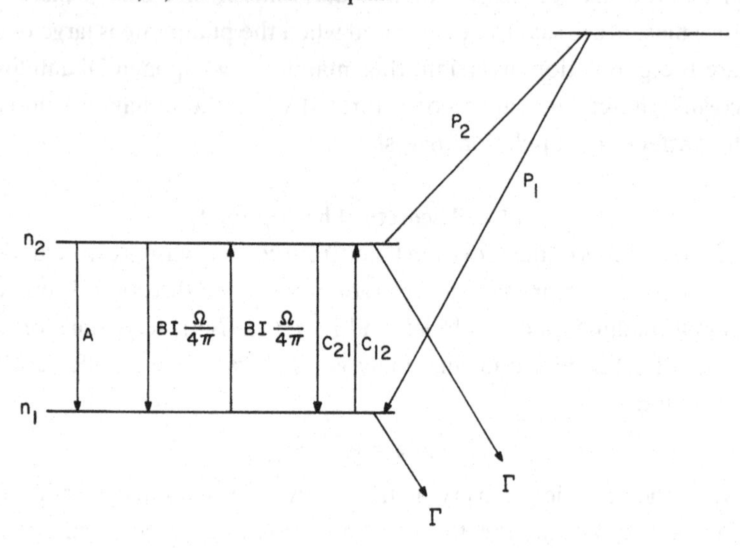

per molecule from the maser levels, J is the intensity averaged over angle and frequency

$$J = \frac{1}{4\pi}\iint I_\nu \phi_\nu \,d\nu d\Omega \approx \frac{I\Omega}{4\pi} \tag{23.6}$$

Ω is the beam angle of the maser emission at any point in the amplifying medium, and I is the intensity at the line center. The quantity BJ is the microwave rate, W, an important parameter in many calculations. Since $A \ll \Gamma$, A can be ignored in the rate equations. Also, $C_{21} \approx C_{12} \equiv C$, for reasonable temperatures, and C can be absorbed into Γ. The sum of Equations (23.5a) and (23.5b) yields

$$P = n_{12}\Gamma \tag{23.7}$$

where $P = P_1 + P_2$ and $n_{12} = n_1 + n_2$, and the difference yields

$$\Delta n = \Delta P/(2BJ + \Gamma) \tag{23.8}$$

where $\Delta n = n_2 - n_1$ and $\Delta P = P_2 - P_1$. The microwave intensity affects the population inversion when the microwave rate exceeds Γ and the maser begins to saturate. The saturation intensity I_s, obtained by setting $2BJ$ equal to Γ, is

$$I_s = 2\pi\Gamma/B\Omega \tag{23.9}$$

The population inversion that can be maintained when $2BJ \ll \Gamma$ is

$$\Delta n_0 = \Delta P/\Gamma \tag{23.10}$$

which is called the unsaturated population inversion (any quantity with subscript 0 denotes its value in the unsaturated regime). The pump efficiency, η, is defined as

$$\eta = \Delta P/P = \Delta n_0/n_{12} \tag{23.11}$$

The excitation temperature, defined by

$$n_2/n_1 = \exp(-h\nu/kT_x) \tag{23.12}$$

will be negative when $n_2 > n_1$, and for small population inversions can be approximated by

$$T_x \approx -\frac{h\nu}{2k}\frac{n_{12}}{\Delta n} \tag{23.13}$$

which in the unsaturated regime is

$$T_{x_0} = -\frac{h\nu}{2k}\frac{P}{\Delta P} \tag{23.14}$$

The unsaturated gain coefficient at line center is, from Equations (23.3) and (23.10),

$$\varkappa_0 = A\lambda^2\Delta P/8\pi\Delta\nu_D\Gamma \tag{23.15}$$

since ϕ_ν at line center is about equal to $\Delta\nu_D^{-1}$. Thus, the equation of radiative transfer, at line center, can be written

$$\frac{dI}{dz} = \frac{\varkappa_0 I}{1 + I/I_s} \tag{23.16}$$

where the spontaneous emission term has been neglected (frequency dependent quantities are written without the subscript ν to denote their value at line center). The general solution to Equation (23.16) is

$$\ln\left[\frac{I(z)}{I(0)}\right] + \frac{I(z) - I(0)}{I_s} = \varkappa_0 z \tag{23.17}$$

where $I(0)$ is the intensity at the input of the maser. The two limiting solutions are

$$I(z) = I(0)\exp(\varkappa_0 z) \quad I \ll I_s \tag{23.18a}$$

$$I(z) = \varkappa_0 I_s z + I_1 \quad I \gg I_s \tag{23.18b}$$

where I_1 is a constant. These expressions can be converted to brightness temperatures with the use of the Rayleigh–Jeans law,

$$I = 2kT_B/\lambda^2 \tag{23.19}$$

If we retain the stimulated emission term, then for $I < I_s$, the solution to Equation (23.2) in terms of brightness temperature is

$$T_B(z) = [T_B(0) - T_x]\exp(\varkappa_0 z) \tag{23.20}$$

The input signal to a maser is a combination of a background source, $T_B(0)$, and spontaneous emission, represented by T_x. In general, the more saturated the maser, the higher $|T_x|$ is, and the less importance a background source has as an input source. The brightness temperature at which a maser becomes saturated, T_s, can be found from Equations (23.9) and (23.19) to be

$$T_s = \frac{h\nu}{2k}\frac{\Gamma}{A}\frac{4\pi}{\Omega} \tag{23.21}$$

This equation is a fundamental result for the discussion of maser saturation.

In general cases of more complicated geometry, it is sometimes easy to calculate the intensity of an unsaturated maser because the radiation field does not affect the level populations. The case of partial saturation as described above is very difficult to analyze because radiation beams in different directions compete for the pump photons, and a three-dimensional solution to the equation of radiative transfer is required except in cases of simple geometry. An interesting and useful case is that of the fully saturated maser, where every part of the amplifying medium is saturated. In this situation, the total emitted power can be easily calculated because every net pump photon is converted to a maser photon. Thus, the power radiated is

$$\mathcal{L} = h\nu\Delta PV = \eta h\nu\Gamma n_{12}V \tag{23.22}$$

where V is the maser volume. A fully saturated maser is the most efficient maser because each net pump cycle results in a maser photon. It is difficult to explain the strength of the strongest masers with current pump theories, so the high conversion efficiency of saturated masers is appealing. We shall now discuss some specific maser geometries.

23.3.1. Specific geometries and their observed properties

Let us consider the simple geometries of a long thin tube or filament and a sphere that are uniformly pumped, in the limiting cases of no saturation and complete saturation, as shown in Figure 23.5 (see Goldreich and Keeley (1972)). It is important to distinguish among the various solid angles that appear in the analysis. Ω is the beam angle of the emission at any point; one is usually concerned with the beam angle at the surface of the maser. Ω_m is the effective solid angle into which the maser radiates, which may be different from Ω. The apparent isotropic luminosity $\mathcal{L}^*$ of a maser, which is usually reported by observers, is related to the true luminosity by

$$\mathcal{L}/\mathcal{L}^* = \Omega_m/4\pi \tag{23.23}$$

The apparent angular size of the source as seen by the observer is Ω_s ($\sim\theta_s^2$). This angle can be measured by very long baseline interferometry (VLBI). The flux density at line center measured by an observer is

$$F = I\Omega_s \tag{23.24}$$

and the brightness temperature inferred from measurements of F and Ω_s is

$$T_B = \lambda^2 F/2k\Omega_s \tag{23.25}$$

A thin tube is essentially a one-dimensional structure whose properties can be evaluated from the considerations of the previous section. Let its length be L and cross sectional radius R, where $R/L \ll 1$. Most of the emission emerges from the ends of the tube. An observer will see the maser only if closely aligned with the axis of the tube. The maser beam angle is (see Figure 23.5(c))

$$\Omega = \Omega_m \sim \pi(R/L)^2 \tag{23.26}$$

In the unsaturated limit, with no background input signal (see Equations (23.19) and (23.20))

$$I = -\frac{2kT_{x0}}{\lambda^2}\exp(\varkappa_0 L) \tag{23.27}$$

Figure 23.5. Simple maser geometries and the beaming of the maser emission. (*a*) Unsaturated sphere; the beam angle is given by Equation (23.33). (*b*) Saturated sphere; the beam angle, given by Equation (23.37), depends on the size of the central unsaturated core. (*c*) Tube; the beam angle is given by Equation (23.26).

(*a*) UNSATURATED SPHERE

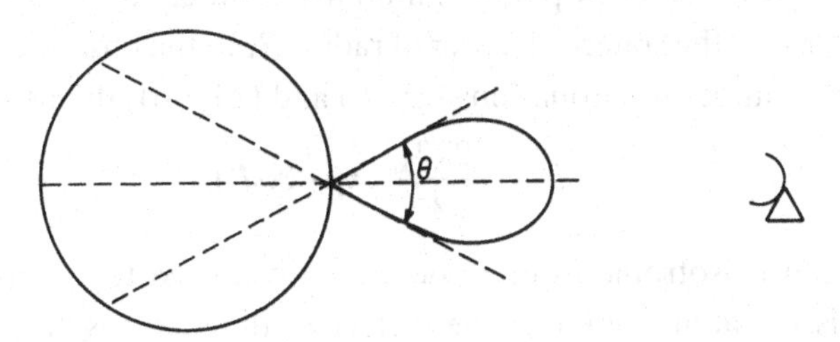

(*b*) SATURATED SPHERE

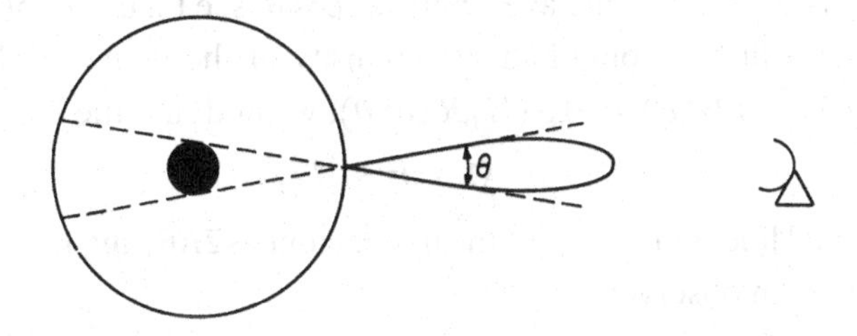

(*c*) TUBE

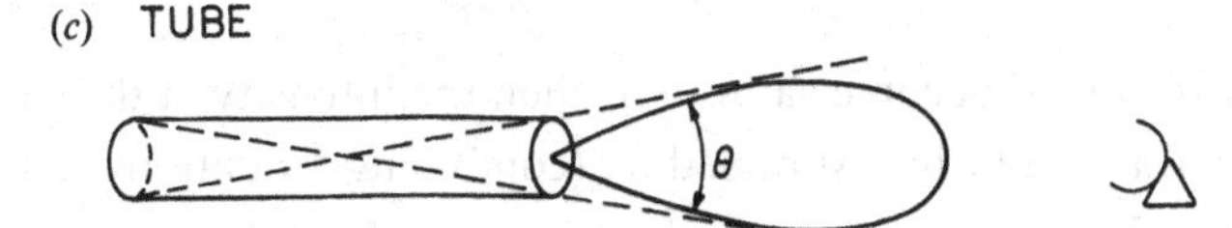

whereas in the fully saturated limit (Equation (23.18b))

$$I = \varkappa_0 I_s L \tag{23.28}$$

The total luminosity can be estimated by the general relation

$$\mathcal{L} = I\Omega\mathcal{A}\Delta\nu \tag{23.29}$$

where $\mathcal{A}$ is the area of the end caps of the tube, $2\pi R^2$.

For the fully saturated case, the substitution of Equations (23.9), (23.15), (23.26), and (23.28) into Equation (23.29) gives the general result of Equation (23.22). The observed flux density is $I\Omega_s$ where

$$\Omega_s = \pi R^2/D^2 \tag{23.30}$$

and D is the distance to the source. Substitution of Equations (23.15), (23.28), and (23.30) into Equation (23.24) yields the result

$$F = \frac{1}{2} h\nu \frac{\Delta P}{\Delta\nu} \frac{L^3}{D^2} \tag{23.31}$$

Note that the flux density in the tubular maser increases as L^3, not L, because the beam angle varies as L^{-2}. The tubular model has the flexibility that L is not a directly observable parameter and it can be made large to account for the observed flux density. Another way of viewing this result is that the beam angle is not an observable, and it can be made small in order to keep the luminosity below any prescribed value set by pump limitations or other considerations.

Next, consider the spherical maser of radius R. In the unsaturated case, with no background source (see Equations (23.19) and (23.20)), the peak intensity is

$$I = -\frac{2kT_{x0}}{\lambda^2} \exp(2\varkappa_0 R) \tag{23.32}$$

The emission is isotropic from an observer's view, so $\Omega_m = 4\pi$. However, the maser emission at any point on the surface of the maser is beamed in the radial direction, because the longest gain paths are the ones that pass through the center of the sphere. To calculate the beam angle, consider a ray making an angle θ to the diametrical one, which will have length $2R\cos\theta$ (see Figure 23.5(a)). Choose θ so that the ray will have only half the intensity of the diametrical one. From the relation $0.5\exp(2\varkappa_0 R) = \exp(2\varkappa_0 R\cos\theta)$, we find that the beam angle is

$$\Omega \sim \theta^2 \sim 1/\varkappa_0 R \tag{23.33}$$

The observed linear diameter of the maser cloud is $2R\theta$, and the solid-angular size measured by an observer is

$$\Omega_s \sim R/D^2\varkappa_0 \tag{23.34}$$

The maser begins to become saturated when the intensity at the surface exceeds I_s. This condition can be expressed as (combining Equations (23.19), (23.21), (23.32), and (23.33))

$$\frac{\exp(2\varkappa_0 R_s)}{2\varkappa_0 R_s} \sim \frac{\Delta P}{P}\frac{\Gamma}{A} \tag{23.35}$$

where R_s is the saturation radius. From symmetry, it is clear that a partially saturated spherical maser will have an unsaturated core surrounded by a saturated shell. A fully saturated maser retains a distinct core, a fact first recognized by Litvak (1971). The size of this core is (Goldreich and Keeley 1972)

$$a = 1.3\left[\frac{P}{\Delta P}\frac{A}{\Gamma}\right]^{1/4} R \tag{23.36}$$

The strongest rays are those that pass through this core. Hence the fully saturated maser is highly beamed with (see Figure 23.5(b))

$$\Omega \sim \pi a^2/R^2 \tag{23.37}$$

and an observer sees primarily the core of the maser cloud that subtends an angle

$$\Omega_s \sim \frac{\pi}{4}\theta_s^2 \sim \pi\frac{a^2}{D^2} \tag{23.38}$$

The flux density observed can be calculated by noting that since $\Omega_m = 4\pi$, $F = \mathscr{L}/4\pi D^2 \Delta\nu$. For a fully saturated maser, use of Equation (23.22) for $\mathscr{L}$ gives

$$F \cong \frac{1}{3} h\nu \frac{\Delta P}{\Delta\nu}\frac{R^3}{D^2} \tag{23.39}$$

which can be expressed in terms of observable quantities if Equations (23.10), (23.11), (23.36), (23.38), and (23.39) are combined as

$$D^{-1}\Delta\nu\theta_s^{-3}F \cong (h\nu/8)A^{-3/4}n_{12}(\Gamma\eta)^{7/4} \tag{23.40}$$

A comparison of Equations (23.31) and (23.39) shows that tubes and spheres have more in common than one might at first think. One can model a maser either as a tube of length $2R$ or a sphere of radius R. The tube will have highly anisotropic emission ($\Omega_m \sim R^2/L^2 \ll 1$), and its total luminosity will be smaller than that of the sphere by a factor $\Omega_m/4\pi$. This anisotropy can alleviate the pumping requirements for a specific maser cloud. However, if a source consists of a large number of randomly oriented tubes, then only a small fraction, $\Omega_m/4\pi$, will be observed. Hence, the overall luminosity requirements for a collection of filamentary masers and spherical masers are about the same.

Parameters can be estimated for a specific case. Consider the very strong interstellar H_2O maser in the Orion-KL region whose flux density versus time is shown in Figure 23.6 (Garay, Moran and Haschick 1989). Its parameters in April 1980 were: $F = 2 \times 10^6$ Jy; $\Delta\nu = 47$ kHz, $\theta_s = 1.0$ mas and $D = 0.5$ kpc. For the H_2O transition, $\nu = 2.2 \times 10^{10}$ Hz ($\lambda = 1.35$ cm), $A = 1.9 \times 10^{-9}$ s^{-1} and Γ is expected to be ~1 s^{-1}, the transition rate for rotational transitions from the 6_{16} and 5_{23} levels. The saturation brightness temperature (Equation (23.21)), for $\Omega =$

Figure 23.6. The flux density versus time for the strong flaring H_2O maser at 8 km s^{-1} in the Orion-KL region. From Garay *et al.* (1989).

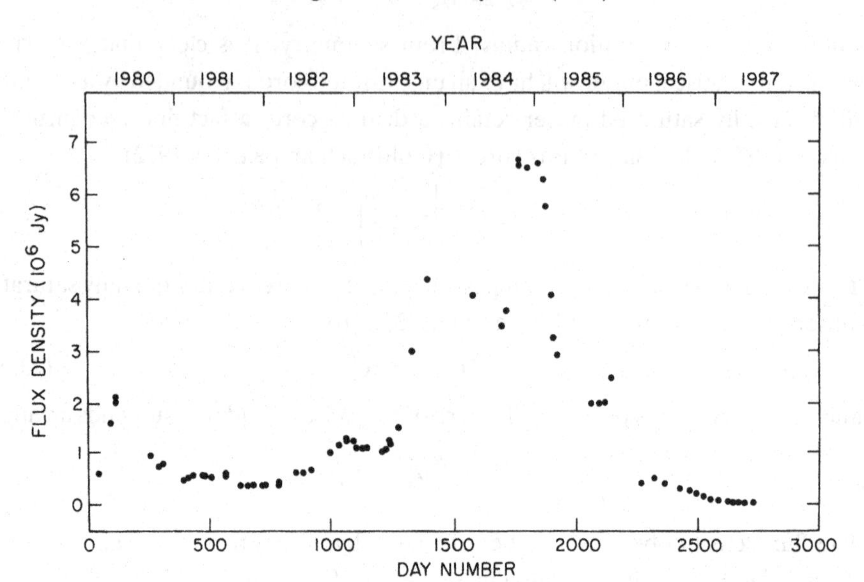

10^{-3} (justified below) is 3×10^{12} K. The observed brightness temperature (Equation (23.25)) is 7×10^{15} K, so the maser is undoubtedly well saturated. Equation (23.40) can be expressed in convenient units for the values of A and Γ given above as

$$0.7\left(\frac{D}{\text{kpc}}\right)^{-1}\left(\frac{\Delta\nu}{50\ \text{km s}^{-1}}\right)\left(\frac{\theta_s}{\text{mas}}\right)^{-3}\left(\frac{F}{10^6\ \text{Jy}}\right) = \left(\frac{n_{12}}{10^5\ \text{cm}^{-3}}\right)\left(\frac{\eta}{0.01}\right)^{7/4} \tag{23.41}$$

The observed parameters require that $(n_{12}/10^5\ \text{cm}^{-3})(\eta/0.01)^{7/4} = 2.6$. Hence, if $\eta = 0.01$, then $n_{12} = 2.6 \times 10^5$ cm^{-3}. The required hydrogen density can be expressed as

$$n_{H_2} = n_{12}\left(\frac{n_{H_2O}}{n_{12}}\right)\left(\frac{n_{H_2}}{n_{H_2O}}\right) \tag{23.42}$$

where n_{H_2O} is the density of water vapor. For temperatures of a few hundred Kelvin, $n_{H_2O}/n_{12} \sim 10^2$, and we adopt $n_{H_2}/n_{H_2O} \sim 10^4$. Hence, $n_{H_2} \sim 2.6 \times 10^{11}$ cm^{-3}. The ratio of the cloud radius to the core radius is (Equation (23.36)) 37, and hence $\Omega \sim 2 \times 10^{-3}$. The apparent or core radius of the maser is (Equation (23.38)) 4×10^{12} cm and the true radius is 1.5×10^{14} cm. The luminosity is 3×10^{32} erg s^{-1} ($0.0007\ \mathcal{L}_\odot$), corresponding to a photon rate of 2×10^{47} s^{-1}. The microwave rate, $W = BJ$, can be written as

$$W = A\frac{kT_B}{h\nu}\frac{\Omega}{4\pi} \tag{23.43}$$

For this example, $W \sim 2500\ \mathrm{s}^{-1}$, which is only a factor of 20 less than $\Delta\nu$. The corresponding tubular model with the same density would have a length about equal to the diameter of a spherical cloud, and an emission angle $\Omega_\mathrm{m} = \Omega \sim 2 \times 10^{-3}$. The luminosity of the tubular maser, which is less than that of the sphere by the factor $\Omega_\mathrm{m}/4\pi$, is about 5×10^{28} erg s^{-1}.

It is difficult to decide whether cosmic masers more closely resemble tubes or spheres. Alcock and Ross (1985a,b) examined the general case of anisotropic geometries by use of a two-dimensional four-stream radiative transfer model. They found that, even in a fully saturated maser, only a slight elongation was necessary to increase greatly the intensity along the major axis. The various radiation beams compete for pump photons, and the beam along the major axis is the most successful. This result suggests that tubular geometry may be more realistic than spherical geometry. The boundary of a masing cloud is determined by the velocity coherence. In the dynamic envelope of stars, the masing clouds are most likely to be filamentary structures within which the velocity is constant to within the linewidth.

23.3.2. Line shapes

The linewidth of maser signals is affected by various mechanisms. The line profile function appropriate for a thermal gas at temperature T_k is a Gaussian function with width $\Delta\nu_\mathrm{D} = (8 \ln 2)^{1/2}(\nu_0/c)(kT_\mathrm{k}/M)^{1/2}$, where M is the molecular mass. In an unsaturated maser, the line strength will increase much faster at line center than in the line wings so the line will narrow. In this case the line shape is expected to be

$$I_\nu = I_\nu(0) \exp \{\varkappa_0 L \exp [-4 \ln 2(\nu - \nu_0)^2/\Delta\nu_\Delta^2]\} \tag{23.44}$$

For $\varkappa_0 L \gg 1$, the line profile of Equation (23.44) is very nearly Gaussian with a linewidth

$$\Delta\nu \cong \frac{\Delta\nu_\mathrm{D}}{(\varkappa_0 L)^{1/2}} \tag{23.45}$$

A maser with a gain factor, $\varkappa_0 L$, of 25 would have its linewidth reduced by a factor of 5, and the apparent kinetic temperature would be $T_\mathrm{k}/\varkappa_0 L$ or $T_\mathrm{k}/25$. The dependence of linewidth on intensity is

$$\Delta\nu = \Delta\nu_\mathrm{D}\{\ln [I/I(0)]\}^{-1/2} \tag{23.46}$$

Since maser features can usually only be tracked over a small dynamic range ($\sim 10^2$), this weak dependence is difficult to detect. However, in some cases, much stronger dependences of $\Delta\nu$ on I have been observed (e.g., Mattila *et al.* (1985), Rowland and Cohen (1986)).

Along the amplification path of a one-stream maser, the linewidth narrows during exponential growth and rebroadens to the thermal value during saturated growth. The rebroadening can be inhibited by the trapping of radiation at wavelengths (usually infrared) involved in the pump cycle (Goldreich and Kwan, 1974a). Such trapping is important if the cross relaxation rate, $\gamma \sim kTA_p/30h\nu_p$, ($A_p$ and ν_p are the Einstein coefficient and frequency of the pump transition) exceeds the stimulated emission rate, i.e., $\gamma > W$. For H_2O masers, this mechanism does not seem to be important since $\gamma \sim 1\ s^{-1}$ and $W \sim 100\ s^{-1}$. In H_2O masers, the linewidths are $\sim$1 km s^{-1} and correspond to reasonable kinetic temperatures, ~200 K. For OH masers, linewidths as small as 0.05 km s^{-1}, corresponding to temperatures of ~5 K, have been observed, and cross relaxation may be important.

The broadening due to the Stark effect is small for $T_B < 10^{17}$ K, and there is no evidence for it. Magnetic fields may cause broadening through Alfven turbulence and the Zeeman effect, which is discussed in the next section.

23.3.3. Polarization

Cosmic masers are often strongly polarized. Most of this polarization is probably due to the Zeeman effect, although anisotropic pumping can also produce linear polarization (e.g., Western and Watson (1983)). Because of the high densities in masers (10^6–10^{11} cm^{-3}), magnetic fields in the milligauss range are reasonable. The theory of maser line formation in the presence of magnetic fields is discussed by Goldreich, Keeley and Kwan (1973a,b) and Western and Watson (1984). The only paramagnetic molecule with well-studied maser properties is OH (e.g., Davies (1973)). The magnetic field required to separate the Zeeman components by more than a linewidth in the 1665 MHz transition is ~1 mG. Interstellar OH masers exhibit the greatest degree of polarization of any of the known cosmic masers. OH masers generally display many features that are 100% circularly polarized. The general separation between left and right circularly polarized features among the various transitions in W3(OH) is indicative of a magnetic field strength of ~5 mG (see Figure 23.1). However, no complete Zeeman pattern has ever been seen. For example, the Zeeman pattern for the 1665 MHz OH transition consists of a π component flanked by two σ components that are offset by about 0.3 km s^{-1} mG^{-1}. In a longitudinal field, the σ components are 100% right and left circularly polarized, and the π component is absent. In a transverse field, all three components are linearly polarized. Many spatially coincident features of opposite circular polarization have been identified, which are almost certainly Zeeman pairs. π components that might be associated with σ components have never been seen (Garcia-Barreto *et al.* 1988).

A possible explanation for this lack of linear polarization is internal Faraday rotation, which effectively reduces the gain path for linearly polarized components. σ components often have unequal amplitudes, and isolated circularly polarized features frequently appear. This condition may be a result of gradients and turbulence in the velocity and magnetic fields. In an extreme case, one could imagine a velocity gradient along the line of sight and a magnetic field gradient that produces a frequency shift that exactly matches the Doppler shift in the σ^+ line. Therefore, the σ^+ line would have a long coherent gain path and the σ^- line a short path.

For nonparamagnetic molecules such as H_2O, SiO, and CH_3OH, the Zeeman splitting is generally 1000 times less than for paramagnetic molecules. For H_2O, a magnetic field of about 50 G is required to produce a splitting equal to the typical linewidth of 50 kHz. However, strong linear polarization can be produced under certain circumstances. The production of linear polarization requires that (Goldreich *et al.* 1973a)

$$\left.\begin{array}{ll} W \ll \Gamma & \text{(saturated)} \\ W \ll \gamma & \text{(cross relaxation negligible)} \\ \text{and} & \\ W \ll \Delta\nu_z \ll \Delta\nu & \end{array}\right\} \quad (23.47)$$

where $\Delta\nu_z$ is the frequency of the Zeeman splitting. Deguchi and Watson (1986) have shown with simple model calculations that the degree of polarization in H_2O masers can approach 70% when the angle between the magnetic field and the line of sight is small, i.e., ~25°. Complete modeling of the polarization properties of masers in high rotational states is a prodigious task. Garay *et al.* (1989) found that the maser flare in Orion-KL region has at times reached 70% linear polarization and that the polarization decreased as the flux density increased, suggesting that the limit $W \sim \Delta\nu_z$ had been reached. They estimate the magnetic field to be about 30 mG.

A very small circular polarization can be measured in the case where $\Delta\nu_z \ll \Delta\nu$. The very small shift between the left and right circularly polarized spectra means that their difference is approximately proportional to the derivative of the individual profiles. The peak amplitude ΔI of the difference spectrum is

$$\Delta I/I \sim 1.4\Delta\nu_z/\Delta\nu \quad (23.48)$$

H_2O masers with $\Delta I/I \sim 10^{-4}$ have been detected corresponding to $B \sim 30$ mG (Fiebig and Güsten 1988).

23.3.4. Pumping mechanisms

Pump mechanisms that produce the population inversion require non-LTE conditions and rely on at least two energy reservoirs at different temperatures. For example, a cloud where hot dust and cold gas coexist can provide the necessary conditions. Another example is a cloud that is optically thin at the infrared wavelengths involved in the pump cycle, so that photons can leak out of the cloud. Two broad classes of maser pumps are called *collisional* and *radiative*, depending on whether the primary source of pump energy is from collisions or radiation. Strelnitskij (1984) classified pumps in more detail by the nature of their energy source and energy sink (R for radiative and C for collisional) as well as by the transitions involved in the pump cycle. One version of the hot dust/cold gas model mentioned above operates through the 6 μm vibrational (v) transition in H_2O masers and is classified as an RCv pump. Only the most general thermodynamic constraints shall be discussed here (e.g., Goldreich and Kwan (1974b)). These can be used in practice to determine if there are adequate energy sources in or near the maser to account for its luminosity. Detailed discussions of pump mechanisms can be found in Strelnitskij (1984), Kylafis (1988), and other review papers cited in the bibliography.

In a radiative pump, the pump photons can be supplied by an external source or an internal source such as hot dust. At least one pump photon is required for every maser photon. The maser photon rate is

$$N_{\mathrm{m}} = 4\pi D^2 F_{\mathrm{m}} \frac{\Delta\nu_{\mathrm{m}}}{h\nu_{\mathrm{m}}} \frac{\Omega_{\mathrm{m}}}{4\pi} \tag{23.49}$$

where F_{m}, $\Delta\nu_{\mathrm{m}}$, and ν_{m} are the flux density, linewidth, and frequency of the maser. The number of pump photons available from an external source is

$$N_{\mathrm{p}} = 4\pi D^2 F_{\mathrm{p}} \frac{\Delta\nu_{\mathrm{p}}}{h\nu_{\mathrm{p}}} \frac{\Omega_{\mathrm{p}}}{4\pi} \tag{23.50}$$

where F_{p} is the flux density of the pump source at the earth (assuming no absorption), Ω_{p} is the solid angle subtended by the maser at the pump, and $\Delta\nu_{\mathrm{p}}$ and ν_{p} are the pump linewidth and frequency. Hence, the condition that $N_{\mathrm{p}} > N_{\mathrm{m}}$ leads to the pump requirement

$$F_{\mathrm{p}} > \frac{\Omega_{\mathrm{m}}}{\Omega_{\mathrm{p}}} F_{\mathrm{m}} \tag{23.51}$$

or, converting to brightness temperature with the Rayleigh–Jeans relation,

$$T_{\mathrm{p}} > \left(\frac{\Omega_{\mathrm{m}}}{\Omega_{\mathrm{p}}}\right)\left(\frac{\nu_{\mathrm{m}}}{\nu_{\mathrm{p}}}\right)^2 T_{\mathrm{m}} \tag{23.52}$$

The most viable pump cycle in H_2O masers involves the 6 μm vibrational transition, where $(\nu_{\mathrm{p}}/\nu_{\mathrm{m}})^2 \sim 5 \times 10^6$. Hence, a distant external radiative source at a temperature $<10^3$ K cannot pump the brightest H_2O masers (10^{15} K).

A collisional pump with a radiative sink requires that at least one sink photon be removed from the pump cycle for every maser photon emitted. The 'used' pump photons must escape from the maser cloud or be absorbed by cold dust. This constraint is sometimes called the thermodynamic limit. The maximum rate at which these photons can be emitted is

$$N_p = B(T_p) \frac{\Delta\nu_p}{h\nu_p} \mathcal{A} \tag{23.53}$$

where $B(T_p)$ is the Planck function at the temperature of the pump transition, T_p, and $\mathcal{A}$ is the area of the maser cloud. If the maser is spherical with radius R, then the requirement that $N_p > N_m$ provides a limit on the cloud size of

$$R > \left(\frac{N_m \lambda_p^2}{8\pi\Delta\nu_p}\right)^{1/2} [\exp(h\nu_p/kT_p) - 1]^{1/2} \tag{23.54}$$

Most pump models do not operate when the density is too high because collisions thermalize the populations and quench the maser. Critical densities are about 10^{11} cm^{-3} for H_2O and CH_3OH masers and 10^9 cm^{-3} for OH masers. Collisional pump models that do not have the normal density limit are possible if two species of particles having different temperatures are present (Strelnitskij 1984). A possible model of this type is the hot electron/cold gas model of Kylafis and Norman (1987). Such a model may be able to explain the strongest H_2O masers. However, it requires very high magnetic fields to raise the energy of the electrons.

The study of pump models usually requires solving rate equations for many levels in the masing molecules. Some pumps take advantage of special relations among the energy levels of the molecule. The discovery of a misassignment of the Λ doublet states of OH has invalidated many OH pumping schemes (Andresen 1986).

23.4. Stellar masers

Maser emission often originates in the dusty circumstellar envelopes of evolved stars. These are usually M giant ($\mathcal{L} \sim 10^4 \mathcal{L}_\odot$) or supergiant ($\mathcal{L} \sim 10^6 \mathcal{L}_\odot$) stars, which are undergoing mass loss at a rate of 10^{-7}–10^{-5} $M_\odot$ yr^{-1} (e.g., Netzer and Knapp (1987)). The optical luminosities of these stars vary, either regularly (long period variables) or irregularly. More than 400 stellar masers have been identified from radio searches of known stars (Engels 1979). The characteristic spectrum of OH masers has been used to identify many infrared sources as evolved stars. Many stellar masers have been found among IRAS sources (e.g., Lewis, Eder and Terzian (1985)).

The general picture of the maser emission in the envelopes of late type stars is basically simple. Dust grains condense in the extended atmospheres of these stars and are driven outwards by radiation pressure. Through collisions with the dust, the gas is also accelerated outwards and forms molecules as it cools. At a radial distance of $\sim 10^{16}$ cm, the outflow reaches a terminal velocity of 10–50 km s^{-1}. The temperature of the photosphere is ~2000 K, but the envelope absorbs most of the radiation and reemits in the infrared at a typical color temperature of ~500 K.

Detailed models for OH and H_2O maser emission in late type stars can be found in Elitzur, Goldreich and Scoville (1976) and Cooke and Elitzur (1985). An example of the maser emission from an evolved star is shown in Figure 23.7. The OH maser emission at 1612 MHz has a readily identifiable spectrum, which consists of emission in two discrete velocity intervals. The mean of these two intervals is the stellar velocity and the total range of velocities is twice the expansion velocity. The ground-state 1612 MHz OH emission occurs in the outer envelope beyond the acceleration zone in a shell of radius $\sim 10^{16}$ cm. In a shell of constant velocity, the longest path length for maser amplification is along a ray passing through the star. Hence, the emission observed in the two spectral peaks arises from 'caps' in front of and in back of the star. The emission from the limbs, which appears at lower Doppler velocities, is much weaker because of the shorter coherent path lengths.

H_2O maser emission, which requires an excitation temperature of $\sim$600 K, exhibits spectral features over a much smaller range of velocities and is confined to a radius of $\sim 10^{14}$ cm. Finally, SiO emission occurs on the smallest range of velocity and smallest radius of $\sim 10^{13}$ cm. Maser emission in 12 different rotational lines of SiO has been identified, which requires excitation temperatures of up to 5300 K (Jewell *et al.* 1987). These characteristics are all depicted in the spectra and maps of masers associated with the star VX Sgr shown in Figure 23.7. One unconfirmed aspect of this picture is the precise alignment among the maser images and the position of the star. Current astrometric techniques are not good enough to establish the absolute positions to sufficient accuracy for precise comparison.

Circular polarization measurements of OH maser emission suggest that the magnetic field strength is $\sim$1 mG at a radius of $\sim 10^{16}$ cm in evolved stars (Reid *et al.* 1979, Cohen *et al.* 1987). Magnetic fields of 10–100 G at radii of $\sim 10^{14}$ cm have been estimated from the detection of circular polarization in SiO masers (Barvainis, McIntosh and Predmore 1987). These measurements suggest a magnetic field at the stellar surface as large as 100 G and a dependence with radius of r^{-2}–r^{-3}.

The luminosities of the masers vary in synchronism with the optical and infrared luminosity of the star. This behavior has been confirmed for OH (e.g., Harvey *et al.* (1974)), H_2O (e.g., Schwartz, Harvey and Barrett (1974)) and SiO masers (e.g., Nyman and Olofsson (1986)). The regular variation of the OH maser emission offers a method of measuring directly the linear dimension of the maser shell. Since the red and blueshifted maser emission features are presumably excited in phase by the stellar radiation or shock waves, the time lag between their light curves is the light travel time across the shell. This 'phase lag' technique has been exploited by Schultz, Sherwood and Winnberg (1978), Jewell, Webber and Snyder (1980), and Diamond *et al.* (1985). The measured time shifts are about 10 days implying shell diameters of about 10^{16} cm. These measurements form the

Figure 23.7. The spectra of the maser emission from the supergiant star VX Sgr and the angular distribution of the emission. The OH emission at 1612 MHz is only shown for two low velocities that reveal the shell structure of the maser: 6.5 km s^{-1} (solid line) and 11.9 km s^{-1} (dashed line). The H_2O, SiO, and OH (1665 MHz) emission occurs closer to the star in the acceleration zone of the wind. The stellar velocity is 5 km s^{-1}. At a distance of 1.7 kpc, 1″ corresponds to 2.5×10^{16} cm. From Chapman and Cohen (1986).

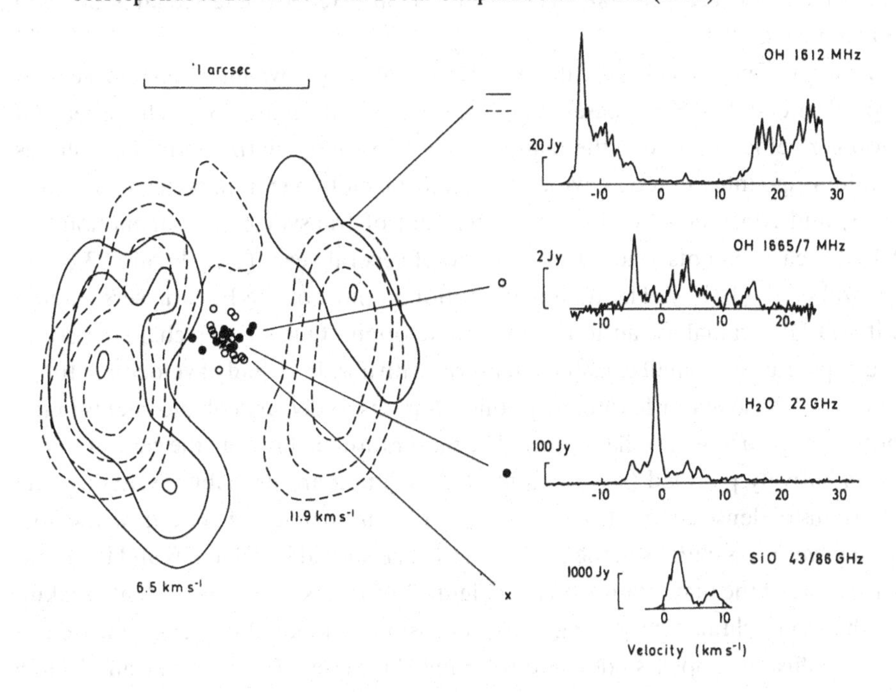

basis of a technique for estimating the distance to the star. The angular size of the OH maser shell can be measured by interferometry (e.g., with the VLA or the MERLIN array) from the weak low velocity emission that arises from the gas at the limb of the shell moving transverse to the line of sight. The distance estimate is the ratio of the 'phase lag' diameter to the angular diameter. The distances to the stars can also be estimated from their velocities and the Schmidt model of the galaxy. Herman *et al.* (1985) made such comparative measurements for 15 stars and found the best agreement when R_0, the distance to the center of the galaxy, is 9.2 ± 1.2 kpc. Their error bars do not include systematic effects. However, this technique is very promising.

23.5. Interstellar masers

Strong masers are firmly associated with the last stages of pre-main-sequence development of OB and perhaps lower mass stars. These stages, which are thought to last $\sim 10^5$ yr, are marked by the appearance of molecular outflows and compact HII regions. Various schemes for relating masers to evolutionary development of stars have been developed (e.g., Genzel and Downes (1977)). H_2O masers are generally associated with the youngest objects, which have strong outflows, while OH masers are generally associated with compact HII regions smaller than 10^{17} cm. Interstellar masers are found mostly near the galactic plane, and their observed angular sizes are often affected by interstellar scattering (Gwinn *et al.* 1988).

Braz and Epchtein (1983) catalogued many of the known interstellar masers. A typical interstellar H_2O maser, Sgr B2, is shown in Figure 23.8. The spread of Doppler velocities among the features of $\sim\pm 50$ km s^{-1} in this and other sources was long ago interpreted as evidence of ballistic motion (Strelnitskij and Sunyaev 1973) and confirmed by direct measurement of transverse angular motions by VLBI measurements made over a period of several years (e.g., Figure 23.8 and Genzel *et al.* (1981)). The dominant motion in most of the H_2O masers studied with VLBI is radial expansion from a single origin. One striking characteristic is the appearance of small clusters of features with a wide spread of velocities (10–30 km s^{-1}). These maser features are condensations or density enhancements in the outflow from a young stellar object. The masers are too far from the central star to be radiatively pumped (see Equation (23.52)) and are probably excited by the collisions of dense clumps ($\sim 10^{11}$ cm^{-3}, $\sim 10^{29}$ g total mass) of material in the flow with lower mass clumps in the ambient gas. The spread in velocities and the rapid variability of the maser emission (see Figure 23.6) attests to the turbulent breakup of the cloud clumps producing maser emission. The kinetic energy released in these collisions appears adequate to pump the masers (e.g., Tarter and Welch (1986)). H_2O masers are probably closely related to H–H objects, another manifestation of mass loss in early stellar evolution. The luminosity of H_2O masers is roughly proportional to the far-infrared luminosity of the parent star ($\mathscr{L}_{H_2O} \sim 10^{-8}\mathscr{L}_{FIR}$) over a wide range of at least four orders of magnitude, suggesting that the H_2O masers are associated with forming stars as massive as O7 down to at least B3.

An important application of the measurement of proper motion of interstellar H_2O masers is that their distances can be estimated. The outflow in the Sgr B2 maser of ~ 50 km s^{-1} creates transverse motions of up to 1.2 mas yr^{-1}, which can easily be measured with VLBI arrays having resolutions of ~ 300 mas (see Figure 23.8). The distance can be found by the method of statistical parallax, as the ratio of the rms deviation of the line-of-sight Doppler velocities and the rms deviation of the transverse angular velocities. In the case where the flow has a simple geometry, e.g., constant expansion velocity, the distance can be derived implicitly from the comparison of transverse angular velocities and line-of-sight velocities as one parameter of a least mean square analysis. In both cases, the fractional

Figure 23.8. (*a*) The spectrum of the H_2O maser towards Sgr B2(N). (*b*) The transverse motions of the masers determined from VLBI observations are indicated by the arrows on the maser spots. The dominant motion is radial expansion from an origin marked by an '×'. From Reid *et al.* (1988).

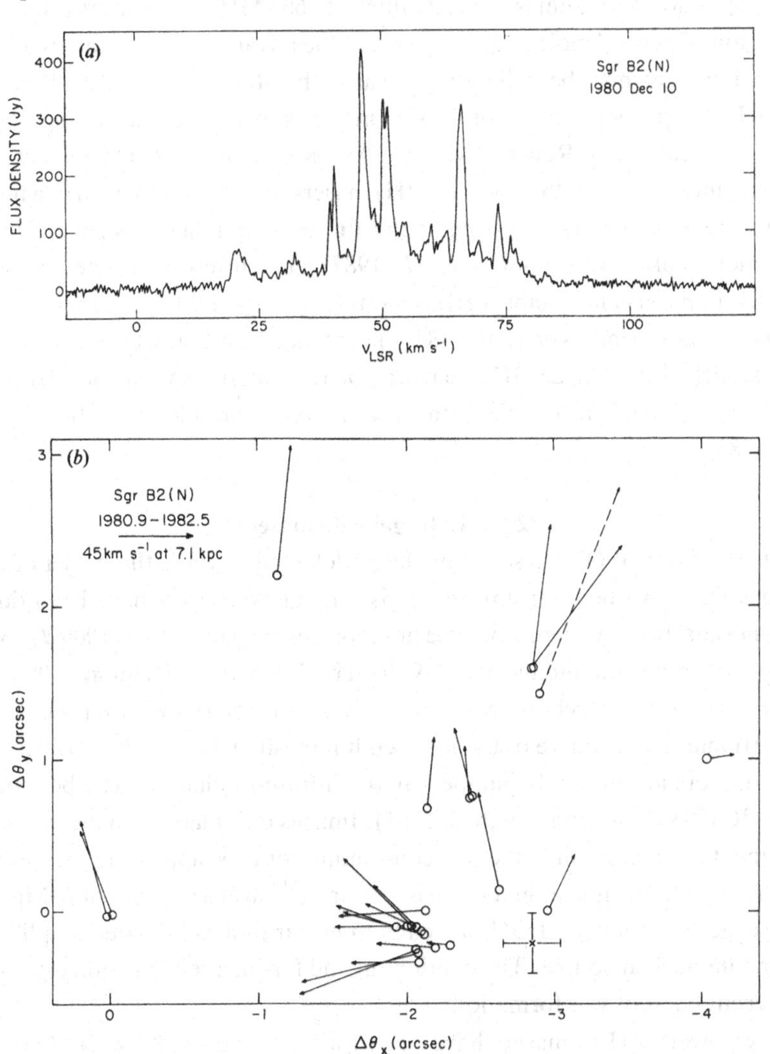

accuracy in the distance estimate is about equal to $(2N)^{-1/2}$, where N is the number of maser features measured. The distance to the maser near the galactic center measured by the modeling method is 7.1 ± 1.5 kpc (Reid *et al.* 1988). This method can be extended to masers in nearby galaxies.

OH masers are generally closely associated with compact HII regions (e.g., Gaume and Mutel (1987)). The strongest emission usually comes from the 1665 MHz transition. A well-studied example is W3(OH), where the OH features are spread over the face of the compact HII region (see Figure 23.3). The stellar velocity can be measured from radio recombination lines, and in many cases the masers are redshifted with respect to the star (Garay, Reid and Moran 1985). Since the radio continuum is optically thick at 1665 MHz, the masers probably lie in a region of neutral molecular gas, outside the advancing ionization front of the HII region, that may be collapsing towards the star. However, in the case of W3(OH), the proper motions of the OH masers suggest that the masing region is expanding (Bloemhof, Reid and Moran 1989). Magnetic fields are typically 5 mG.

Strong interstellar methanol (CH_3OH) masers have been reported in a number of transitions, and work is proceeding in this area at a vigorous pace. The 12.1 GHz methanol masers (Batrla *et al.* 1987) are closely associated with OH interstellar masers and compact HII regions (e.g., Figure 23.3), and some of them are linearly polarized (Koo *et al.* 1988). There may be a distinct class of methanol masers, such as those at 25 GHz, that are not associated with compact HII regions but that are probably associated with an earlier phase of stellar evolution (Menten *et al.* 1986).

23.6. Extragalactic masers

About 15 powerful OH masers have been detected towards the nuclei of active galaxies that have high far-infrared emission. These masers have been dubbed 'megamasers' because their isotropic luminosities range from 1 to 1000 $\mathscr{L}_\odot$, which is up to 10^8 times the luminosity of W3(OH) (Baan 1985, Diamond 1988). The maser luminosity, which appears primarily at 1665 and 1667 MHz, is roughly proportional to the square of the infrared luminosity ($\mathscr{L}_{OH} \propto \mathscr{L}_{IR}^2$) (Baan 1989). These masers are probably pumped by the infrared radiation. The best studied case is IC4553 (Baan and Haschick 1984). Images of it made with the VLA show that the OH maser and nuclear continuum sources appear to be perfectly superimposed. The linear dimension is $\sim 1''$ or 10^{21} cm at a distance of 70 Mpc. The data suggest that inverted OH in a circumnuclear molecular cloud amplifies the continuum nuclear source. The continuum and far-infrared emission may be due to a recent burst of star formation.

Very powerful H_2O masers have been found towards the nuclei of six other galaxies. These are active galaxies that have lower infrared and radio luminosities than those with OH megamasers. The isotropic luminosities in the maser lines are as high as 500 $\mathscr{L}_\odot$. NGC3079 is the best studied example of this class. The maser features are unresolved by VLBI and are smaller than 1 mas (2×10^{16} cm at 17 Mpc) (Haschick *et al.* 1989), and the emission is confined to a region of 10^{17} cm in diameter. The maser is superimposed on a nuclear continuum source. Because the maser volume is so small, it is difficult to understand how this maser is pumped unless its radiation is highly beamed. The collisional pumping schemes involving two streams of particles of Strelnitskij (1984) or Kylafis and Norman (1987) may help to explain these masers.

H_2O masers similar to those in our galaxies ($\mathcal{L}_{H_2O} < 10\,\mathcal{L}_\odot$) have been found in the spiral arms of several nearby galaxies (<1 Mpc) such as M33 (Huchtmeier, Eckart and Zensus 1988). It may be possible to estimate the distance to these masers and hence their host galaxies by the methods of statistical parallax described above. Proper motions within one maser complex are about 15 μas yr^{-1} (50 km s^{-1} at 0.7 Mpc), and the orbital proper motions about the galactic nucleus are about 120 μas yr^{-1}.

I am grateful to M. Elitzur, L. J. Greenhill, K. M. Menten, M. J. Reid, L. F. Rodríguez and V. Strelnitskij for helpful discussions.

RECOMMENDED READING (REVIEW ARTICLES)

Diamond, P. J. (1988). In *The Impact of VLBI on Astrophysics and Geophysics, IAU Symposium 129*, eds M. J. Reid and J. M. Moran, p. 213. Dordrecht: Kluwer Academic Publishers.

Downes, D. (1985). In *Birth and Infancy of Stars, Les Houches Sess. XLI*, eds R. Lucas, A. Omont, and R. Stora, p. 560. Amsterdam: North-Holland.

Elitzur, M. (1982). *Reviews in Modern Physics*, **54**, 1225.

Genzel, R. (1985). In *Masers, Molecules and Mass Outflows in Star Forming Regions*, ed. A. D. Haschick, p. 233. Westford: Haystack Observatory.

Goldreich, P. (1974). In *Atomic and Molecular Physics and the Interstellar Matter, Les Houches Sess. XXVI*, eds R. Balian, P. Encrenaz, and J. Lequeux, p. 413. Amsterdam: North-Holland.

Kylafis, N. D. (1988). In *The Impact of VLBI on Astrophysics and Geophysics, IAU Symposium 129*, eds M. J. Reid and J. M. Moran, p. 233. Dordrecht: Kluwer Academic Publishers.

Litvak, M. M. (1974). *Annual Review of Astronomy and Astrophysics*, **12**, 97.

Moran, J. M. (1976). In *Frontiers of Astrophysics*, ed. E. H. Avrett, p. 385. Cambridge, MA: Harvard University Press.

Moran, J. M. (1982). In *Handbook of Laser Science and Technology*, Vol. 1, ed. M. Weber, p. 483. Boca Raton, Florida: CRC Press.

Reid, M. J., and Moran, J. M. (1981). *Annual Review of Astronomy and Astrophysics*, **19**, 231.

Reid, M. J., and Moran, J. M. (1988). In *Galactic and Extragalactic Radio Astronomy*, eds G. L. Verschuur and K. I. Kellermann, p. 255. New York: Springer-Verlag.

REFERENCES

Alcock, C., and Ross, R. R. (1985a). *Astrophysical Journal*, **290**, 433.

Alcock, C., and Ross, R. R. (1985b). *Astrophysical Journal*, **299**, 763.

Andresen, P. (1986). *Astronomy and Astrophysics*, **154**, 42.

Baan, W. (1985). *Nature*, **315**, 26.

Baan, W. (1989). *Astrophysical Journal*, **338**, 804.

Baan, W., and Haschick, A. D. (1984). *Astrophysical Journal*, **279**, 541.

Barvainis, R., McIntosh, G., and Predmore, C. R. (1987). *Nature*, **329**, 613.

Batrla, W., Matthews, H. E., Menten, K. M., and Walmsley, C. M. (1987). *Nature*, **326**, 49.
Bloemhof, E. E., Reid, M. J., and Moran, J. M. (1989). In *Proceedings of the Conference on the Physics and Chemistry of Interstellar Molecular Clouds*, eds G. Winnewisser and J. T. Armstrong. New York: Springer-Verlag. In press.
Braz, M. A., and Epchtein, E. (1983). *Astronomy and Astrophysics Supplement*, **54**, 167.
Chapman, J. M., and Cohen, R. J. (1986). *Monthly Notices of the Royal Astronomical Society*, **220**, 513.
Cohen, R. J., Downs, G., Emerson, R., Grimm, M., Gulkis, S., Stevens, G., and Tarter, J. (1987). *Monthly Notices of the Royal Astronomical Society*, **225**, 491.
Cooke, B., and Elitzur, M. (1985). *Astrophysical Journal*, **295**, 175.
Davies, R. D. (1973). In *Galactic Radio Astronomy, IAU Symposium 60*, eds F. J. Kerr, and S. C. Simonson, p. 275. Dordrecht: Reidel Publishing.
Deguchi, S., and Watson, W. D. (1986). *Astrophysical Journal*, **302**, 750.
deNoyer, L. K., and Dodd, J. G. (1989). *Astrophysical Journal,* in press.
dePater, I., Palmer, P., and Snyder, L. E. (1986). *Astrophysical Journal (Letters)*, **304**, L33.
Diamond, P. J. (1988). In *The Impact of VLBI on Astrophysics and Geophysics, IAU Symposium 129*, eds M. J. Reid and J. M. Moran, p. 213. Dordrecht: Kluwer Academic Publishers.
Diamond, P. J., Norris, R. P., Rowland, P. R., Booth, R. S., and Nyman, L. A. (1985). *Monthly Notices of the Royal Astronomical Society*, **212**, 1.
Elitzur, M., Goldreich, P., and Scoville, N. (1976). *Astrophysical Journal*, **205**, 384.
Engels, D. (1979). *Astronomy and Astrophysics Supplement*, **36**, 337.
Evans, N. J., Hill, S. R. E., Rydbeck, O. E. H., and Kollberg, E. (1972). *Physical Review A,* **6**, 1643.
Fiebig, D., and Güsten, R. (1988). *Astronomy and Astrophysics,* **214**, 333.
Garay, G., Moran, J. M., and Haschick, A. D. (1989). *Astrophysical Journal,* **338**, 244.
Garay, G., Reid, M. J., and Moran, J. M. (1985). *Astrophysical Journal*, **289**, 681.
Garcia-Barreto, J. A., Burke, B. F., Reid, M. J., Moran, J. M., Haschick, A. D., and Schilizzi, R. T. (1988). *Astrophysical Journal*, **326**, 954.
Gaume, R. A., and Mutel, R. L. (1987). *Astrophysical Journal Supplements (Series)*, **65**, 193.
Genzel, R., and Downes, D. (1977). *Astronomy and Astrophysics Supplement*, **30**, 145.
Genzel, R., Reid, M. J., Moran, J. M., and Downes, D. (1981). *Astrophysical Journal*, **244**, 884.
Goldreich, P., and Keeley, D. A. (1972). *Astrophysical Journal*, **174**, 517.
Goldreich, P., Keeley, D. A., and Kwan, J. Y. (1973a). *Astrophysical Journal*, **179**, 111.
Goldreich, P., Keeley, D. A., and Kwan, J. Y. (1973b). *Astrophysical Journal*, **182**, 55.
Goldreich, P., and Kwan, J. (1974a). *Astrophysical Journal*, **190**, 27.
Goldreich, P., and Kwan, J. (1974b). *Astrophysical Journal*, **191**, 93.
Guilloteau, S. (1982). *Astronomy and Astrophysics*, **116**, 101.
Gwinn, C. R., Moran, J. M., Reid, M. J., and Schneps, M. H. (1988). *Astrophysical Journal*, **330**, 817.
Harvey, P. M., Bechis, K. P., Wilson, W. J., and Ball, J. A. (1974). *Astrophysical Journal Supplements (Series)*, **27**, 331.
Haschick, A. D., Baan, W. A., Schneps, M. H., Reid, M. J., Moran, J. M., and Güsten, R. (1989). In preparation.

Herman, J., Baud, B., Habing, H. J., and Winnberg, A. (1985). *Astronomy and Astrophysics*, **143**, 122.
Huchtmeier, W. K., Eckart, A., and Zensus, A. J. (1988). *Astronomy and Astrophysics*, **200**, 26.
Jewell, P. R., Dickinson, D. F., Snyder, L. E., and Clemens, D. P. (1987). *Astrophysical Journal*, **323**, 749.
Jewell, P. R., Webber, J. C., and Snyder, L. E. (1980). *Astrophysical Journal (Letters)*, **242**, L29.
Koo, B. C., Williams, D. R. W., Heiles, C., and Backer, D. C. (1988). *Astrophysical Journal*, **326**, 931.
Kylafis, N. D. (1988). In *The Impact of VLBI on Astrophysics and Geophysics, IAU Symposium 129,* eds M. J. Reid and J. M. Moran, p. 223. Dordrecht: Kluwer Academic Publishers.
Kylafis, N. D., and Norman, C. A. (1987). *Astrophysical Journal*, **323**, 346.
Lewis, B. M., Eder, J., and Terzian, Y. (1985). *Nature*, **313**, 200.
Litvak, M. (1971). *Astrophysical Journal*, **170**, 71.
Mattila, K., Holsti, N., Torisera, M., Anttila, R., and Malkamaki, L. (1985). *Astronomy and Astrophysics*, **145**, 192.
Menten, K. M., Reid, M. J., Moran, J. M., Wilson, T. L., Johnston, K. J., and Batrla, W. (1988). *Astrophysical Journal (Letters)*, **333**, L83.
Menten, K. M., Walmsley, C. M., Henkel, C., and Wilson, T. L. (1986). *Astronomy and Astrophysics*, **157**, 318.
Moran, J. M. (1981). *Bulletin of the American Astronomical Society*, **13**, 508.
Moran, J. M. (1982). In *Handbook of Laser Science and Technology*, Vol. 1, *Lasers and Masers*, ed. M. J. Weber, p. 483. Boca Raton: CRC Press.
Mumma, M. J., Buhl, D., Chin, G., Drake, D., Espenak, F., and Kostiuk, T. (1981). *Science*, **212**, 45.
Netzer, N., and Knapp, G. R. (1987). *Astrophysical Journal,* **323**, 734.
Nyman, L. A., and Olofsson, H. (1986). *Astronomy and Astrophysics*, **158**, 67.
Reid, M. J., Moran, J. M., Leach, R. W., Ball, J. A., Johnston, K. J., Spencer, J. H., and Swenson, G. W. (1979). *Astrophysical Journal (Letters)*, **227**, L89.
Reid, M. J., Schneps, M. H., Moran, J. M., Gwinn, C. R., Genzel, R., Downes, D., and Rönnäng, B. (1988). *Astrophysical Journal*, **330**, 809.
Rowland, P. R., and Cohen, R. J. (1986). *Monthly Notices of the Royal Astronomical Society*, **220**, 223.
Schultz, G. V., Sherwood, W. A., and Winnberg, A. (1978). *Astronomy and Astrophysics*, **63**, L5.
Schwartz, P. R., Harvey, P. M., and Barrett, A. H. (1974). *Astrophysical Journal*, **187**, 491.
Strelnitskij, V. S. (1984). *Monthly Notices of the Royal Astronomical Society*, **207**, 339.
Strelnitskij, V. S., and Sunyaev, R. A. (1973). *Soviet Astronomy*, **16**, 579.
Tarter, J. C., and Welch, W. J. (1986). *Astrophysical Journal*, **305**, 467.
Western, L. R., and Watson, W. D. (1983). *Astrophysical Journal*, **274**, 195.
Western, L. R., and Watson, W. D. (1984). *Astrophysical Journal*, **285**, 158.

24

Chemistry in the circumstellar envelopes around mass-losing red giants

M. JURA

Department of Astronomy, University of California, Los Angeles, USA

24.1. Introduction

When low and intermediate mass stars evolve off the main sequence, they become red giants. During the first ascent up the red giant branch, the stars lose mass at a relatively modest rate ($\sim 10^{-9}$ $M_\odot$ yr^{-1}. Dupree (1986)). However, during the second ascent up the red giant branch, the asymptotic giant branch (AGB), stars lose mass at a much greater rate; up to 10^{-4} $M_\odot$ yr^{-1} (Zuckerman 1980, Jura 1986a, Olofsson 1985). These outflows from AGB stars are very cold ($T < 1000$ K), and they contain large amounts of dust and molecules. The chemistry in these outflows is the topic of this chapter (see also Olofsson (1987), Omont (1987), Millar (1988)).

Stars derive most of their luminosity from nuclear reactions that occur in their interiors. In the red giant phase, material from the interior is mixed with that at the surface and the composition of the photosphere of a red giant can be very different from the star's initial composition. In most cases, it appears that in the atmospheres of the red giants, the major elements are still hydrogen and helium, even though material is mixed from the interiors to the surface. However, the next most common elements, carbon, nitrogen and oxygen, may have very nonstandard compositions, and this can greatly affect the chemistry. In particular, the ratio [C]/[O] is critical. In the usual thermodynamic description for material in cool stellar atmospheres, we expect CO, because of its very high binding energy, to contain as much carbon and oxygen as possible. If $[C]/[O] < 1$, there is excess oxygen and the material is oxygen-rich while if $[C]/[O] > 1$, there is excess carbon and the material is carbon-rich. In the sun, $[O] \approx 2[C]$, and the material is oxygen-rich. However, AGB stars come in a wide variety of types including both oxygen-rich and carbon-rich stars and S stars (objects with $[C] \approx [O]$). These differences are manifestations of a heterogeneous stellar population with different progenitor main sequence masses and surface compositions.

In the Yale Bright Star Catalog of the 10 000 optically brightest stars in the sky, there are only about 20 carbon-rich stars, and it would seem that these objects are quite rare. However, among the red giants that are losing large amounts of mass, at least half are carbon-rich (Knapp and Morris 1985) as is also the case for planetary nebulae (Zuckerman and Aller 1986), the evolutionary descendants of the mass-losing AGB stars (Iben and Renzini 1983). Even S stars are not remarkably exotic; there are about $\frac{1}{3}$ as many mass-losing S stars as there are mass-losing carbon stars (Wing and Yorka 1977, Jura 1988).

There are now known to be a few stars which have carbon-rich atmospheres, but oxygen-rich circumstellar material (Little-Marenin 1986, Nakada *et al.* 1987, Willems and de Jong 1986). This may occur because of a star's rapid evolution from being oxygen-rich to being carbon-rich (Willems and de Jong 1986), but this is questionable (Claussen *et al.* 1987). Another possibility is that there are complicated binary interactions which can account for this phenomenon (Benson and Little-Marenin 1987).

There are some red giants that are much more massive and luminous than asymptotic giant branch stars. Objects like NML Cyg with an initial main sequence mass of 40 $M_\odot$ and $L \gg 10^5$ $L_\odot$ (Morris and Jura 1983a) have a different evolutionary history than most of the stars discussed in this chapter. The maximum luminosity of AGB stars is 6×10^4 $L_\odot$ (Iben and Renzini 1983), and any red giant star that is more luminous than this must have an initial mass larger than ~8 $M_\odot$. These mass-losing red supergiants are much less common than the AGB stars and have received less attention.

Mass loss during the AGB is of central importance in determining the ultimate fate of a star. Unless the final mass of a star is less than 1.4 $M_\odot$, it cannot become a white dwarf; instead it must become a neutron star, a black hole and/or a supernova. Typically, however, it seems that in the Milky Way, stars with main sequence masses as large as about 6 $M_\odot$ ultimately become white dwarfs. The evidence is: (i) White dwarfs are observed in open clusters where the mass at the turn-off of the main sequence is about 6 $M_\odot$ (Weidemann and Koester 1983). (ii) The rate of supernovae in the galaxy is consistent with the death rate of stars with masses larger than ~8 $M_\odot$ (Miller and Scalo 1979). If all the stars with initial masses larger than 1.4 $M_\odot$ become supernovae, the rate of these explosions would be an order of magnitude larger than observed.

Dust grain cores are formed in the outflows from red giants. While there is considerable evolution of this material once it leaves the red giant and escapes into the interstellar medium, there is evidence that at least the grain cores can retain their identity for a considerable time. For example, very small grain inclusions in some meteorites have such unusual and distinctive isotopic compositions that they cannot be general solar system material. Instead, they appear to have been transported into the early solar system from a red giant wind, while having managed to retain much of their compositional integrity during their passage through the interstellar medium (Zinner *et al.* 1986, Zinner and Epstein 1987).

During the past 20 years, the great advances in the instrumentation available to do radio and infrared astronomy have opened up the field of study of the circumstellar matter around mass-losing red giants. Molecules are studied by both emission and absorption in the radio and infrared while the dust is primarily studied by its infrared emission. With extensive sky surveys available at 2 μm (Neugebauer and Leighton 1969) and at 12 μm, 25 μm, 60 μm and 100 μm (the IRAS survey), it is possible to describe systematically the spatial distribution and physical properties of the different mass-losing stars. We write for the local space density of carbon stars, n, that:

$$n = n_0 \exp(-z/z_0) \tag{24.1}$$

where z is measured vertical to the galactic plane. Within about 1.5 kpc of the sun, the space density of carbon stars n_0 = 100 kpc^{-3} with an exponential scale height of z_0 = 200 pc above the galactic plane (Claussen *et al.* 1987). All these stars are losing $\geq 10^{-7}$ $M_{\odot}$ yr^{-1}. Although the analysis is not yet finished, it seems that there is more mass injected into the interstellar medium from the relatively rare stars with high mass loss rates than the much more numerous stars with low mass loss rates (Knapp and Morris 1985).

The S stars display the same spatial distribution as the carbon stars, but there appear to be about $\frac{1}{3}$ as many S stars and they typically seem to lose about $\frac{1}{3}$ as much mass (Jura 1988). A full analysis of the infrared properties of oxygen-rich mass-losing stars in the solar neighborhood is not yet completed.

In addition to conducting surveys, it is also instructive to study more intensively a few prototypes. In Table 24.1, we list some representative mass-losing stars in the solar neighborhood. Probably the single most famous object is the carbon-rich star, IRC+10216. This is the brightest 10 μm source in the sky outside the solar system, and it has a bolometric flux at the earth of 2×10^{-8} W m^{-2} (see, for example, Sopka *et al.* (1985)). If it were not surrounded by an envelope of dust, this would be one of the ~10 brightest optical stars in the sky. However, it is barely detectable at optical wavelengths with even the largest telescopes (Cohen and Schmidt 1982), and it was not even discovered until the 2 μm sky survey was performed. Even though this object is very well studied, its distance is uncertain by a substantial factor; it probably lies between 100 and 200 pc from the sun (Zuckerman, Dyck and Claussen 1986).

Table 24.1. *Representative mass-losing stars*

Class	Prototype	d (pc)	dM/dt ($M_\odot$ yr^{-1})	v (km s^{-1})	Reference
Oxygen-rich Mira	Mira	77	6(−7)	5	1, 2
Carbon-rich Mira	V Hya	340	1(−6)	15	1, 3, 4, 5
Dusty carbon star	IRC+10216	150?	2(−5)	15	6
OH/IR star	OH 26.5 + 0.6	1800	7(−5)	15	7, 8
Optical supergiant	α Ori	200	4(−5)	15	9, 10
Dusty supergiant	NML Cyg	2000	6(−5)	23	11
Post AGB	CRL 2688	1000	7(−5)	19	12, 13
Low mass post AGB	U Mon	600	8(−6)	10?	14
Young planetary	NGC 7027	940	5(−5)	22	15, 16

References: (1) Knapp and Morris (1985); (2) Zuckerman and Dyck (1986a); (3) Zuckerman and Dyck (1986b); (4) Kahane, Maizels and Jura (1988); (5) Tsuji *et al.* (1988); (6) Kwan and Linke (1982); (7) Werner *et al.* (1980); (8) Bowers, Johnston and Spencer (1983); (9) Glassgold and Huggins (1986); (10) Huggins (1987); (11) Morris and Jura (1983b); (12) Ney *et al.* (1975); (13) Kawabe *et al.* (1987); (14) Jura (1986c); (15) Jura (1984); (16) Masson (1986).

Besides systematic infrared surveys of the sky, there have been extensive surveys of the molecular emission from mass-losing stars (see Table 24.2). H_2 is only detectable in special circumstances – such as shocks formed in the expansion of the ionized gas around planetary nebulae; in most cases H_2 is not directly measured. Approximately 200 stars are CO sources, and they are detected by emission in either the $J = 1$–0 or $J = 2$–1 rotational lines at wavelengths of 0.26 and 0.13 cm, respectively, that are readily studied by ground-based millimeter telescopes. Other molecules such as OH, H_2O, HCN, and SiO have been studied in a wide variety of sources, while some molecules have only been detected in the unusually bright object, IRC+10216. The number of molecules known in circumstellar envelopes has increased dramatically during the past few years.

24.2. Physical conditions in the outflows

In order to understand the chemistry in the outflows from mass-losing red giants, it is necessary to describe the density and temperature as a function of radius around the star. These quantities then constrain the chemical and dust composition in the envelope.

As a first approximation, we assume a spherically symmetric outflow with a mass loss rate of dM/dt. Some stars clearly display a bipolar asymmetry in their outflow in the sense that there is much more mass in a disk than along the poles of the outflow (Morris 1981). In extreme cases, $dM/dt(\theta)$ may vary by a factor of 10 as a function of θ (Yusef-Zadeh, Morris and White 1984). However, for these

Table 24.2. *Molecules in circumstellar shells*

Molecule	Location	Representative reference
H_2	Planetary nebulae	1
CO	Most mass losers	2, 3, 4, 5, 6, 7, 8
OH	OH/IR stars	9, 10
SiO	Many mass losers	11
SiS	IRC+10216	12, 13
CS	IRC+10216	14
CN	IRC+10216	15
SO	IRC+10216	16
NaCl	IRC+10216	17
AlCl	IRC+10216	17
KCl	IRC+10216	17
HCN	IRC+10216	18, 19
HNC	Oxygen-rich; carbon-rich	20, 21
C_2H	IRC+10216	15
SiC_2	IRC+10216	22
H_2O	OH/IR stars	10, 23
H_2S	Various stars	24
HCO^+	OH231.8+4.2	20
OCS	OH231.8+4.2	20
SO_2	Oxygen-rich stars	16
C_2S	IRC+10216	14
C_2H_2	IRC+10216	25
NH_3	CRL 2688	26
C_3N	IRC+10216	21
C_3S	IRC+10216	14
C_3H	IRC+10216	27
HC_3N	Carbon stars	28
CH_4	IRC+10216	29
SiH_4	IRC+10216	30
HC_4	IRC+10216	21
C_4H	IRC+10216	31, 32
C_3H_2	IRC+10216	33
CH_3CN	IRC+10216	21
C_5H	IRC+10216	34
C_2H_4	IRC+10216	35
HC_5N	Carbon stars	28
C_6H	IRC+10216	32
HC_7N	IRC+10216	28
HC_9N	IRC+10216	36
$HC_{11}N$	IRC+10216	37
PAHs	Planetary nebulae	38

References: (1) Zuckerman and Gatley (1988); (2) Zuckerman and Dyck (1986a); (3) Zuckerman and Dyck (1986b); (4) Zuckerman, Dyck and Claussen (1986); (5) Knapp and Morris (1985); (6) Olofsson, Eriksson and Gustafsson (1987); (7) Nguyen-Q-Rieu *et al.* (1987); (8) Olofsson, Eriksson and Gustafsson (1988); (9) Bowers *et al.* (1983); (10) Engels (1979); (11) Jewell *et al.* (1985); (12) Henkel, Matthews and Morris (1983); (13)

cases, we can treat the mass loss per unit solid angle, $dM/dt(\theta)$ as if the mass loss were locally equivalent to a spherically symmetric mass loss (Jura 1983), so, for many purposes, the simplifying assumption of spherical symmetry does not affect our understanding of the chemistry in the outflow.

From the equation of continuity, we may write for the mass density of the material, ρ, that:

$$\rho = (dM/dt)/(4\pi r^2 v) \qquad (24.2)$$

where r is the distance from the center of the star and v is the outflow velocity. Typically, v reaches a terminal value of ~ 15 km s^{-1} within 10^{15} cm of the central star.

For most mass-losing red giants, we expect the bulk of the hydrogen to be molecular rather than atomic (Glassgold and Huggins 1986). However, at least for mass loss from the warmer stars, much of the hydrogen that flows out of the photosphere may be atomic, and there may be production of atomic hydrogen by photodissociation of the molecular material in the outflow (Morris and Jura 1983a, Glassgold and Huggins 1983). There are one or two stars where 21 cm emission from atomic hydrogen has been detected in the circumstellar shell (Bowers and Knapp 1987), but mostly there are only upper limits (Knapp and Bowers 1983, Zuckerman, Terzian and Silverglate 1980).

The characteristic dynamic time scale for the outflow, t_{flow}, is r/v. To describe the chemistry, it is necessary to compare t_{flow} with the local time for chemical reactions to occur which, of course, depends upon the local density and temperature. If we ignore the complications described below, the density is given by Equation (24.2). Because the density near the star must be high and then decrease significantly further from the star, one can naturally imagine the chemistry as having three main regimes (Lafont, Lucas and Omont 1982, Omont 1987, Millar 1988). (i) There is the region near the star where the density is sufficiently high that three-body chemical reactions frequently occur in a time short compared to t_{flow}. In this regime, we expect the conditions to approach thermodynamic equilibrium. (ii) Further away from the star, there is a 'freeze-out' of the products

Table 24.2 References contd.
Turner (1987); (14) Cernicharo *et al.* (1987); (15) Huggins, Glassgold and Morris (1984); (16) Guilloteau *et al.* (1986); (17) Cernicharo and Guélin (1987); (18) Bieging, Chapman and Welch (1984); (19) Ziurys and Turner (1986); (20) Morris *et al.* (1987); (21) Johansson *et al.* (1984); (22) Thaddeus, Cummins and Linke (1984); (23) Lane *et al.* (1987); (24) Ukita and Morris (1983); (25) Betz (1981); (26) Nguyen-Q-Rieu, Winnberg and Bujarrabal (1986); (27) Thaddeus *et al.* (1985); (28) Jewell and Snyder (1984); (29) Hall and Ridgwa (1978); (3) Goldhaber and Betz (1984); (31) Guélin *et al.* (1986); (32) Guélin *et al.* (1987); (33) Matthews and Irvine (1985); (34) Cernicharo *et al.* (1986); (35) Goldhaber, Betz and Ottusch (1987); (36) Matthews, Friberg and Irvine (1985); (37) Bell *et al.* (1982); (38) Léger and Puget (1984).

of the three-body reactions. In this region two-body reactions dominate the chemistry, and this is an environment similar to that within interstellar clouds. (iii) Finally, the density becomes sufficiently low, that only 'one-body' reactions occur. In this case, the only significant chemical processing results from the absorption of ambient interstellar ultraviolet photons by the molecules flowing out from the central star.

In addition to the density, it is necessary to know the temperature as a function of r. The temperature of the gas and the temperature of the grains are decoupled from each other (Goldreich and Scoville 1976). If T_{gr} denotes the temperature of the grains, we expect that their temperature is mainly controlled by a balance between absorbing photons emitted by the central star, and then reradiating this energy in the infrared. For grains that are further from the star, the rate of heating is lower, and T_{gr} decreases radially from the star. With the very simple approximation that the grain emissivity as a function of frequency is a power law such that the ε_ν varies as ν^q, with perhaps $q = 1$ (Jura 1986b), we expect:

$$T_{gr} \sim r^{-2/5} \tag{24.3}$$

A convenient discussion of estimates of the grain temperature vs radius is given in Sopka *et al.* (1985). Very near the star where the grains are formed, we expect T_{gr} to be ~1000 K, and then to fall as a function of radius according to Equation (24.3).

The gas temperature is also determined by a balance between heating and cooling rates, but the processes are quite different for the gas than for the dust grains. The main source of heating of the gas is the supersonic streaming of the grains through the gas. The radiation from the central star drives the grains through the gas with a drift velocity, v_{dr}, of (Kwan and Linke 1982):

$$v_{dr} = [vQL/(c\mathrm{d}M/\mathrm{d}t)]^{1/2} \tag{24.4}$$

In this expresion, v is the outflow velocity of the gas, L is the luminosity of the star and Q is the ratio, averaged over frequency, of the momentum transfer cross section (for radiation) to the geometric cross section of the grains. Typical values of v_{dr} are 2–20 km s^{-1} (Papoular and Pegourie 1986).

The gas heating rate from gas–grain collisions, $\mathrm{d}q_{dust}/\mathrm{d}t$, depends upon the rate of collisions and the kinetic energy transfer per collision. Therefore, (Goldreich and Scoville 1976, Kwan and Linke 1982):

$$\mathrm{d}q_{dust}/\mathrm{d}t = (n_{gr}\sigma_{gr}v_d)\tfrac{1}{2}\rho v_d^2 \tag{24.5}$$

The cooling of the gas is controlled both by the adiabatic expansion of the gas and radiation. We write (Goldreich and Scoville 1976):

$$\begin{aligned}\mathrm{d}T_{ga}/\mathrm{d}r = {} & (2-2\gamma)(T_{ga}/r) - (\gamma-1)(\mathrm{d}q_{cool})(knv)^{-1} \\ & + (\gamma-1)(\mathrm{d}q_{dust})(knv)^{-1}\end{aligned} \tag{24.6}$$

In the above expression, γ is the ratio of specific heats, k is Boltzmann's constant and $n = \rho/\mu$ where μ is the mean molecular weight of the outflowing material.

The gas temperature as a function of radius can be sensitive to such parameters as the grain size, the dust to gas ratio and the total mass loss rate. In particular, the drift velocity of the grains depends upon the amount of resistance provided by the gas. In stars with low mass loss rates, the density in the circumstellar envelope is lower and therefore the drift velocity should be appreciably higher than in stars with high mass loss rates. Since the heating varies as the drift velocity cubed (see Equation (24.5)), we expect higher temperatures in the circumstellar shells of stars with lower mass loss rates. Although the full consequences of the different heating and cooling rates in different circumstellar envelopes have not yet been fully explored (see, for example, Tielens (1983), Jura, Kahane and Omont (1988)), it seems that near the stars, temperatures in excess of 1000 K are possible. Far from the star, the adiabatic cooling dominates and the temperature may fall close to 2.7 K, the temperature of the microwave background.

The dust and gas ratio and the size of the dust grains are important in the physics and chemistry of the outflows from mass-losing stars. Indeed, radiation pressure on the dust may play a key role in driving the material into the interstellar medium (Kwok 1975, Lucy 1976, Jura 1986a). Unfortunately, the theory for the nucleation and growth of dust in stellar envelopes is still very uncertain (Donn and Nuth 1985). Therefore, we must rely upon observations to provide insight into the nature of the circumstellar dust. It appears that at least in carbon-rich stars and in S stars, much if not most of the carbon is contained within dust (Jura 1986b, 1988) so that the dust to gas ratio by mass is typically ~ 0.01. The grain sizes also are not well known, and like the dust grains in the interstellar medium (Mathis, Rumpl and Nordsieck 1977), many of the circumstellar grains are typically 0.01–0.1 μm in size. However, there may be both larger and smaller grains produced in the outflows (Kroto *et al.* 1987).

24.3. Chemistry

24.3.1. The near zone – three-body reactions

The typical radius of a mass-losing red giant star is $\sim 10^{13}$ cm. The near zone, that region within roughly 10 stellar radii or about 10^{14} cm of the star, is complex and not well studied with the observational probes that are currently available.

The large majority of mass-losing red giants are pulsating (Jura 1986a); some mass-losing stars exhibit changes in their radii of >30% during their pulsations (Richichi, Salinari and Lisi 1988). These large scale oscillations of the stellar photosphere lead to shocks which can levitate matter to regions well above the nominal photosphere.

Once the matter is far from the star, it cools, grains can form, and radiation pressure can dramatically accelerate the matter away from the star (Dyck *et al.* 1984, Jura 1986b). However, flows in the region near the star can be quite complex; observations indicate some material near the star actually flowing back on to the surface (Jura 1984). Also, the outflow may not be spherically symmetric because, for example, of the presence of a close companion star to the red giant. Small condensations may form in the outflows (Alcock and Ross 1986), perhaps from thermal instabilities (Muchmore, Nuth and Stencel 1987, Stencel, Carpenter and Hagen 1986). Therefore, it is very difficult to compute the density as a function of radius in this near zone. From Equation (24.2), for a mass loss rate of 10^{-6} $M_\odot$ yr^{-1} at a speed of 10 km s^{-1}, the particle density at 10^{14} cm from the star is 10^8 cm^{-3}. The flow time, r/v, is then 10^8 s while the three-body chemical reaction time is $\sim 10^8$ s. Therefore, three-body reactions dominate the chemistry, consistent with the picture that grains form in this region. We also expect that in most circumstances, the molecules H_2 and CO are formed with large abundances. If the star is oxygen-rich, there should be large amounts of H_2O while if it is carbon-rich, C_2H_2 should be abundant (Salpeter 1977, McCabe, Smith and Clegg 1979, Lafont *et al.* 1982).

Because we are considering a dynamic environment where both small molecules and large grains are formed, it is quite possible that large numbers of intermediate-size species, such as PAHs or C_{60}, are also created in large abundances (Kroto *et al.* 1987). There is indirect evidence for a substantial production of PAHs in the outflows from carbon stars, since carbon-rich planetary nebulae often display emission in broad infrared features, for example, at 3.28 μm, which are considered to be signatures of PAHs (Léger and Puget 1984). (There may also be some oxygen-rich planetaries which display this emission (Martin 1987), but this is not confirmed by all observers.)

24.3.2. Intermediate distances – two-body reactions

Beyond about 10^{14} or 10^{15} cm from the central star, the density drops sufficiently that the time scale for three-body reactions to occur is longer than the outflow time. However, two-body reactions may still be important. The typical outflow time, r/v, is 10^9 s at 10^{15} cm from the central star. For a mass loss rate of 10^{-6} $M_\odot$ yr^{-1} and an outflow speed of 10 km s^{-1}, the density at 10^{15} cm from the star is 10^6 cm^{-3}. If the reaction rate for a neutral–neutral reaction is 10^{-15} cm^3 s^{-1}, then the time between collisions is also 10^9 s. Ion–molecule reactions proceed much more rapidly, and they can be quite important in the synthesis of some species.

There have been a number of studies of the importance of two-body reactions in the outflows from mass-losing red giants. These calculations are natural extensions of previous work that has been directed towards understanding the gas phase chemistry within interstellar clouds. While all the models do not agree with each other in detail, there are clearly some trends that are important. For example, it is possible to synthesize minor constituents which are not predicted to be abundant in the simple thermodynamic 'freeze-out' model. Abundances of such species as HCN in oxygen-rich outflows, and HCO^+ in all envelopes are computed and agree reasonably well with the available data (Glassgold, Lucas and Omont 1986, Glassgold *et al.* 1987, Mamon, Glassgold and Omont 1987, Nejad and Millar 1987).

Another sort of two-body reaction that can be important in the outflows is the collisions between gas and solid dust grains. The amount of material in grains is sufficiently large (typically 0.01 of the total mass in the outflow) that much of the material heavier than helium must be incorporated into grains. The relative weakness of the thermal SiO emission in the low lying rotational levels, which is produced in an extended circumstellar envelope (as opposed to SiO maser emission which is produced very near the photosphere of the star) indicates that perhaps 99% of all the silicon is incorporated into the dust grains (Morris *et al.* 1979). Much of the grain composition is determined by what happens in the inner zone of the envelope where three-body reactions dominate. However, even in the outer envelope, there are gas–grain collisions that occur that are important for the chemistry as well as for heating the gas.

Once the dust is far from the star, it becomes cool (see Equation (24.3)). On a sufficiently cool grain, the time scale for the thermal evaporation of an adsorbed molecule may be longer than the dynamic outflow time; in this case, the grain may grow a mantle of adsorbed molecules. Although, for example, the binding energy of CO onto a grain is sufficiently low that we do not expect much solid CO to be on grain surfaces, other molecules, such as H_2O can reside on relatively cool grains for a long time. Jura and Morris (1985) have shown that in some outflows, ice–grain mantles should be present, and their calculations roughly agree with observations which show that a few circumstellar envelopes do contain solid ice as well as gas phase H_2O (see, for example, Forveille *et al.* (1987)).

24.3.3. The outer envelopes – photochemistry

Far from the star, the density falls sufficiently that there are very few collisions among the different molecules. However, the material is still subject to processing by the ambient interstellar ultraviolet radiation field. In the absence of shielding, the typical lifetime for a molecule against photodestruction in the average interstellar ultraviolet radiation field is $\sim 10^{10}$ s (Lee 1984). Therefore, photodissociation begins to be important for molecules once they are $\sim 10^{16}$ cm from the star.

A particularly noteworthy example of photodissociation is OH in oxygen-rich red giants. Near the photosphere, most of the oxygen is contained either within

CO or H_2O. As the matter flows away from the star, the H_2O is photodissociated into OH (Goldreich and Scoville 1976, Huggins and Glassgold 1982). The OH that is thus produced is often detected from its maser emission which characteristically is significantly displaced from the central star.

In considering the photoprocesses in the outflows, it is important to include the effects of shielding either from dust or, for very abundant molecules, from self-shielding. In outflows from stars losing a large amount of mass, there may be so much circumstellar dust that the ultraviolet light from the interstellar medium does not penetrate deeply into the envelope. If we write τ as the optical depth, measured radially from position r to infinity, then:

$$\tau = \chi(\mathrm{d}M/\mathrm{d}t)/(4\pi r v) \tag{24.7}$$

In equation (24.7), χ is the opacity of the material. For the usual dust to gas ratios in the outflows from red giants and dust properties of the material that condenses into solids, $\chi_{UV} \sim 1000\ \mathrm{cm^2\ gm^{-1}}$ (Jura 1986b, Le Bertre 1987, Martin and Rogers 1987). When one averages over all directions, it is possible to show that the ultraviolet radiation in a spherically symmetric outflow is diminished by an average optical depth, $\tau_{av} \approx 1.4\tau$ (Jura and Morris 1981). In this case, if I_0 denotes the photodestruction rate by ultraviolet photons in the absence of shielding, the actual destruction rate inside a circumstellar envelope, I, is given by:

$$I = I_0 \exp(-\tau_{av}) \tag{24.8}$$

With this simple form of the photodestruction rate, it is straightforward to derive simple analytic expressions for the density as a function of radius around the star (Jura and Morris 1981, Huggins and Glassgold 1982). It is also possible to consider the complications in nonspherically symmetric outflows (Jura 1983).

H_2 and CO are sufficiently abundant in many outflows that self-shielding by these species of themselves can be important (Morris and Jura 1983b, Glassgold and Huggins 1983, Mamon, Glassgold and Huggins 1988). That is, there is so much H_2 and CO present in the outflows, that all the ultraviolet photons that can destroy these species are consumed in the outer circumstellar envelope so that none of these photons penetrate into the inner regions. These effects can be quite important, for example, in studying the different isotopes of CO.

24.4. Conclusions and prospects for future studies

While optical studies enabled astronomers to recognize that mass loss from red giants is occurring (Deutsch 1956, Weymann 1962), almost everything we know about mass-losing red giants has been learned from radio and infrared studies. The past 20 years have shown a remarkable advance in this field, and we now have made good progress in determining the mass loss rates, the outflow velocities, the chemical compositions and other physical properties of the mass-losing stars. We have a much better quantitative understanding of the role that the mass loss plays in stellar evolution and the chemical evolution of the Milky Way galaxy as a whole. There is good evidence that mass-losing stars occur in other galaxies as well (Impey, Wynn-Williams and Becklin 1986, Jura *et al.* 1987, Soifer *et al.* 1986) and play a key role in determining the evolution of those systems.

Of course, there are still very many uncertainties in our understanding of the mass-losing red giants, and many of the unsolved problems are quite central to all of astronomy. For example, a few mass-losing carbon stars are very lithium-rich, and it is possible that much if not most of the galactic lithium is produced in these unusual carbon-rich red giants (Scalo 1976). This could profoundly affect the use of the measured Population I lithium abundance to estimate the amount of primordial lithium in the galaxy and thus determine the baryon/photon ratio in the early universe (Delbourgo-Salvador *et al.* 1985). Future studies of mass-losing red giant stars will provide essential insights into the nature of the universe in which we live.

This work has been partly supported by NASA.

REFERENCES

Alcock, C., and Ross, R. R. 1986, *Ap. J.*, **310**, 835.
Bell, M. B., Feldman, P. A., Kwok, S., and Matthews, H. E. 1982, *Nature*, **295**, 389.
Benson, P. J., and Little-Marenin, I. R. 1987, *Ap. J. (Letters)*, **316**, L37.
Betz, A. L. 1981, *Ap. J. (Letters)*, **244**, L103.
Bieging, J. H., Chapman, B., and Welch, W. J. 1984, *Ap. J.*, **285**, 656.
Bowers, P. F., and Knapp, G. R. 1987, *Ap. J.*, **315**, 305.
Bowers, P. F., Johnston, K. J., and Spencer, J. H. 1983, *Ap. J.*, **274**, 733.
Cernicharo, J., and Guélin, M. 1987, *Astr. Ap.*, **183**, L10.
Cernicharo, J., Guélin, M., Hein, H., and Kahane, C. 1987, *Astr. Ap.*, **181**, L9.
Cernicharo, J., Kahane, C., Gomez-Gonzalez, J., and Guélin, M. 1986, *Astr. Ap.*, **164**, L1.
Claussen, M. J., Kleinmann, S. G., Joyce, R. R., and Jura, M. 1987, *Ap. J. Suppl.*, **65**, 385.
Cohen, M., and Schmidt, G. D. 1982, *Ap. J.*, **259**, 693.
Delbourgo-Salvador, P., Gry, C., Malinie, G., and Audouze, J. 1985, *Astr. Ap.*, **150**, 53.
Deutsch, A. J. 1956, *Ap. J.*, **123**, 210.
Donn, B., and Nuth, J. A. 1985, *Ap. J.*, **288**, 187.
Dupree, A. K. 1986, *Ann. Rev. Astr. Ap.*, **24**, 377.
Dyck, H. M., Zuckerman, B., Leinert, Ch., and Beckwith, S. 1984, *Ap. J.*, **287**, 801.
Engels, D. 1979, *Astr. Ap. Suppl.*, **36**, 337.
Forveille, T., Morris, M., Omont, A., and Likkel, L. 1987, *Astr. Ap.*, **176**, L13.
Glassgold, A. E., and Huggins, P. J. 1983, *MNRAS*, **203**, 517.
Glassgold, A. E., and Huggins, P. J. 1986, *Ap. J.*, **306**, 605.
Glassgold, A. E., Lucas, R., and Omont, A. 1986, *Astr. Ap.*, **157**, 35.
Glassgold, A. E., Mamon, G. A., Omont, A., and Lucas, R. 1987, *Astr. Ap.*, **180**, 183.
Goldhaber, D. M., and Betz, A. L. 1984, *Ap. J. (Letters)*, **279**, L55.
Goldhaber, D. M., Betz, A. L., and Ottusch, J. J. 1987. *Ap. J.*, **314**, 356.

Goldreich, P., and Scoville, N. 1976, *Ap. J.*, **205**, 149.
Guélin, M., Cernicharo, J., Kahane, C., and Gomez-Gonzalez, J. 1986, *Astr. Ap.*, **157**, L17.
Guélin, M., Cernicharo, J., Kahane, C., Gomez-Gonzalez, J., and Walmsley, C. M. 1987, *Astr. Ap.*, **175**, L5.
Guilloteau, S., Lucas, R., Nguyen-Q-Rieu, and Omont, A. 1986, *Astr. Ap.*, **165**, L1.
Hall, D. N. B., and Ridgway, S. T. 1978, *Nature*, **273**, 281.
Henkel, C., Matthews, H. E., and Morris, M. 1983, *Ap. J.*, **267**, 184.
Huggins, P. J. 1987, *Ap. J.*, **313**, 400.
Huggins, P. J., and Glassgold, A. E. 1982, *Astr. J.*, **87**, 1828.
Huggins, P. J., Glassgold, A. E., and Morris, M. 1984, *Ap. J.*, **279**, 284.
Iben, I., and Renzini, A. 1983, *Ann. Rev. Astr. Ap.*, **21**, 271.
Impey, C. D., Wynn-Williams, C. G., and Becklin, E. E. 1986, *Ap. J.*, **309**, 572.
Jewell, P. R., and Snyder, L. E. 1984, *Ap. J.*, **278**, 176.
Jewell, P. R., Walmsley, C. M., Wilson, T. L., and Snyder, L. E. 1985, *Ap. J. (Letters)*, **298**, L55.
Johansson, L. E. B., *et al.* 1984, *Astr. Ap.*, **130**, 227.
Jura, M. 1983, *Ap. J.*, **275**, 683.
Jura, M. 1984, *Ap. J.*, **282**, 200.
Jura, M. 1986a, *Irish Astr. J.*, **17**, 322.
Jura, M. 1986b, *Ap. J.*, **301**, 624.
Jura, M. 1986c, *Ap. J.*, **309**, 732.
Jura, M. 1988, *Ap. J. Suppl.*, **66**, 33.
Jura, M., and Morris, M. 1981, *Ap. J.*, **251**, 181.
Jura, M., and Morris, M. 1985, *Ap. J.*, **292**, 487.
Jura, M., Kahane, C., and Omont, A. 1988, *Astr. Ap.*, **201**, 80.
Jura, M., Kim, D. W., Knapp, G. R., and Guhathakurta, P. 1987, *Ap. J. (Letters)*, **312**, L11.
Kahane, C., Maizels, C., and Jura, M. 1988, *Ap. J. (Letters)*, **328**, L25.
Kawabe, R., Ishiguro, M., Kasuga, T., Morita, K.-I., Ukita, N., Kobayashi, H. J., Okumura, S., Fomalont, E., and Kaifu, N. 1987, *Ap. J.*, **314**, 322.
Knapp, G. R., and Bowers, P. F. 1983, *Ap. J.*, **266**, 701.
Knapp, G. R., and Morris, M. 1985, *Ap. J.*, **292**, 640.
Kroto, H. W., Heath, J. R., O'Brien, S. C., Curl, R. F., and Smalley, R. E. 1985, *Nature*, **318**, 162.
Kroto, H. W., Heath, J. R., O'Brien, S. C., Curl, R. F., and Smalley, R. E. 1987, *Ap. J.*, **314**, 352.
Kwan, J., and Linke, R. A. 1982, *Ap. J.*, **254**, 587.
Kwok, S. 1975, *Ap. J.*, **198**, 583.
Lafont, S., Lucas, R., and Omont, A. 1982, *Astr. Ap.*, **106**, 201.
Lane, A. P., Johnston, K. J., Bowers, P. F., Spencer, J. H., and Diamond, P. J. 1987, *Ap. J.*, **323**, 756.
Le Bertre, T. 1987, *Astr. Ap.*, **176**, 107.
Lee, L. C. 1984, *Ap. J.*, **282**, 172.
Léger, A., and Puget, J.-L. 1984, *Astr. Ap.*, **137**, L5.
Little-Marenin, I. R. 1986, *Ap. J. (Letters)*, **307**, L15.
Lucy, L. B. 1976, *Ap. J.*, **205**, 482.
Mamon, G. A., Glassgold, A. E., and Huggins, P. J. 1988, *Ap. J.*, **328**, 797.
Mamon, G. A., Glassgold, A. E., and Omont, A. 1987, *Ap. J.*, **323**, 306.
Martin, W. 1987, *Astr. Ap.*, **182**, 290.

Martin, P. G., and Rogers, C. 1987, *Ap. J.*, **322**, 374.
Masson, C. R. 1986, *Ap. J. (Letters)*, **302**, L27.
Mathis, J. S., Rumpl, W., and Nordsieck, K. H. 1977, *Ap. J.*, **217**, 425.
Matthews, H. E., and Irvine, W. M. 1985, *Ap. J. (Letters)*, **298**, L61.
Matthews, H. E., Friberg, P., and Irvine, W. M. 1985, *Ap. J.*, **290**, 609.
McCabe, E. M., Smith, R. C., and Clegg, R. E. S. 1979, *Nature*, **281**, 263.
Millar, T. J. 1988, in *Rate Coefficients in Astrochemistry*, eds. T. J. Millar and D. A. Williams (Reidel: Dordrecht), in press.
Miller, G. E., and Scalo, J. M. 1979, *Ap. J. Suppl.*, **41**, 513.
Morris, M. 1981, *Ap. J.*, **249**, 572.
Morris, M., and Jura, M. 1983a, *Ap. J.*, **267**, 179.
Morris, M., and Jura, M. 1983b, *Ap. J.*, **264**, 546.
Morris, M., Guilloteau, S., Lucas, R., and Omont, A. 1987, *Ap. J.*, **321**, 888.
Morris, M., Redman, R., Reid, M. J., and Dickinson, D. F. 1979, *Ap. J.*, **229**, 257.
Muchmore, D. L., Nuth, J. A., and Stencel, R. E. 1987, *Ap. J. (Letters)*, **315**, L141.
Nakada, Y., Izumiura, H., Onaka, T., Hashimoto, O., Ukita, N., Deguchi, S., and Tanabe, T. 1987, *Ap. J. (Letters)*, **323**, L77.
Neugebauer, G., and Leighton, R. B. 1969, *Two Micron Sky Survey* (NASA: SP-3047).
Nejad, L. A. M., and Millar, T. J. 1987, *Astr. Ap.*, **183**, 279.
Ney, E. P., Merrill, K. M., Becklin, E. E., Neugebauer, G., and Wynn-Williams, C. G. 1975, *Ap. J. (Letters)*, **198**, L129.
Nguyen-Q-Rieu, Epchtein, N., Truong-Bach, and Cohen, M. 1987, *Astr. Ap.*, **180**, 117.
Nguyen-Q-Rieu, Winnberg, A., and Bujarrabal, V. 1986, *Astr. Ap.*, **165**, 204.
Olofsson, H. 1985, in *Workshop on Submillimeter Astronomy*, eds. P. A. Shaver and K. Kjar (Garching: European Southern Observatory) p. 535.
Olofsson, H. 1987, in *Late Stages of Stellar Evolution*, eds. S. Kwok and S. R. Pottasch (Dordrecht: Reidel), p. 149.
Olofsson, H., Eriksson, K., and Gustafsson, B. 1987, *Astr. Ap.*, **183**, L13.
Olofsson, H., Eriksson, K., and Gustafsson, B. 1988, *Astr. Ap.*, in press.
Omont, A. 1987, in *Astrochemistry, IAU Symp. No. 120*, eds. M. S. Vardya and S. P. Tarafdar (Dordrecht: Reidel), p. 357.
Papoular, R., and Pegourie, B. 1986, *Astr. Ap.*, **156**, 199.
Richichi, A., Salinari, P., and Lisi, F. 1988, *Ap. J.*, **326**, 791.
Ridgway, S. T., Hall, D. N. B., Kleinmann, S. G., Weinberger, D. A., and Wojslaw, R. S. 1976, *Nature*, **264**, 345.
Salpeter, E. E. 1977, *Ann. Rev. Astr. Ap.*, **15**, 267.
Scalo, J. M. 1976, *Ap. J.*, **206**, 795.
Soifer, B. T., Rice, W. L., Mould, J. R., Gillett, F. C., Rowan-Robinson, M., and Habing, H. J. 1986, *Ap. J.*, **304**, 651.
Sopka, R. J., Hildebrand, R., Jaffe, D. T., Gatley, I., Roellig, T., Werner, M., Jura, M., and Zuckerman, B. 1985, *Ap. J.*, **294**, 242.
Stencel, R. E., Carpenter, K. G., and Hagen, W. 1986, *Ap. J.*, **308**, 859.
Thaddeus, P., Cummins, S. E., and Linke, R. A. 1984, *Ap. J. (Letters)*, **283**, L45.
Thaddeus, P., Gottlieb, C. A., Hjalmarson, A., Johansson, L. E. B., Irvine, W. M., Friberg, P., and Linke, R. A. 1985, *Ap. J. (Letters)*, **294**, L49.
Tielens, A. G. G. M. 1983, *Ap. J.*, **271**, 702.
Tsuji, T., Unno, W., Kaifu, N., Izumiura, H., Ukita, N., Cho, S., and Koyama, K. 1988, *Ap. J. (Letters)*, **327**, L17.
Turner, B. E. 1987, *Astr. Ap.*, **183**, L23.
Ukita, N., and Morris, M. 1983, *Astr. Ap.*, **121**, 15.

Weidemann, V., and Koester, D. 1983, *Astr. Ap.*, **121**, 77.
Werner, M. W., Beckwith, S., Gatley, I., Sellgren, K., Berriman, G., and Whiting, D. L. 1980, *Ap. J.*, **239**, 540.
Weymann, R. 1962, *Ap. J.*, **136**, 844.
Willems, F. J., and de Jong, T. 1986, *Ap. J. (Letters)*, **309**, L39.
Wing, R. F., and Yorka, S. B. 1977, *MNRAS*, **178**, 383.
Yusef-Zadeh, F., Morris, M., and White, R. L. 1984, *Ap. J.*, **278**, 186.
Zinner, E., and Epstein, S. 1987, *Earth and Planetary Sci. Letters*, **84**, 359.
Zinner, E. K., Fahey, A. J., Goswami, J. N., Ireland, T. R., and McKeegan, K. D. 1986, *Ap. J. (Letters)*, **311**, L103.
Ziurys, L. M., and Turner, B. E. 1986, *Ap. J. (Letters)*, **300**, L19.
Zuckerman, B. 1980, *Ann. Rev. Astr. Ap.*, **18**, 263.
Zuckerman, B., and Aller, L. H. 1986, *Ap. J.*, **301**, 772.
Zuckerman, B., and Dyck, H. M. 1986a, *Ap. J.*, **304**, 394.
Zuckerman, B., and Dyck, H. M. 1986b, *Ap. J.*, **311**, 345.
Zuckerman, B., and Gatley, I. 1988, *Ap. J.*, **324**, 501.
Zuckerman, B., Dyck, H. M., and Claussen, M. 1986, *Ap. J.*, **304**, 401.
Zuckerman, B., Terzian, Y., and Silverglate, P. 1980, *Ap. J.*, **241**, 1014.

25

Atoms and molecules in Supernova 1987A

RICHARD McCRAY

Joint Institute for Laboratory Astrophysics, University of Colorado and National Bureau of Standards, Boulder, Colorado, USA

25.1. Introduction

It's a great pleasue for me to contribute to this volume in honor of Alex Dalgarno's 60th birthday. I first met Alex when I came to Harvard as an assistant professor in 1968. From these turbulent times up to now, Alex has been a very good friend and mentor to me. From Alex I gained an interest in atomic and molecular processes that has influenced my research ever since. My writing skills improved considerably by virtue of working with him. Most important, I am grateful to Alex for setting an example of what a good professor should be, not only as a scholar and teacher, but also as a generous and loyal friend to his students and colleagues.

With Supernova 1987A (February 23, 1987), nature has provided some birthday fireworks that will be a festive reminder of Alex's many important contributions to astrophysics. The brightest supernova since SN1604 (Kepler's supernova), SN1987A is the first one that has been observed in every electromagnetic wavelength band and it is the first one that will remain observable for several years as the debris clears away to allow a detailed view of its interior. Thus, SN1987A offers an unprecedented opportunity to infer details of supernova explosion dynamics and nucleosynthesis. This task presents fascinating and challenging problems in atomic and molecular astrophysics because, as I will describe, SN1987A is remarkably cool ($\lesssim$7000 K) throughout its interior and there is good evidence that CO and SiO molecules have already formed there. SN1987A is now a hypersonically expanding dense molecular cloud illuminated from within by gamma rays and possibly also by a compact X-ray source.

Arnett *et al.* (1989), Dopita (1988), and McCray and Li (1988) have written comprehensive reviews of current observations and theory of SN1987A. I will therefore concentrate here on those aspects concerning atomic and molecular processes in the expanding envelope during the 'nebula phase,' in which the optical/infrared spectrum is dominated by emission lines. The nebula phase began in late 1987 and continues today (August 1988), as we begin to sort out the details of the physical processes and mechanisms of spectrum formation. It is clear that the interior of SN1987A is an exciting environment in which to study molecular astrophysics, but we have only begun to explore the molecular processes. Therefore, in this chapter I will be able to say more about the physical environment in SN1987A than I will about the molecular processes themselves. In Section 25.2 I will briefly summarize the essential observations and physical model that form the basis for understanding the nebular spectrum. Then, in Section 25.3 I will discuss the nebular spectrum and its interpretation. In Section 25.4 I will briefly discuss the issues raised by the observations of CO and SiO molecular emission bands. Finally, in Section 25.5 I will speculate on the future of SN1987A.

25.2. Basic observations and model

The first signal from SN1987A to reach Earth was a burst of neutrinos that was detected simultaneously in Japan and Michigan. Since the temperature ($\approx$4 MeV), duration (few seconds), and energy ($E_\nu \sim$ few $\times 10^{53}$ erg) of the neutrino burst had values in just the range expected for the formation of a neutron star, we are virtually certain that the supernova contains a neutron star at its center and that the explosion resulted from the rebound shock that follows the collapse of the core ($\sim$1.4 $M_\odot$) of a massive star. Models for the early light curve and spectrum of the supernova, as well as observations of the progenitor star, require that the expanding envelope of SN1987A has mass in the range $10 \lesssim M_{env} \lesssim 20\ M_\odot$ and kinetic energy $E_{kin} \approx (1\text{–}3) \times 10^{51}$ erg.

Figure 25.1 shows the light curve of SN1987A, which was unusual in that it took three months to reach maximum, in contrast to that of a typical Type II supernova, which begins to fade after a few days. We now understand that this difference results from the fact that the progenitor of SN1987A was a blue giant star (with radius $R_* \sim 3 \times 10^{12}$ cm) rather than a red giant ($R_* \sim 10^{14}$ cm) as is the case with typical SN IIs. Because the blast energy was deposited in a smaller star, it radiated less effectively at first and a greater fraction of the interior radiation field was converted adiabatically to kinetic energy in the nearly opaque expanding envelope. Indeed, after the first six weeks the light from the SN1987A was dominated by radiation produced by radioactive elements in the interior and not by the initial blast. Note that the net radiative energy output from the supernova, $\int L(t)dt \sim 10^{49}$ erg, is much less than the blast energy $E_0 \sim$ few $\times 10^{51}$ erg, which in turn is much less than the energy of the neutrino burst, $E_\nu \sim$ few $\times 10^{53}$ erg.

As the supernova expanded, the interior radiation temperature cooled rapidly from an initial value $T_0 \sim 10^6$ K as a result of adiabatic losses in the interior and radiation from the photosphere. If we approximate the supernova spectrum by a blackbody with color temperature T_c, we can define an 'effective photospheric radius,' from the Stefan–Boltzmann formula $L = 4\pi\sigma R_p^2 T_c^4$. After the first few days, the color temperature stabilized at $T_c \approx 5500$ K, and it has remained fairly constant ever since. We find from the observations of $L(t)$ and $T_p(t)$ that R_p increased for about three months, reaching a maximum value of $R_p \sim 10^{15}$ cm, and decreased rapidly thereafter.

This behavior can be understood as follows. For the first three months, the photosphere radius increased because the envelope was opaque and expanding. The timescale for radiation to diffuse from the core was greater than the age of the supernova. However, the strong radiative cooling caused the matter temperature of the photosphere to drop to ~5500 K, at which temperature most atoms recombined. The (mostly neutral) matter outside the photosphere became transparent to optical radiation, and the photosphere moved inward relative to the expanding matter of the supernova envelope. The relative velocity of this inward motion of the photosphere accelerated and by $t = 3$ months exceeded the expansion velocity. Then the photosphere shrank rapidly toward the center and the optical luminosity dropped rapidly. By $t = 4$ months the supernova envelope had become transparent in the optical and infrared bands almost to its center.

Figure 25.1. Optical light curve (equivalent V magnitude) of SN1987A from the FES camera on the *International Ultraviolet Explorer* satellite. Deficit from exponential decay (dashed line) is due to escape of X-rays and gamma rays (courtesy of Dr Nino Panagia).

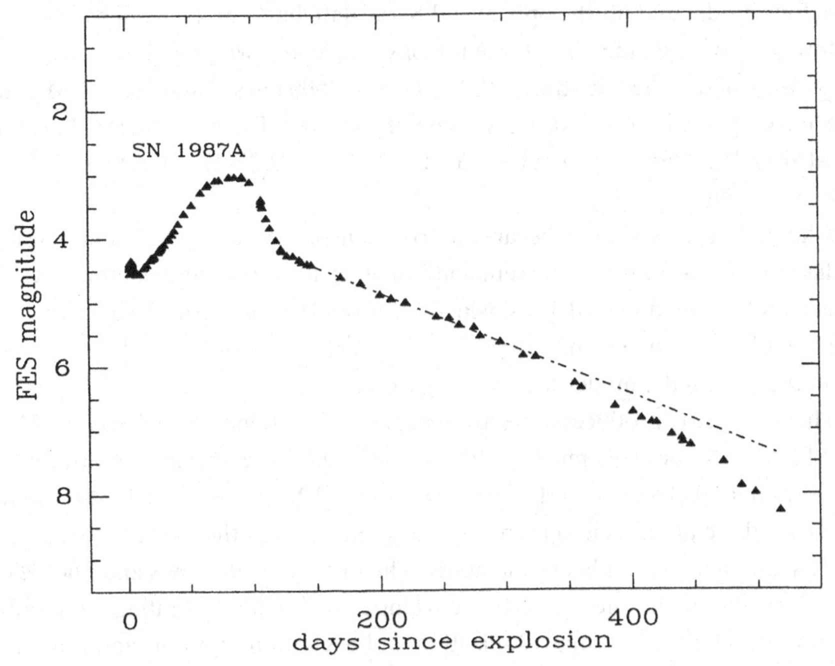

For $t \gtrsim 4$ months, the diffusion time for optical radiation to escape from the supernova is much less than the supernova age, so the optical light curve faithfully tracks the instantaneous energy injection rate. We can see from Figure 25.1 that for 4 months $\lesssim t \lesssim 9$ months, this light curve is almost a perfect exponential, with a decay constant very close to the value $\lambda \approx (111.3\ \mathrm{d})^{-1}$ appropriate for the radioactive decay of ^{56}Co (which deposits ≈3 MeV per decay). Therefore, we can infer that the supernova explosion produced ≈0.07 $M_\odot$ of ^{56}Ni, which decayed quickly ($\lambda \approx (8.8\ \mathrm{d})^{-1}$) into ^{56}Co, which in turn decays into ^{56}Fe. For $t \gtrsim 9$ months, the light curve drops below the exponential law; this occurs because an increasing fraction of the radioactive energy leaks out as hard X-rays and gamma rays without being converted into optical radiation, as will be discussed below. By $t \approx$ 16 months this fraction had increased to approximately 25%.

Although the supernova blast energy $E_0 \sim (1–3) \times 10^{51}$ erg was initially divided roughly equally between the internal radiation energy and kinetic energy of the expanding matter, most of the trapped radiation energy was converted to kinetic energy of expansion and by $t \gtrsim 1$ week the supernova envelope was coasting freely in hypersonic expansion. (During the first week ^{56}Ni radioactivity releases $\sim 10^{49}$ erg, which may be dynamically significant and lead to some mixing in the core.)

Thus, the envelope dynamics at later times can be described by homologous free expansion. Co-moving coordinates have radius $r(t) = vt$ with fixed velocity v. The density obeys the law $\rho(r, t) = (t/t_0)^{-3}\rho(v, t_0)$, where t_0 is some fiducial time (say, 1 yr). Figure 25.2 shows the density profile and elemental composition of an envelope model that fits the optical and X-ray data fairly well, as will be discussed below. It may be divided into three regions: the 'core,' with $v < 1400\ \mathrm{km\ s^{-1}}$ and mass $M_c = 4\ M_\odot$; the 'mantle,' with $1400 < v < 3500\ \mathrm{km\ s^{-1}}$ and $M_m = 7\ M_\odot$; and the 'envelope,' with $v > 3500\ \mathrm{km\ s^{-1}}$ and $M_e = 3\ M_\odot$. The net masses ($M_\odot$) of the most abundant elements are: H – 4.5, He – 7.6, C – 0.22, O – 1.3, Si – 0.26, S – 0.17, and ^{56}Ni – 0.075.

Although the envelope became fairly transparent to optical and infrared radiation within 4 months, it remained opaque to X-rays and gamma rays for much longer. The optical depths, which are proportional to column density, obey the homologous expansion law $\tau(\varepsilon, t) = (t/t_0)^{-2}\tau(\varepsilon, t_0)$, where $\tau(1\ \mathrm{MeV}, 1\ \mathrm{yr}) \approx 7$ from the core in the model shown in Figure 2.

The ^{56}Co decay produces a spectrum of gamma rays, mainly in lines at 847 keV and 1238 keV. These gamma rays diffuse out through the envelope, scattering off free and bound electrons and degrading into hard X-rays as a result of Compton recoil until the photon energies become low enough that they are photoabsorbed by K-shell ionization of heavy elements. The fast recoil electrons and photoelectrons deposit their energy in the envelope as Coulomb heating of thermal electrons and electron impact ionization and excitation of atoms and ions. This process has been described in detail by Shull and van Steenberg (1985) and Fransson and Chevalier (1989), following pioneering work by Dalgarno and Lejeune (1971).

For the first several months, the supernova envelope is so opaque that virtually all of the gamma ray energy is deposited in the envelope. However, as the envelope thins out, a spectrum of hard X-rays from down-Comptonized gamma rays begins to emerge, and, somewhat later, unscattered gamma ray line photons appear. Several authors (McCray, Shull, and Sutherland 1987, Grebenev and Sunyaev 1987a,b, Xu *et al.* 1988, Ebisuzaki and Shibazaki 1988a) predicted that these X-rays and gamma rays would be detected in late 1987, on the basis of supernova models in circulation at the time. In July 1987, the Japanese team operating the *Ginga* spacecraft (Dotani *et al.* 1987) and the Soviet team operating the *Mir* spacecraft (Sunyaev *et al.* 1987) detected hard X-rays from SN1987A with a spectrum and intensity nearly as predicted.

However, the X-rays were detected some four months earlier than predicted, and this result forced a revision of the supernova model. The predictions were based on models in which the element composition was highly stratified, with the newly synthesized ^{56}Ni all concentrated at the center. In such a model, the envelope is too opaque to allow detectable fluxes of X-rays or gamma rays to

Figure 25.2. Radial dependence of density and elemental composition for Model 10HM constructed by Pinto and Woosley (1988) to fit the optical and X-ray light curve of SN1987A. The curve marked 'Ni' denotes the density of ^{56}Ni and its decay products.

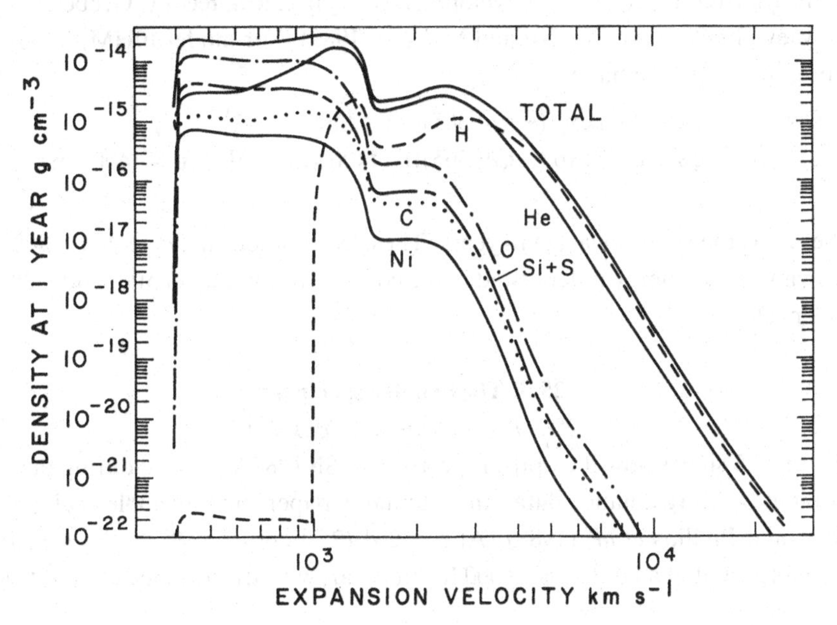

emerge before $t \approx 8$ months. For detectable fluxes to emerge at $t \approx 5$ months, some of the ^{56}Ni must be located nearer the surface, and therefore, several authors (Grebenev and Sunyaev 1987a,b, Shull and Xu, 1987, Itoh *et al.* 1987, Ebisuzaki and Shibazaki 1988b, Kumagai *et al.* 1988, Pinto and Woosley 1988) suggested 'mixed models,' in which the ^{56}Ni and the other heavy elements are homogenized with the hydrogen and helium throughout some substantial fraction of the envelope, say, the inner 5 $M_\odot$. Pinto and Woosley's (1988) model 10HM, shown in Figure 25.2, is one such mixed model that gives a satisfactory fit to both the X-ray and optical light curves, and we shall employ it here to discuss the atomic and molecular processes. However, one should keep in mind that there are other possible explanations for the early emergence of X-rays and gamma rays. For example, the supernova envelope might be filamentary, or have deep holes.

Early evidence for the emergence of X-rays and gamma rays was provided by the appearance of an absorption line at 1.08 μm due to the HeI 2^3S–2^3P in the infrared spectrum of SN1987A (Graham 1988, MacGregor 1988). The metastable 2^3S levels lies some 20 eV above the HeI ground state and cannot be populated thermally at the ~5000 K photospheric temperature; instead, the level is populated by impact excitation and recombination following ionization by nonthermal electrons produced by gamma ray degradation outside the photosphere. This absorption line appeared at $t = 40$ d.

To construct a theory for the emission line spectrum of SN1987A, we must first know the distribution of radioactive energy deposition in the envelope. This function, which is a by-product of the Monte Carlo simulations of gamma ray and X-ray transfer through the envelope, has been calculated by Grebenev and Sunyaev (1987a,b) and by Xu and McCray (1989). For model 10HM, it may be represented approximately by

$$\begin{aligned} \dot{\varepsilon} &\approx (10^8\,\mathrm{erg\,s^{-1}\,g^{-1}})\exp(-t_y/0.305), \quad v < 1400\,\mathrm{km\,s^{-1}} \\ &\approx (10^8\,\mathrm{erg\,s^{-1}\,g^{-1}})\exp(-t_y/0.305)[v/(1400\,\mathrm{km\,s^{-1}})]^{-2}, \quad v > 1400\,\mathrm{km\,s^{-1}}, \end{aligned} \tag{25.1}$$

where t_y is the supernova age in years. These calculations indicate that ~80% of the gamma ray energy is deposited in the core, ~18% in the mantle, and ~2% in the envelope.

25.3. The nebular spectrum

25.3.1. Spectrum development

Figure 25.3 shows how the optical spectrum of SN1987A evolved during the first six months. More detailed data can be found in papers by Catchpole *et al.* (1987, 1988) and Phillips *et al.* (1988). At $t = 2$ d (25 February 1987), the spectrum resembled that of a hot ($T_c \approx 15\,000$ K) blue star, with strong blueshifted P-Cygni lines in the Balmer lines and HeIλ5876, indicating a rapidly expanding ($v_p \sim$ 15 000 km s^{-1}) photosphere. At that time the spectrum had a bright ultraviolet continuum ($1500 < \lambda < 4000$ Å), but this continuum vanished by $t = 3$ d. By $t = 50$ d (14 April 1987), the continuum had become much cooler ($T_c \approx 5000$ K) and absorption lines from singly ionized metals and a strong P-Cygni line of metastable CaII had appeared. The Balmer absorption lines had become narrower and had shifted to the red, with $v_p \approx 5000$ km s^{-1}, indicating that the photosphere had moved deeper into the envelope.

From September 1987 until today (August 1988), the supernova spectrum has resembled an emission nebula more than a star, with bright emission lines of Hα and CaII atop a ragged quasi-continuum showing a few metallic absorption lines. The emission lines are broad (FWHM $\Delta V \approx 3500$ km s^{-1}) and their centroid

Figure 25.3. Evolution of the optical spectrum of SN1987A (courtesy of Dr Mark Phillips).

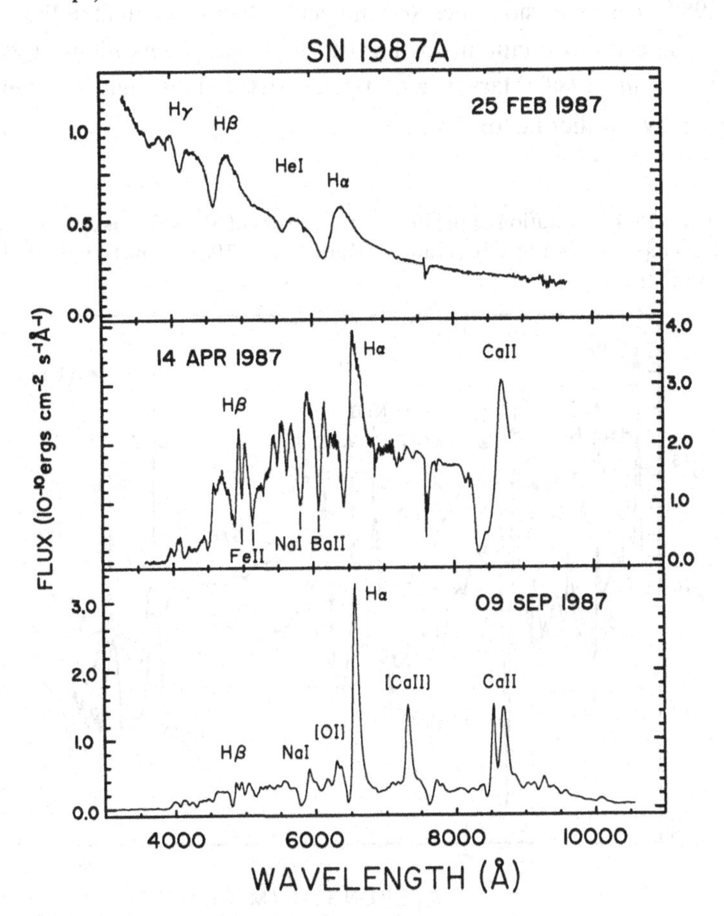

wavelengths correspond roughly to the barycenter velocity of the LMC, indicating that the evelope has become almost transparent and that most of the emission comes from the core and mantle. Curiously, there is almost no Hβ emission despite the bright Hα line. There is no evidence in the optical spectrum of any elements ionized more than once, indicating that the supernova is relatively cool ($T \lesssim 7000$ K) and neutral throughout. In November 1987, ~65% of the bolometric luminosity was in the optical quasi-continuum, ~25% was in the Hα and CaII emission lines, and ~10% was in the infrared band.

SN1987A is most transparent at infrared wavelengths, and observations with the Kuiper Airborne Observatory have provided the most revealing view to date of its interior. Figure 25.4 shows the time development of the spectrum in the 1.5–12 μm range. In mid-April 1987, it was dominated by a continuum, with $F_\lambda \propto \lambda^{-2}$. By mid-November 1987, the continuum had faded by a factor ~4 and the spectrum was dominated by hydrogen recombination lines, emission lines of NiII, ArII, and CoII, and CO and SiO ($\Delta v = 1$) molecular emission bands. By mid-April 1988, these emission lines and molecular bands dominated the 1.5–12 μm spectrum and the continuum was hard to find, but observations at 20–100 μm (Moseley *et al.* 1988, Harvey and Lester 1988) show that its intensity had decreased by another factor ~5.

Figure 25.4. Evolution of the infrared spectrum of SN1987A as observed with the Kuiper Airborne Observatory (Rank *et al.* (1988), courtesy of Dr Fred Witteborn).

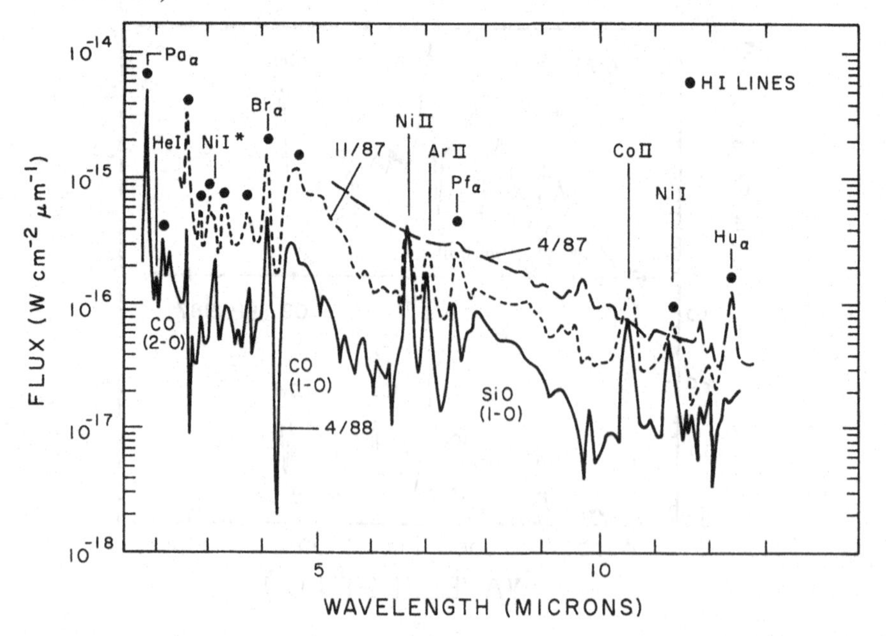

The infrared continuum has a flat frequency spectrum, $F_\nu \propto \nu^0$, suggestive of optically thin bremsstrahlung. (There is no evidence for thermal infrared emission by dust, which should have a very different spectrum (Dwek 1988).) Assuming bremsstrahlung to be the emission mechanism, we may infer an emission integral, $EM(+) = \int n_e n^+ dV \approx (3 \times 10^{63}\ \mathrm{cm}^{-3}) F_\nu(\mathrm{Jy})$, for a distance $D = 50$ kpc and an interior temperature $T \approx 6000$ K (Moseley *et al.* 1988). At $t \approx 9$ months (mid-November 1987), the observed flux $F_\nu \approx 10$ Jy implies that $EM(+) \approx 3 \times 10^{64}$ cm^{-3}. From the observed infrared hydrogen recombination line intensities we may infer (see below) an analogous emission integral, $EM(\mathrm{H}^+) \approx \int n_e n(\mathrm{H}^+) dV$ $\approx 6 \times 10^{63}\ \mathrm{cm}^{-3}$. Comparing $EM(+)$ with $EM(\mathrm{H}^+)$ tells us that, in November 1987, a substantial fraction, ~0.2, of the positive ions in the interior of SN1987A were H^+ ions.

We may compare these emission integrals to an analogous quantity $EM(0) = \int n_b^2 dV$, where n_b is the baryon number density in a hydrodynamic model. For model 10HM, $n_b \approx 1.2 \times 10^{10}\, t_y^{-3}\ \mathrm{cm}^{-3}$ in the core and $EM(0) = 8 \times 10^{67}\, t_y^{-3}\ \mathrm{cm}^{-3}$, where t_y is the age in years. Thus, we find (in November 1987) the ratio $EM(+)/EM(0) \approx 1.6 \times 10^{-5}$. If we assume for simplicity that the ionized fraction is constant throughout the envelope, we find $n_+/n_b \approx 4 \times 10^{-3}$. However, if the interior structure of SN1987A is well approximated by model 10HM, we would expect that $EM(+)$ is dominated by C^+, Si^+, and S^+ in the core, while $EM(\mathrm{H}^+)$ is dominated by H^+ in the mantle. If so, we may infer that the ionized fraction of hydrogen in the mantle, $n(\mathrm{H}^+)/n(\mathrm{H}) \approx 0.1$.

According to the homologous expansion model, $EM(0) \propto t^{-3}$. The infrared continuum observations imply that $EM(+)$ decreased faster than this, $EM(+) \propto t^{-4}$, between November 1987 and April 1988. This result shows that the supernova interior is slowly recombining – $n_e \propto t^{-3.5}$.

We may also estimate the optical depth for Compton scattering by free electrons, $\tau_C \approx 0.16 t_y^{-2.5}$. For $t \lesssim 1$ year, optical and infrared photons emitted from the core are substantially redshifted as they are scattered by electrons in the expanding envelope (Fransson and Chevalier 1989, Woosley, Pinto, and Weaver 1988). This effect likely accounts for the redshifts (~200–1200 km s^{-1}) of infrared spectral lines observed by Oliva, Moorwood, and Danziger (1987), Erickson *et al.* (1988), and MacGregor (1988). Detailed analysis of high resolution line profiles will be a valuable technique for inferring the distribution of free electrons with expansion velocity.

The most exciting feature in the infrared spectrum is the [CoII] 10.5 μm emission line, which probably comes mostly from the radioactive $^{56}\mathrm{Co}$. If so, we might expect the intensity of this line to decay as $\exp(-t/111.3\ \mathrm{d})$. However, it does not; indeed, the line became brighter between April 1987 and November 1987 (Aitken *et al.* 1988). The reason is that the line was optically thick, so that its strength was not simply proportional to the mass of ^{56}Co. In order to interpret the optical and infrared emission line spectrum of SN1987A, it is essential to understand the physics of spectral line transfer in the envelope; this is discussed below.

23.3.2. Spectral line trapping

At present, virtually all of the important ultraviolet, optical, and infrared resonance line transitions in SN1987A are optically thick. As a result, a line photon that is emitted somewhere in the expanding envelope scatters many times locally before it diffuses far enough in frequency to move out of resonance and propagate freely. While the line photon is trapped, it may be lost as a result of continuum absorption or collisional de-excitation of the excited level, or it may be converted into other lines as a result of other decay branches of the excited level. Thus, line trapping can in effect change a permitted transition into a forbidden transition.

Fortunately, it is not difficult to describe these processes. The Sobolev formalism (Castor 1970, Kirshner and Kwan 1975) is an excellent approximation for line transfer in a diverging hypersonic flow such as the envelope of SN1987A. The scattering optical depth is given by (cf. Rybicki and Lightman (1979)) $\tau = n\sigma(\nu)\Delta r$, where $\sigma(\nu) = (\pi e^2/m_e c) f_{lu}\phi(\delta\nu)$, and $\phi(\delta\nu)$ is the line profile function, f_{lu} is the oscillator strength between the lower and upper level, and $\delta\nu = (\nu - \nu_0)$. The Sobolev approximation can be understood heuristically as follows: write the line profile function $\phi(\delta\nu) = 1/\Delta\nu$, where $\Delta\nu$ is the line width. In the envelope of SN 1987A, the local radius is given by the homologous expansion law $r = vt$, so $\Delta r = \Delta vt$. The Doppler shift associated with $\Delta\nu$ is $\Delta\nu = \nu_0\Delta v/c = \Delta r/(\lambda_0 t)$. For a thermally broadened line, $\Delta\nu \approx \nu_0 a_s/c$, where $a_s = (2kT/m)^{1/2}$ is the sound speed, the resonance scattering at a given frequency is confined to a shell of fractional thickness $\Delta r/r \sim a_s/v(r)$. Thus, we may write the optical depth averaged over the line profile

$$\bar{\tau} = \frac{\pi e^2}{m_e c} f_{lu}\lambda_0 t n_l \left[1 - \frac{g_l n_u}{g_u n_l}\right] = \frac{\lambda_0^3 t g_u A_{ul} n_l}{8\pi g_l}\left[1 - \frac{g_l n_u}{g_u n_l}\right] \tag{25.2}$$

where the square brackets account for induced emission from the upper state.

According to the Sobolev theory, the probability that a line photon will escape from the scattering layer is given by

$$P_{esc}(\bar{\tau}^{-1}[1 - \exp(-\bar{\tau})] \tag{25.3}$$

and the local line emission coefficient is given by

$$j_\nu = \frac{h\nu_0}{4\pi} n_u A_{ul} P_{esc}\phi(\Delta\nu) \tag{25.4}$$

Thus, the effect of line trapping is to reduce the effective decay rate $A_{\rm ul}$ by the factor $P_{\rm esc}$.

Optical depths are large for many emission lines in SN1987A. For example, in the mantle of model 10HM we have $n({\rm H}) \sim 10^8 t_{\rm y}^{-3}$ cm^{-3} and so $\bar{\tau}({\rm Ly}\alpha) \gtrsim 4 \times 10^8 t_{\rm y}^{-2}$, and n(CaII) $\sim 10^5\, t_{\rm y}^{-3}$ cm^{-3}, so for the CaII$\lambda\lambda$3933, 3968 (H,K) doublet, $\bar{\tau}({\rm H,K}) \sim 10^7 t_{\rm y}^{-2}$. Similarly, we may estimate from Equation (25.2) that many infrared transitions are optically thick; for example, the [CoII] 10.5 μm line (with $A_{\rm ul} \approx 0.02$ s^{-1}) has $\tau(\lambda 10.5\ \mu{\rm m}) \sim 40 t_{\rm y}^{-2} \exp(-t_{\rm y}/0.305)$ in the core, where $n(^{56}{\rm Co}) \sim (5 \times 10^6\ {\rm cm}^{-3}) t_{\rm y}^{-3} \exp(-t_{\rm y}/0.305)$. This result explains why the [CoII] line actually increased in strength between April 1987 and November 1987.

In the limit that $\bar{\tau} \gg 1$, the emission coefficient becomes

$$j_\nu = \frac{\nu_0}{ct\Delta\nu}\frac{2h\nu_0^3}{c^2}\left(\frac{g_{\rm u}n_{\rm l}}{g_{\rm l}n_{\rm u}} - 1\right)^{-1}. \tag{25.5}$$

If the atom is in local thermodynamic equilibrium (LTE), with $(g_{\rm l}n_{\rm u})/(g_{\rm u}n_{\rm l}) = \exp(-h\nu/kT)$, Equation (25.5) gives the reasonable result (cf. Colgan and Hollenbach (1988)) that the specific intensity emerging from an expanding shell of thickness $\Delta r = \Delta v t = ct\Delta\nu/\nu_0$ is $I_\nu = j_\nu \Delta r = B_\nu(T)$, i.e., that the specific intensity of an emission line with $\bar{\tau} \gg 1$ is the Planck function, $B_\nu(T)$.

Likewise, line trapping causes the population of an excited state to reach LTE at a lower density. For a two-level atom excited by electron collisions in a gaseous nebula, the excited state population is given by

$$\frac{n_{\rm u}}{n_{\rm l}} = \frac{g_{\rm u}}{g_{\rm l}} \exp(-h\nu/kT)(1 + n_{\rm cr}/n_{\rm e})^{-1} \tag{25.6}$$

where, without line trapping, the critical electron density, $n_{\rm cr} = A_{\rm ul}/C_{\rm ul}$, and $C_{\rm ul}$ is the rate coefficient for de-excitation of the upper state by electron impact. For allowed optical transitions, $A_{\rm ul} \sim 10^9\,{\rm s}^{-1}[\lambda/(1000\ {\rm \AA})]^{-3}$ and $C_{\rm ul} \sim 10^{-7}$ cm^3 s^{-1} at $T \approx 5000$ K (cf. Osterbrock (1974)), so LTE requires $n_{\rm e} \gtrsim n_{\rm cr} \sim 10^{16}\,{\rm cm}^{-3}[\lambda/(1000\ {\rm \AA})]^{-3}$. However, if line trapping is important ($\bar{\tau} \gg 1$), we may estimate from Equations (25.2) and (25.6) that $n_{\rm cr} = A_{\rm ul}/(C_{\rm lu}\bar{\tau}) \sim 10^{16}\,{\rm cm}^{-3}[\lambda/(1000\ {\rm \AA})]^{-3} n_{\rm l}^{-1} t_{\rm y}^2$, independent of oscillator strength, $f_{\rm lu}$.

Thus, for example, in the mantle the critical electron density for hydrogen Lyman transitions is $n_{\rm cr} \approx 10^8$ cm$^{-3} t_{\rm y}^2$. For $t \lesssim 1$ yr this is comparable to the electron density $n_{\rm e} \sim 10^8$ cm^{-3} $t_{\rm y}^{-3.5}$ that we infer from the infrared continuum flux. Since the two-photon decay rate of H(2^{2}S) is also less than the electron impact de-excitation rate, we may conclude that the population of H*(n = 2) is approximately in LTE for $t \lesssim 1$ yr: $n({\rm H}^*)/n({\rm H}) \approx 4 \exp(-10.2\ {\rm eV}/kT) \approx 2 \times 10^{-7}$ at $T = 7000$ K.

The density n(H*) in SN1987A is great enough to make the higher Balmer lines and the Balmer continuum (BaC) opaque. For example, we estimate that (for t_y = 1 and T = 7000 K) $\tau(H\alpha)$, $\tau(H\beta)$, and $\tau(H\gamma) \approx 1200$, 150, and 50, respectively, and $\tau(BaC) \approx 8[\lambda/(3646\ Å)]^3$ in the mantle of SN1987A. (These values are very sensitive to the assumed temperature; they would each decrease by a factor 17 if T were 6000 K.) However, the Paschen and higher series lines are optically thin. These results have important implications for the interpretation of the hydrogen line spectrum of SN1987A, which is an example of a 'case C' recombination spectrum. (cf. Osterbrock (1974) for a description of cases A and B.) When the recombination cascade results in the emission of a higher Balmer line, say Hβ, the line photon is absorbed and re-emitted several times until the upper (n = 1) level branches to Hα + Pα, and so forth. This mechanism explains why we see no Hβ emission line in the spectrum of SN1987A (Figure 25.3) despite the strong Hα line.

Furthermore, a recombination directly from the continuum to the n = 2 level is ineffective, because the resulting Balmer continuum photon is photoabsorbed locally and ionizes another H*(n = 2) atom. Thus, the effective hydrogen recombination coefficient, $\alpha^{(3)}(T) = (2.5 \times 10^{-13}\ \mathrm{cm^3\ s^{-1}})\ [T/(7000\ \mathrm{K})]^{-0.9}$, is a factor ~0.7 less than the effective recombination coefficient for case B, $\alpha^{(2)}(T)$, which is the usual situation for emission nebulae (cf. Osterbrock (1974)). Since in case C every higher Balmer line is converted into infrared lines plus Hα, the emissivity of each of these lines is substantially greater than in case B. For example, the emissivities of Hα, Pα, and Brα are increased by factors 1.7, 2.5, and 1.9 respectively, times the case B results. We used the case C results to infer the emission integral $EM(H^+)$ in Section 25.3.1 from the observed line strengths.

Similarly, resonant trapping of the CaII (H,K) lines accounts for the strong emission in the infrared triplet CaII$\lambda\lambda$8600 (Figure 25.4) and the forbidden [CaII]$\lambda\lambda$7300 doublet. These lines are discussed by Kirshner and Kwan (1975). The $\lambda\lambda$8600 lines come from the same upper 4p(^{2}P^0) level as the (H,K) lines, but their branching ratio is small, $A(8600)/A(3933) = 0.073$. However, since the (H,K) lines are trapped, when the ^{2}P^0 level is populated (by electron impact excitation), it has no alternative but to emit $\lambda\lambda$8600.

The Balmer continuum opacity probably accounts for the paucity of ultraviolet emission from the interior of SN1987A. A significant fraction, $\gtrsim$3% of the ^{56}Co decay luminosity must be converted into ultraviolet emission lines in the core as a result of impact excitation by nonthermal electrons, and this fraction should increase to ~10–30% as the interior recombines further (cf. Shull and Van Steenberg (1985), Fransson and Chevalier (1989). However, such ultraviolet emission is not observed; indeed, the observed ultraviolet luminosity of SN1987A is $L_{UV} \lesssim 0.01 L_{bol}$ (Kirshner 1988). The ultraviolet emission from the core is probably stopped by Balmer continuum absorption in the mantle and by scattering in several hundred resonance lines of FeII and other metals in the spectral range 1600–3200 Å (cf. Nussbaumer and Storey (1980), Phillips (1979)). This trapped ultraviolet radiation is converted into optical emission by resonance fluorescence in much the same manner as occurs in QSOs and active galactic nuclei (Phillips 1978, Wills, Netzer, and Wills 1985).

25.3.3. Temperature and ionization

The Balmer continuum opacity is very sensitive to temperature, and so we expect that the supernova will rapidly become transparent in the ultraviolet when its temperature drops below $T \sim 6000$ K. Then, we might expect SN 1987A to have an 'ultraviolet renaissance,' in which a significant fraction of L_{bol} will appear as ultraviolet emission lines (McCray *et al.* 1987). Unfortunately, we are not yet able to estimate accurately the time when this ultraviolet renaissance will occur, because its onset is so sensitive to the temperature. To calculate the temperature accurately, we need to construct a detailed microscopic theory for the radiative cooling of the interior, a task that is not yet completed. The outstanding problem is to identify the radiative transitions that are responsible for the optical quasi-continuum.

In the meantime, we can illustrate the possible evolution of the interior temperature with the aid of a crude model as follows. Assume that the local radiative cooling (erg s^{-1} cm^{-3}) is given by an expression of the form $C \approx n_e n_b \Lambda(T)$, where $\Lambda(T) \propto T^{\alpha}$, where $1 \lesssim \alpha \lesssim 3$ for $T \sim 7000$ K. We know the time-dependence of the local heating, $H \propto n_b \exp(-t_y/0.305)$. Setting $H = C$, and assuming $T \approx 7000$ K for $t_y = 1.5$ (roughly the present time), we obtain $T \approx (7000\ \mathrm{K})[t_y/1.5]^{(3/\alpha)} \exp[-(t_y - 1.5)/(0.305\alpha)]$, if n_e/n_b is approximately constant. The result is that the temperature will drop below 6000 K when $t_y = 1.63$ (September 1988) if $\alpha = 1$ and when $t_y = 1.83$ (November 1988) if $\alpha = 3$. We expect the Balmer continuum to become transparent when $T \lesssim 6000$ K, and so we can predict that the ultraviolet renaissance will occur sometime in late 1988, depending on the value of the exponent α. We can make such a prediction despite the great uncertainty in the radiative cooling because the radioactive heating has such a strong time-dependence. Note that the temperature drops faster as α becomes smaller. This is a manifestation of the thermal instability that occurs (see Section 25.5) when the radiative cooling function $\Lambda(T)$ becomes flat.

The inferred ionization of hydrogen, $n(H^+)/n(H) \approx 0.1$ in November 1987, presents an interesting puzzle. A mantle temperature $T \approx 4800$ K would be sufficient to account for this ionization level in LTE, according to the Saha equation; but LTE would require the radiation field in the mantle to be a 4800 K blackbody, and this is clearly not the case. A simple argument (cf. Moseley *et al.* (1988)) shows that the observed hydrogen ionization cannot be caused by nonthermal electrons resulting from gamma ray degradation. If so, the total hydrogen ionization rate, I_H(atoms s^{-1}) = $\eta_m L_{SN}/\zeta$, where L_{SN} is the total supernova luminosity, $\eta_m \approx 0.18$ is the fraction of L_{SN} deposited in the mantle, and ζ is the energy cost per ionization by a nonthermal electron. If we equate $I_H = R_H$, where the effective case C hydrogen recombination rate is given by $R_H = \alpha^{(3)}(T) \int n_e n(H^+) dV$, we obtain $\zeta \approx 3$ eV (for $T = 7000$ K). However, the calculations of Shull and Van Steenberg (1985) show that $\zeta \approx 80$ eV in a medium with $n(H^+)/n(H) \approx 0.1$. Therefore, we conclude that ionization by nonthermal electrons fails by a factor ~30 to account for the hydrogen ionization in the mantle.

The fact that $\zeta \approx 3$ eV is a strong hint that the hydrogen is ionized from the $n = 2(H^*)$ state rather than from the ground state. The most likely mechanism is photoionization of H^* by near-ultraviolet lines ($3.4 < h\nu \lesssim 5$ eV) produced by thermal electron impact excitation of metals, e.g. MgIIλ2800 (cf. Chugai (1987)) and FeII. This indirect ionization mechanism is more effective than direct electron impact ionization of H^* because the metal ions are more abundant than H^* by factors $\gtrsim 10^3$ and the cross sections for electron impact excitation of metals are greater than the cross section for electron impact ionization of H^*. It is possible that the near-ultraviolet lines that ionize H^* in the mantle are produced in the metal-enriched core as well as the mantle; if so, $\eta \approx 1$ and $\zeta \approx 15$ eV, so there is sufficient thermal energy available to ionize the H^*.

The observed ionization of trace elements results from charge transfer reactions (Dalgarno and Butler 1978). Multiply-ionized species are not observed because they are removed by rapid exothermic charge transfer reactions with OI in the core and HI in the mantle and envelope, while most trace elements with ionization potential <13.6 eV are once-ionized as a result of charge transfer reactions with H^+ and O^+. (See Chapter 26 on the effects of charge transfer on the ionization structures of nebulae.)

The temperature in the interior of SN1987A is certainly not uniform. The observed ionization of hydrogen and the high Balmer continuum opacity seem to require a rather high temperature, $T \approx 7000$ K, in the mantle and envelope, but other diagnostics indicate a lower temperature in the core. For example, the brightness and profile of a 26.0 μm line that was observed at $t = 254$ d can be interpreted as an optically thick [FeII] emission line from the core with $T = 3500 \pm 1500$ K (Erickson *et al.* 1988). Moreover, as will be discussed below, the observations of CO molecular bands suggest temperatures $T \lesssim 3000$ K in the core. This apparent temperature inversion is not surprising, and was in fact predicted in models by Fransson and Chevalier (1987, 1989). It occurs because the metal-enriched core has a much greater radiative efficiency than the envelope.

For $t \gtrsim 450$ d, neutral atoms began to appear in the core, as evidenced by the appearance of a strong [OI]$\lambda\lambda$6300,6364 emission feature in the optical spectrum

(Whitelock *et al.* 1988) and lines of [CoI], [NiI], and CII in the infrared spectrum (Aitken 1988).

25.4. Molecules

SN1987A is an exciting laboratory for molecular astrophysics. Gas densities are sufficiently high ($\sim 10^7$–10^{11} cm^{-3}) that binary gas phase reactions are fast, but not so high that three-body reactions are important. The interior is partially ionized as a result of the gamma ray illumination, it is cooling rapidly, and as yet there is no evidence for dust. Consequently, the chemistry is dominated by gas phase ion–molecule reactions that have been studied extensively by Dalgarno and his co-workers (cf. Dalgarno (1987)). However, the chemical abundances in the interior of SN1987A are very different from those in the interstellar medium and so the reaction networks may be substantially different. The study of the chemistry of SN1987A is only just beginning; the discussion below is derived in part from work by Lepp, Dalgarno, and McCray (1989).

The dominant spectral features in the infrared spectrum of SN1987A (Figure 25.4) are the $\Delta v = 1$ emission bands of COλ4.6 μm and SiOλ8.1 μm (Rank *et al.* 1989). Ground-based observations (Oliva, Moorwood, and Danziger 1987, Spyromilio *et al.* 1988) also show emission from the COλ2.3 μm ($\Delta v = 2$) overtone band. These emission bands surely come from the interior of SN1987A. The observed broadening of the 2.3 μm band head, $\Delta v \gtrsim 1000$ km s^{-1}, is typical of the supernova expansion but much greater than interstellar velocities, and the band strength and profile are evolving rapidly.

If one assumes that the vibration–rotation populations of the CO molecules are in LTE and that the 2.3 μm band is optically thin, one can estimate the mass and temperature of the CO from its intensity and profile. The results are $M(\mathrm{CO}) \approx 10^{-4}$ M$_\odot$ and that the temperature decreased from $T \approx 3000$ K at $t = 192$ d to $T \approx 1800$ K at $t = 255$ d (Spyromilio *et al.* 1988). If so, the fraction of carbon that is in CO is $n(\mathrm{CO})/n(\mathrm{C}) \sim 10^{-3}$. Evidently, CO is not in chemical equilibrium, because that equation would require $T \approx 4000$ K, substantially greater than the temperatures inferred from the band profiles.

However, these inferred core temperatures are suspect because the assumptions of LTE and optically thin emission may be invalid. For the rotational levels of CO($v = 2$) to be in LTE, the electron density in the core should exceed a critical density $n_{\mathrm{cr}} = A_{20}/C_{20} \approx 10^{10}$ cm^{-3}, where $A_{20} \approx 1$ s^{-1} is the radiative decay rate of CO($v = 2$) (Scoville 1984) and $C_{20} \approx 10^{10}$ cm^{-3} s^{-1} (cf. Lane (1980)) is the rate for de-excitation by electron impact. We estimated in Section 25.3.1 that $n_e \sim 10^8 t_y^{-3.5}$ cm^{-3}, so $n_e < n_{\mathrm{cr}}$. Indeed, satisfactory LTE fits of the 2.3 μm bands cannot be found for $t \gtrsim 255$ d (Spyromilio *et al.* 1988). Moreover, the CO bands have significant optical depths. One can estimate from Equation (25.2) and the oscillator strengths tabulated by Kirby-Docken and Liu (1978) that the CO(1–0) band has optical depth $\bar{\tau}(4.6\,\mu m) \approx 100t_y^{-2}$ and that the CO(2–0) band has $\bar{\tau}$(2.3 μm) $\approx 0.5t_y^{-2}$.

It is not surprising that the CO bands are the most prominent features in the infrared spectrum of SN1987A. With a dissociation energy of 11.09 eV, CO is the most durable molecule that can be formed from abundant elements in SN1987A. The gas density in SN1987A is low enough that collisional dissociation of CO is suppressed by the radiative stabilization mechanism discussed by Roberge and Dalgarno (1982). The CO is probably formed by the radiative association reaction

$$C^+ + O \rightarrow CO^+ + h\nu \tag{25.7}$$

followed by the charge transfer reaction

$$CO^+ + O \rightarrow CO + O^+ \tag{25.8}$$

The CO^+ is also removed by photodissociation and by the reaction

$$CO^+ + e \rightarrow C + O \tag{25.9}$$

The CO is probably destroyed by ultraviolet photons in the range 912–1118 Å, which can cause discrete transitions to predissociating states (van Dishoeck and Black 1988). Photoionization of CO by harder ($h\nu > 14$ eV) ultraviolet radiation does not lead to a net destruction because the CO is recovered by Reaction (25.8). The source of the 912–1118 Å photons which can dissociate the CO is probably the two-photon decay of metastable 2^1S helium atoms which have been produced by the nonthermal electrons. The resulting CO abundance is far from chemical equilibrium and is a sensitive function of the density and gamma ray intensity.

There is evidence for CO^+ in the near infrared spectrum of SN1987A. Spyromilio *et al.* find that (in November 1987) the 2.3 μm band and an additional emission feature at 2.26 μm can be fitted much better with a mixture of CO and CO^+ in the ratio $(CO^+/CO) \approx 0.16$.

25.5. The future

Although SN1987A is fading fast, it is still a bright object and we can expect that it will continue to yield a rich harvest of data for a long time. Figure 25.5 illustrates a likely scenario for the future evolution of its ultraviolet/optical/infrared light. At present, the light curve is dropping below the 111.3 d exponential because an increasing fraction of the ^{56}Co luminosity is escaping as hard X-rays and gamma rays. However, as Pinto, Woosley, and Ensman (1988) have pointed out, the supernova explosion must have produced other radioisotopes besides ^{56}Ni, so one should expect that the energy deposition will eventually be dominated by isotopes with longer mean lives than ^{56}Co, in particular ^{57}Co, which has a mean lifetime λ^{-1} = 392 d, and ^{44}Ti, (λ^{-1} = 68 yr). Indeed, if SN1987A produces ^{57}Ni with an abundance ratio ^{57}Ni/^{56}Ni = 0.024, the cosmic abundance ratio of ^{57}Fe/^{56}Fe, the resulting ^{57}Co will dominate the radioactive energy deposition for $t \gtrsim 3$ yr. If a significant fraction of this power is converted to optical and UV emission lines,

SN1987A should have $V \approx 19^m$ at $t \approx 6$ yr; this is still bright enough for ground-based optical spectroscopy and ultraviolet spectroscopy with the Hubble Space Telescope.

Actually, SN1987A is likely to remain much brighter than this because of heating and ionization by a compact X-ray source at its center. The observations of a neutrino pulse make it almost certain that a neutron star is there (Section 25.2). If the neutron star is magnetized and spinning fast, it may produce nonthermal X-rays with luminosity comparable to or greater than that of the Crab Nebula ($L_x \approx 10^{38}$ erg s^{-1}). If it is not magnetized, it can produce $L_x \approx 2 \times 10^{38}$ erg s^{-1} (the Eddington limit for a neutron star) by accreting gas at a very small rate $\dot{M} \sim 10^{-8}$ M$_\odot$ yr^{-1}. Thus, in either case we might expect a compact X-ray source with $L_x \sim 10^{38}$ erg s^{-1} at the center of SN1987A. If so, the heating due to this source should dominate the optical/infrared luminosity of SN1987A for $t \gtrsim 2.5$ yr and the light curve should level out at $V \approx 13$–14^m, as illustrated in Figure 25.5. It should remain bright for many years, because the luminosity (mainly soft X-rays) of the pulsar or accreting neutron star should not decrease rapidly and the envelope should remain opaque to these X-rays for more than 10 yr, unless the envelope is fragmented or has deep holes. Ionization of the core by a compact X-ray source will also be manifested by the appearance of emission lines due to FeIII, FeVII, and FeVIII in the infrared spectrum (Colgan and Hollenbach 1988).

The interior temperature of SN1987A is dropping fast and we may soon expect to see the thermal instability, or 'infrared catastrophe,' predicted by Axelrod

Figure 25.5. A possible scenario (Pinto, Woosley, and Ensman 1988) for the future optical light curve of SN1987A, including contributions from the radioactive energy deposition by ^{56}Co (0.07 M$_\odot$), ^{57}Co (1.7×10^{-3} M$_\odot$), ^{44}Ti (3×10^{-5} M$_\odot$), and a compact X-ray source with luminosity $L_x = 10^{38}$ erg s^{-1}.

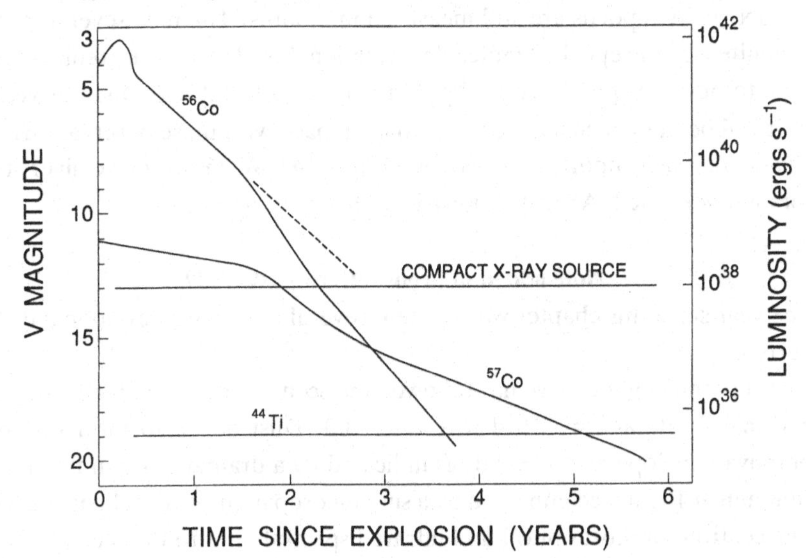

(1980), in which the core temperature drops to $\lesssim$100 K. Fransson and Chevalier (1987) estimate that this event will occur at $t \sim 2$ yr, while Colgan and Hollenbach (1988) estimate $t \sim 3$ yr. However, these estimates must be revised to include the important effects of late radioactive energy sources and cooling by molecules.

As the interior temperature continues to decrease, molecules less durable than CO and SiO should survive and a more complex chemistry should ensue. The observed CO and SiO infrared emission bands probably come from the core, where $T \lesssim 4000$ K, and other molecular species should soon form there. We also expect that the temperature of the mantle will soon decrease to $T \lesssim 3000$ K, and then a hydrogen dominated, gas phase, ion–molecule chemistry rather similar to interstellar chemistry should commence. As the temperature of the mantle drops, the Balmer continuum opacity should vanish and we might observe molecular emission from SN1987A in the ultraviolet as well as the infrared. Early evidence for this 'ultraviolet renaissance' might be the appearance of H_2 Lyman band emission in the range 1100–1600 Å.

It is also possible that the supernova chemistry will progress to form dust grains. Such an event would be bad news for spectroscopists because the dust would cause the supernova to become black at optical and infrared wavelengths (Dwek 1988). Such transitions have been observed to occur in some but not all novae (Gehrz and Ney 1987), but as yet there is no evidence for dust formation in SN1987A (Aitken 1988). Whether dust grains will form depends on whether the gas temperature in the supernova drops below the condensation temperature before the density is too low for nucleation to occur. If dust does form in SN1987A, it would probably be silicates rather than graphite and PAH molecules, because the supernova envelope is believed to be oxygen-rich.

For most of astronomy, 'epochs' are measured in millions or billions of years; but in SN1987A, epochs are still measured in months. The next several months will definitely be the epoch of molecule formation. SN1987A should remain bright enough to observe spectroscopically in the infrared for at least another year or two. It will be quite a challenge to try to keep pace with these observations.

This work was supported in part by Grant NAGW-766 to the University of Colorado under the NASA Astrophysical Theory Program.

Note added in proof – 16 August, 1989

In the year since this chapter was written, several important developments have occurred.

The supernova light curve has followed the scenario illustrated in Figure 25.5: the V magnitude at $t = 850$ d was $V = 13.3$. Dust began to form within the supernova envelope at $t \approx 500$ d, as indicated by a dramatic rise in the infrared continuum at 10 μm accompanied by a sudden drop in the optical light and a shift of the centroid of most optical and infrared spectral lines to the blue (caused by obscuration of the receding side of the supernova envelope). At t = 850 d the bolometric luminosity of the supernova, $L_{bol} = 5 \times 10^{38}$ erg s, continued to follow the exponential decay of ^{56}Co and was distributed as follows: optical and near infrared – 15%; far infrared continuum – 30%, X-rays and gamma rays – 55%. There is still no evidence for a significant ($\gtrsim$20%) contribution to the luminosity from ^{57}Co or a compact object.

The temperature in the core of SN1987A has dropped below 2000 K, as inferred from the ratio of optical and infrared lines of FeII and the predicted 'infrared catastrophe' is probably now underway.

The 2.26 μm emission feature in the infrared spectrum continued to remain strong despite the rapid decay of the 2.3 μm band of CO. This behavior casts doubt on its identification as emission by CO^+. The feature may be a recombination line of SI (E. Oliva, private communication).

Recent calculations by A. Dalgarno, M.-L. Du, and J.-H. You (1989, *Ap. J.*, in press) indicate that the rate coefficient of Reaction (25.7) is too low for it to contribute significantly to the observed abundance of CO in SN1987A. Lepp, Dalgarno, and McCray (1989) now believe that the CO is formed in the core of SN1987A primarily by the radiative association reaction C + O → CO + h and is destroyed primarily by the reaction $He^+ + CO \rightarrow He + C^+ + O$. The resulting abundance of CO is sensitive to the degree of microscopic mixing of the elements and to the unknown rate coefficients for charge transfer reactions of He with neutral species such as O.

REFERENCES

Aitken, D. K. *et al.* 1988, *Australian J. Phys., MNRAS,* **235**, 19P.

Arnett, W. D., Bahcall, J. N., Kirshner, R. P., and Woosley, S. E. 1988, *Ann. Rev. Astr. Ap.*, **27**, 629.

Axelrod, T. 1980, Ph.D. Thesis, University of California at Santa Cruz.

Castor, J. I. 1970, *MNRAS,* **149**, 111.

Catchpole, R. M. *et al.* 1987, *MNRAS,* **229**, 15p.

Catchpole, R. M. *et al.* 1988, *MNRAS,* **231**, 75p.

Chugai, N. N. 1987, *Soviet Astron. Lett.*, **13(4)**, 282.

Colgan, S. W. J., and Hollenbach, D. J. 1988, *Ap. J. (Letters)*, **329**, L25.

Dalgarno, A. 1987, in *Physical Processes in Interstellar Clouds*, eds. G. E. Morfill and M. Scholer (Dordrecht: Reidel), p. 219.

Dalgarno, A., and Butler, S. E. 1978, *Comments Atomic & Molecular Phys.*, **7**, 129.

Dalgarno, A., and Lejeune, G. 1971, *Planetary Space Sci.*, **19**, 1653.

Dopita, M. A. 1988, *Space Sci. Rev.,* **46**, 225.

Dotani, T. *et al.* 1987, *Nature*, **330**, 230.

Dwek, E. 1988, *Ap. J.*, **329**, 814.

Ebisuzaki, T., and Shibazaki, N. 1988a, *Ap. J.*, **328**, 699.

Ebisuzaki, T., and Shibazaki, N. 1988b, *Ap. J. (Letters)*, **327**, L5.

Erickson, E. F. *et al.* 1988, *Ap. J. (Letters)*, **330**, L39.

Fransson, C., and Chevalier, R. A. 1987, *Ap. J. (Letters)*, **322**, L15.

Fransson, C., and Chevalier, R. A. 1989, *Ap. J.,* **343**, 323.

Gehrz, R. D., and Ney, E. P. 1987, *Proc. Natl. Acad. Sci. USA,* **84**, 6961.
Graham, J. R. 1988, *Ap. J. (Letters)*, **335**, L53.
Grebenev, S. A., and Sunyaev, R. A. 1987a,b, *Soviet Astr. Lett.*, **13**, 945; **13**, 1042.
Harvey, P., and Lester, D. 1988, private communication.
Itoh, M. *et al.* 1987, *Nature*, **330**, 233.
Kirby-Docken, K., and Liu, B. 1978, *Ap. J. Suppl.*, **36**, 359.
Kirshner, R. P. 1988, in *A Decade of UV Astronomy with the IUE Satellite*, ed. Y. Kondo (European Space Agency), in press.
Kirshner, R. P., and Kwan, J. 1975, *Ap. J.*, **197**, 415.
Kumagai, S., Itoh, M., Shigeyama, T., Nomoto, K., and Nishimura, J. 1988, *Astr. Ap.*, **197**, L7.
Lane, N. 1980, *Rev. Mod. Phys.*, **52**, 29.
Lepp, S., Dalgarno, A., and McCray, R. 1989, in preparation.
MacGregor, P. J. 1988, *Proc. Astr. Soc. Australia,* **7**(4), 450.
McCray, R., and Li, H. W. 1988, in *Structure and Evolution of Galaxies,* ed. L. Z. Fang (Singapore: World Scientific Publishing Co.), pp. 8–46.
McCray, R., Shull, J. M., and Sutherland, P. 1987, *Ap. J. (Letters)*, **317**, L73.
Moseley, S. H., Dwek, E., Silverberg, R. F., Glaccum, W. J., Graham, J. R., and Loewenstein, R. F. 1988, in preparation.
Nussbaumer, H., and Storey, P. J. 1980, *Astr. Ap.*, **89**, 308.
Oliva, E., Moorwood, A. F. M., and Danziger, I. J. 1987, *ESO Messenger*, **50**, 18.
Osterbrock, D. E. 1974, *Astrophysics of Gaseous Nebulae* (San Francisco: W. H. Freeman & Co.).
Phillips, M. M. 1978, *Ap. J.*, **226**, 736.
Phillips, M. M. 1979, *Ap. J. Suppl.*, **39**, 377.
Phillips, M. M., Heathcote, S. R., Hamuy, M., and Navarete, M. 1988, *Astr. J.*, **95**, 1087.
Pinto, P. A., and Woosley, S. E. 1988, *Ap. J.*, **329**, 820.
Pinto, P. E., Woosley, S. A., and Ensman, L. M. 1988, *Ap. J. (Letters)*, **331**, L101.
Rank, D., Bregman, J., Cohen, M., Pinto, P., Witteborn, F., and Wooden, D. 1989, in preparation.
Roberge, W., and Dalgarno, A. 1982, *Ap. J.*, **255**, 176.
Rybicki, G. B., and Lightman, A. P. 1979, *Radiative Processes in Astrophysics* (New York: John Wiley & Sons, Inc.).
Scoville, N. Z. 1984, in *Galactic and Extragalactic Infrared Spectroscopy*, eds. M. F. Kessler and J. P. Phillips (Dordrecht: Reidel), p. 167.
Shull, J. M., and Van Steenberg, M. E. 1985, *Ap. J.*, **298**, 268.
Shull, J. M., and Xu, Y. 1987, *Proc. George Mason Conference on Supernova 1987A*, ed. M. Kafatos (Cambridge: Cambridge University Press).
Spyromilio, J., Meikle, W. P. S., Learner, R. C. M., and Allen, D. A. 1988, *Nature*, **334**, 327.
Sunyaev, R. *et al.* 1987, *Nature*, **330**, 227.
van Dishoeck, E., and Black, J. H. 1988, *Ap. J.*, **334**, 711.
Whitelock, P. A. *et al.* 1988, *MNRAS,* **234**, 5P.
Wills, B. J., Netzer, H., and Wills, D. 1985, *Ap. J.*, **288**, 94.
Woosley, S. E., Pinto, P. A., and Weaver, T. A. 1988, *Proc. Astr. Soc. Australia,* **7**(4), 355.
Xu, Y., Sutherland, P., Ross, R. R., and McCray, R. 1988, *Ap. J.*, **327**, 197.

VIII

Moderately ionized gas and chemistry at large redshifts

26

Charge transfer in astrophysical nebulae

GREGORY A. SHIELDS

Department of Astronomy, University of Texas, Austin

26.1. Introduction

The importance of charge transfer in astrophysics has been known for several decades. The last ten years, however, have seen a great increase in the number and quality of rate coefficients available and the variety of applications. These include the interstellar medium (ISM), the solar atmosphere, HII regions and planetary nebulae (PN), supernova remnants (SNR), the Galactic corona, and active galactic nuclei (AGN).

An early application of charge transfer was made by Chamberlain (1956). The reactions

$$H^+ + O^0 \rightarrow H^0 + O^+ \tag{26.1a}$$

and

$$H^0 + O^+ \rightarrow H^+ + O^0 \tag{26.1b}$$

have large rate coefficients at thermal collision energies because of the near resonance of the ionization potentials. Using cross sections for $H^0 + H^+$ charge transfer by Dalgarno and Yadav (1953), Chamberlain estimated a rate coefficient $k = 10^{-8}\ cm^3\ s^{-1}$ for Reactions (26.1a) and (26.1b). The ratio of probabilities per second that an O^+ ion will be neutralized by charge transfer or by radiative recombination is

$$R = k(T)N(H^0)/[\alpha(T)N_e] \tag{26.2}$$

where α is the radiative recombination coefficient and $N(H^0)$ and N_e are the densities of H^0 atoms and free electrons (cm^{-3}). Typical values of α are $\sim 10^{-12}$ $cm^3\ s^{-1}$; and therefore charge transfer dominates for $N(H^0)/N_e > 10^{-4}$. Chamberlain (1956) used the thermodynamic relation between the rate coefficients for Reactions (26.1a) and (26.1b), derived by A. Dalgarno, and noted that when charge transfer dominates both the ionization and recombination, the O^+/O ratio is governed by the H^+/H^0 ratio,

$$N(O^+)/N(O^0) \approx (8/9)[N(H^+)/N(H^0)] \tag{26.3}$$

This relation Chamberlain applied to an analysis of the spectrum of the Cas A SNR, and it has been widely used in subsequent work. Bates and Moisewitsch (1954) had found a large rate for $Si^{2+} + H^0$ (confirmed by McCarroll and Valiron 1976), as did Dalgarno (1954) for other multiply-charged ions reacting with H^0; but many years passed before astrophysicists gained a full appreciation of the significance of charge transfer for multiply-charged ions in nebulae.

The importance of charge transfer in the ISM was studied by Field and Steigman (1971), who estimated rates for Reactions (26.1a) and (26.1b) and showed that the ionization of oxygen in the ISM is governed by Equation (26.3). In the case of PN, Williams (1973) showed that observed intensities of the [OI] $\lambda\lambda$6300, 6363 emission lines could be understood in terms of charge transfer. The flowering of the subject of charge transfer in nebulae, however, can be traced to IAU Symposium No. 76 on *Planetary Nebulae* at Cornell University. There Dalgarno (1978) reviewed charge transfer reactions of astrophysical significance. Harrington (1978) noted the underprediction of O^+ in nebular models and the possible importance of charge transfer; and Pequignot (1978) and Perinotto (1978) proposed empirical charge transfer rates to account for discrepancies between observed and predicted abundances of O^+ and other ions. Although this proposal met with some skepticism at the symposium, work during the next few years confirmed the importance of charge transfer for many ions, in particular the reaction $O^{2+} + H^0$ (Butler, Heil and Dalgarno 1980). Charge transfer quickly became a standard ingredient in models of ionized nebulae, SNR, and AGN.

26.2. Charge transfer rate coefficients

Astrophysical applications of charge transfer generally involve temperatures $T \approx 10^4$ K. The corresponding collision energies of up to a few electron volts are such that the relative collision velocity is small compared with the orbital velocity of the valence electrons. Therefore the molecular model may be used to describe the charge transfer process. The computational techniques are reviewed by McCarroll, Valiron and Opradolce (1983), with an emphasis on multiply-ionized target ions. Potential energy curves of the molecular states are determined for the entry and all possible exit channels. The charge transfer cross section is governed by the internuclear separation R_x at avoided crossings and the energy separation $\Delta E(R_x)$. The cross sections are sensitive to the details of these parameters and differ by orders of magnitude from one ion to another.

Many of the rate coefficients of astrophysical interest are due to Dalgarno and collaborators. Full quantal calculations were carried out by Butler *et al.* (1980) for $(C, N, O, Ne)^{2,3+}$ reacting with H^0; O^{2+} and C^+ with He^0; and C^+ with He^+. The rate coefficient for $O^{2+} + H^0$ was found to have the large value 0.77×10^9 cm^3 s^{-1} at 10^4 K, in contrast to earlier theoretical expectations. Also large is the $N^{2+} + H^0$ rate, whereas $Ne^{2+} + H^0$ is much slower. The results are consistent with observations of nebulae. Butler and Dalgarno (1980) gave Landau–Zener calculations of charge transfer rates for numerous multiple-charged ions reacting with H^0 and He^0 and for ionization reactions involving H^+ and He^+. Potential curves were calculated for many ions and an interpolation procedure suggested. Comparison of the quantal and Landau–Zener results shows good agreement for astrophysical purposes except in the case of $O^{3+} + H^0$. However, McCarroll *et al.* (1983) question the general reliability of approximation procedures for charge transfer. Butler and Dalgarno (1980) also defined Type I charge transfer processes to be one-electron reactions in which electron capture takes place without any change in the configuration of the ionic core. Type II processes involve a rearrangement of the core configuration, e.g., $O^{2+}(2s^22p^2)^3P + H(1s)^2S \rightarrow O^+(2s2p^4)^4P + H^+$.

Although emphasis has been on charge transfer recombination, ionization by charge transfer is important in some cases. For example, Baliunas and Butler (1980) gave rate coefficients for $Si^+ + H^+ \rightarrow Si^{2+} + H^0$ and $Si^{2+} + He^+ \rightarrow Si^{3+} + He^0$. These processes strongly affect the ionization of silicon in the solar corona and the SiIII]λ1892 emissivity in the quiet sun transition region. (The half bracket indicates an intercombination line.) Neufeld and Dalgarno (1987) found that charge transfer ionization $Fe^+ + H^+ \rightarrow Fe^{2+} + H$ and the reverse reaction are important at nebular temperatures, leading to a relation analogous to Equation (26.3). On the other hand, they found that while $Ni^{2+} + H^0$ is rapid, the ionization reaction is slow; and this suggests the possibility of comparing the ionization of Fe and Ni to probe the abundance of H^0.

26.3. The physics of ionized nebulae

Astrophysical nebulae are exemplified by PN and HII regions. A typical PN consists of a central star with effective temperature $35\,000 \lesssim T_{eff} \lesssim 200\,000$ K, whose ionizing continuum excites a surrounding shell of gas of radius $\sim 10^{-1}$ pc and mass $\sim$0.1–1 $M_\odot$, density $\sim 10^4$ cm^{-3}, and electron temperature 10^4 K. The gas may be concentrated in small clouds or filaments. HII regions are ionized by one or more luminous main sequence stars with $35\,000 \lesssim T_{eff} \lesssim 60\,000$ K, typically with dimensions of several parsecs, densities $N \approx 10^3$–10^4 cm^{-3}, and masses of many solar masses. These objects have a rich spectrum of optically thin emission lines from a variety of elements and ionization stages and therefore are rewarding subjects for spectroscopic analysis.

The basic physics of gaseous nebulae is surveyed in the book by Osterbrock (1974). For an ion A^{i+} with photoionization cross section a_ν (cm^2), the probability per second of photoionization at distance R from the central star is

$$\Gamma_{\mathrm{pi}} = \int_{\nu_{\mathrm{T}}}^{\infty} \frac{L_\nu a_\nu \exp(-\tau_\nu)}{4\pi R^2 h\nu} \, d\nu \tag{26.4}$$

where L_ν is the ionizing luminosity of the central star (erg s^{-1} Hz^{-1}), τ_ν is the optical depth, and ν_{T} is the threshold frequency. (For simplicity we ignore here the nebular diffuse radiation.) The equilibrium ionization ratio is then

$$\frac{N(\mathrm{A}^{(i+1)+})}{N(\mathrm{A}^{i+})} = \frac{\Gamma_{\mathrm{pi}} + k_{\mathrm{i}} N(\mathrm{H}^+)}{(\alpha_{\mathrm{r}} + \alpha_{\mathrm{di}}) N_{\mathrm{e}} + k_{\mathrm{r}} N(\mathrm{H}^0)}, \tag{26.5}$$

where $k_{\mathrm{r}} = k(A^{(i+1)+} + \mathrm{H}^0)$, $k_i = k(A^{i+} + \mathrm{H}^+)$, and α_{di} is the dielectronic recombination coefficient (see Storey (1983)). For most HII regions and PN, $N(\mathrm{H}^0)/N(\mathrm{H}^+)$ is very small except near the Strömgren radius R_{s}, where the gas rapidly becomes neutral. R_{s} is given by equating the number of recombinations of H in the nebula to the number of ionizing photons emitted per second by the star,

$$(4\pi/3) R_{\mathrm{s}}^3 \alpha_{\mathrm{B}} N_{\mathrm{e}} N(\mathrm{H}^+) = Q(\mathrm{H}^0) = \int_{\nu_{\mathrm{H}}}^{\infty} L_\nu (h\nu)^{-1} \, d\nu, \tag{26.6}$$

where α_{B} is the recombination coefficient to states $n \geq 2$. The degree of ionization of H and other elements at a representative point inside R_{s} is determined by the 'ionization parameter',

$$U \equiv Q(\mathrm{H}^0)/4\pi R_{\mathrm{s}}^2 N c \tag{26.7}$$

where N is the gas density and c is the speed of light. From Equations (26.4) and (26.5), the ratio $\Gamma_{\mathrm{pi}}/\alpha N_{\mathrm{e}}$ is proportional to U, so larger U means higher ionization where τ_ν is modest.

For $T_{\mathrm{eff}} \lesssim 75\,000$ K, the star emits few photons with $\nu > 4\nu_{\mathrm{H}}$, and the nebular interior contains H^+, He^+, O^{2+}, Ne^{2+}, etc. Near R_{s}, as τ_ν increases, the ionization drops. O^+ predominates in a shell at and just inside R_{s} (the $\mathrm{H}^+ \rightarrow \mathrm{H}^0$ transition zone) and O^0 prevails further out. In PN with $T \gtrsim 100\,000$ K, there is an inner zone with He^{2+}, where oxygen is in the form of O^{3+} and possibly higher stages.

Line emission from H and He in nebulae consists of permitted lines formed by radiative recombination. Most observed lines of heavier elements are formed by collisional excitation by thermal electrons. The physical conditions N_{e} and T can be derived from the intensity ratios of certain pairs of emission lines. Given accurate collision strengths, one can then derive ionic abundance ratios $N(\mathrm{A}^{i+})/N(\mathrm{H}^+)$ from the emission lines and compare the results with computer models of the ionization equilibrium in the nebula (Harrington (1983), and references therein). The underprediction of the abundances of O^+ and N^+ in early models eventually was resolved by the inclusion of charge transfer.

Because $N(H^0)/N_e$ is very small in the interiors of most steady state photoionized nebulae, charge transfer is most important in the transition zone. In the case of nebulae that are undergoing progressive recombination with time because the ionizing source has diminished, the H^0 fraction builds up rapidly because there is only one stage of ionization to recombine through ($H^+ \rightarrow H^0$). Harrington and Marionni (1976) discuss this process for FG Sagitae. Once $N(H^0)/N_e$ exceeds $\sim 10^{-3}$, charge transfer overwhelms radiative recombination and greatly accelerates the recombination of many ions such as O^{2+} and N^{2+}. This stage is reached in a small fraction of the hydrogen recombination time, $t_r = (\alpha_B N_e)^{-1}$. Similarly, in a shock wave, the gas initially reaches a state of ionization determined by the postshock temperature, in turn fixed by the shock velocity. The gas then enters a time-dependent cooling and recombination phase, in which H^0 soon reaches a concentration sufficient for charge transfer to dominate the recombination of many ions (Butler and Raymond 1980).

26.4. Photoionized nebulae

We now review some results in the application of charge transfer to the ionization structure and line emission of ionized nebulae, and the use of nebulae in turn to test theoretical predictions of charge transfer rates.

26.4.1. Charge transfer recombination and ionic abundances

Early photoionization models of PN and HII regions showed that the predicted intensities of [OII] λ3727 and λ7325, [NII] $\lambda\lambda$6584, and [OI] λ6300 were less than observed. Inclusion of dense clouds, optically thick in the Lyman continuum, helped to raise these line intensities but was inadequate (Harrington 1969, Williams 1973). Williams (1973) achieved adequate [OI] emission in models with dense condensations and O–H charge transfer. Kirkpatrick (1972) found too little O^+ in his model of NGC 7662 and considered *ad hoc* modifications of the stellar continuum. Balick (1975) found that internal reddening of the ionizing continuum in NGC 7027 did not sufficiently increase the [OII] intensity; and for the same object, Shields (1978) found that density condensations and dust were not successful. Shields and Oke (1975) underpredicted the O^+ abundance in photoionization models of the Seyfert galaxy NGC 1068, and Boksenberg and Netzer (1977) found the same problem for NGC 3516. The ubiquitous 'O^+ problem' in photoionized nebulae appeared to call for a fundamental addition to the models.

Blint, Watson and Christensen (1976) and Christensen, Watson and Blint (1977) calculated rates for $C^{3+} + H^0$ and $N^{3+} + H^0$, respectively, and found that these processes profoundly reduce the abundances of C^{3+} and N^{3+} in the interstellar medium. These rates were used in models of the PN NGC 7662 by

Bohlin, Harrington and Stecher (1978) to give improved agreement with observed ionic abundances.

Pequignot, Aldrovandi and Stasinska (1978) proposed empirical charge transfer rate coefficients to reconcile differences between photoionization models and observation of NGC 7027. These included $O^{2+} + H^0$, $N^{2+} + H^0$, and $Ne^{2+} + H^0$. This proposal was vindicated for $O^{2+} + H^0$ and $N^{2+} + H^0$ by the quantal calculations of Butler *et al.* (1980), but the theoretical rate for $Ne^{2+} + H^0$ is much smaller than required by Pequignot *et al.* (1978). Evidently, when the effect of charge transfer is large enough to exceed the uncertainties in nebular modeling, a meaningful empirical rate can be derived. On the other hand, the rate for $Ne^{2+} + H^0$ is too slow to be measured by nebular model fitting, which runs the risk of attributing to charge transfer discrepancies arising from other causes. However, the low rate for $N^+ + H^0$ derived empirically by Pequignot *et al.* agrees with the quantal calculation by Butler and Dalgarno (1979).

As soon as the list of charge transfer rate coefficients by Butler *et al.* (1980) and Butler and Dalgarno (1980) became available, their inclusion in nebular models, along with other theoretical or empirical values, became routine. Shields *et al.* (1981) analyzed the effect of charge transfer in models of the planetary nebula NGC 2440. They noted that charge transfer recombination governs the boundaries of the radial zones in which various ions predominate and thereby the fractional abundance $\langle X(A^{i+})\rangle$ of an ion averaged over the nebula. For example, $\langle X(O^+)\rangle$ varies as $[k(O^{2+} + H^0)]^{0.35}$. Agreement with observation was improved by inclusion of H^0 charge transfer with C^{3+}, O^{2+}, N^+, Ar^{3+}, and Ar^{4+}. On the other hand, the models suggested that charge transfer of N^{3+}, O^{3+}, and Ne^{4+} with H^0 may proceed less rapidly than estimated by Butler and Dalgarno (1980) from the Landau–Zener approximation.

A recent discussion of charge transfer in PN, containing references to earlier work, is that of Allan *et al.* (1988). They calculate cross sections and rate coefficients for the ionization process $Mg + H^+ \rightarrow Mg^+ + H^0$. The process occurs to the 4s and 3p states, the latter coupling to an intermediate molecular state. The process is endothermic to the states involved. The total rate coefficient is fit by $k_{tot} = (1.74 \times 10^{-9}\ cm^3\ s^{-1}) \exp(-2.210/t)$ for $0.8 < t \leq 2$, where $t = T/10^4$ K. The process has an important effect in reducing the intensity of MgI] $\lambda\lambda$4571, 4562 (Clegg *et al.* 1987) in nebular models of NGC 3918, NGC 7662, IC 418, and IC 4997. The derived Mg/O abundance ratios from MgI, MgII, and MgIV agree within a factor of 2, when allowance is made for interstellar absorption of MgII λ2800. The mean Mg/O value of 0.024 is only a factor 2 below solar, indicating that Mg suffers less severe depletion onto grains than does Fe.

26.4.2. Spatial variations of ionization in nebulae

The preceding discussion has dealt with models and observations of ionic abundances in nebulae as a whole. A graphic illustration of the operation of charge transfer is provided by the spatial variations of ionization in the Ring Nebula, NGC 6720. Hawley and Miller (1977) observed the emission line spectrum as a function of distance from the central star. The [NeIII] λ3869 line, relative to Hβ, remains strong to the edge of the nebula whereas [OIII] gives way to [OII] and [OI]. Harrington (1983) gives a photoionization model that explains this in terms of the large rate of $O^{2+} + H^0$ charge transfer and the very small rate for $Ne^{2+} + H^0$. Toward the edge of the nebula, the H^0 fraction becomes sufficient to lower the ionization of O to O^+ and then O^0, whereas Ne remains Ne^{2+}. The Ring Nebula is an excellent laboratory for study of charge transfer because of its low ionization parameter, leading to a relatively large H^0 abundance.

26.4.3. Charge transfer line emission

Modeling of ionic abundances in nebulae provides information about the total charge transfer rate k_{tot} $(A^{i+} + H^0)$ but not about the specific exit channels. Transfer to excited states in some cases gives rise to weak but observable emission lines whose intensities provide a test of the calculated rates. A list of excited states populated by charge transfer and the resulting allowed emission line wavelengths is given by Shields, Dalgarno and Sternberg (1983). Dalgarno, Heil and Butler (1981) called attention to the possibility of emission in OIIIλ5592 resulting from charge transfer of O^{3+} with H^0 to populate $2s^2p3p$, ^1P. Shields *et al.* (1983) noted that this line is seen in PN with HeIIλ4686 emission, indicating the presence of an He^{2+} zone containing O^{3+}; but it is absent from the spectrum of PN that lacks an He^{2+} zone. This is qualitatively consistent with the charge transfer emission process. Using a detailed model of NGC 2440, they derived from the observed intensity of λ5592 an empirical rate coefficient $k(^1P) = 2.0 \times 10^{-9}$ cm^3 s^{-1} at 3000 K. The theoretical value is 1.6×10^{-9} cm^3 s^{-1} (Dalgarno *et al.* 1981). Similarly good agreement was found for the PN NGC 7027. Shields *et al.* (1983) also suggested that the OIII$\lambda\lambda$3774, 3757 lines of PN arise from the charge transfer reaction O^{3+} $(2s^22p) + H^0 \rightarrow O^2(2s^22p3p, {}^3D_1) + H^+$. For two PN, they derive an empirical rate coefficient of 2.1×10^{-9} cm^3 s^{-1}, compared with a theoretical value into all the ^3D_J levels of 6.2×10^{-9} cm^3 s^{-1} at 10^4 K (Dalgarno *et al.* 1981). Dalgarno and Sternberg (1982) showed that the intensities of $\lambda\lambda$3757, 3774 in two PN, relative to λ5592, were consistent with charge transfer. Likkel and Aller (1986) similarly found agreement with predicted OIII line intensities in a number of PN. With the aid of rates for charge transfer onto individual fine structure levels by Roueff and Dalgarno (1988), Sternberg, Dalgarno and Roueff (1988) found improved agreement for PN and argued that charge transfer may be contributing to the observed OIII lines of X-ray sources.

Charge transfer emission in $\lambda5592$ was also studied by Clegg and Walsh (1985). Using detailed models to analyze the spectrum of two PN, NGC 7662 and NGC 3916, they find empirical rate coefficients $k\,(^1P) = 0.9 \times 10^{-9}$ and 2.3×10^{-9} cm^3 s^{-1}, respectively. They also argue that dielectronic recombination is not an important source of $\lambda5592$ and that radiative recombination and fluorescent excitation by the stellar continuum each fall short by a factor ~30 of explaining the observed intensities. These results suggest that empirical rates can be derived to factor of 2 accuracy and that they agree with quantal calculations at this level of accuracy.

NeIII$\lambda\lambda2590, 2678$ emission can result from the charge transfer reaction $Ne^{3+} + H^0 \rightarrow Ne^{2+} + H^+$. From observations and a model of NGC 3918, Clegg, Harrington and Storey (1986) find empirical rate coefficients $k\,(^5P) = 3.5 \times 10^{-9}$ and $k\,(^3P) = 2.8 \times 10^{-9}$ cm^3 s^{-1}, compared with theoretical values at 14 000 K of 4.7×10^{-9} and 1.5×10^{-9}, respectively (Butler *et al.* 1980). Again, agreement at the factor 2 level is found. Deeper spectra could provide an accurate measure of the ratio $k\,(^5P)/k(^3P)$. Similarly, sensitive spectra of PN could provide accurate measurements of $k(^3D)/k(^1P)$ and the relative rates form the ^3D fine structure levels of OIII populated by $O^{3+} + H^0$ charge transfer.

26.5. Supernova remnants

Supernova explosions leave expanding remnants that produce shock waves as they advance into the ambient interstellar gas. Gas passing through shocks of velocity $v_s \approx 10^2$ km s^{-1} is heated to temperatures sufficient to ionize hydrogen and multiply ionize heavier elements. As time goes on, a given volume of gas undergoes radiative cooling and progressive recombination as it moves further downstream from the shock front. The rising H^0 abundance leads to the dominance of charge transfer recombination for many ions. Shull and McKee (1979) computed models of interstellar shocks including a variety of charge transfer reactions with H^0 and He^0 (Dalgarno and Butler 1978). They noted the strong effect of charge transfer in reducing the Si^{2+} abundance and the importance of $H^0 + H^+$ charge transfer in reducing the downstream penetration of neutral atoms. Butler and Raymond (1980) systematically explored the effects of including charge transfer in models of shocks with $v_s = 60$ or 120 km s^{-1}. For 120 km s^{-1} shocks, $N(H^0)/N(H^+)$ rises to 10^{-3} by the time the temperature has fallen to 36 000 K; and for the 60 km s^{-1} case $N(H^0)/N(H^+)$ always exceeds 10^{-3}. Charge transfer substantially reduces the postshock column densities of N^+, $O^{1,2+}$, and Si^{2+}; but the $Si^+ + H^+$ ionization process offsets $Si^{2+} + H^0$ recombination. The O^+/O^0 and Si^{2+}/Si^+ ratios are locked to H^+/H^0 in the manner of Equation (26.3);

but N^+/N^0, while affected by charge transfer, is not locked to H^+/H^0. Those collisionally excited lines of a given ion with higher excitation potentials are predominantly emitted in warmer zones closer to the shock, compared with lines of lower excitation potential. Because $N(H^0)/N_e$ increases as T drops, the lines of weaker temperature sensitivity are more affected by charge transfer. Thus, Butler and Raymond found for $v_s = 120$ km s^{-1} that [OIII] $\lambda5007$ was reduced by charge transfer whereas $\lambda4363$ was not; and this affects the electron temperature that would be inferred from the line ratio. In general, ultraviolet lines are less affected by charge transfer than optical lines. For $v_s = 120$ km s^{-1}, the lines of [NI] $\lambda5200$ and [OI] $\lambda6300$ are increased by factors ~2 by charge transfer; but SiII] $\lambda2325$, SiIII] $\lambda1892$, and SiIV] $\lambda1400$ are changed by more modest factors.

A different role of charge transfer is discussed by Chevalier, Kirshner and Raymond (1980) for the remnant of Tycho's supernova. The spectrum shows only emission lines of Hα, Hβ, and Hγ with a composite line profile having a narrow core and broad wings of 1800 km s^{-1} full width at half maximum. This can be explained in terms of a shock with $v_s \approx 2300$ km s^{-1}. The broad wings result from fast moving ions that undergo $H^+ + H^0$ charge transfer recombination into excited states; and the resulting stationary ions produce the narrow line cores. A similar explanation for the broad and narrow lines of Seyfert galaxies was proposed by Ptak and Stoner (1973) but is not widely accepted. Chevalier *et al.* (1980) use charge transfer cross sections by Bates and Dalgarno (1953) to estimate that a fraction 0.02–0.04 of all charge transfers result in Hα photons. A similar discussion for the Cygnus Loop SNR is given by Raymond *et al.* (1980).

26.6. The Galactic corona

Charge transfer has also played a role in discussions of photoionized gas in the Galactic corona. The key observations are measurements of ultraviolet absorption lines of CIV and SiIV at high Galactic latitudes (Savage (1987) and references therein). This gas appears to be concentrated several kiloparsecs above the Galactic plane, with column densities $N(C^{4+}) \approx 10^{14}$ cm^{-3} and $N(C^{4+})/N(Si^{4+}) \approx 5$. There are two competing models for the support of the Galactic corona above the plane. In the 'Galactic fountain' model (Shapiro and Field 1976, Bregman 1980), interstellar gas heated to 10^6 K by supernovae tries to rise to its pressure scaleheight $Z \approx 7$ kpc; but radiative cooling causes thermal instabilities and the formation of clouds that orbit ballistically back to the disk. Alternatively, cooler gas could be supported by cosmic ray pressure (Hartquist and Morfill (1986), and references therein). Chevalier and Fransson (1984) and Hartquist, Pettini and Tallant (1984) consider models for the CIV, SiIV gas involving photoionization by the extragalactic radiation field produced by AGN. The ionization balance favors doubly-ionized over triply-ionized species, and the fractional abundances of C^{3+}

and Si^{3+} are very sensitive to the ionization parameter. For U below a critical value, $N(C^{3+})/N(C) \propto U^2$, because charge transfer dominates the recombination of C^{3+} to C^{2+} and $N(H^0)/N_e \propto U$. Sufficient columns of C^{3+} occur only if there is a low density gas at 10^4 K filling much of the coronal volume; and cosmic ray support then may be necessary. Bregman and Harrington (1986) include ionizing radiation from OB stars, planetary nebulae, AGN, and the Galactic soft X-ray background. Their best agreement with observed values of $N(C^{3+})$ and $N(C^{3+})/N(Si^{3+})$ occurs for gas of density 0.01–0.0003 cm^{-3}. They note that for high enough U, radiative recombination dominates charge transfer and $N(C^{3+})/N(C) \propto U^{-1}$. For a given photon density $N(\gamma)$, the abundance of C^{3+} (cm^{-3}) is independent of the total gas density N, and thus the C^{3+} column over a fixed path length is independent of N. One may note that, in this regime, the C^{3+} is directly proportional to $N(\gamma)$ and thus measures $N(\gamma)$ as long as U is above the critical value that makes charge transfer unimportant. (Of course, the C^{3+} column is also proportional to the filling factor of the 10^4 K gas.) Thus there is great interest in comparing $N(\gamma)$ determined this way with values inferred from the Hα surface brightness of high velocity clouds (Reynolds 1987).

26.7. Summary

Charge transfer has already played a major role in the understanding of the ISM, HII regions, PN, SNR, and the Galactic corona. In AGN and other objects ionized by hard radiation, there typically is a large zone of partially ionized gas in which charge transfer is crucial; and inclusion of all available rates is standard in AGN models (see review by Ferland and Shields (1985)). Another potentially important class of objects is old nova shells, where the charge transfer line emission by C, N, O, Ne will be enhanced by the extreme enrichment of these elements and the low electron temperature ($\lesssim 10^3$ K) suppresses collisionally excited emission (Williams 1982). Advances in astronomical observations and atomic physics are sure to widen the role of charge transfer in astrophysical research.

The author is grateful to J. P. Harrington, J. C. Raymond, and P. R. Shapiro for valuable discussions. This work was supported in part by Research Grant F-910 from the Robert A. Welch Foundation.

REFERENCES

Allan, R. J., Clegg, R. E. S., Dickinson, A. S., and Flower, D. R. 1988, *Monthly Not. Roy. Astron. Soc.*, in press.
Balick, B. 1975, *Astrophys. J.*, **201**, 705.
Baliunas, S. L., and Butler, S. E. 1980, *Astrophys. J. (Letters)*, **235**, L45.
Bates, D. R., and Dalgarno, A. 1953, *Proc. Phys. Soc. A*, **66**, 972.
Bates, D. R., and Moisewitsch, B. L. 1954, *Proc. Phys. Soc. A*, **67**, 805.

Blint, R. J., Watson, W. D., and Christensen, R. B. 1976, *Astrophys. J.*, **205**, 634.
Bohlin, R. C., Harrington, J. P., and Stecher, T. P. 1978, *Astrophys. J.*, **219**, 575.
Boksenberg, A., and Netzer, H. 1977, *Astrophys. J.*, **212**, 37.
Bregman, J. N. 1980, *Astrophys. J.*, **236**, 577.
Bregman, J. N., and Harrington, J. P. 1986, *Astrophys. J.*, **309**, 833.
Butler, S. E., and Dalgarno, A. 1979, *Astrophys. J.*, **234**, 765.
Butler, S. E., and Dalgarno, A. 1980, *Astrophys. J.*, **241**, 838.
Butler, S. E., Heil, T. G., and Dalgarno, A. 1980, *Astrophys. J.*, **241**, 442.
Butler, S. E., and Raymond, J. C. 1980, *Astrophys. J.*, **240**, 680.
Chamberlain, J. W. 1956, *Astrophys. J.*, **124**, 390.
Chevalier, R. A., and Fransson, C. 1984, *Astrophys. J.* (*Letters*), **279**, L43.
Chevalier, R. A., Kirshner, R. P., and Raymond, J. C. 1980, *Astrophys. J.*, **235**, 186.
Christensen, R. B., Watson, W. C., and Blint, R. J. 1977, *Astrophys. J.*, **213**, 712.
Clegg, R. E. S., Harrington, J. P., Barlow, M. J., and Walsh, J. R. 1987, *Astrophys. J.*, **314**, 551.
Clegg, R. E. S., Harrington, J. P., and Storey, P. J. 1986, *Monthly Not. Roy. Astron. Soc.*, **221**, 61P.
Clegg, R. E. S., and Walsh, J. R. 1985, *Monthly Not. Roy. Astron. Soc.*, **215**, 323.
Dalgarno, A. 1954, *Proc. Phys. Soc. A.*, **67**, 1010.
Dalgarno, A. 1978, *IAU Symposium No. 76, Planetary Nebulae,* ed. Y. Terzian (Dordrecht: D. Reidel), p. 139.
Dalgarno, A., and Butler, S. E. 1978, *Comments Atom. Mol. Phys.*, **7**, 129.
Dalgarno, A., Heil, T. G., and Butler, S. E. 1981, *Astrophys. J.*, **245**, 793.
Dalgarno, A., and Sternberg, A. 1982, *Monthly Not. Roy. Astron. Soc.*, **200**, 77P.
Dalgarno, A., and Yadav, H. N. 1953, *Proc. Phys. Soc. A,* **66**, 173.
Ferland, G. J., and Shields, G. A. 1985, in *Astrophysics of Active Galaxies and Quasistellar Objects,* ed. J. S. Miller (Mill Valley CA: University Science Books), pp. 157–184.
Field, G. B., and Steigman, G. 1971, *Astrophys. J.*, **166**, 59.
Harrington, P J. 1969, *Astrophys. J.*, **156**, 903.
Harrington, J. P. 1978, *IAU Symposium No. 76, Planetary Nebulae,* ed. Y. Terzian (Dordrecht: D. Reidel), p. 151.
Harrington, J. P. 1983, *IAU Symposium No. 103, Planetary Nebulae,* ed. D. R. Flower (Dordrecht: D. Reidel), p. 219.
Harrington, P. J., and Marionni, P. A. 1976, *Astrophys. J.*, **206**, 458.
Hartquist, T. W., and Morfill, G. E. 1986, *Astrophys. J.*, **311**, 518.
Hartquist, T. W., Pettini, M., and Tallant, A. 1984, *Astrophys. J.*, **276**, 519.
Hawley, S. A., and Miller, J. S. 1977, *Astrophys. J.*, **212**, 94.
Kirkpatrick, R. 1972, *Astrophys. J.*, **176**, 381.
Likkel, L., and Aller, L. H. 1986, *Astrophys. J.*, **301**, 825.
McCarroll, R., and Valiron, P. 1976, *Astron. Astrophys.*, **53**, 83.
McCarroll, R., Valiron, P., and Opradolce, L. 1983, *IAU Symposium No. 103, Planetary Nebulae,* ed. D. R. Flower (Dordrecht: D. Reidel), p. 187.
Neufeld, D. A., and Dalgarno, A. 1987, *Phys. Review A,* **35**, 3142.
Osterbrock, D. E. 1974, *Astrophysics of Gaseous Nebulae* (San Francisco: Freeman).
Pequignot, D. 1978, *IAU Symposium No. 76, Planetary Nebulae,* ed. Y. Terzian (Dordrecht: D. Reidel), p. 162.
Pequignot, D., Aldrovandi, S. M. U., and Stasinska, G. 1978, *Astron. Astrophys.*, **63**, 313.
Perinotto, M. 1978, *IAU Symposium No. 76, Planetary Nebulae,* ed. Y. Terzian (Dordrecht: D. Reidel), p. 161.

Ptak, R., and Stoner, R. E. 1973, *Astrophys. J.*, **185**, 121.
Raymond, J. C., Davis, M., Gull, T. R., and Parker, R. A. R. 1980, *Astrophys. J.* (*Letters*), **238**, L21.
Reynolds, R. J. 1987, *Astrophys. J.*, **323**, 553.
Roueff, E., and Dalgarno, A. 1988, *Phys. Review A,* in press.
Savage, B. D. 1987, *Interstellar Processes,* eds. D. J. Hollenbach and H. A. Thronson, Jr (Dordrecht: D. Reidel), p. 123.
Shapiro, P. R., and Field, G. B. 1976, *Astrophys. J.*, **205**, 762.
Shields, G. A. 1978, *Astrophys. J.*, **219**, 565.
Shields, G. A., Allen, L. H., Czyzak, S. J., and Keyes, C. D. 1981, *Astrophys. J.*, **248**, 569.
Shields, G. A., Dalgarno, A., and Sternberg, A. 1983, *Phys. Review A,* **28**, 2137.
Shields, G. A., and Oke, J. B. 1975, *Astrophys. J.*, **197**, 5.
Shull, J. M., and McKee, C. F. 1979, *Astrophys. J.*, **227**, 131.
Steigman, G. J. 1975, *Astrophys. J.*, **199**, 642.
Sternberg, A., Dalgarno, A., and Roueff, E. 1988, in press.
Storey, P J. 1983, *IAU Symposium No. 103, Planetary Nebulae,* ed. D. R. Flower (Dordrecht: D. Reidel), p. 199.
Williams, R. E. 1973, *Monthly Not. Roy. Astron. Soc.*, **164**, 111.
Williams, R. E. 1982, *Astrophys. J.*, **261**, 170.

27

Molecules at early epochs

JOHN H. BLACK

Steward Observatory, University of Arizona, Tucson, USA

27.1. Introduction

In this chapter, we consider the earliest chemistry and the most distant and primitive material that can be probed by means of molecular spectra. The formation of the first chemical bonds probably occurred around the epoch of recombination in the early development of the Universe. Although the first molecules may have had profound effects on the thermal properties of matter, they are not expected to be directly observable. Sometime later, after stars, galaxies, and quasi-stellar objects (QSOs or quasars) have arisen, molecules in intergalactic clouds and in the interstellar clouds of distant, disk-type galaxies may become observable. Molecules like H_2 provide excellent probes of such material.

27.2. Pregalactic molecules

Figure 27.1 displays the thermal history of a homogeneous, expanding Universe as a function of the red shift z for a particular 'standard model' (Harrison 1973, Zel'dovich and Novikov 1983). The present epoch is at $z = 0$, corresponding to a time $t \approx 4 \times 10^{10}$ yr after the beginning of the expansion in the model displayed. As the Universe expands, the temperature of the cosmic background radiation decreases as $T_r \propto (1 + z)$; its current value is $T_r = 2.7$ K. The normalization of a particular model depends upon the mass density parameter Ω and the current expansion rate or Hubble parameter, H_0. For the model shown, the number density of hydrogen nuclei varies as

$$n_H = 212.5\left(\frac{1+z}{1000}\right)^3\left(\frac{\Omega}{0.1}\right)^4\left(\frac{H_0}{50}\right)^8 \text{ cm}^{-3} \qquad (27.1)$$

when H_0 is in km s^{-1} Mpc^{-1}. The composition of the baryonic matter is fixed by the equilibrium prevailing in the very early nuclear era. Relative abundances by

Figure 27.1. Thermal and chemical history of the Universe for a standard Big Bang model with $\Omega = 0.1$ and $H_0 = 50$ km s^{-1} Mpc^{-1}. (*a*) The matter and radiation temperatures as functions of redshift z. (*b*) and (*c*) Fractional abundances relative to the number density of hydrogen nuclei for the principal forms of the most abundant elements. (From the work of Latter (1989).)

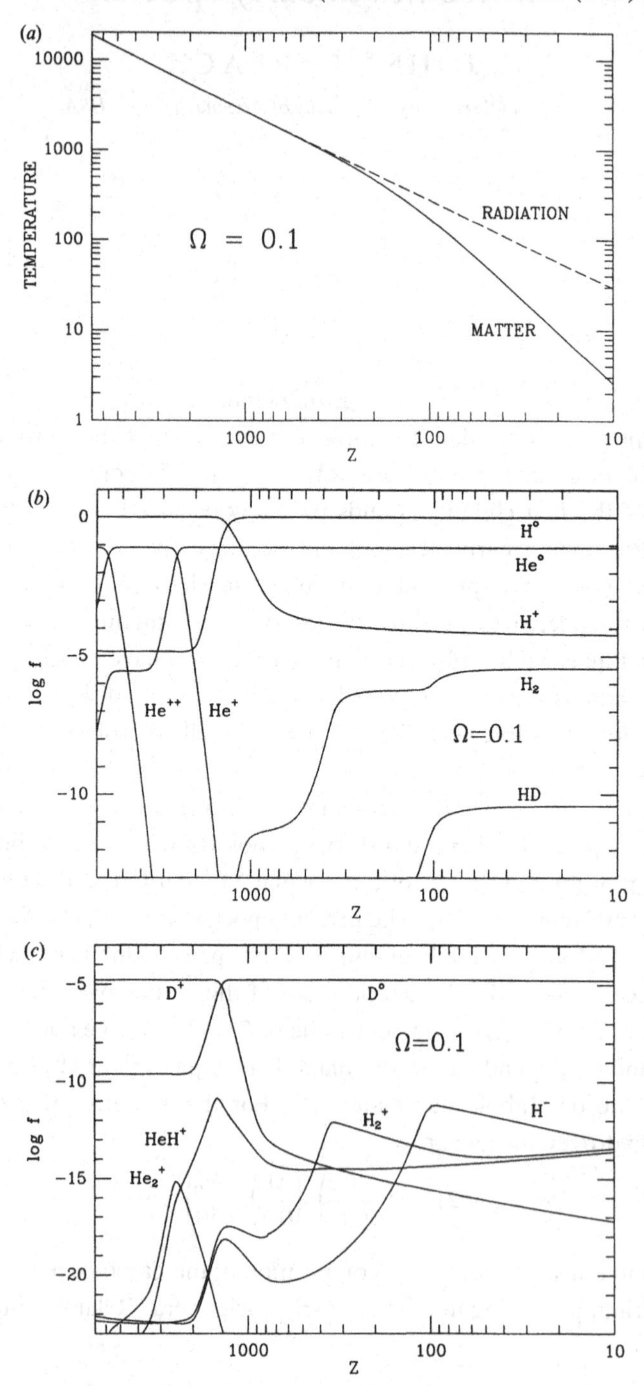

number of nuclei for the model illustrated are H : He : D = 10^5 : 8053 : 1.701, with smaller traces of Li, Be, and B. At the end of the nuclear era, the matter is almost completely ionized. As the temperatures of the matter, T_m, and the radiation drop, fewer photons and electrons possess enough energy to ionize He^+, He, and H in competition with radiative recombination processes like

$$H^+ + e \rightarrow H + h\nu \quad (27.2)$$

the rate of which increases with time, since $\alpha \propto T_m^{-0.61}$ at low temperature. Thus the ions He^{2+}, He^+, and H^+ recombine in succession and the Universe approaches an almost fully neutral state. Equal abundance fractions of H^+ and H, $f(H^+) = f(H)$, occur at $z = 1340$ when $T_r \approx T_m = 3630$ K in the model of Figure 27.1. This recombination epoch happens to coincide approximately with the time at which the matter and radiation lose close contact with each other. Initially, $T_m = T_r$, but eventually the rate of momentum transfer by Compton scattering can no longer keep pace with the decrease in density due to cosmic expansion; subsequently the matter evolves according to $T_m \propto (1 + z)^2$. It is also at this time that a detailed statistical balance among microscopic processes replaces thermodynamic equilibrium as the appropriate description of the state of the matter. Eventually, the expansion will overcome Process (27.2) in controlling the decrease of the proton and electron densities, with the result that the ionization fraction will approach a non-zero, asymptotic value ($n(e)/n_H \approx 5 \times 10^{-5}$ in Figure 27.1).

That molecules might form in the early Universe was noted by Saslaw and Zipoy (1967) and Peebles and Dicke (1968). Molecular abundances and their effects on cooling rates at early times have been discussed by Lepp and Shull (1984) and the entire subject has been reviewed recently by Dalgarno and Lepp (1987). Chemistry evidently begins with the formation of the helium-bearing molecular ions He_2^+ and HeH^+ by radiative association reactions, although their fractional abundances are expected to be very low. Owing to symmetry properties that exclude any allowed radiative transitions, the hydrogen molecule cannot form directly by radiative association of ground-state atoms. At the onset of recombination there is, however, a source of H_2 from excited atom association

$$H(n = 1) + H(n = 2) \rightarrow H_2^* \rightarrow H_2 + h\nu \quad (27.3)$$

which is estimated to have a rate coefficient of the order of 10^{-13} cm^3 s^{-1} at T_m = 5000 K (Latter 1989). The special circumstances that optimize the product of the n = 2 population and the overall neutral fraction are short lived and are expected to be rare in other astronomical phenomena.

The principal routes to H_2 are by association to form H_2^+

$$H^+ + H \rightarrow H_2^+ + h\nu \quad (27.4)$$

followed by charge transfer

$$H_2^+ + H \rightarrow H_2 + H^+ \quad (27.5)$$

and by radiative attachment to produce H^-

$$H + e \rightarrow H^- + h\nu \quad (27.6)$$

followed by associative detachment

$$H^- + H \rightarrow H_2 + e \quad (27.7)$$

Both sources of H_2 behave like catalytic sequences, the protons and electrons being returned upon formation of H_2. The first sequence can be interrupted by photodissociation

$$H_2^+ + h\nu \rightarrow H^+ + H \quad (27.8)$$

and by dissociative recombination

$$H_2^+ + e \rightarrow H + H \quad (27.9)$$

The rates of both Processes (27.8) and (27.9) are sensitive to the distribution of H_2^+ among its long-lived, excited vibrational states. A minor sink of excited H_2^+ $(v > 3)$

$$H_2^+(v > 3) + He \rightarrow HeH^+ + H \quad (27.10)$$

is also a major source of HeH^+, the principal, primordial, polar molecule with a strongly allowed rotational and vibrational spectrum. Processes that govern the abundance of HeH^+ are treated in detail by Roberge and Dalgarno (1982).

The sequence of Reactions (27.6) and (27.7) can be interrupted by photodetachment

$$H^- + h\nu \rightarrow H + e \quad (27.11)$$

and by mutual neutralization

$$H^- + H^+ \rightarrow H + H \quad (27.12)$$

Reaction (27.11) is especially effective around the time of recombination when the radiation field still contains a high density of photons with energies above the threshold for detachment, $h\nu > 0.75$ eV.

The net result of these processes is that the matter is endowed with a non-zero asymptotic value of the molecular fraction as $z \rightarrow 0$, $f(H_2) \approx 10^{-6}$–10^{-5}, depending on the density parameter Ω. The presence of even this small amount of H_2 can increase the cooling rate of the gas by orders of magnitude at low temperatures. Collisional excitation of excited vibrational and rotational transitions of hydride molecules is effective at cooling gas in the temperature range $T_m \approx 10^2$–10^3 K where even the lowest excited states in atomic H and He are energetically inaccessible. Under some circumstances, the less abundant molecules like HD, HeH^+, and even LiH, can contribute significantly to the cooling. In comparison with H_2, their rotational transitions at lower frequencies cool effectively at lower temperatures, and their more probable dipole-allowed radiative transitions partly compensate for their lower abundances. The cooling function, in turn, affects the characteristic value of the Jeans mass

$$M_J \approx 60 T_m^{3/2} n_H^{-1/2} \, M_\odot \tag{27.13}$$

which is the smallest mass for which gravitational contraction can overcome gas pressure. In the model of Figure 27.1, the Jeans mass at recombination is 9×10^5 $M_\odot$ and still exceeds 10^4 $M_\odot$ at $z \approx 10$. Efficient, low-temperature cooling in the first condensations can allow the Jeans mass to be reduced, permitting fragmentation and contraction to occur on smaller, stellar mass scales. Once dense condensations arise, additional molecular processes will be important. In particular, the H_2 abundance in primordial protostars can be enhanced by three-body association reactions

$$H + H + H \rightarrow H_2 + H \tag{27.14}$$

at high densities, $n_H \geq 10^9$ cm^{-3}, as shown by Palla, Salpeter and Stahler (1983).

Molecules formed in the early Universe are likely to be destroyed by the ultraviolet photons and hot gas that accompany the first generation of stars, and thus will probably have no directly observable consequences. Even so, they may affect the formation of the first stars, clusters and galaxies, and their study is useful for understanding the processes by which molecules form in dust-free material of primordial composition.

27.3. Molecules in QSO spectra

At somewhat more contemporary epochs, $z \approx 2$, absorption lines of stable, possibly abundant molecules like H_2 and CO become observable in the spectra of high-redshift QSOs. While the violent environs of a QSO itself may be inhospitable to molecules, the light of the QSO has a non-zero probability of passing through a molecule-containing cloud while travelling more than halfway across the Universe to reach us. Although a number of possible identifications of H_2 have been suggested on the basis of coincidences between QSO absorption lines and H_2 line positions at plausible redshifts (see review by Varshalovich and Levshakov (1982)), only recently have spectra been obtained that are of sufficiently high resolution and signal/noise ratio to settle the issue of the presence of H_2 definitively.

The spectrum of PHL 957, a relatively bright, high-redshift QSO, illustrates the spectroscopic problem (Figure 27.2). The redshift z now has an empirical meaning as the connection between the observed (Doppler shifted) and rest wavelength of a spectral line: $\lambda_{obs} = (1 + z)\lambda_{rest}$. By analogy with interstellar clouds in our Galaxy, H_2 is expected to appear where the column density of atomic hydrogen is large, as in the $z = 2.309$ redshift system of PHL 957, which has strong, damped HI

Lyman lines that indicate a column density $N(\mathrm{H}) = 2.5 \times 10^{21}$ cm^{-2}. The strongest lines of H_2 lie at $\lambda_{rest} = 912$–1110 Å: it is evident from Figure 27.2 that there will be a severe problem of confusion between the expected lines of H_2 and the plethora of HI Lyman lines at other redshifts. PHL 957 has been regarded as one of the best cases for the identification of H_2; however, detailed comparison of the spectrum of Figure 27.2 with simulated H_2 spectra for a variety of excitation conditions shows that H_2 is not present. The molecular fraction at $z = 2.309$ is quite low, $f(H_2) = 2N(H_2)/N_H < 4 \times 10^{-6}$, in comparison with that found in Galactic interstellar clouds of comparable column density of H (Black, Chaffee and Foltz 1987).

More recently, spectra have been obtained of the QSO PKS 0528–250 that are in harmony with the identification of H_2 lines at $z = 2.811$ (Foltz, Chaffee and Black 1988). The detailed comparison of observed and simulated spectra places

Figure 27.2. A spectrum of PHL 957 obtained with the Multiple Mirror Telescope at 1 Å resolution. The abscissa is the vacuum wavelength in the observer's frame of reference. The lower curve represents the 1σ noise level. The identified absorption lines arise in the predominant absorbing region at a redshift of $z = 2.309$. The broad peak in the spectrum near 4500 Å is the HI Lyman α emission of the QSO itself ($z_{em} = 2.69$, $\lambda_{rest} = 1215.6701$ Å). Most of the myriad of sharp absorption features that lie shortward of the Lyα emission, i.e. at $z < z_{em}$, are probably HI Lyman lines that arise in numerous small clouds along the line of sight to the QSO. (See Black, Chaffee and Foltz (1987) for details.)

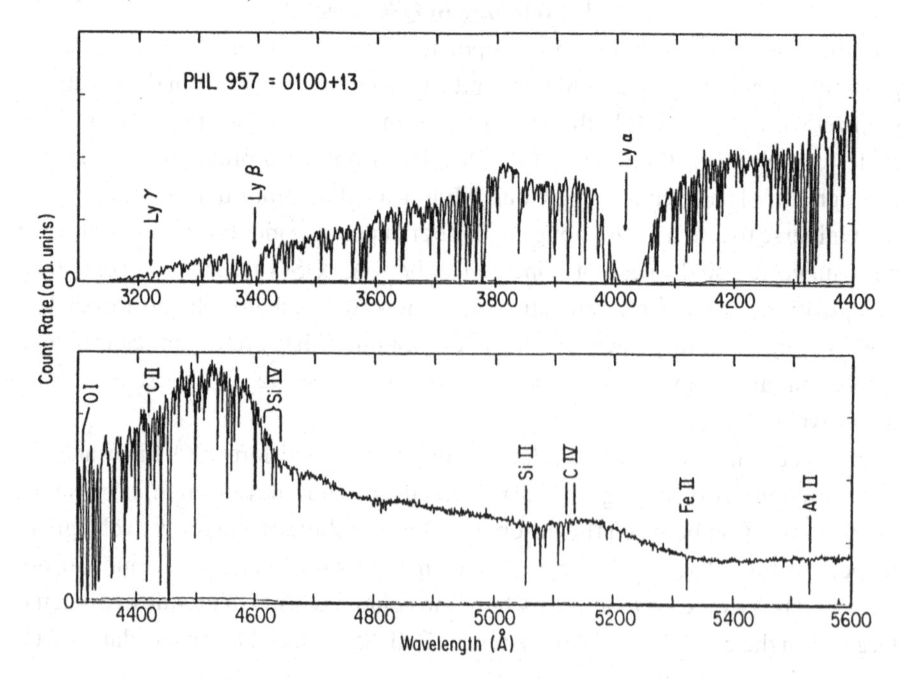

Table 27.1. *Properties of high-z clouds and a Galactic cloud*

Property	Cloud			
	MC 1331+170	PHL 957	PKS 0528−250	ζ Oph
z	1.776	2.309	2.811	0
$N(H)$ (cm^{-2})	1.5×10^{21}	2.5×10^{21}	1.3×10^{21}	5.2×10^{20}
$N(H_2)$ (cm^{-2})	—	$<5 \times 10^{15}$	10^{18}	4.2×10^{20}
$N(C^+)$ (cm^{-2})	5.9×10^{16}	7.8×10^{15}	—	9.3×10^{16}
$N(CO)$ (cm^{-2})	$<1.1 \times 10^{13}$	$<4 \times 10^{13}$	$<5 \times 10^{16}$	2.0×10^{15}
[C]/[H]	$(0.3–4) \times 10^{-4}$	4×10^{-6}	—	3×10^{-4}

rather narrow constraints on the molecular column density, $N(H_2) \approx 10^{18}$ cm^{-2}, on the Doppler line broadening parameter $b \approx 5$ km s^{-1}, and on the rotational excitation temperature of H_2, $T_{ex} \approx 100$ K. The implied molecular fraction in this redshift system, $f(H_2) \approx 10^{-3}$, appears to be too large to result from gas phase processes alone as described in the preceding paragraphs. If H_2 forms on the surfaces of dust particles in this system, then it may eventually be possible to estimate the dust content of the absorbing region. A reliable quantitative assessment will be possible, however, only when additional diagnostic information on the gas density and the intensity of dissociating radiation becomes available.

If the prominent QSO absorption systems with large column densities of H are attributed to interstellar clouds in disk galaxies at high redshift (Wolfe *et al.* 1986), then the absorption lines probe the composition of and physical conditions in galaxies at a very early stage in their development. For example, the H_2 molecules at $z = 2.811$ toward PKS 0528–250 absorbed the QSO light at a time 87% of the age of the Universe into the past, if a low-density cosmology like that of Figure 27.1 applies. Such observations probe conditions over time intervals that are long compared with the lifetimes of stars that enrich the matter in heavy elements through nucleosynthesis. In this way, the chemical evolution of the cosmos is being opened further to direct investigation. Table 27.1 summarizes some of the observed properties of three QSO absorption line regions and of a well-studied diffuse interstellar cloud in the Galaxy. The column densities of H and H_2 and the limits on CO show that the molecular fractions can vary over a wide range. The low molecular fractions at early epochs can be attributed to such causes as: (1) less efficient formation of H_2 on grain surfaces due to smaller, less abundant, or hotter grains; and (2) more intense ultraviolet starlight, perhaps due to a very active first generation of star formation. The overall carbon abundance, [C]/[H], is an indicator of the extent of enrichment of the gas through stellar nucleosynthesis. In the case of PHL 957, the low carbon abundance clearly suggests the effect of chemical evolution between $z = 2.3$ and the current epoch. As the blanks and inconclusive limits in the table indicate, there is a continuing need for more and better observations aided by more detailed theoretical analysis.

The investigation of molecules in the distant past provides a very striking example of the unavoidable connection between the microscopic world of atoms and molecules and the nature of the very largest phenomena.

REFERENCES

Black, J. H., Chaffee, F. H., Jr, and Foltz, C. B. (1987). *Astrophys. J.* **317**, 442–9.
Dalgarno, A., and Lepp, S. (1987). In *Astrochemistry,* eds. M. S. Vardya and S. P. Tarafdar, pp. 109–20. Dordrecht: D. Reidel.
Foltz, C. B., Chaffee, F. H., Jr, and Black, J. H. (1988). *Astrophys. J.* **324**, 267–78.
Harrison, E. R. (1973). *Ann. Rev. Astron. Astrophys.* **11**, 155–86.
Latter, W. B. (1989). Ph.D. Thesis, University of Arizona, in preparation.
Lepp, S., and Shull, J. M. (1984). *Astrophys. J.* **280**, 465–9.
Palla, F., Salpeter, E. E., and Stahler, S. W. (1983). *Astrophys. J.* **271**, 632–41.
Peebles, P. J. E., and Dicke, R. H. (1968). *Astrophys. J.* **154**, 891–908.
Roberge, W., and Dalgarno, A. (1982). *Astrophys. J.* **255**, 489–96.
Saslaw, W. C., and Zipoy, D. (1967). *Nature* **216**, 976–8.
Varshalovich, D. A., and Levshakov, S. A. (1982). *Comments Astrophys.* **9**, 199–209.
Wolfe, A. M., Turnshek, D. A., Smith, H. E., and Cohen, R. D. (1986). *Astrophys. J. Suppl.* **61**, 249–304.
Zel'dovich, Ya. B., and Novikov, I. D. (1983). *Relativistic Astrophysics* Volume 2: *The Structure and Evolution of the Universe.* Chicago: University of Chicago Press.

INDEX